Clofeta   III

Check-list of the fishes
of the eastern tropical Atlantic

Clofeta

Catalogue des poissons
de l'Atlantique oriental tropical

Volume III

Editors/Rédacteurs :
J. C. Quéro,
J. C. Hureau, C. Karrer,
A. Post & L. Saldanha

UNESCO

JNICT-Portugal

Published in 1990 by:

Junta Nacional de Investigação Científica e Tecnológica
Avenida D. Carlos I 126, 1200 Lisbon, Portugal
European Ichthyological Union
43 rue Cuvier, 75005 Paris, France
United Nations Educational,
Scientific and Cultural Organization
7 place de Fontenoy, 75700 Paris, France

Composed by C. Métivier and J. Sánchez-Jaimes

Printed by Gráfica Europam, Lda.
Apartado 28, 2726 Mem Martins Codex, Portugal

Publié en 1990 par :

Junta Nacional de Investigação Científica e Tecnológica
Avenida D. Carlos I 126, 1200 Lisbonne, Portugal
Union européenne d'ichtyologie
43 rue Cuvier, 75005 Paris, France
Organisation des Nations Unies pour
l'éducation, la science et la culture
7 place de Fontenoy, 75700 Paris, France

Composition : C. Métivier et J. Sánchez-Jaimes

Impression : Gráfica Europam, Lda.
Apartado 28, 2726 Mem Martins Codex, Portugal

ISBN 92-3-002620-4

Contents   Table des matières

Bibliography
Bibliographie

ABBOTT, C. C. 1861. Description of new species of apodal fishes in the Museum of the Academy of Natural Sciences of Philadelphia. Proc. Acad. nat. Sci. Philad., 12:475-479.

--. 1868. Catalogue of vertebrate animals of New Jersey. Rep. St. Geol., New Jersey, Appendix E:751-830.

--. 1899. The marine fishes of Peru. Proc. Acad. nat. Sci. Philad., 51: 324-364.

ABE, T. 1953. New, rare or uncommon fishes from Japanese waters. II. Records of rare fishes of the families Diretmidae, Luvaridae, and Tetragonuridae, with an appendix (description of a new species, Tetragonurus pacificus, from off the Solomon Islands). Jap. J. Ichthyol., 3(1):39-48, fig. 1-7.

--. 1955. Notes on the adult of Cubiceps gracilis from the western Pacific. J. oceanogr. Soc. Japan, 11:75-80.

--. 1957. Notes on fishes from the stomachs of whales taken in the Antarctic. I. Xenocyttus nemotoi, a new genus and new species of zeomorph fish of the subfamily Oreosomatinae Goode and Bean, 1895. Scient. Rep. Whales Res. Inst., Tokyo, (12):225-233, 2 pl.

--. 1961-1962. Notes on some fishes of the sub-family Braminae, with the introduction of a new genus, Pseudotaractes. Jap. J. Ichthyol., 8: 92-99, 101-114, fig. 1-31.

--. 1963. Colour illustrated book of fishes with keys. Tokyo, v + 358p., 780 col; fig. (in Japanese).

--. 1967. Records from northern Japan of two females of Ceratias holboelli each parasitized by a male. Proc. Japan Acad., 43:797-799, fig. 1, 2.

--. 1975. Notes on some fishes collected by the Fisheries Research Vessel "Kaiyomaru" in the South China Sea. III. Bull. biogeogr. Soc. Japan, 30(2):31-34, fig. 1.

ABE, T.; HOTTA, H. 1963. Description of new deep-sea fish of the genus Rondeletia from Japan. Jap. J. Ichthyol., 10 (2/6):43-48, 9 fig.

ABE, T.; KOJIMA, S.; KOSAKAI, T. 1963. Description of a new nomeid fish from Japan. Jap. J. Ichthyol., 11:31-35.

ABE, T.; MARUMO, R.; KAWAGUCHI, K., 1965. Description of a new alepocephalid fish from Suruga Bay. Jap. J. Ichthyol., 13(1-3):69-72, 4 fig.

ABE, T.; MARUYAMA, K. 1963. A record of Barbourisia rufa Parr from off the Kurile Islands. Jap. J. Ichthyol., 10(2/6):49-50, 1 pl.

ABE, R.; NAKAMURA, C. 1954. A record of an adult female with a supposedly parasitic male of Cryptopsaras couesii from the Pacific coast of northern Japan. Jap. J. Ichthyol., 3:96-97.

ABE, T.; SUZUKI, M. 1981. Works on some fishes associated with the Antarctic Krill. II. On Xenocyttus nemotoi Abe, and again on Neopagetopsis ionah Nybelin. Antarctic Rec., 71:121-129, 3 fig., 2 tab.

ABEL, E. 1959. Über die Verbreitung von Tripterygion minor Kolomb. im Mittelmeer, sowie ergänzende Beschreibung des Fisches. Pubbl. Staz. zool. Napoli, 31:98-108.

ABOUSSOUAN, A. 1964. Contribution à l'étude des oeufs et larves péla-

giques des poissons téléostéens dans le golfe de Marseille. Recl
Trav. Stn mar. Endoume, 32(48):87-178, 30 fig.

--. 1965. Oeufs et larves de téléostéens de l'ouest africain. II.
Distribution verticale. Bull. Inst. fr. Afr. noire, (A)27(4):1504-
1521.

--. 1966. Oeufs et larves de téléostéens de l'ouest africain. III.
Larves de Monacanthus hispidus (L.) et de Balistes forcipatus Gm.
Bull. Inst. fond. Afr. noire, (A)28(1):276-282, fig. 1-6.

--. 1966. Oeufs et larves de téléostéens de l'ouest africain. IV.
Galeoides polydactylus (Vahl) (Polynemidae). Bull. Inst. fond. Afr.
noire, (A)28(3):1037-1040, fig. 1-5.

--. 1968. Oeufs et larves de téléostéens de l'ouest africain. VII.
Larves de Syacium guineensis (Blkr.) (Bothidae). Bull. Inst. fond.
Afr. noire, (A)30(3): 1188-1197, 7 fig.

--. 1968. Oeufs et larves de téléostéens de l'ouest africain. VIII.
larves de Bregmaceros macclellandi Thompson (Bregmacerotidae). Bull.
Inst. fond. Afr. noire, (A)30(4):1590-1602.

--. 1971. Oeufs et larves de téléostéens de l'ouest africain. IX. Sur
la migration des larves de Sardinella aurita Val. à proximité de la
presqu'île du Cap Vert. Bull. Inst. fond. Afr. noire, (A)33:227-236.

--. 1972. Oeufs et larves de téléostéens de l'ouest africain. XI. Larves
serraniformes. Bull. Inst. fond. Afr. noire, (A)34(1):485-502, 32
fig.

--. 1972. Oeufs et larves de téléostéens de l'ouest africain. XII. Les
larves d'Heterosomata récoltés aux environs de l'île de Gorée
(Sénégal). Bull. Inst. fond. Afr. noire, (A)34(4):974-1003, 19 fig.

ABRAMOV, A. A. 1982. On the range of some species from the family
Chlorophthalmidae. p. 62-63. In: VNIRO Fishery investigations in the
eastern tropical Atlantic (E.V. Vladimirskaya, ed.). (Moscow).

ADAMICKA, P. 1972. Funktionsanatomische Untersuchungen am Kopf von Akan-
thopterygiern (Pisces, Teleostei): Sarda sarda, Pterois volitans,
Chrysophrys aurata. Teil I: Einleitung und Sarda sarda. Zool. Jb.
(Anat.), 89(2):301-331.

ADAMS, J. A. 1960. A contribution to the biology and postlarval
development of the Sargassum fish, Histrio histrio (Linnaeus), with a
discussion of the Sargassum complex. Bull. mar. Sci. Gulf Caribb.,
10(1):55-82.

AGASSIZ, L. 1829-1831. In: J.B. von Spix. Selecta genera et species
piscium quos in itinere per Brasiliam annos 1817-20. Monachii, 138
p., 101 pl. (1, 1829:1-82, pl. 1-48; 2, 1831:83-138, pl. 49-101).

--. 1833-1844. Recherches sur les poissons fossiles. Neuchâtel, 5 vol.
text + 5 vol. atlas. 1:xlix + 188 p.; 2:310 + 336 p.; 3:390 + 32 p.;
4:296 + 22 p.; 5:122 + 160 p.

--. 1845. Nomenclator zoologicus continens nomina systematica generum
piscium tam viventium quam fossilium secundum ordinem alphabeticum
disposita, adjectis auctoribus, libris, in quibus reperiuntur, anno
editionis, etymologia et familiis, ad quas pertinent in singulis
classibus. Soloduri, 1846(1845):VI + 69 + 8 p.

--. 1846. Nomina systematica generum piscium,... p. vi + 69 + 8. In:
Nomenclator zoologicus,... Soloduri, 1846(1845).

--. 1847. Nomenclatoris Zoologici. Index universalis continens nomina
systematica classium, ordinum, familiarum et generum animalium
omnium, tam viventium quam fossilium. Soloduri, 1846(1847):viii + 393
p.

--. 1872. Fish-nest in the sea-weed of the Sargasso Sea. Am. J. Sci.
(3)3:154-156.

--. 1888. Three cruises of the United States Coast and Geodetic Survey
Steamer "BLAKE" in the Gulf of Mexico, in the Caribbean Sea, and

along the Atlantic coast of the United States from 1877 to 1880. 2. Boston, 220 p.

AGASSIZ, A.; WHITMAN, C. O. 1885. The development of osseus fishes. I. The pelagic stages of young fishes. Mem. Mus. comp. Zool. Harv., 14(1), pt. 1:1-56, pl. 1-19.

AHL, E. 1923. Zur Kenntniss der Knochenfischfamilie Chaetodontidae insbesondere der Unterfamilie Chaetodontinae. Arch. Naturgesch., (A)89(5):1-205, pl. 1-2.

--. 1925. Über eine neue Grundel aus Afrika. Zool. Anz., 64:51.

--. 1935. Beschreibungen zweier neuer Süsswasserfische aus West-Afrika. Zool. Anz. 110:251-253.

AHL, J. N. 1789. In: Thunberg, C.P. & J.N. Ahl. Specimen ichthyologicum de muraena et ophichtho, quod... Praeside C.P. Thunberg... offert J.N. Ahl. Upsaliae. Inaug. Dissert., 14 p., 2 pl.

AHLSTROM, E. H. 1969. Mesopelagic and bathypelagic fishes in the California Current region. Rep. Calif. coop. oceanic Fish. Invest., 13:39-44.

--. 1971. Kinds and abundance of fish larvae in the eastern tropical Pacific, based on collections made on EASTROPAC I. Fishery Bull., 69 (1):3-77.

AHLSTROM, E. H.; BUTLER, J. L.; SUMIDA, B. Y. 1976. Pelagic stromateoid fishes (Pisces, Perciformes) of the eastern Pacific: kinds, distributions, and early life histories and observations on five of these from the northwest Atlantic. Bull. mar. Sci., 26:285-402, 49 fig., 52 tab.

AHLSTROM, E. H.; COUNTS, R. 1958. Development and distribution of Vinciguerria lucetia and related species in the eastern Pacific. Fishery. Bull. Fish. Wildl. Serv. U.S., 58(139):359-416, 29 fig.

AHLSTROM, E. H.; MOSER, H. G.; O'TOOLE, M. J. 1976. Development and distribution of larvae and early juveniles of the commercial lanternfish Lampanyctodes hectoris (Günther) of the west coast of southern Africa with a discussion of phylogenetic relationships of the genus. Bull. Sth. Calif. Acad. Sci., 75(2):138-152.

AHMAD, M. F. 1970. Redescription of Abudefduf luridus (Cuvier), (Pisces: Pomacentridae). Senckenberg. biol., 51(5/6):311-316.

AKAZAKI, M. 1962. Studies on the spariform fishes - anatomy, phylogeny, ecology and taxonomy. Spec. Rep. Misaki mar. biol. Inst., (1):1-368, fig.

--. 1974. In: TOMIYAMA, T., Fisheries in Japan. Sea Bream. Tokyo, 181 p., num. col. phot.

ALAEJOS, L. 1931. La pesca maritima en el puerto de Santander. Notas Resum. Inst. esp. Oceanogr., (2)(56):1-43.

ALAEJOS SANZ, L. 1923. Provincia de Santander: La pesca maritima en España en 1920. Inst. Estud. Cient. Estad. Pesca, Madrid, 2:159-210, 20 fig. (= Boln Pescas, 1922:35-86, 20 fig.).

ALBARET, J.-J.; CHARLES-DOMINIQUE, E. 1983. Exposé synoptique des données biologiques sur l'ethmalose (Ethmalosa fimbriata S. Bowdich, 1825). Revue Hydrobiol. trop., 15(4):373-397.

ALBARET, J. J.; GERLOTTO, F. 1976. Biologie de l'ethmalose (Ethmalosa fimbriata Bowdich) en Côte d'Ivoire. 1. Description de la reproduction et des premiers stades larvaires. Docums. scient. Centre Rech. oceanogr., Abidjan, 7(1):113-133.

ALBERT, PRINCE OF MONACO. 1898. Oceanography of the North Atlantic. Geogrl J., 12(5):445-469, fig.

ALBUQUERQUE, R. M. 1954-1956. Peixes de Portugal e ilhas adjacentes. Chavas para a sua determinaçâo. Port. Acta biol. (B), 5:xvi, 1-1167, fig. 1-445.

ALCOCK, A. W., 1889. Natural history notes from H.M. Indian Marine

Survey Steamer "Investigator", Commander Alfred Carpenter, R.N., D.S.O., commanding. No. 13. On the bathybial fishes of the Bay of Bengal and neighbouring waters, obtained during the seasons 1885-1889. Ann. Mag. nat. Hist., (6)4:376-399 + 450-461 (same title).

--. 1890. Natural history notes from H. M. Indian Marine Survey Steamer "Investigator", Commander R. F. Hoskyn, R. N., commanding. No 16. On the bathybial fishes collected in the Bay of Bengal during the season 1889-1890. Ann. Mag. nat. Hist., (6)6 :197-222, pl. 8-9.

--. 1890. Natural history notes from H.M. Indian marine survey steamer "Investigator", Commander R.F. Hoskyn, R.N., commanding. No 18. On the bathybial fishes of the Arabian Sea, obtained during the season 1889-1890. Ann. Mag. nat. Hist., (6) 6(34):295-311.

--. 1890. Natural history notes from H.M. Indian Marine Survey Steamer "Investigator" Commander R.F. Hoskyn, R.N. commanding. No 20. On some undescribed shorefishes from the bay of Bengal. Ann. Mag. nat. Hist., (6)6:425-443.

--. 1892. On the bathybial fishes collected during the season of 1891-92. Ann. Mag. nat. Hist., (6)10(59):345-365, 2 fig., 1 pl.

--. 1893. Natural history notes from H. M. Indian Marine Survey Steamer "Investigator", Commander C. F. Oldham, R. N., commanding. Ser. II, No 9. An account of the deep-sea collection made during the season of 1892. J. Asiat. Soc. Beng., 62:170-184.

--. 1894. Natural history notes from H. M. Indian Marine Survey Steamer "Investigator", Commander C. F. Oldham, R.N., commanding. Series II., No 11. An account of a recent collection of bathybial fishes from the bay of Bengal and from the Laccadive sea. J. Asiat. Soc. Beng., 63(2):115-137, pl. 6-7.

--. 1896. A supplementary list of the marine fishes of India, with descriptions of 2 new genera and 8 new species. J. Asiat. Soc. Beng., 65(2):301-338.

--. 1897. Illustrations of the zoology of the Royal Indian Marine Survey Steamer "Investigator" under the command of Commander C.F. Oldham, R. N. Fishes, pt. 4. Calcutta, pl. 17.

--. 1898. Natural history notes from H. M. Indian Survey Steamer "Investigator", Commander T. H. Heming, R. N., commanding. Ser. II, No 25. A note on the deep-sea fishes, with descriptions of some new genera and species, including another probably viviparous ophidioid. Ann. Mag. nat. Hist., (7)2:136-156.

--. 1899. A descriptive catalogue of the Indian deep-sea fishes in the Indian Museum. Being a revised account of the deep-sea fishes collected by the Royal Indian Marine Survey Ship "Investigator". Calcutta, i-iii + 211 + i-viii p., 1 map (pl. published separately).

ALCOCK, A. W.; MCARDLE, A. F. 1900. Illustrations of the zoology of the Royal Indian Marine Survey Ship "Investigator", under the command of Comm. T.H. Heming. Fishes, pt. 7, pl. 27-35. Calcutta.

ALCOCK, A.W.; MACGILCHRIST, A. C. 1905. Illustrations of the zoology of the Royal Indian Marine Survey Ship "Investigator". Fishes, pt. 8, pl. 36-38. Calcutta.

ALDEBERT, Y. 1970. Répartition bathymétrique et géographique des poissons hétérosomes récoltés par la "Thalassa" en 1962 au Rio de Oro et en Mauritanie. Rapp. P.-v. Réun. perm. int. Explor. Mer, 159:213-217.

ALDRIN, J.F.; MARCHAL, E. 1964. Un nouveau poisson intéressant l'industrie de la conserve: le "sardineau" des côtes d'Afrique. 1er Congr. Int. Indust. Aliment. Agric., Abidjan, Côte d'Ivoire, 14-19 Déc. 1964:1199-1204.

ALDRIN, J. F; NOYER, O.; BREGEAT, D. 1972. Les poissons de mer à Abidjan. Républ. Côte d'Ivoire, Min. Prod. anim., Lab. Dir. Pêch. marit.

lagun., Abidjan (53), 4 + 110 p + indexes, 1 + 111 fig. (mimeo).

ALDROVANDI, U. 1613. De piscibus libri v, et de cetis liber unus. Bononiae, 732 p.

ALEXANDER, E. C. 1961. A contribution to the life history, biology and geographical distribution of the bonefish, Albula vulpes (L.). Dana Rep., (53):1-51, 16 fig., 12 tab.

ALLEN, E. J. 1917. Post-larval Teleosteans collected near Plymouth. J. mar. biol. Ass. U.K., 11(2):207-250, 8 fig., 45 tab.

ALLEN, G. R., 1981. Drepanidae, Ephippidae, Kuhliidae, Lutjanidae, Monodactylidae, Polynemidae, in Fischer, W., Bianchi, G. & W. B. Scott (ed.), FAO species identification sheets for fishery purposes. Eastern Central Atlantic; fishing areas 34, 47 (in part). Canada Funds-in-Trust. Ottawa, Department of Fisheries and Oceans Canada, by arrangement with the Food and Agriculture Organization of the United Nations, vols. 1-7: pag. var.

ALLEN, G. R.; HOESE, D. F.; PAXTON, J. R.; RANDALL, J. E.; RUSSELL, B. C.; STARCK II, W. A.; TALBOT, F. H.; WHITLEY, G. P. 1976. Annotated checklist of the fishes of Lord Howe Island. Rec. Aust. Mus., 30:365-454.

ALLEYNE, H. G.; MACLEAY, W. 1877. The ichthyology of the Chevert Expedition. Proc. Linn. Soc. N.S.W., 1:261-281, 321-359, pl. 1-9, 10-17.

ALLUE, C.; BORRUEL, C.; LLORIS, D.; RUCABADO, J. 1984. Datos pesqueros de la campana "Benguela II". Datos informativos. Inst. Investnes Pesc., Barcelona, 9:85-190.

ALMEIDA, A. J.; GOMES, J. A. 1978. Quelques poissons nouveaux pour la faune du Portugal. Mems Mus. mar. (Zool.), 1:1-23.

ALMEIDA, A. J.; GOMES, J. A.; RE, P. 1980. Trois Blenniidae nouveaux pour la faune du Portugal (Pisces: Perciformes). Tethys, 9(3):235-241.

ALONCLE, H. 1967. Remarques sur une variété locale de Blennius goreensis Valenciennes, 1836, espèce nouvelle pour les côtes du Maroc atlantique. Bull. Inst. Pêch. marit. Maroc, (15):105-109, 1 fig.

--. 1968. Note sur la présence de Lutjanus goreensis (Valenciennes, 1830) dans les eaux du littoral atlantique marocain. Bull. Inst. Pêch. marit. Maroc, (16):97-99, 1 fig.

--. 1968. Catalogue des types de poissons téléostéens en collection au Muséum de La Rochelle. Bull. Mus. natn. Hist. nat., Paris, (2) 40 (4):683-691.

--. 1972. Sélaciens Hypotrèmes: raies & formes affines. In: COLLIGNON J. & ALONCLE, H. Catalogue raisonné des poissons des mers marocaines. Bull. Inst. Pêch. marit. Maroc, (19):87-159, fig. 27-52, photos 16-34.

AMANIEU, M.; CAZAUX, C. 1962. Animaux rares observés dans la région d'Arcachon en 1961-1962.. P.-v. Soc. linn. Bordeaux, 99:1-7, fig. 1-2.

AMAOKA, K.; ABE, K. 1977. Description of a new alepocephalid fish, Bajacalifornia erimoensis, and a second record of Alepocephalus umbriceps off Japan. Jap. J. Ichthyol., 23(4):185-191, 4 fig.

ANCONA, U. d'. 1928. Murenoidi (Apodes) del Mar Rosso del Golfo di Aden. Materiali raccolti dal Prof. Luigi Sanzo nella Campagna della R.N. "Ammiraglio Magnaghi" 1923-24. Memorie R. Com. talassogr. ital., (146):1-146, 5 pl.

--. 1928. Sulla possibilità di ordinare sistematicamente la specie larvali dei murenoidi. Atti Accad. naz. Lincei Rc., (6)7(6):516-520.

--. 1929-1931. Anguillidae, Muraenidae, Congridae, Echelidae, Nettastomidae, Ophichthyidae. Faune ichthyol. Atlant. N., Cahier 1, 1929:130, 134, 138, 139; Cahier 2, 1929:129; Cahier 4, 1930:135, 136; Cahier 7, 1931:137.

--. 1931. Clupeoidei, Heteromi, Apodes, Synentognathi. In: Uova, larve e stadi giovanili di Teleostei (ed. U. d'Ancona). _Fauna Flora Golfo Napoli_, 38:1-21, 93-177, fig. 1-23, 59-166, pl. 1-2, 8-11.

--. 1933. Gadidae, Berycoidei, Lampridae. _In_: Uova, larve e stadi giovanili di Teleostei. (ed. U. d'Ancona). _Fauna Flora Golfo Napoli_, 38:178-255, 280-306, fig. 167-226, 240-244, pl. 12-15, 17-18.

ANCONA, U. d'; CAVINATO, G., 1965. The fishes of the family Bregmacerotidae. _Dana Rep._, (64):1-92, 58 fig.

ANDERSON, M. E. 1984. On the anatomy and phylogeny of Zoarcidae (Teleostei: Perciformes). Ph. D. Thesis, College of William and Mary, Williamsburg, Va., 254 p.

--. 1985. Zoarcidae, p. 401-404. _In_: Fischer, W.& Hureau, J.C. (ed.). FAO Species identification sheets for fishery purposes. Southern Ocean (Fishing Areas 48, 58 and 88). Prepared and printed with the support of the Commission for the Conservation of Antarctic Marine Living Resources (CCAMLR), vol.1. Rome.

ANDERSON, M. E.; HUBBS, C. L. 1981. Redescription and osteology of the northeastern Pacific fish _Derepodichthys alepidotus_ (Zoarcidae). _Copeia_, 1981(2):341-352, 10 fig.

ANDERSON, W. D. 1967. Field guide to the snappers (Lutjanidae) of the western Atlantic. _Circ. U.S. Fish Wild. Serv., Wash._, (252):14 p., 29 fig.

--. 1971. Investigations on the gray snapper _Lujanus griseus_. Copeia, 1971, (4):764 (reviews and comments).

ANDERSON, W. D.; CALDWELL, D. K.; MCKINNEY, J. F.; FARMER, C. H., 1972. Morphological and ecological data on the priacanthid fish _Cookeolus boops_ in the western North Atlantic. _Copeia_, 1972(4):884-885.

ANDERSON, W. W.; GEHRINGER, J. V.; BERRY, F. H. 1966. Fam. Synodontidae. _Mem. Sears Fdn mar. Res._, 1(5):30-102, fig. 8-35.

ANDRIASHEV, A. P. 1954. Ryby severnykh morei SSSR. Moskva, Leningrad, 1-566, 300 fig. (English trans. 1964, IPST, Jerusalem, 617 p., 300 fig.).

--. 1958. List of ichthyological stations with preliminary characterisation of the hauls. _Inf. Byull. sov. antarkt. Eksped. D/E OB 1955-1956:_ 199-204 (in Russian).

--. 1959. On the systematic position of the south African lycodid fish (_Lycodes agulhensis_, sp. n.) confused with the Arctic species, _Lycodes frigidus_ Collett. _Zool. Zh._, 38(3):465-468, 3 fig. (in Russian with English summary).

--. 1960. Families of fishes new for the Antarctic. 1. _Paradiplospinus antarcticus_ gen. et sp. nov. (Pisces, Trichiuridae). _Zool. Zh._, 39(2):244-249, 2 fig. (in Russian).

--. 1960. Families of fishes new to the Antarctic. 2. Pearleyes fishes (Scopelarchidae). _Zool. Zh._, 39(4):563-566, 2 fig. (in Russian).

--. 1961. Review of genus _Artediellus_ of the Bering Sea. _Vop. Ikhtiol._, (19)=1(2):231-242, 6 fig. (in Russian).

--. 1962. Bathypelagic fishes of the Antarctic. 1. Family Myctophidae. _Issled. Fauny Morei_, 1:216-294, fig. 1-36 (in Russian) = Biol. Res. Sov. Antarctic Exped. (1955-1958), 1.

--. 1965. A general review of the Antarctic fish fauna. _Monographiae biol._, 15:491-550, 18 fig.

--. 1973. Zoarcidae, p. 540-547, _in_ Hureau, J. C. & Monod, Th. (ed.), Check-list of the fishes of the north-eastern Atlantic and of the Mediterranean/Catalogue des poissons du nord-est Atlantique et de la Méditerranée (Clofnam). Paris, Unesco, 2 vol.

--. 1977. Some additions to schemes of the vertical zonation of marine bottom fauna. p. 351-360, 5 fig. _In_: G.A. Llano (ed.) Adaptions within Antarctic ecosystems. _Proc. 3rd SCAR Symp. Antarct. Biol._, Washington, xliv + 1252 p.

--. 1979. O nekotorych voprosach vertikalnoj zonalnosti morskoj donnoj fauny. p. 117-138, 6 fig. In: Biologiceskie resoursy mirovogo Okeana. (ed. S.A Studeneskij). Moskva.

--. 1985. Liparididae, p. 289-291. In: Fischer, W.& Hureau, J.C. (ed.). FAO Species identification sheets for fishery purposes. Southern Ocean (Fishing Areas 48, 58 and 88). Prepared and printed with the support of the Commission for the Conservation of Antarctic Marine Living Resources (CCAMLR). Vol.2, Rome.

--. 1986. Zoarcidae, in P.J.P. Whitehead, M.L. Bauchot, J.C. Hureau, J. Nielsen & E. Tortonese. Fishes of the North-eastern Atlantic and the Mediterranean/Poissons de l'Atlantique du Nord-Est et de la Méditerranée, Paris, Unesco, 3:1130-1150, 27 + 10 fig.

--. 1986. Agonidae, in P.J.P. Whitehead, M.L. Bauchot, J.C. Hureau, J. Nielsen & E. Tortonese. Fishes of the North-eastern Atlantic and the Mediterranean/Poissons de l'Atlantique du Nord-Est et de la Méditerranée, Paris, Unesco, 3:1265-1268, 3 + 3 fig.

Anonymous. 1798. Allg. Literatur-Zeitung. Sept. 24. Berlin.

--. 1948. Fiche ichthyologique N° 2: Ethmalosa fimbrata (Bowdich). Cybium (2):5-7.

--. 1948. Report on fisheries investigations 1942-48 (Sess. Pap. 23/- 1948). Lagos, 63 p.

--. 1957. Interim report on a survey of the stock of Ethmalosa fimbriata (Bowdich) in the Sierra Leone river. Oceanography and sea fisheries on the West African coast, CCTA/CSA Symposium (20-27 November 1957, Luanda), Paper No 63 (mimeo).

--. 1961. (Summarized translations of Soviet research on Sardinella off tropical West Africa: Nigeria Federal Fisheries Service). Occ. Pap. Fed. Fish. Serv., Lagos, (3):1-14.

--. 1964. International Code of Zoological Nomenclature adopted by the XV International Congress of Zoology. London, 176 p.

--. 1967. Opinion 809: Thunnus South, 1845 (Pisces): validated under the plenary powers. Bull. zool. Nom., 24(2):85-86.

--. 1971. Grammicolepis brachiusculus. Mar Pesca, (64): p. (63)DB-321, fig.

--. 1972. Colored illustrations of bottomfishes collected by Japanese trawlers. (ed. Far Seas Fish. Res. Lab.) Tokyo, (1):v + 145 p., fig. (in Japanese).

--. 1973. Sedentary, migratory and intermingling species, their habitat and distribution. FAO Fish. Circ., (48)(Revision 1), FAO, Rome, 44 p.

--. 1976. Colored illustrations of bottomfishes collected by Japanese trawlers. (ed. Far Seas Fish. Res. Lab.) Tokyo, 2:iv + 188 p., fig. (in Japanese).

--. 1976. Rapport du groupe de travail sur la sardinelle (S. aurita) des eaux ivoiro-ghanéennes. Abidjan, 28 June - 3 July. ORSTOM, Abidjan, 40 p.

--. 1979. Report of the ad hoc working group on West African coastal pelagic fish from Mauritania to Liberia (26°N to 5°N). Dakar-Thiaroye, 19-24 June 1978, CECAF/ECAF Ser. 78/10 (En), UNDP/FAO, Rome, 161 p.

ANSLIJN, N. 1828. Systematische Beschrijving der voor ons meest belangrijke visschen. Door N. Anslijn, N.Z. D. du Mortier en zoon, Leyden, (ii):1-345, 98 col. pl.

ANSA-EMMIN, M. 1982. Fisheries in the CINECA region. Rapp. P.-v. Réun. Cons. perm. int. Explor. Mer, 182:405-422.

ANTIPA, G. 1906. Die Clupeiden des westlichen Teiles des Schwarzen Meeres und der Donaumündungen. Denkschr. Akad. Wiss. Wien, 78:1-57, 6 fig., 3 pl.

APOSTOLIDES, N. C. 1883. La pêche en Grèce. Athens, 87 p.

ARATA, G. F., Jr. 1954. A contribution to the life history of the swordfish, _Xiphias gladius_ Linnaeus, from the South Atlantic coast of the United States and the Gulf of Mexico. _Bull. mar. Sci. Gulf Caribb._, 4(3):183-243, 19 fig.

ARCHEY, G. 1922. A second specimen of _Idiacanthus niger_ Regan from New Zealand. _N.Z. Jl Sci. Technol._, 5(1):295-296.

ARIOLA, V. 1912. Nuovo pesce abissale del Golfo di Genova (_Cubiceps capensis_ Smith). _Riv. mens. Pesca, Pavia_, 7:185-192, 1 pl.

--. 1904. Pesci nuovi o rari per il Golfo di Genova. _Annali Mus. civ. Stor. nat. Genova_, (3)1:153-168.

ARNOLD, D. C. 1956. A systematic revision of the fishes of the teleost family Carapidae (Percomorphi, Blennioidea), with descriptions of two new species. _Bull. Br. Mus. nat. Hist._, (Zool.), 4(6):247-307.

ARNOULT, J.; AUBENTON, F. d'; BAUCHOT, M. L.; BLANC, M. 1966. Poissons téléostéens. _In_: Résultats scientifiques des Campagnes de la "Calypso". _Annls Inst. océanogr., Monaco_, 44:1-22.

ARNOUX, J. 1976. Les poissons de mer de St. Louis. Répertoire synonymique. Publication du Centre d'Etude des Pêches de Guet N'Dar, 1956; réédition Centre de Recherches Océanographiques de Dakar. Thiaroye, novembre 1976, multicopié.

ARON, W. 1962. The distribution of animals in the eastern North Pacific and its relationships to physical and chemical conditions. _J. Fish. Res. Bd Can._, 19(2):271-314.

ARON, W.; GOODYEAR, R. H., 1969. Fishes collected during a midwater trawling survey of the Gulf of Elat and the Red Sea. _Israel J. Zool._, 18:237-244.

ARRUDA, L. M., 1977. Morphological comparisons between the Mediterranean and northeastern Atlantic populations of _Chromis chromis_ L. (Pisces: Pomacentridae). _Archos Mus. Bocage_, (2)6(1):207-217.

ARTE, P., 1952. Notas ictiológicas. I. Peces raros o nuevos para el litoral gallego (N.W. de España). _Publnes Inst. Biol. apl., Barcelona_, 10:93-103, fig. 1-6.

ARTEDI, P., 1738. Ichthyologia, sine opera omnia de piscibus scilicet... Omnia in hoc opere perfectiora, quam antea ulla. Posthuma vindicavit, recognovit, coeptavit et editit Carolus Linnaeus. Pars III. Genera piscium. Lugduni Batavorum, vi + 88 p.

--. 1789-1793. P. Artedi renovati... Ichthyologica. Cura J. J. Walbaumii. (5 pt in 3 vol.). Grypeswaldiae.

--. 1793. Synonymia nominum piscium fere omnium... Ichthyologiae. Pars IV (ed. II). Grypeswaldiae, 3 + 140 p.

ASANO, H. 1958. Description of a new genus _Japonoconger_ typed by _Arisoma sivicola_ Matsubara and Ochiai, with consideration to the related genera. _Zool. Mag., Tokyo_, 67(10):316-321, fig. 1-5.

ASCANIUS, P. 1767-1772. Icones rerum naturalium, ou figures enluminées d'histoire naturelle du Nord. Copenhague, Cah. (1), 1767:22 p., pl. 1; Cah. (2), 1772:8 p., pl. 11-20.

ASSO Y DEL RIO, I. J. 1801. Introducción à la ichthyología oriental de España. _An. Cienc. nat._, 4:28-52 (also separate, Madrid, 28 p.).

ATZ, J. W. 1950. Strange animal lures. _Anim. Kingd._, 53(4):110-113.

--. 1951. Strange fish lures. _Aquar. J._, 22(4):70-72.

AWERINZEW, S. 1913. Ergebnisse der Untersuchungen über parasitische Protozoen der tropischen Region Afrikas. III. _Zool. Anz._, 42(4):151-156.

AYRES, W. O. 1855. _Notorhynchus maculatus_ Ayres. _Proc. Calif. Acad. Sci._ 1:76-77.

--. 1857. Description of a new species of mackerel, _Scomber diego_, Ayres. _Proc. Calif. Acad. Sci._, 1:92.

--. 1860. Descriptions of new species of fishes. _Proc. Calif. Acad._

Sci., 2:73-86, fig. 19-23.

AZEVEDO, C. J. C. de. 1971. Antenarídeos de Angola. Publiçao Cienc. biol. Fac. Cienc. Univ. de Luanda, Angola, 1971:89-92, 4 pl.

BAADER, F. 1875. Beiträge zur Kenntnis der Fische von Marocco. Ber. senckenb. naturf. Ges., 1875:179-182.

BACCI, G.; RAZZAUTI, A. 1958. Protogynous hermaphroditism in Coris julis L. Nature. Lond., 181(4606):432-433, 1 fig.

BACKUS, R. H.; CRADDOCK, J. E.; HAEDRICH, R. L.; ROBINSON, B. H. 1977. Atlantic mesopelagic zoogeography. Mem. Sears Fdn mar. Res., 1 (7): 266-287.

BACKUS, R. H.; CRADDOCK, J. E.; HAEDRICH, R. L.; SHORES, D. L. 1969. Mesopelagic fishes and thermal fronts in the western Sargasso Sea. Mar. Biol., Berl., 3(2):87-106, 9 fig.

--. 1970. The distribution of mesopelagic fishes in the equatorial and western North Atlantic. J. mar. Res., 28(2):179-201, fig. 1-17.

BACKUS, R. H.; MEAD, G. W.; HAEDRICH, R. L.; EBELING, A. W. 1965. The mesopelagic fishes collected during Cruise 17 of the R/V "Chain", with a method for analysing faunal transects. Bull. Mus. comp. Zool. Harv., 134(5):139-15, fig. 1-9.

BADCOCK, J. 1969. Colour variation in two mesopelagic fishes and its correlation with ambient light conditions. Nature. Lond., 221(5178): 383-385, 2 fig.

--. 1970. The vertical distribution of mesopelagic fishes collected on the SOND Cruise. J. mar. biol. Ass. U.K., 50:1001-1044, 10 fig.

--. 1977. Aspects of the development and biology of postlarval Valenciennellus tripunctulatus (Esmark) and Bonapartia pedaliota Goode and Bean (Pisces, Stomiatoidei). p.537-552, fig. 1-4. In: M. Angel (ed.) A voyage of discovery: George Deacon 70th anniversary, Oxford.

--. 1982. A new species of the deep-sea fish genus Cyclothone Goode & Bean (Stomiatoidei, Gonostomatidae) from the tropical Atlantic. J. Fish Biol., 20:197-211, fig. 1-6.

--. 1984. Gonostomatidae, in Whitehead, P.J.P., Bauchot, M.L, Hureau, J.C., Nielsen, J. & E. Tortonese, Fishes of the North-eastern Atlantic and the Mediterranean/Poissons de l'Atlantique du Nord-Est et de la Méditerranée, Paris, Unesco, 1:284-301, 17+10 fig., 17 maps.

--. 1984. Sternoptychidae, in Whitehead, P.J.P., Bauchot, M.L., Hureau, J.C., Nielsen, J. & E. Tortonese, Fishes of the North-eastern Atlantic and the Mediterranean/Poissons de l'Atlantique du Nord-Est et de la Méditerranée, Paris, Unesco, 1:302-317, 11+9 fig., 11 maps.

--. 1984. Photichthyidae, in Whitehead, P.J.P., Bauchot, M.L., Hureau, J.C., Nielsen, J. & E. Tortonese, Fishes of the North-eastern Atlantic and the Mediterranean/Poissons de l'Atlantique du Nord-Est et de la Méditerranée, Paris, Unesco, 1:318-324, 6+1 fig., 6 maps.

BADCOCK, J.; BAIRD, R.C. 1979. Remarks on systematics, development, and distribution of the hatchetfish genus Sternoptyx (Pisces, Stomiatoidei). Fishery Bull., 77(4):803-820, 11 fig.

BADCOCK, J.; LARCOMBE, R. A. 1980. The sequence of photophore development in Xenodermichthys copei (Pisces: Alepocephalidae). J. mar. biol. Ass. U. K., 60:277-294.

BADCOCK, J.; MERRETT, N. R. 1972. On Argyripnus atlanticus Maul 1952 (Pisces, Stomiatoidei), with a description of post-larval forms. J. Fish Biol., 4:277-287, 3 fig.

--. 1976. Midwater fishes in the eastern North Atlantic. 1. Vertical distribution and associated biology in 30°N, 23°W, with developmental notes on certain myctophids. Prog. Oceanogr., 7(1):3-58, fig. 1-25.

--. 1977. On the distribution of midwater fishes in the eastern North

Atlantic. p.249-282, fig. 1-11. In: Andersen, N.R. & B.J. Zahuranec (ed.). Ocean Sound Scattering Prediction. New York.

BADIA, P.; LORENZO, A. 1982. Preliminary studies on transmural potential and intensity of the short circuit current in intestine of Gobius maderensis. Rev. esp. Fisiol., 38:221-226.

BAGNIS, R.; MAZELLIER, P.; BENNETT, J.; CHRISTIAN, E. 1972. Fishes of Polynesia. Melbourne, 368 p., many col. photo. and fig.

BAGUET, F; CHRISTOPHE, B.; MARECHAL, G. 1980. Luminescence of Argyrope-lecus photophores electrically stimulated. Comp. Biochem. Physiol., (A)67(3):375-381.

BAILEY, R. M.; FITCH, J. E.; HERALD, E. S.; LACHNER, E. A.; LINDSEY, C. C.; ROBINS, C. R.; SCOTT, W. B. 1970. A list of common and scientific names of fishes from the United States and Canada. (3rd ed.) Spec. Publ. Am. Fish. Soc., (6):1-150.

BAILLY, P. 1930. Description d'un Stomiatide nouveau de la région des Iles Canaries. Bull. Mus. natn. Hist. nat., Paris, (2)2:378-380.

BAINBRIDGE, V. 1957. Eggs and larvae of Ethmalosa dorsalis. Oceanography and sea fisheries on the West African coast. CCTA/CSA Symposium, Luanda (20-27 November 1957), (mimeo).

--. 1957. Food of Ethmalosa dorsalis (Cuvier and Valenciennes). Nature, Lond., 179:874-875.

--. 1960. The plankton of inshore waters off Freetown, Sierra Leone. Fishery Publs. colon. Off., (13), 48 p.

--. 1961. The early life history of the Bonga, Ethmalosa dorsalis (Cu-vier & Valenciennes). J. Cons. perm. int. Explor. Mer, 26:347-353.

--. 1962. The larvae of Pellonula vorax (Günther) (Clupeidae) in Sierra Leone coastal waters. Bull. Inst. fr. Afr. noire, (A)24(1):262-269.

--. 1963. The food, feeding habits and distribution of the Bonga Ethma-losa dorsalis (Cuvier & Valenciennes). J. Cons. perm. int. Explor. Mer., 28(2):270-284.

BAIRD, R. C. 1973. Sternopthychidae, p. 123-125, in Hureau, J.C. & Th. Monod (ed.), Check-list of the fishes of the north-eastern Atlantic and of the Mediterranean/Catalogue des poissons du nord-est Atlanti-que et de la Méditerranée (Clofnam). Paris, Unesco, 1973, 2 vol.

--. 1971. The systematics, distribution and zoogeography of the marine hatchetfishes (Family Sternoptychidae). Bull. Mus. comp. Zool. Harv., 142(1):1-128, 80 fig.

BAIRD, R. C.; HOPKINS, T. L.; WILSON, D. F. 1975. Diet and feeding chro-nology of Diaphus taaningi (Myctophidae) in the Cariaco Trench. Cope-ia, 1975:356-365.

BAIRD, R. C.; WILSON, D.; MILLIKEN, D. 1973. Observations on Bregmaceros nectabanus Whitley in the anoxic, sulfurous water of the Cariaco Trench. Deep Sea Res., 20:503-504.

BAIRD, R. C.; WILSON, D. F.; BECKETT, R. C.; HOPKINS, T. L. 1974. Dia-phus taaningi Norman, the principal component of a shallow sound-scattering layer in the Cariaco Trench, Venezuela. J. mar. Res., 32:301-312.

BAIRD, S.F. 1873. Natural history of some of the more important foodfis-hes of the south shore of New England II. The bluefish Pomatomus saltatrix. Rep. U.S. Commnr Fish. (1871-72), 1:228-252.

BAIRD, S. F.; GIRARD, C. 1855. Descriptions of new species of fishes collected in Texas, New Mexico and Sonora... Proc. Acad. nat. Sci. Philad., (1854):24-29.

BALASUBRAHMANYAN, K.; BUSHANA RAO, K. S. P.; SUBBA RAO, R. C. 1969. Larval and juvenile stages of the flying fish Exocoetus volitans Linn. from the Bay of Bengal. Bull. natn. Inst. Sci. India, 38:876-884, fig. A-L.

BALDUCCI, E. 1915. Notizie e osservazioni sul Bathophilus nigerrimus

Gigl. _Annls_ _Inst._ _océanogr._, _Monaco_, 7(3):1-16.
BALLARD, W. W. 1969. Normal embryonic stages of _Gobius_ _niger_ _jozo_. _Pub-_
_bl._ _Staz._ _zool._ _Napoli_, 37:1-17.
BANARESCU, P. 1964. Pisces. Osteichthyes. _Fauna_ _Repub._ _pop._ _rom._, 13:1-
962, 402 fig.
BANE, G. W., Jr. 1965. Spawning of the margined flyingfish, _Cypselurus_
cyanopterus (Valenciennes) in the Gulf of Guinea. _Copeia_, 1965:382.
--. 1970. The food of the Yellowfin tuna in the Gulf of Guinea (Pisces:
Scombridae). _Wasmann_ _J._ _Biol._, 28:207-218.
BARANENKOVA, A. S. 1955. The catch of _Paralepis_. _Priroda_, _Mosk._, 7:118-
119, 1 fig.
BARBER, R. T.; HAEDRICH, R. L. 1969. Gobies associated with a scattering
layer off southwest Africa. _Deep_ _Sea_ _Res._, 16:105-106.
BARBOUR, T. 1905. Notes on Bermudian fishes. _Bull._ _Mus._ _comp._ _Zool._
Harv., 46(7):109-134, pl; 1-4.
--. 1941. Note on pediculate fishes. _Proc._ _New_ _Engl._ _zool._ _Club_, 19:7-
14, pl. 2-7.
--. 1942. On two Cuban deep sea fish. _Proc._ _New_ _Engl._ _zool._ _Club_, 19:41-
44.
--. 1942. The northwestern Atlantic species of frog fishes. _Proc._ _New_
Engl. _zool._ _Club_, 19:21-40, pl. 8-17.
--. 1942. More concerning ceratioid fishes. _Proc._ _New_ _Engl._ _zool._ _Club_,
21:77-86.
--. 1942. More Cuban deep-sea fishes. _Proc._ _New_ _Engl._ _zool._ _Club_, 19:45-
50, 73.
BARKOVA, N. A. 1978. Particularités de la répartition et état des stocks
de la sardine (_Sardina_ _pilchardus_) dans la région de l'Afrique du
Nord-Ouest. _In_: Rapport du groupe de travail _ad_ _hoc_ sur les poissons
pélagiques côtiers ouest-africains de la Mauritanie au Libéria (26°N
à 5°N). COPACE/PACE. Ser. 78/10 (mimeo).
BARNARD, K. H. 1925-1927. A monograph of the marine fishes of South
Africa. _Ann._ _S._ _Afr._ _Mus._, 21:1-1065, fig; 1-32, pl. 1-37.
--. 1927. Diagnoses of new genera and species of South African marine
fishes. _Ann._ _Mag._ _nat._ _Hist._, (9)20:66-79.
--. 1937. Further notes on South African marine fishes. _Ann._ _S._ _Afr._
Mus., 32(2):41-67, 4 fig.
--. 1940. (Report of) Department of fishes and marine invertebrates. _In_:
Rep. _S._ _Afr._ _Mus._ (1939), 19 p., 1 pl.
--. 1947. A pictorial guide to South African fishes marine and freshwater. Cape Town, xvii + 226, 2 fig., 25 pl.
--. 1948. Further notes on South African marine fishes. _Ann._ _S._ _Afr._
Mus., 36:341-406, 17 fig., pl. 9-13.
BARNARD, K. H.; BONDE, C. von 1944. An adult specimen of _Diretmus_ (Bery-
comorphi). _Ann._ _Mag._ _nat._ _Hist._, (11)11:236-240, fig.
BARNETT, M. A.; GIBBS, R. H., Jr. 1968. Four new stomiatoid fishes of
the genus _Bathophilus_ with a revised key to the species of _Bathophi-_
lus. _Copeia_, 1968 (4):826-832.
BARNEVILLE, B. de. 1847. Note sur un nouveau genre d'Anguilliformes.
Revue _zool._, 1847:219-220.
BARNHART, P. S. 1936. Marine fishes of southern California. Berkeley,
209 p., 290 fig.
BARON, J. C. 1971. Les transferrines de deux espèces de sardinelles:
Sardinella _aurita_ (C.V.) et _Sardinella_ _eba_ (C.V.). _Cah._ _ORSTOM_, _Océa-_
nogr., 9(1):85-96.
--. 1973. Note sur les protéines sériques de _Sardinella_ _aurita_ Val.,
1847 in C.V. _Cah._ _ORSTOM_, _Océanogr._, 11(2):133-170.
--. 1973. Les esterases du sérum de _Sardinella_ _aurita_ Valenciennes, 1847
in C.V. Application à l'étude des populations. _Cah._ _ORSTOM_, _Océa-_

nogr., 11(4):389-418.

--. 1974. Preliminary studies on the blood of Sardinella from the West African coast. Europ. Conf. Anim. Blood grp biochem. Polymorph., 12:593-595.

BARSUKOV, V. V. 1979. Subspecies of the Atlantic Helicolenus dactylopterus (De la Roche, 1809) (in Russian). Vop. Ikhtiol., 19(4):579-595, 2 fig.

--. 1986. Anarhichadidae, in P.J.P. Whitehead, M.L. Bauchot, J.C. Hureau, J. Nielsen & E. Tortonese. Fishes of the North-eastern Atlantic and the Mediterranean/Poissons de l'Atlantique du Nord-Est et de la Méditerranée, Paris, Unesco, 3:1113-1116, 3 + 3 fig.

BARTLETT, M.R.; BACKUS, R.H. 1962. A catch of the rare gempylid Lepidocybium flavobrunneum (Smith) in the Bahamas. Copeia, 1962(4):845-847, fig.

BARTLETT, M. R.; HAEDRICH, R. L. 1968. Neuston nets and South Atlantic larval blue marlin (Makaira nigricans). Copeia, 1968(3):469-474, 3 fig.

BAS, C. 1974. Distribución de especies demersales recogidas durante la expedición oceanogràfica "Sahara I". Result. Exped. cient. B/O Cornide de Saavedra, Madrid, (3):187-247, 37 fig.

BAS, C.; ARIAS, A.; GUERRA, A. 1976. Pesca efectuadas durante la campaña "Altor V" (C. Bojador - C. Blanco, abril-mayo 1974). Características y tratamiento de las capturas. Result. Exped. cient. B/O Cornide de Saavedra, Madrid, (5):161-172, 3 tab.

BASS, A. J.; d'AUBREY, J. D.; KISTNASAMY, N. 1973. Sharks of the east coast of southern Africa. I. The genus Carcharhinus (Carcharhinidae). Investl Rep. (S. Afr. Ass. mar. biol. Res.), (33):1-168.

--. 1975. Sharks of the east coast of southern Africa. II. The families Scyliorhinidae and Pseudotriakidae. Investl Rep. (S. Afr. Ass. mar. biol. Res.), (37):1-64.

--. 1975. Sharks of the east coast of southern Africa. IV. The families Odontaspididae, Scapanorhynchidae, Isuridae, Cetorhinidae, Alopiidae, Orectolobidae, and Rhiniodontidae. Investl Rep. (S. Afr. Ass. mar. biol. Res.), (39):1-102.

--. 1975. Sharks of the east coast of southern Africa. V. The families Hexanchidae, Chlamydoselachidae, Heterodontidae, Pristiophoridae, and Squatinidae. Investl Rep. (S. Afr. Ass. mar. biol. Res.), (43):1-50.

--. 1976. Sharks of the east coast of southern Africa. VI. The families Oxynotidae, Squalidae, Dalatiidae, and Echinorhinidae. Investl Rep. (S. Afr. mar. biol. Res.), (45):1-103.

BASSINDALE, R. 1961. On the marine fish fauna of Ghana. Proc. zool. Soc. Lond., 137(4):481-510.

BAST, D.; BERGERARD, P.; LAMBERT, K.; MAIGRET, J.; RICHER de FORGES, B.; SOUKHOVERSHINE, V.; WEISS, R. 1983. Exploration, par chalutage, du talus continental des côtes mauritaniennes (missions du N/R "Ernst Haeckel" et du "Walter Barth" - Mars-Avril, 1982). Bull. Cent. natn. Rech. océanogr. Pêch. Nouadhibou, 11(1):157-215, 10 fig., 34 tab.

BATALIANTS, K. Y., 1960. (Characteristics of gametogenesis in Sardinella aurita in the Dakar and Takoradi regions, in relation to reproduction - in Russian). Trudy balt. nauchno-issled. Inst. morsk. ryb. Khoz. Okeanogr., 5:118-126.

--. 1978. (Peculiarities of population composition of fish species and specific features of reproduction - in Russian). Trudy atlant. nauchno-issled. Inst. ryb. Khoz. Okeanogr., (74):18-28.

--. 1979. Composition par taille et par poids des captures à la senne tournante de certaines espèces pélagiques dans l'Atlantique centre-est. Annexe 17, p. 133-161. In: Anon., 1979.

BATH, H. 1965. Hypleurochilus phrynus n. sp. Erstmaliger Nachweis der

Gattung Hypleurochilus Gill im Mittelmeer (Pisces, Blennioidea, Blenniidae). Senckenberg. biol., 46(4):251-255.

--. 1966. Erstmaliger Nachweis von Blennius vandervekeni Poll, 1959 im Mittelmeer (Pisces: Blennioidea: Blenniidae). Senckenberg. biol., 47 (5):411-417.

--. 1968. Untersuchung von Blennius zvonimiri Kolombatovic und Beschreibung von Blennius incognitus n. sp. aus dem Mittelmeer (Pisces, Blennioidea, Blenniidae). Senckenberg. biol., 49(5): 367-386.

--. 1970. Vergleichend-morphologische, taxonomische und zoogeographische Untersuchungen an den Schleimfischarten Blennius cristatus, crinitus und nuchifilis (Pisces: Blennioidea: Blenniidae). Senckenberg. biol., 51 (5/6):287-306.

--. 1973. Wiederbeschreibung und neuer Nachweis von Tripterygion melanurus. Senckenberg. biol., 54:47-56.

--. 1973. Blenniidae, 1:519-527; 2:321; in Hureau, J.C. & Th. Monod (ed.), Check-list of the fishes of the north-eastern Atlantic and of the Mediterranean/Catalogue des poissons du nord-est Atlantique et de la Méditerranée (Clofnam). Paris, Unesco, 2 vol.

--. 1977. Revision der Blenniini (Pisces: Blenniidae). Senckenberg. biol., 57(4/6):167-234.

--. 1979. Geographische Variation der Körperfäbung und Flossenformel von Coryphoblennius galerita (Linnaeus, 1758) (Pisces: Blenniidae). Senckenberg. biol., 59(5/6):317-324.

--. 1981. Der Holotypus von Blennius ponticus Slastenenko 1934 und seine taxonomische Stellung (Pisces: Blenniidae). Senckenberg. biol., 61 (5/6):337-348, 12 fig., 4 tab.

--. 1981. Blenniidae, in Fischer W., Bianchi G. & W. B. Scott (ed.), FAO species identification sheets for fishery purposes. Eastern Central Atlantic; fishing areas 34, 47 (in part). Canada Funds-in-Trust. Ottawa, Department of Fisheries and Oceans Canada, by arrangement with the Food and Agriculture Organization of the United Nations, vol. 1-7: pag. var.

--. 1982. Beitrag zur Revalidation von Parablennius ruber (Valenciennes, 1836) mit kritischen Bermerkungen zur Gültigkeit der Gattung Pictiblennius Whitley, 1930 (Pisces: Blenniidae). Senckenberg. biol., 62 (4/6):211-224, 19 fig.

BATH, H.; WIRTZ, P., 1981. Rediscovery of Hypleurochilus aequipinnis (Günther, 1861) in West Africa (Pisces: Blenniidae). Senckenberg. biol., 61 (5/6):349-355,6 fig., 1 tab.

BAUCHOT-BOUTIN, M. L., 1953. Révision synoptique du genre Serrivomer (Anguilliformes). Bull. Mus. natn. Hist. nat., Paris, (2)25(4):365-367.

BAUCHOT-BOUTIN, M.L., 1954. Identification de Serrivomer beani Gill et Ryder (Téléostéen Anguilliforme). Bull. Mus. natn. Hist. nat., Paris, (2) 26:303-306, 2 fig.

BAUCHOT-BOUTIN, 1955. Détermination des Serrivomer du "Michael Sars" et de l'"Armauer Hansen". p. 11-12. In: E. Koefoed, Iniomi (Myctophidae exclusive), Lyomeri, Apodes from the Michael Sars North Atlantic Deep-sea Expedition, 1910. Rep. scient. Results Michael Sars. N. Atlant. deep Sea Exped., 4(2)(4):1-15, 2 pl.

BAUCHOT, M. L. 1959. Etude des larves leptocéphales du groupe Leptocephalus lanceolatus, Strömman et identification à la famille des Serrivomeridae. Dana Rep., (48):1-148, 105 fig., 2 pl.

--. 1963. Catalogue critique des types de poissons du Muséum national d'Histoire naturelle. I: Famille des Labridae. II: Famille des Chaetodontidae, Scatophagidae, Toxotidae, Monodactylidae, Ephippidae, Scorpidae, Pempheridae, Kyphosidae, Girellidae. Publs divers. Mus. natn. Hist. nat., Paris, (20):1-195.

--. 1966. Poissons marins de l'Est Atlantique Tropical - Téléostéens Perciformes. II. Percoidei (3e p.). III. Acanthuroidei. IV. Balistoidei. V. Blennioidei. Atlantide Rep., (9): I-IV:7-43, 8 fig. 5 pl.; V:63-91, 4 fig. 3 pl.

--. 1967. Catalogue critique des types de poissons du Muséum national d'Histoire naturelle. Blennioidei. Publs divers. Mus. natn. Hist. nat., Paris, (21):1-70.

--. 1969. Les poissons de la collection de Broussonet au Muséum National d'Histoire Naturelle de Paris. Bull. Mus. natn. Hist. nat., Paris, (2)41(1):125-143.

--. 1969. Etude d'une collection récoltée au cours des campagnes de chalutage dans le Golfe de Guinée (G.T.S. 1963-64). Bull. Mus. natn. Hist. nat., Paris, (2)41(2):410-425, 8 fig.

--. 1970. Catalogue critique des types de poissons du Muséum national d'Histoire naturelle (suite). (Lampridiformes, Stephanoberyciformes, Beryciformes, Zeiformes, Coryphaeniformes). Publs divers. Mus. natn. Hist. nat., Paris, (24):1-55.

--. 1973. Catalogue critique des types de poissons du Muséum national d'Histoire naturelle (suite). (Familles des Centracanthidae, Dipterygonotidae et Emmelichthyidae). Bull. Mus. natn. Hist. nat., Paris, (3)(143) (Zool. 107):881-995.

--. 1973. Saccopharyngidae, Eurypharyngidae, p. 216-219, in Hureau, J.C. & Th. Monod (ed.), Check-list of the fishes of the north-eastern Atlantic and of the Mediterranean/Catalogue des poissons du nord-est Atlantique et de la Méditerranée (Clofnam). Paris, Unesco, 2 vol.

--. 1986. Anguillidae, in P.J.P. Whitehead, M.L. Bauchot, J.C. Hureau, J. Nielsen & E. Tortonese. Fishes of the North-eastern Atlantic and the Mediterranean/Poissons de l'Atlantique du Nord-Est et de la Méditerranée, Paris, Unesco, 2:535-536, 1 fig.

--. 1986. Muraenidae, in P.J.P. Whitehead, M.L. Bauchot, J.C. Hureau, J. Nielsen & E. Tortonese. Fishes of the North-eastern Atlantic and the Mediterranean/Poissons de l'Atlantique du Nord-Est et de la Méditerranée, Paris, Unesco, 2:537-544, 7 + 8 fig.

--. 1986. Serrivomeridae, in P.J.P. Whitehead, M.L. Bauchot, J.C. Hureau, J. Nielsen & E. Tortonese. Fishes of the North-eastern Atlantic and the Mediterranean/Poissons de l'Atlantique du Nord-Est et de la Méditerranée, Paris, Unesco, 2:548-550, 2 + 9 fig.

--. 1986. Ophichthidae, in P.J.P. Whitehead, M.L. Bauchot, J.C. Hureau, J. Nielsen & E. Tortonese. Fishes of the North-eastern Atlantic and the Mediterranean /Poissons de l'Atlantique du Nord-Est et de la Méditerranée, Paris, Unesco, 2:577-585, 8 + 7 fig.

BAUCHOT, M. L.; BAUCHOT, R. 1983. Les Pagellus de l'Océan Indien (Pisces, Perciformes, Sparidae). Bull. Mus. natn. Hist. nat., Paris, (4) 5 A (4):1123-1138.

BAUCHOT, M. L.; BLACHE, J. 1979. Présence d'Ariosoma balearicum (de la Roche, 1809) en mer Rouge (Pisces, Teleostei, Congridae). Bull. Mus. natn. Hist. nat., Paris, (4)1 A (4):1131-1137, fig. 1-2.

BAUCHOT, M. L.; BLANC, M. 1961. Catalogue des types de Scombroidei (Poissons Téléostéens Perciformes) des collections du Muséum National d'Histoire Naturelle de Paris. Bull. Mus. natn. Hist. nat., Paris, (2)33(4):369-379.

--. 1961. Poissons marins de l'Est Atlantique tropical - Téléostéens Perciformes. I. Labroidei. II. Percoidei (le partie). Atlantide Rep., (6), I:43-64, 6 fig.; II: 65-100, 4 fig., 1 pl.

--. 1962. Sur deux espèces de Labridae (Poissons Téléostéens Perciformes) des côtes occidentales d'Afrique. Bull. Mus. natn. Hist. nat., Paris, (2)34(1):67-71.

--. 1963. Poissons marins de l'Est Atlantique tropical. II. Percoidei,

(Téléostéens, Perciformes). 2e partie. Atlantide Rep., 7:37-61, 3 fig.

BAUCHOT, M. L.; DAGET, J. 1967. Les Lutjanus des côtes occidentales d'Afrique. Réhabilitation de L. endecacanthus Bleeker, 1863 (Poissons, Perciformes). Bull. Mus. natn. Hist. nat., Paris, (2)39(2): 260-264.

--. 1971. Les Diplodus (Pisces, Sparidae) du groupe cervinus-fasciatus. Cah. ORSTOM. Océanogr., 9(3): 319-338, 6 fig.

BAUCHOT, M. L.; HUREAU, J. C. 1978. Sparus pagrus miqueli n. ssp., nouvelle sous-espèce des côtes tropicales ouest-africaines (Téléostéens, Sparidae). Cybium, (3)(4):97-98.

--. 1980. Sparus pagrus miqueli Bauchot et Hureau, 1978, synonyme junior de Pagrus africanus Akasaki, 1962. Cybium, (3)(8):99.

--. 1986. Sparidae, in P.J.P. Whitehead, M.L. Bauchot, J.C. Hureau, J. Nielsen & E. Tortonese. Fishes of the North-eastern Atlantic and the Mediterranean/Poissons de l'Atlantique du Nord-Est et de la Méditerranée, Paris, Unesco, 2:883-907, 25 + 18 fig.

BAUCHOT, M.L.; PRAS, A. 1980. Guide des poissons marins d'Europe. Lausanne, Paris, 427 p., 34 fig., pl. 1-24 + 1-40.

BAUCHOT, M. L.; QUIGNARD, J. P. 1973. Labridae, p. 426-443, in Hureau, J.C. & Th. Monod (ed.), Check-list of the fishes of the north-eastern Atlantic and of the Mediterranean/Catalogue des poissons du nord-est Atlantique et de la Méditerranée (Clofnam). Paris, Unesco, 2 vol.

BAUCHOT, M. L.; SALDANHA, L. 1973. Serrivomeridae, p. 229-230, Cyemidae, p. 234, in Hureau, J.C. & Th. Monod (ed.), Check-list of the fishes of the northeastern Atlantic and of the Mediterranean/Catalogue des poissons du nord-est Atlantique et de la Méditerranée (Clofnam). Paris, Unesco, 2 vol.

--. 1986. Muraenesocidae, in P.J.P. Whitehead, M.L. Bauchot, J.C. Hureau & E. Tortonese. Fishes of the North-eastern Atlantic and the Mediterranean/Poissons de l'Atlantique du Nord-Est et de la Méditerranée, Paris, Unesco, 2:559-561, 2 + 4 fig.

--. 1986. Congridae, in P.J.P. Whitehead, M.L. Bauchot, J.C. Hureau, J. Nielsen & E. Tortonese. Fishes of the North-eastern Atlantic and the Mediterranean/Poissons de l'Atlantique du Nord-Est et de la Méditerranée, Paris, Unesco, 2:567-574, 7 + 6 fig.

--. 1986. Derichthyidae, in P.J.P. Whitehead, M.L. Bauchot, J.C. Hureau, J. Nielsen & E. Tortonese. Fishes of the North-eastern Atlantic and the Mediterranean/Poissons de l'Atlantique du Nord-Est et de la Méditerranée, Paris, Unesco, 2:575-576, 2 + 2 fig.

BAUCHOT, M. L.; DESOUTTER, M.; ALLEN, G. R. 1978. Catalogue critique des types de Poissons du Muséum national d'Histoire naturelle (suite) (Famille des Pomacentridae). Bull. Mus. natn. Hist. nat., Paris, (3) suppl. 1:1-54 p.

BAUCHOT, M. L.; HUREAU, J. C.; MIQUEL, J. C. 1981. Sparidae, in Fischer, W., Bianchi, G. & W.B. Scott (ed.), FAO species identification sheets for fishery purposes. Eastern Central Atlantic; fishing areas 34, 47 (in part). Canada Funds-in-Trust. Ottawa, Department of Fisheries and Oceans Canada, by arrangement with the Food and Agriculture Organization of the United Nations, vol. 1-7: pag. var.

BAUCHOT, M. L.; WHITEHEAD, P. J. P.; MONOD, T. 1982. Date of publication and authorship of the fish names in Eydoux & Souleyet's Zoology of "La Bonîte", 1841-1852. Cybium, 6(3):59-73.

BAUCHOT, M. L.; DAGET, J.; HUREAU, J. C.; MONOD, Th. 1970. Le problème des "auteurs secondaires" en taxionomie. Bull. Mus. natn. Hist. nat. Paris, 42(2):301-304.

BAUCHOT, M. L.; IWAMOTO, T.; GEISTDOERFER, P.; RANNOU, M. 1971. Etude critique des résultats des expéditions scientifiques du "Travailleur"

et du "Talisman". Nouvel examen des Macrouridae (Téléostéens, Gadiformes). Bull. Mus. natn. Hist. nat., Paris, (3)(14) (Zool. 14):653-
669.

BAUGHMAN, J. L. 1943. The lutianid fishes of Texas. Copeia, 1943(4):212-
215.

--. 1950. Random notes on Texas fishes. II. Texas J. Sci., 2:242-263.

--. 1955. The oviparity of the whale shark, Rhincodon typus, with records of this and other fishes in Texas water. Copeia, 1955(1):54-55.

BAUZA-RULLAN, J. 1955. Contribución al conocimiento de la ictiología
actual y fósil de España. Boln R. Soc. esp. Hist. nat. (Geol.),
(1954) 52:63-71, 3 pl.

--. 1957. Nueva contribución al estudio de los otolitos de peces actua-
les y fósiles de España. Mems Comun. Inst. geol., Barcelona, 16:33-
44, 5 pl.

--. 1960. Contribución al conocimiento de los otolitos de peces. Boln R.
Soc. esp. Hist. nat. (Biol.), (1959) 57:89-118, pl. 1-8.

--. 1961. Contribuciones al conocimiento de los otolitos de peces actua-
les. Boln R. Soc. esp. Hist. nat. (Biol.), 59:153-168.

--. 1969. Contribución al conocimiento de los otolitos de peces actua-
les. Boln R. Soc. esp. Hist. nat. (Biol.), 66:105-114, 68 fig.

BAYAGBONA, E. O. 1963. Biometric study of two species of Pseudotolithus
from Lagos trawling grounds. Bull. Inst. fr. Afr. noire, (A)25:238-
264.

--. 1965. The effect of fishing effort on croakers in the Lagos fishing
grounds. Bull. Inst. fr. Afr. noire, (A)27(1):334-338.

BEAN, B. A. 1905. Fishes of the Bahama Islands. p. 293-325, pl. 52-61.
In: The Bahama Islands. A report of the Bahama expedition sent out by
the Geographical Society of Baltimore (ed. G.B. Shattuck). New York.

--. 1925. Description of a new species of luminous lizard fish, Scopelo-
saurus smithii, from off the coast of Brazil. Proc. biol. Soc. Wash.,
38:13-14.

BEAN, T. H. 1885. Description of a new species of Plectromus (P. crassi-
ceps) taken by the United States Fish Commission. Proc. U. S. natn.
Mus., 8(5):73-74.

--. 1887. Description of a new species of Thyrsitops (T. violaceus) from
the fishing banks off the New England coast. Proc. U.S. natn Mus.,
10:513-514.

--. 1890. Scientific results of explorations by the U.S. Fish Commission
steamer Albatross, no 11. New fishes off the coast of Alaska and the
adjacent region southward. Proc. U.S. natn. Mus., 13:37-45.

--. 1903. Catalogue of the fishes of New York. Bull. N.Y. St. Mus.,
(60): 784 p.

--. 1906. A catalogue of the fishes of Bermuda, with notes on a collection made in 1905 for the Field Museum. Publs Field Mus. nat. Hist.
(Zool., 108), (2)7:21-89, 14 fig.

--. 1912. Description of new fishes of Bermuda. Proc. biol. Soc. Wash.,
25:121-126.

BEAN, T. H.; DRESSEL, H. L. 1884. A catalogue of fishes received from
the public museum of the Institute of Jamaica, with description of
Pristipoma approximans and Tylosurus euryops, two new species. Proc.
U.S. natn. Mus., 7(418):151-170.

BEARDSLEY, G. L. Jr.; MERRETT, N. R.; RICHARDS, W. J. 1975. Synopsis of
the biology of the sailfish, Istiophorus platypterus (Shaw and Nodder, 1791); NOAA tech. Rep. NMFS SSRF, 675 (Part 3):95-120, 9 fig.

BEAUFORT, L. F. de. 1913. Fishes of the eastern part of the Indo-Australian Archipelago with remarks on its zoogeography. Bijdr. Dierk., 19:
95-163, 8 fig., 2 pl.

BEAUFORT, L. F. de; BRIGGS, J. C. 1962. The fishes of the Indo-Australian Archipelago. Scleroparei.. Pediculati.. Plectognathi.. Xenopterygii. Leiden, 11:xi + 481 p., 100 fig.

BECKLEY, L. E. 1984. The ichtyofauna of the Sundays Estuary, South Africa, with particular reference to the juvenile marine component. Estuaries, 7(3):248-258.

BEEBE, W. 1926. A new ceratioid fish. Preliminary description of a new genus and species. Bull. N.Y. zool. Soc., 29(2):80, fig.

--. 1929. Deep-sea fishes of the Hudson Gorge. Zoologica, N.Y., 12(1):1-19, 1 fig.

--. 1932. Nineteen new species and four postlarval deep-sea fish. Zoologica, N.Y., 13(4):47-107, fig. 1-31.

--. 1933. Preliminary account of deep sea dives in the bathysphere with especial reference to one of 2200 feet. Proc. natn. Acad. Sci. U.S.A., 19:178-188.

--. 1933. Deep-sea isospondylous fishes. Two new genera and four new species. Zoologica, N.Y., 13(8):159-167, fig. 39-42.

--. 1933. Deep-sea fishes of the Bermuda Oceanographic Expeditions. (Introduction, family Alepocephalidae, family Argentinidae). Zoologica, N.Y. 16(1-3):1-147, 46 fig.

--. 1933. New data on the deep sea fish Stylophthalmus and Idiacanthus. Science, N.Y., 78(2026):390.

--. 1933. Deep-sea stomiatoid fishes. One new genus and eight new species. Copeia, 1933(4):160-175.

--. 1934. Deep-sea fishes of the Bermuda Oceanographic Expeditions 1931: Family Idiacanthidae. Zoologica, N.Y. 16(4):149-241, fig. 47-81.

--. 1935. Deep-sea fishes of the Bermuda Oceanographic Expeditions: Family Nessorhamphidae. Zoologica, N.Y., 20(2):25-51, fig. 10-22.

--. 1937. Preliminary list of Bermuda deep-sea fish, based on the collections from fifteen hundred metre net hauls, made in an eight-mile circle south on Nonsuch Island, Bermuda. Zoologica, N.Y., 22(3) (14): 197-208.

--. 1941. Eastern Pacific Expedition of the New York Zoological Society. XXVII. A study of young sailfish (Istiophorus). Zoologica, N.Y., 26 (3):209-227, 9 fig., 9 pl.

BEEBE, W.; CRANE, I. 1937. Deep-sea fishes of the Bermuda Oceanographic Expeditions. Family Serrivomeridae. Part. II: Genus Platuronides. Zoologica, N.Y., 22(26):331-348, 14 fig.

--. 1939. Deep-sea fishes of the Bermuda Oceanographic Expeditions. Family Melanostomiatidae. Zoologica, N.Y., 24(6):65-238, 77 fig.

--. 1946. Eastern Pacific Expeditions of the New York Zoological Society. XXXVII. Deep-sea ceratioid fishes. Zoologica, N.Y., 31(2):151-182, 19 fig., 3 pl.

BEEBE, W.; PYL, M. van der 1944. Eastern Pacific Expeditions of the New York Zoological Society. XXXIII. Pacific Myctophidae. (Fishes). Zoologica, N.Y., 29(2):59-95.

BEEBE, W.; TEE-VAN, J. 1928. The fishes of Port-au-Prince Bay, Haiti. Zoologica, N.Y., 10:1-279, 268 fig., pl. A.

--. 1932. New Bermuda fish, including six new species and forty-three species hitherto unrecorded from Bermuda. Zoologica, N.Y., 13(5):109-120.

--. 1933. Field book of the shore fishes of Bermuda. New York, London, xiv + 337 p., 343 fig.

--. 1936. Systematic notes on Bermudian and West Indian tunas of the genera Parathunnus and Neothunnus. Zoologica, N.Y., 21(3):177-194.

--. 1938. Eastern Pacific Expeditions of the New-York Zoological Society, 15: Seven new marine fishes from Lower California. Zoologica, N.Y., 23(3):299-312, 5 text-fig., 3 pl.

BEKKER, V.E. 1963. New data on the lantern fish genera Electrona and

Protomyctophum (Pisces, Myctophidae) of the southern hemisphere. Vop. Ikhtiol., 3(1):15-28 (in Russian).

--. 1964. Slender tailed myctophids (genera Loweina, Tarletonbeania, Gonichthys and Centrobranchus) of the Pacific and Indian Ocean. Systematics and distribution. Trudy Inst. Okeanol., 73:11-75, 25 fig. (in Russian).

--. 1964. On the temperate-cold water complex of myctophids (Myctophidae, Pisces). Okeanologija, 4 (3): 469-475 (in Russian).

--. 1965. The lanternfishes of the genus Hygophum (Myctophidae, Pisces). Systematics and distribution. Trudy Inst. Okeanol., 80:62-103, 11 fig. (in Russian).

--. 1967. The lanternfishes (Myctophidae) from the "Petr Lebedev" Atlantic Expeditions, 1961-1964. Trudy Inst. Okeanol., 84:84-124, 13 fig. (in Russian).

--. 1967. Luminous anchovies (Fam. Myctophidae). Pacif. Ocean (Biol.), 7(3):145-181 (in Russian).

--. 1968. New data on mesopelagic fishes of the genus Opisthoproctus Vaillant. Probl. Ichthyol., 8(2):189-193 (Engl. transl. from Vop. Ikhtiol., 8(2):241-246).

BEKKER, V. E.; BORODULINA, O. D. 1968. Lantern fishes of the genus Ceratoscopelus Günther. Systematic and distribution. Vop. Ikhtiol., 8(5):779-798 (in Russian; Engl. transl.: Probl. Ichthyol., 8(5):625-640, 5 fig.).

--. 1978. "Myctophum asperum" species-group with description of a new species and Myctophum selenops Tåning (Myctophidae, Osteichthyes). Taxonomy and distribution. Trudy Inst. Okeanol., 111:108-125 (in Russian).

BEKKER, V. E.; SHCHERBACHEV, Y. N.; TCHUVASOV, V. M. 1975. Deep-sea pelagic fishes of the Caribbean, Gulf of Mexico and the Puerto Rico Trench. Trudy Inst. Okeanol., 100:289-336.

BELL, R. 1859. On the natural history of the Gulf of St. Lawrence, and the distribution of the Mollusca of eastern Canada. Can. Naturalist Geologist, 4(3):197-220.

BELLOC, G. 1926. Note préliminaire sur un poisson nouveau du genre Centropholis. Bull. Mus. natn. Hist. nat., Paris, (2)32:271-274.

--. 1927. Remarques sur le Germon. Part 2 (Biologie). Rapp. P.-v. Réun. Cons. perm. int. Explor. Mer, 44:92-97.

--. 1934. Catalogue illustré des poissons comestibles de la côte occidentale d'Afrique (du Cap Spartel au Cap Vert). le partie, poissons cartilagineux. Revue Trav. Off. (scient. tech.) Pêch. marit., 7(2): 117-195, 38 fig.

--. 1937. Sur la capture dans le bassin d'Arcachon d'un squale nouveau pour la faune de France, Euprotomicrus sarmenti Noronha. Bull. Mus. natn. Hist. nat., Paris, (2)9:370-372.

--. 1938. Liste des poissons pélagiques et bathypélagiques capturés au cours de la cinquième croisière avec diagnoses préliminaires de deux espèces nouvelles. Revue Trav. Off. (scient. Tech.) Pêch. marit., 11:281-313, 29 fig.

--. 1949. Catalogue des types de poissons du Musée Océanographique de Monaco. Bull. Inst. océanogr., Monaco, (958):1-23, 3 pl.

--. 1954. Les thons de la Mediterranée. Première note: germon, pélamide et melva. Proc. tech. Pap. gen. Fish. Coun. Mediterr., 2(49):283-318.

--. 1954. Historia natural del atûn, Thunnus thynnus (L.). Boln Inst. esp. Oceanogr., 67:1-88.

--. 1954. Rapport préliminaire sur le Neothunnus albacora (Lowe), avec des indications sur le N. macropterus (Temm. et Schl.). Rapp. présenté au "Comité Atlantique" Cons. int. Explor. Mer Copenhagen. 136 p.

BELLON L. ; BARDAN de BELLON, E. 1949. Algunos datos sobre los "Thunni-

dae" de Canarias. Boln Inst. esp. Oceanogr., 19:1-28.
BELLON URIARTE, L.; BARDAN MATEU, E. 1931. Nota sobre los peces Elasmo-
branquios de Canarias. Notas Resúm. Inst. esp. Oceanogr., (2)(53):1-
39, fig. 1-9.
BELLOTTI, C., 1878. Note ittiologiche. Osservazioni fatte sulla collezione ittiologica del Museo civico di Storia Naturale in Milano. Atti
Soc. ital. Sci. nat., 20(1):53-60.
--. 1883. Note ittiologiche: osservazioni fatte sulla collezione ittiologica del Museo civico di Storia Naturale in Milano. VII. I lepto-
cefali del mare di Messina. Atti Soc. ital. Sci. nat., 26:165-181,
fig.
--. 1891. Appunti all'opera del Dottor Emilio Moreau: "Histoire naturelle des poissons de la France", e al relativo supplemento. Atti Soc.
ital. Sci. nat., 33(2):107-143, 1 pl.
--. 1892. Note al manuale d'ittiologia francese del Dott. Emilio Moreau.
Atti Soc. ital. Sci. nat., 34:19- 35.
--. 1903. Di un nuovo pteraclide giapponese. Atti Soc. ital. Sci. nat.,
42:136-139.
BELVEZE, H. 1971. Premières observations sur le stock sardinier de l'A-
tlantique marocain. Bull. Inst. Pêch. marit. Maroc, (18):3-35.
--. 1972. Observations complémentaires sur le stock sardinier de l'A-
tlantique marocain de 1968 à 1971 et essais d'estimatíon de quelques
paramètres. Bull. Inst. Pêch. marit. Maroc, (20):5-55.
--. 1974. L'état des stocks de la sardine marocaine à la fin de l'année
1973. Trav. Docums Dév. Pêch. marit. Casablanca, (6):1-17 (mimeo).
--. 1975. Pluviométrie et pêche de la sardine au Maroc depuis 1936 mise
en évidence de cycles parallèles. Trav. Docums Dév. Pêch. marit.
Casablanca, (10):1-8, (mimeo).
BELVEZE, H.; BRAVO de LAGUNA, J. 1980. Les ressources halieutiques de
l'Atlantique centre-est, deuxième partie: les ressources de la côte
ouest-africaine entre 24°N et le détroit de Gibraltar. FAO Docums
Tech. Pêch., (182.2): 64 p. (mimeo).
BELVEZE, H.; RAMI, M. 1978. Détermination de l'âge de la sardine marocaine (Sardina pilchardus Walb.) éléments de croissance comparés.
Bull. Inst. Pêch. marit. Maroc, (23):57-81.
BELYANINA, T. N. 1974. Development, taxonomy and distribution of fishes
of the family Bregmacerotidae. Trudy Inst. Okeanol., 96:143-188.
--. 1977. Materials on development of chauliodontid fishes (Chauliodontidae, Pisces). Trudy Inst. Okeanol., 109:113-132, 8 fig. (in Russian).
--. 1982. Larvae of midwater fishes in the western tropical Pacific
Ocean and the seas of the Indo-Australian Archipelago. Trudy Inst.
Okeanol., 118:3-54.
BENNETT, E. T. 1827. Observations on the fishes contained in the collection of the Zoological Society. Zool. J., 3(11)(37):371-378, pl. 9.
--. 1831. (1 Sept. 1831). (A small collection of Fishes, formed during
the voyage of H.M.S. Chanticleer...). Proc. Comm. Sci. Corresp. zool.
Soc. Lond., pt. 1 (26 July 1831):112. (in the index quoted as: Observations on a collection of Fishes, formed during the voyage of H.M.S.
Chanticleer, with characters of two new species.)
--. 1831. (6 December 1831). (A collection of Fishes was exhibited...
presented to the Society by Captain Belcher, R.N...). Proc. Comm.
Sci. Corresp. zool. Soc. Lond., 1 (8 Nov. 1831):145-148.(in the
index quoted as: Characters of new genera and species of fishes of
the Atlantic coast of northern Africa presented by Capt. Belcher,
R.N... p. 146-148.)
--. 1832. (2 March 1832). (The remaining portion of the collection of
Fishes formed at the Mauritius by Charles Telfair...). Proc. Comm.

Sci. Corresp. zool. Soc. Lond., 1 (27 December 1831):165-169. (in the index quoted as: Observations on a collection of Fishes from the Mauritius, presented by Mr. Telfair, with characters of new genera ans species.)

--. 1835. Characters of several previously undescribed fishes from Trebizond, collected by Keith Abbott. Proc. zool. Soc. Lond., 1835, (3):91-92.

BENNETT, F. D. 1840. Narrative of a whaling voyage round the globe, from the year 1833 to 1836. 2, London, vii + 1-395 p., text illust.

BENNETT, G. 1860. Gatherings of a naturalist in Australasia. London, xii + 1-456 p., text illust., 8 pl.

BEN-TUVIA, A. 1953. Mediterranean fishes of Israel. Bull. Sea Fish. Res. Stn Israel, (8):1-40, 20 fig.

--. 1956. Eels of the family Heterenchelidae. Ann. Mag. nat. Hist., (12)9:401-408, 4 fig.

--. 1960. Synopsis of biological data on Sardinella aurita of the Mediterranean and other waters. FAO Proc. Wld scient. Meet. Biol. Sardines, 2:287-312 (Synopsis 14).

--. 1960. Synopsis of the systematics and biology of Sardinella maderensis (Lowe). FAO Proc. Wld scient. Meet. Biol. Sardines, 2:497-519.

--. 1963. Variation in vertebral number of young Sardinella aurita in relation to temperature during spawning season. Rapp. p.-v. Réun. Commn int. Explor. scient. Mer Méditerr., 17(2):313-318, 2 fig.

--. 1966. Red Sea fishes recently found in the Mediterranean. Copeia, 1966(2):254-275, 2 fig.

--. 1971. Revised list of the Mediterranean fishes of Israel. Israel J. Zool., 20:1-39.

--. 1981. Mullidae. In Fischer, W., Bianchi G. & W.B. Scott (ed.), FAO species identification sheets for fishery purposes. Eastern Central Atlantic; fishing areas 34, 47 (in part). Canada Funds-in-Trust. Ottawa, Department of Fisheries and Oceans Canada, by arrangement with the Food and Agriculture Organization of the United Nations, vol. 1-7: pag. var.

--. 1986. Holocentridae, in P.J.P. Whitehead, M.L. Bauchot, J.C. Hureau, J. Nielsen & E. Tortonese. Fishes of the North-eastern Atlantic and the Mediterranean/Poissons de l'Atlantique du Nord-Est et de la Méditerranée, Paris, Unesco, 2:753-755, 2 fig.

--. 1986. Teraponidae, in P.J.P. Whitehead, M.L. Bauchot, J.C. Hureau, J. Nielsen & E. Tortonese. Fishes of the North-eastern Atlantic and the Mediterranean/Poissons de l'Atlantique du Nord-Est et la Méditerranée, Paris, Unesco, 2:797-799, 2 fig.

--. 1986. Sillaginidae, in P.J.P. Whitehead, M.L. Bauchot, J.C. Hureau, J. Nielsen & E. Tortonese. Fishes of the North-eastern Atlantic and the Mediterranean/Poissons de l'Atlantique du Nord-Est et de la Méditerranée, Paris, Unesco, 2:875-876, 1 fig.

--. 1986. Siganidae, in P.J.P. Whitehead, M.L. Bauchot, J.C. Hureau, J. Nielsen & E. Tortonese. Fishes of the North-eastern Atlantic and the Mediterranean/Poissons de l'Atlantique du Nord-Est et de la Méditerranée, Paris, Unesco, 2:964-966, 2 fig.

--. 1986. Sphyraenidae, in P.J.P. Whitehead, M.L. Bauchot, J.C. Hureau, J. Nielsen & E. Tortonese. Fishes of the North-eastern Atlantic and the Mediterranean/Poissons de l'Atlantique du Nord-Est et de la Méditerranée, Paris, Unesco, 3:1194-1196, 3 fig.

--. 1986. Mugilidae, in P.J.P. Whitehead, M.L. Bauchot, J.C. Hureau, J. Nielsen & E. Tortonese. Fishes of the North-eastern Atlantic and the Mediterranean/Poissons de l'Atlantique du Nord-Est et de la Méditerranée, Paris, Unesco, 3:1197-1204, 8 + 2 fig.

BEN-TUVIA, A.; DICKSON, W. (ed.). 1969. Proceedings of the FAO Conference on fish behaviour in relation to fishing techniques and tactics (19-27 October 1957, Bergen). FAO Fishery Rep., (62)3(Experience papers):463-884.

BEN-TUVIA, A.; MCKAY, R. 1986. Haemulidae, in P.J.P. Whitehead, M.L. Bauchot, J.C. Hureau, J. Nielsen & E. Tortonese. Fishes of the Northeastern Atlantic and the Mediterranean/Poissons de l'Atlantique du Nord-Est et de la Méditerranée, Paris, Unesco, 2:858-869, 6 fig.

BERDAR, A. 1970. Bathophilus nigerrimus Gigl. raccolto sulla spiaggia di Ginzirri (Messina). Atti Soc. Pelor. Sci. fis. mat. nat., 16(1-2):75-86, 7 fig.

BERG, C. 1895. Enumeración sistemática y sinonímica de los peces de las costas argentina y uruguaya. An. Mus. nac. Hist. nat. B. Aires, 4:1-120, 1 pl.

BERG, L.S. 1913. A review of the clupeoid fishes of the Caspian Sea, with remarks on the herring-like fishes of the Russian Empire. Ann. Mag. nat. Hist., (8)11:472-480.

--. 1915. Predvaritel'nyi otchet o sel'dyakh, sobrannikh v Caspyiskom moreye Caspyiscoy ekspeditziey 1913. Mater. Pozn. russk. rybolov., 4(6):3-8, 2 pl.

--. 1940. Classification of fishes both recent and fossil. Trudy zool. Inst., Leningr., 5(2):87-517, fig. 1-190. (Russian with complete English translation).

--. 1947. Classification of fishes both recent and fossil. Ann Arbor, Mich., 517 p., 187 fig.

--. 1955. Classification of fishes, both recent and fossil. Trudy zool. Inst., Leningr.1, 20:1-286, fig. 1-263.

BERGERARD, P.; DOMAIN, F.; RICHER de FORGES, B. 1983. Evaluation, par chalutages, des ressources démersales du plateau continental mauritanien. Bull. Cent. natn. Rech. océanogr. Pêch. Nouadhibou, 11(1):217-250, 6 fig., 13 tab.

BERKELEY, S. A.; HOUDE, E. D. 1978. Biology of two exploited species of halfbeaks, Hemiramphus brasilienis and H. balao from southeast Florida. Bull. mar. Sci., 28(4):624-644.

BERKENHOUT, J. 1789. Synopsis of the natural history of Great Britain and Ireland, 1. London, 380 p.

BERNARD, F. 1958. Plancton et benthos observés durant trois plongées en Bathyscaphe au large de Toulon. Annls Inst. océanogr. Monaco, (n.s.) 35(4):287-326, 2 pl., 8 fig.

BERRY, F. H. 1958. A new species of fish from the western North Atlantic, Dikellorhynchus tropidolepis, and relationships of the genera Dikellorhynchus and Malacanthus. Copeia, 1958:116-125, 6 fig.

--. 1978. Grammicolepidae, in Fischer, W. (ed.) FAO species identification sheets for fishery purposes. Western Central Atlantic (Fishing Area 31). FAO, Rome, Vol. 1-7, pag. var.

BERRY, F. H.; BALDWIN, W. J. 1966. Triggerfishes (Balistidae) of the eastern Pacific. Proc. Calif. Acad. Sci., (4)34(9):429-474.

BERRY, F. H.; COHEN, L. 1974. Synopsis of the species of Trachurus (Pisces, Carangidae). Q. Jl. Fla Acad. Sci., 35(4):177-211.

BERRY, F. H.; PERKINS, H. C. 1966. Survey of pelagic fishes of the California Current area. Fishery Bull., 65(3):625-682, fig. 1-30.

BERRY, F. H.; POLL, M. 1961. Synonymy of the Atlantic Ocean filefish Alutera heudelotii Hollard. Copeia, 1961:360-362.

BERRY, F. H.; ROBINS, C. R. 1967. Macristiella perlucens, a new clupeiform fish from the Gulf of Mexico. Copeia, 1967(1):46-50.

BERRY, F. H; VOGELE, L. E. 1961. Filefishes (Monacanthidae) of the western North Atlantic. Fishery Bull. Fish Wildl. Serv. U.S., 61(181): 57-109, 42 fig., 16 tab.

BERTELSEN, E. 1943. Notes on the deep-sea angler fish Ceratias holbölli

Kr. <u>Vidensk. Meddr</u> dansk <u>naturh. Foren.</u>, 107:185-206, 5 fig.

--. 1951. The ceratioid fishes, ontogeny, taxonomy, distribution and biology. <u>Dana Rep.</u>, (39):276 p., 141 fig., 1 pl.

--. 1958. The argentinoid fish <u>Xenophthalmichthys danae</u>. <u>Dana Rep.</u>, (45):1-8, fig. 1-7.

--. 1980. Notes on Linophrynidae V: A revision of the deepsea angler-fishes of the <u>Linophryne arborifera</u>-group (Pisces, Ceratioidei). <u>Steenstrupia</u>, 6(6):29-70.

--. 1982. Notes on Linophrynidae VIII. A review of the genus <u>Linophryne</u>, with new records and descriptions of two new species. <u>Steenstrupia</u>, 8(3):49-104.

--. 1986. Eutaeniophoridae, <u>in</u> P.J.P. Whitehead, M.L. Bauchot, J.C. Hureau, J. Nielsen & E. Tortonese. Fishes of the North-eastern Atlantic and the Mediterranean/Poissons de l'Atlantique du Nord-Est et de la Méditerranée, Paris, Unesco, 2:522-523, 2 + 2 fig.

--. 1986. Caulophrynidae, <u>in</u> P.J.P. Whitehead, M. L. Bauchot, J.C. Hureau, J. Nielsen & E. Tortonese. Fishes of the North-eastern Atlantic and the Mediterranean/Poissons de l'Atlantique du Nord-Est et de la Méditerranée, Paris, Unesco, 3:1373-1375, 2 + 2 fig.

--. 1986. Melanocetidae, <u>in</u> P.J.P. Whitehead, M.L. Bauchot, J.C. Hureau, J. Nielsen & E. Tortonese. Fishes of the North-eastern Atlantic and the Mediterranean/Poissons de l'Atlantique du Nord-Est et la Méditerranée, Paris, Unesco, 3:1376-1377, 2 + 7 fig.

--. 1986. Himantolophidae, <u>in</u> P.J.P. Whitehead, M.L. Bauchot, J.C. Hureau & Nielsen & E. Tortonese. Fishes of the North-eastern Atlantic and the Mediterranean/Poissons de l'Atlantique du Nord-Est et la Méditerranée, Paris, Unesco, 3:1378-1380, 3 + 1 fig.

--. 1986. Diceratiidae, <u>in</u> P.J.P. Whitehead, M.L. Bauchot, J.C. Hureau, J. Nielsen & E. Tortonese. Fishes of the North-eastern Atlantic and the Mediterranean/Poissons de l'Atlantique du Nord-Est et de la Méditerranée, Paris, Unesco, 3:1381-1382, 1 fig.

--. 1986. Oneirodidae, <u>in</u> P.J.P. Whitehead, M.L. Bauchot, J.C. Hureau, J. Nielsen & E. Tortonese. Fishes of the North-eastern Atlantic and the Mediterranean/Poissons de l'Atlantique du Nord-Est et de la Méditerranée, Paris, Unesco, 3:1383-1399, 18 + 45 fig.

--. 1986. Thaumatichthyidae, <u>in</u> P.J.P. Whitehead, M.L. Bauchot, J.C. Hureau, J. Nielsen & E. Tortonese. Fishes of the North-eastern Atlantic and the Mediterranean/Poissons de l'Atlantique du Nord-Est et de la Méditerranée, Paris, Unesco, 3:1400, 1 + 1 fig.

--. 1986. Centrophrynidae, <u>in</u> P.J.P. Whitehead, M.L. Bauchot, J.C. Hureau, J. Nielsen & E. Tortonese. Fishes of the North-eastern Atlantic and the Mediterranean/Poissons de l'Atlantique du Nord-Est et de la Méditerranée, Paris, Unesco, 3:1401-1402, 1 fig.

--. 1986. Ceratiidae, <u>in</u> P.J.P. Whitehead, M.L. Bauchot, J.C. Hureau, J. Nielsen & E. Tortonese. Fishes of the North-eastern Atlantic and the Mediterranean/Poissons de l'Atlantique du Nord-Est et de la Méditerranée, Paris, Unesco, 3:1403-1405, 2 + 2 fig.

--. 1986. Gigantactinidae, <u>in</u> P.J.P. Whitehead, M.L. Bauchot, J.C. Hureau, J. Nielsen & E. Tortonese. Fishes of the North-eastern Atlantic and the Mediterranean/Poissons de l'Atlantique du Nord-Est et de la Méditerranée, Paris, Unesco, 3:1406-1407, 1 + 1 fig.

--. 1986. Linophrynidae, <u>in</u> P.J.P. Whitehead, M.L. Bauchot, J.C. Hureau, J. Nielsen & E. Tortonese. Fishes of the North-eastern Atlantic and the Mediterranean/Poissons de l'Atlantique du Nord-Est et de la Méditerranée, Paris, Unesco, 3:1408-1414, 7 + 9 fig.

BERTELSEN, E.; MARSHALL, N. B. 1956. The Miripinnati, a new order of teleost fishes. <u>Dana Rep.</u>, (42):1-34, 15 fig., 1 pl.

--. 1958. Notes on Miripinnati (An addendum to Dana Report n° 42), a change of name and further records. Dana Rep., (45):9-10, 1 fig.

BERTELSEN, E.; NIELSEN, J. G. 1986. Saccopharyngidae, in P.J.P. Whitehead, M.L Bauchot, J.C. Hureau, J. Nielsen & E. Tortonese. Fishes of the North-eastern Atlantic and the Mediterranean/Poissons de l'Atlantique du Nord-Est et de la Méditerranée, Paris, Unesco, 2:530-533, 5 fig.

BERTELESEN, E.; PIETSCH, T. W. 1975. Results of the research cruises of FRV "Walther Herwig" to South America. 38. Osteology and relationships of the ceratioid anglerfish genus Spiniphryne (Family Oneirodidae). Arch. FischWiss., 26(1):1-11.

--. 1977. Results of the research cruises of FRV "Walther Herwig" to South America. 47. Ceratioid anglerfishes of the family Oneirodidae collectd by the FRV "Walther Herwig". Arch. FischWiss., 27(3):171-189.

--. 1983. The ceratioid anglerfishes of Australia. Rec. Austr. Mus., 35:77-99.

BERTELSEN, E.; QUERO, J. C. 1981. Capture au large du Maroc de Centrophryne spinulosa Regan et Trewavas, 1932 (Pisces, Lophiiformes, Centrophrynidae), espèce nouvelle pour l'Atlantique nord-est. Cybium, (3)5(2):89-90.

BERTELSEN, E.; STRUHSAKER, P.J. 1977. The ceratioid fishes of the genus Thaumatichthys. Osteology, relationships, distribution and biology. Galathea Rep., 14:7-40, pl. 1-3.

BERTELSEN, E., PIETSCH, T. W.; LAVENBERG, R. J. 1981. Ceratioid anglerfishes of the family Gigantactinidae: morphology, systematics, and distribution. Contr. Sci., (332):1-74, 69 fig., 21 tab.

BERTELSEN, E.; KREFFT, E.; MARSHALL, N. B., 1976. The fishes of the family Notosudidae. Dana-Rep., (86):1-114, fig. 1-67, pl. 1.

BERTIN, L., 1926. Les migrations de l'anus au cours de l'ontogenèse chez les poissons apodes. Bull. Soc. zool. Fr., 51:327-344, 3 fig.

--. 1934. Les poissons apodes appartenant au sous-ordre des Lyomères. Dana Rep., (3):1-56, 2 pl.

--. 1935. Les types de Kaup dans la collection des leptocéphales du Museum. Bull. Mus. natn. Hist. nat., Paris (2)7(2):99-106, 1 fig.

--. 1936. Contribution à l'étude des larves de poissons apodes (les types de Strömman à l'Institut zoologique de l'Université d'Uppsala). Bull. Inst. océanogr. Monaco, (694):1-16, fig. 1-13.

--. 1936. Sur une série de leptocéphales appartenant au genre Saccopharynx Mitchill. - C.r. hebd. Séanc. Acad. Sci., Paris, 203:1540-1542.

--. 1936. Un type primitif de nageoire caudale chez les poissons du genre Cyema Günther. Bull. Soc. zool. Fr., 61:440-446, 5 fig.

--. 1936. Un nouveau genre de poissons apodes caractérisé par l'absence de machoire supérieure. - Bull. Soc. zool. Fr., 61(7):533-540, 4 fig.

--. 1937. Les poissons abyssaux du genre Cyema Günther (anatomie, embryologie, bionomie). Dana Rep., (10):1-30, 24 fig.

--. 1938. Formes nouvelles et formes larvaires de poissons apodes appartenant au sous-ordre des Lyomères. Dana Rep., (15):1-26, 17 fig., 2 pl.

--. 1939. Acquisitions récentes sur la biologie larvaire des poissons apodes. Bull. Soc. zool. Fr., 63:385-395, 2 fig.

--. 1939. Catalogue des types de poissons du Muséum National d'Histoire Naturelle. 1ère Partie: Cyclostomes et Sélaciens. Bull. Mus. natn. Hist. nat., Paris, (2)11:51-98.

--. 1940. Remaniement dans la classification des poissons apodes appartenant à la famille des Serrivoméridés. C.R. hebd. Séanc. Acad. Sci., Paris, 211:76-78.

--. 1940. Catalogue des types de poissons du Muséum National d'Histoire Naturelle. 2e partie: Dipneustes, Chondrostéens, Holostéens, Isospondyles. <u>Bull</u>. <u>Mus</u>. <u>natn</u>. <u>Hist</u>. <u>nat</u>., <u>Paris</u>, (2)12(6):244-322.

--. 1940. Révision des Stomiatiformes (Téléostéens Isospondyles) du Muséum. <u>Bull</u>. <u>Mus</u>. <u>natn</u>. <u>Hist</u>. <u>nat</u>., <u>Paris</u>, (1939) (2)11(4):378-382.

--. 1941. Mise au point sur quelques espèces de Clupéidés. <u>Bull</u>. <u>Soc</u>. <u>zool</u>. <u>Fr</u>., **66**:18-25.

--. 1941. Description anatomique du genre <u>Stemonidium</u>, type peu connu de Serrivoméridés, et définition de cette famille de poissons apodes. <u>Bull</u>. <u>Mus</u>. <u>natn</u>. <u>Hist</u>. <u>nat</u>., <u>Paris</u>, (2)13(6):524-531, 3 fig.

--. 1944. Synopsis ostéologique et synonymie des poissons de la famille des Serrivoméridés (Apodes, Anguilliformes). <u>Bull</u>. <u>Mus</u>. <u>natn</u>. <u>Hist</u>. <u>nat</u>., <u>Paris</u>, (2) **16**:101-108, 2 fig.

--. 1944. La distribution mesogéenne des <u>Elops</u>. <u>C</u>. <u>r</u>. <u>somm</u>. <u>Séanc</u>. <u>Soc</u>. <u>Biogéogr</u>., 21(178):17-23.

--. 1945. Les ouvrages d'ichthyologie et les types des poissons de Risso au Muséum de Paris. <u>Bull</u>. <u>Mus</u>. <u>natn</u>. <u>Hist</u>. <u>nat</u>., <u>Paris</u>, (2)17:373-381.

--. 1947. Notules ichthyologiques:1. Compléments sur le genre <u>Avocetti-nops</u>. <u>Bull</u>. <u>Soc</u>. <u>zool</u>. <u>Fr</u>., **72**:54-56.

BERTIN, L.; ESTEVE, R. 1950. Catalogue des types de poissons du Muséum National d'Histoire Naturelle. 5e partie: Ostariophysaires (Siluri-formes). Paris, 85 p.

BERTIN, L.; ARAMBOURG, C. 1958. Super-ordre des Téléostéens. p. 2204-2500, fig. 1561-1788. <u>In</u>: P.P. Grassé (ed.). Traité de Zoologie, Paris, 13(3):1813-2758, fig. 1303-1889.

BERTOLINI, F. 1933. Apogonidae, Serranidae. <u>In</u>: Uova, larve e stadi giovanili di Teleostei. (U. d'Ancona ed.). <u>Fauna</u> <u>Flora</u> <u>Golfo</u> <u>Napoli</u>, **38**:306-331, fig. 245-246; pl. 15, fig. 14-16; pl. 19-21.

BEST, A. G. C.; BONE, Q. 1976. On the integument and photophores of the alepocephalid fishes <u>Xenodermichthys</u> and <u>Photostylus</u>. <u>J</u>. <u>mar</u>. <u>biol</u>. <u>Ass</u>. <u>U.K</u>., **56**:227-236, 2 fig., 2 pl.

BIANCHI, G. 1984. Fiches FAO d'Identification des espèces pour les be-soins de la pêche. Guide des Resources Halieutiques de l'Atlantique Marocain (Espèces Marines et d'Eaux Saumâtres). Préparé et publié avec le support du PNUD (Project MOR/81/002) Rome, FAO, 151 p.

BIAZ, R. 1978. Mise au point d'une méthode d'identification des popu-lations de sardines par electrophorèse des protéines solubles du noyau cristallin. <u>Bull</u>. <u>Inst</u>. <u>Pêch</u>. <u>marit</u>. <u>Maroc</u>, (23):3-55.

BIDEN, C.L. 1930. Sea angling fishes of the Cape. London, 304 p., 48 pl. (2nd ed. 1948 Cape Town & Johannesburg, same pagination).

BIERBAUM, G. 1914. Untersuchungen über den Bau der Gehörorgane von Tief-seefischen. <u>Z</u>. <u>wiss</u>. <u>Zool</u>., 3(3):281-380, pl. 5, 6.

BIGELOW, H. B.; BARBOUR, T. 1944. A new giant ceratioid fish. <u>Proc</u>. <u>New</u> <u>Engl</u>. <u>zool</u>. <u>Club</u>, **23**:9-15.

BIGELOW, H. B.; SCHROEDER, W. C. 1929. A rare bramid fish (<u>Taractes</u> <u>princeps</u> Johnson) in the northwestern Atlantic. <u>Bull</u>. <u>Mus</u>. <u>comp</u>. <u>Zool</u>. <u>Harv</u>., **69**(2):41-50, 1 pl.

--. 1936. Supplementary notes on fishes of the Gulf of Maine. <u>Bull</u>. <u>Bur</u>. <u>Fish</u>., <u>Wash</u>., 48(20):319-343.

--. 1948. Fishes of the western North Atlantic. Cyclostomes:29-58, fig. 1-5, Sharks:59-576, fig. 6-106. <u>Mem</u>. <u>Sears</u> <u>Fdn</u> <u>mar</u>. <u>Res</u>., 1(1).

--. 1948. New genera and species of batoid fishes. <u>J</u>. <u>mar</u>. <u>Res</u>., 7(3):543-566, fig. 1-9.

--. 1950. New and little known cartilaginous fishes from the Atlantic. <u>Bull</u>. <u>Mus</u>. <u>comp</u>. <u>Zool</u>. <u>Harv</u>., 103:385-408, pl. 1-7.

--. 1951. Three new skates and a new chimaerid fish from the Gulf of Mexico. <u>J</u>. <u>Wash</u>. <u>Acad</u>. <u>Sci</u>., 41(12):383-392, fig. 1-4.

--. 1953. Fishes of the western North Atlantic. Sawfishes, Guitarfishes, Skates and Rays, Chimaeroids. <u>Mem</u>. <u>Sears</u> <u>Fdn</u> <u>mar</u>. <u>Res</u>., 1(2):xv + 588 p., 127 fig.

--. 1953. Fishes of the Gulf of Maine. <u>Fishery</u> <u>Bull.</u> <u>Fish</u> <u>Wildl.</u> <u>Serv.</u> <u>U.S.</u>, **53**:viii + 1-577, fig. 1-288.

--. 1954. Deep water elasmobranchs and chimaeroids from the northwestern Atlantic slope. <u>Bull.</u> <u>Mus.</u> <u>comp.</u> <u>Zool.</u> <u>Harv.</u>, 112(2):37-87, fig. 1-7.

BIGELOW, H. B.; WELSH, W. W. 1925. Fishes of the Gulf of Maine. <u>Bull.</u> <u>Bur.</u> <u>Fish.,</u> <u>Wash.</u>, 40(1):1-567, 278 fig.

BINI, G. 1931. Rapporto sulla Crociera di pesca del tonno compiuta a bordo del Piropeschereccio "Orata" nel' Atlantico. <u>Boll.</u> <u>Pesca</u> <u>Piscic.</u> <u>Idrobiol.</u>, 7(1):57-90, 14 fig., 1 map.

--. 1965. Catalogo dei nomi dei pesci, dei molluschi e dei crostacei di importanza commerciale nel Mediterraneo. Roma, xv + 407 p., 587 fig.

--. 1967-1972. Atlante dei pesci delle Coste italiane. Mondo Sommerso, Milano, 9 vol.:1, 1967, Leptocardi, Ciclostomi, Selaci, 206 p., 66 fig. + 64 col. fig.; 2, 1971, Osteittei (Acipenseriformi, Clupeiformi, Mictofiformi, Anguilliformi), 300 p., 30 fig., 73 col. fig.; 3, 1970, Notacantiformi...Zeiformi, 229 p., 34 fig., 63 col. fig.; 4, 1968, Perciformi (Mugiloidei, Percoidei), 163 p., 34 fig., 49 col. fig.; 5, 1968, Perciformi (Percoidei), 175 p., 22 fig., 56 col. fig.; 6, 1968, Perciformi (Trichiuroidei...Blennioidei), 177 p., 48 fig., 57 col. fig.; 7, 1969, Perciformi (Ofidioidei...Dactilopteroidei), 196 p., 57 fig., 59 col. fig.; 8, 1968, Pleuronettiformi, Echeniformi, Gobiesociformi, Tetraodontiformi, Lofiformi, 164 p., 34 fig., 63 col. fig.; 9, 1972, Introduzione, Parte generale, Aggiornamenti, Indici, 176 p.

BINI, G. & TORTONESE, E. 1955. Missione sperimentale di pesca nel Chile e nel Peru-Pesci marini peruviani. <u>Boll.</u> <u>Pesca</u> <u>Piscic.</u> <u>Idrobiol.</u>, 30(9):1-39

BLACHE, J. 1962. Liste des poissons signalés dans l'Atlantique tropico-oriental sud du Cap des Palmes (4° Lat. N) à Mossamédès (15° Lat. S)(Province guinéo-équatoriale). <u>Trav.</u> <u>Centre</u> <u>océanogr.</u> <u>Pointe-Noire</u>, (2):13-102 (=Cah. ORSTOM, Océanogr).

--. 1963. <u>Ichthyococcus</u> <u>polli</u> nov. sp. (Gonostomidae) poisson téléostéen bathypélagique nouveau de l'Atlantique tropical africain. <u>Bull.</u> <u>Mus.</u> <u>natn.</u> <u>Hist.</u> <u>nat.</u>, Paris, (2) 35(5):468-472, fig. 1-3.

--. 1964. Les poissons du bassin du Tchad et du bassin adjacent du Mayo Kebbi. Etude systématique et biologique. <u>Mém.</u> <u>ORSTOM,</u> <u>Paris</u> (4):1-483, 141 fig.

--. 1964. Note préliminaire sur les larves leptocéphales d'Apodes du Golfe de Guinée (zone sud). <u>Trav.</u> <u>Centre</u> <u>océanogr.</u> <u>Pointe-Noire,</u> (5): 5-23, 32 pl. (=Cah. ORSTOM, Océanogr).

--. 1964. Sur la présence de <u>Luvarus</u> <u>imperialis</u> Raf., 1810 dans l'Atlantique oriental sud révélée par la découverte de deux larves au stade <u>Hystricinella</u> de L. Roule (1924)(Pisces, Teleostei, Perciformi, Luvaroidei, Luvaridae). <u>Cah.</u> <u>ORSTOM,</u> <u>Océanogr.</u>, (5):57-59, 1 fig. (=Trav. Centre océanogr. Pointe-Noire, III).

--. 1964. Contribution à la connaissance des Stomiatidae (Pisces, Teleostei, Clupeiformi, Stomiatoidei) dans l'Atlantique tropical oriental sud. Mise en évidence d'une sous-espèce de <u>Stomias</u> <u>colubrinus</u> Garman, 1899, caractéristique des formes du Golfe de Guinée. <u>Cah.</u> <u>ORSTOM,</u> <u>Océanogr.</u>, (5):61-69 (=Trav. Centre océanogr. Pointe-Noire, III).

--. 1964. Poissons bathypélagiques de la famille des Sternoptychidae (Teleostei, Clupeiformi, Stomiatoidei) provenant des campagnes de l'"Ombango", de la "Reine Pokou" et du "Gérard Tréca" dans les eaux africaines de l'Atlantique tropical. <u>Cah.</u> <u>ORSTOM,</u> <u>Océanogr.</u>, (5):71-87, 6 fig. (=Trav. Centre océanogr. Pointe-Noire, III).

--. 1964. Poissons bathypélagiques rares ou peu connus provenant des eaux de l'Atlantique oriental tropical. 1ère note. <u>Cah.</u> <u>ORSTOM,</u> <u>Océa-</u>

nogr., (5):89-96, 4 fig. (=Trav. Centre océanogr. Pointe-Noire, III).

--. 1964. Le genre _Bathylagus_ dans l'Atlantique tropical oriental sud (Teleostei, Clupeiformi, Opisthoproctoidei, Bathylagidae). _Cah. ORSTOM. Océanogr._, 2(1):7-16, 4 fig.

--. 1964. Sur la validité de _Chauliodus schmidti_ Ege, 1948 (Pisces, Teleostei, Clupeiformi, Chauliodidae) espèce caractéristique de l'Atlantique oriental. _Cah. ORSTOM. Océanogr._, 2(1):33-44.

--. 1964. Les genres _Astronesthes_ et _Borostomias_ dans l'Atlantique tropical oriental sud (Pisces, Teleostei, Clupeiformi, Stomiatoidei, Astronesthidae). _Cah. ORSTOM. Océanogr._, 2(1):17-31, 4 fig. 2 map.

--. 1967. Contribution à la connaissance de poissons anguilliformes de la côte occidentale d'Afrique. 1ère note: _Enchelycore nigricans_ (Bonnaterre, 1788). _Bull. Inst. fond. Afr. noire_, (A)29(1):163-177, 8 fig.

--. 1967. Contribution à la connaissance des poissons anguilliformes de la côte occidentale d'Afrique. 2e note: le genre _Muraena_ (Artedi) Linné, 1758. _Bull. Inst. fond. Afr. noire_, (A)29(1):178-217, 19 fig.

--. 1967. Contribution à la connaissance des poissons anguilliformes de la côte occidentale d'Afrique. 3e note: le genre _Echidna_ Forster, 1788. _Bull. Inst. fond. Afr. noire_, (A)29(2):695-709, 7 fig.

--. 1967. Contribution à la connaissance des poissons anguilliformes de la côte occidentale d'Afrique. 4e note: le genre _Lycodontis_ McClelland, 1845. _Bull. Inst. fond. Afr. noire_, (A)29(3):1122-1187, 29 fig.

--. 1967. Contribution à la connaissance des poissons anguilliformes de la côte occidentale d'Afrique. 5e note: le genre _Gymnothorax_ Bloch, 1795. _Bull. Inst. fond. Afr. noire_, (A)29(4):1695-1705, 5 fig.

-- 1967. Contribution à la connaissance des poissons anguilliformes de la côte occidentale d'Afrique. 6e note: les genres _Anarchias, Uropterygius_ et _Channomuraena_ (Muraenidae). _Bull. Inst. fond. Afr. noire_, (A)29(4):1706-1731, 14 fig.

--. 1967. Sur la présence de _Lycodontis polygonius_ (Poey, 1867) (Pisces, Teleostei, Anguilliformi, Muraenidae) sur les côtes de l'Archipel du Cap Vert. _Archos Mus. Bocage_, (2)1(16):339-350, 4 fig.

--. 1968. Contribution à la connaissance des poissons anguilliformes de la côte occidentale d'Afrique. 7e note: la famille des Muraenesocidae. _Bull. Inst. fond. Afr. noire_, (A)30(2):690-736, fig. 1-31.

--. 1968. Contribution à la connaissance des poissons anguilliformes de la côte occidentale d'Afrique. 8e note: la famille des Echelidae. _Bull. Inst. fond. Afr. noire_, (A)30(4):1501-1539.

--. 1968. Contribution à la connaissance des poissons anguilliformes de la côte occidentale d'Afrique. 9e note: la famille des Heterenchelidae. _Bull. Inst. fond. Afr. noire_, (A)30(4):1540-1581, 20 fig.

--. 1971. Contribution à la connaissance des poissons anguilliformes de la côte occidentale d'Afrique. 11e note: les genres _Mystriophis_ et _Echiopsis_ (Fam. des Ophichthidae). _Bull. Inst. fond. Afr. noire_, (A) 33(1):202-226.

--. 1971. Larves leptocéphales des poissons anguilliformes dans le Golfe de Guinée (zone sud). 1re note: Larves de Muraenidae. _Cah. ORSTOM. Océanogr._, 9(2):203-246, 36 fig.

--. 1972. Larves leptocéphales des poissons anguilliformes dans le Golfe de Guinée (zone sud). 2e note: les espèces adultes de Xenocongridae et leurs larves. _Cah. ORSTOM. Océanogr._, 10(3):219-241, 10 fig.

--. 1972. Contribution à la connaissance des poissons anguilliformes de la côte occidentale d'Afrique. 14e note: la famille des Synaphobranchidae. _Bull. Inst. fond. Afr. noire_, (A)34(4):966-973, 2 fig.

--. 1975. Contribution à la connaissance des poissons anguilliformes de la côte occidentale d'Afrique. 15e note: compléments aux familles des Muraenidae, des Heterenchelyidae et des Ophichthidae. _Bull. Inst._

fond. <u>Afr. noire</u>, (A)37(3):708-740, 13 fig.
--. 1977. Leptocéphales des poissons anguilliformes dans la zone sud du Golfe de Guinée. <u>Faune trop.</u>, 20:1-381, fig. 1-116.
BLACHE, J.; BAUCHOT, M. L. 1972. Contribution à la connaissance des poissons anguilliformes de la côte occidentale d'Afrique. 13e note: les genres <u>Verma, Apterichthus, Ichthyapus, Hemerorhinus, Caecula, Dalophis</u> avec la description de deux genres nouveaux (Fam. des Ophichthidae). <u>Bull. Inst. fond. Afr. noire</u>, (A)34(3):692-773.
--. 1976. Contribution à la connaissance des poissons anguilliformes de la côte occidentale d'Afrique. 16e note: les familles des Congridae et Colocongridae. <u>Bull. Inst. fond. Afr. noire</u>, (A)38(2):369-444, fig. 1-30.
--. 1976. Contribution à une révision générale des Ophichthidae (Pisces, Anguilliformi). I. Le genre <u>Herpetoichthys</u> Kaup, 1856. <u>Bull. Mus. natn. Hist. nat. Paris</u>, (3)(374)(Zool. 262):441-452, 9 fig.
BLACHE, J.; CADENAT, J. 1971. Contribution à la connaissance des poissons anguilliformes de la côte occidentale d'Afrique. 10e note: les genres <u>Myrichthys, Bascanichthys</u> et <u>Callechelys</u> (Fam. des Ophichthidae). <u>Bull. Inst. fond. Afr. noire</u>, (A)33(1):158-201.
BLACHE, J.; DUCROZ, J. 1960. <u>Lepidotrigla laevispinnis</u> n. sp. (Pisces, Perciformi, Scorpaenoidei, Triglidae), poisson nouveau du plateau continental congolais. <u>Bull. Mus. natn. Hist. nat. Paris</u>, (2)32(3): 205-208.
BLACHE, J.; ROSSIGNOL, M. 1961. Sur un poisson bathypélagique nouveau du Golfe de Guinée <u>Astronesthes blanci</u> nov. sp. (Clupeiformes, Stomiatoidei, Astronesthidae). <u>Bull. Mus. natn. Hist. nat. Paris</u>, (2)33(3): 282-284.
--. 1961. <u>Uraspis cadenati</u> nov. sp. (Carangidae) et <u>Xesurus biafraensis</u> nov. sp. (Acanthuridae). Poissons perciformes nouveaux du Golfe de Guinée. <u>Bull. Mus. natn. Hist. nat. Paris</u>, (2)33(5):479-484, 2 fig.
--. 1962. <u>Nansenia atlantica</u> nov. sp. (Teleostei, Clupeiformes, Opisthoproctoidei, Microstomidae) poisson bathypélagique nouveau du Golfe de Guinée. <u>Trav. Centre océanogr. Pointe-Noire</u> 1:101-106, fig.(=Cah. ORSTOM., Océanogr.).
BLACHE, J.; SALDANHA, L. 1972. Contribution à la connaissance des poissons anguilliformes de la côte occidentale d'Afrique. 12e note: les genres <u>Pisodonophis, Ophichtus, Brachysomophis</u> et <u>Ophisurus</u> (fam. Ophichthidae). <u>Bull. Inst. fond. Afr. noire</u>, (A)34(1):127-159.
BLACHE, J.; STAUCH, A. 1964. Contribution à la connaissance des poissons de la famille des Myctophidae dans la partie orientale du Golfe de Guinée (Teleostei, Clupeiformi, Myctophoidei). 1ère note. Les genres <u>Electrona</u> G. & B., 1895, <u>Hygophum</u> (Tån.) Bolin, 1939.<u>Trav. Centre océanogr. Pointe-Noire</u>, (5):97-104, fig. (=Cah. ORSTOM, Océanogr.).
--. 1964. Contribution à la connaisance des poissons de la famille des Myctophidae dans la partie orientale du Golfe de Guinée (Teleostei, Clupeiformi, Myctophoidei). 2ème note. Les genres <u>Benthosema</u> G. & B. 1895, <u>Diogenichthys</u> Bolin, 1939 et <u>Myctophum</u> Raf., 1810. <u>Cah. ORSTOM, Océanogr.</u>, 2(2):61-78, 11 fig.
BLACHE, J.; TORTONESE, E. 1968. Note sur un rare poisson anguilliforme de la Méditerranée: <u>Cynoponticus ferox</u> Costa, 1846 (Fam. Muraenesocidae). <u>Annali Mus. civ. Stor. nat. Giacomo Doria</u>, 77:1-11, 6 fig.
BLACHE, J.; BAUCHOT, M. L.; SALDANHA, L. 1973. Anguillidae, Simenchelidae, Muraenidae, Heterenchelyidae, p. 220-228; Xenocongridae, p. 233; Muraenesocidae, Nettastomatidae, Nettodaridae, Congridae, Derichthyidae, Echelidae, Nessorhamphidae, Ophichthidae, Synaphobranchidae, p. 235-253, <u>in</u> Hureau, J.C & Th. Monod (ed.). Check-list of the fishes of the north-eastern Atlantic and of the Mediterranean/Catalogue des poissons du nord-est Atlantique et de la Méditerranée, Paris, Unesco,

2 vol.

BLACHE, J.; CADENAT, J.; STAUCH, A. 1970. Clés de détermination des poissons de mer signalés dans l'Atlantique oriental entre le 20e parallèle nord et le 15e parallèle sud. _Faune trop._, 18:1-479, fig. 1-1152.

BLACHE, J.; MAUL, G. E.; SALDANHA, L. 1970. Présence d'adultes et de larves de _Nettodarus brevirostris_ et de _Nettodarus_ sp. dans l'Atlantique oriental (Pisces Anguilliformi Nettodaridae). _Archos Mus. Bocage_, (2)2(16):319-331, 15 fig.

BLACKBURN, M. 1957. The relation between the food of the Australian barracouta, _Thyrsites atun_ (Euphrasen), and recent fluctuations in the fisheries. _Aust. J. mar. Freshwat. Res._, 8(1):29-54.

BLACKBURN, M.; GARTNER, P. E. 1954. Population of barracouta, _Thyrsites atun_ (Euphrasen), in Australian waters. _Aust. J. mar. Freshwat. Res._, 5(3):411-468.

BLAINVILLE, H. M. DUCROTAY de, 1816. Prodrome d'une nouvelle distribution systématique du règne animal. _Bull. Soc. philomath., Paris_, 8: 105-124.

--. 1818. Poissons fossiles p. 314. _In_: Nouv. Dict. Hist. nat. (ed. 2), Paris, 27.

--. 1825. Vertébrés. Classe V. Poissons. Sous classe 1ère. Poissons cartilagineux. _In_: P. Vieillot _et al._, Faune française ou histoire naturelle générale et particulière des animaux qui se trouvent en France. Paris, livr. 13-14, 96 p., 22 pl.

BLANC, A. 1949. Recherches ichthyologiques dans le secteur de la Petite Côte. _Bull. Inf. Docum. (Sect. tech. Pêch.)_. Serv. Elevage Industr. anim. Sénégal, Dakar, (3):33-34.

--. 1950. Etude de l'Ethmalose (_Ethmalosa fimbriata_) de la Petite Côte du Sénégal. Serv. Océanogr. Pêch. marit. Dakar (mimeo).

--. 1950. Les Clupéidés de la Petite Côte. _Bull. Inf. Docum. (Sect. tech. Pêch.)_. Serv. Elevage Industr. anim. Sénégal, (6):7-13.

--. 1951. Aperçus biologiques sur _Ethmalosa fimbriata_ (Cobo) de la Petite Côte du Sénégal. _Bull. Inf. Docum. (Sect. tech. Pêch.)_. Serv. Elevage Industr. anim. Sénégal, (11):1-13.

--. 1957. Contribution à la biologie des sardinelles de la côte sud du Sénégal (_Sardinella eba_ et _S. aurita_), p. 43-48. _In_: J. Arnoux (ed.), La pêche maritime au Sénégal. CCTA/CSA Colloque d'Afrique occidentale, Luanda.

--. 1957. La pêche africaine sur la Petite Côte du Sénégal. Oceanography and sea fisheries on the West African coast, CCTA/CSA Symposium (20-27 November 1957, Luanda), Paper N° 27 (7 p., mimeo).

BLANC, M.; BAUCHOT, M. L. 1960. Révision des _Thalassoma_ (Poissons, Téléostéens Labridae) de l'Est Atlantique. _Bull. Mus. natn. Hist. nat., Paris_, (2)32(1):88-96, 5 fig.

--. 1964. Les Scombroidei (Poissons Téléostéens Perciformes) du Muséum National d'Histoire Naturelle de Paris. _Proc. Symp. Scombroid Fishes, mar. biol. Ass. India_, 1(1):443-458, 7 pl.

BLANC, M.; BLACHE, J. 1963. _Rhadinesthes lucberti_ n. sp. (Teleostei, Clupeiformi, Stomiatoidei, Astronesthidae) poisson bathypélagique nouveau de l'Atlantique oriental-sud. _Bull. Mus. natn. Hist. nat., Paris_ (2)35(2):136-139.

BLANC, M.; HUREAU, J. C. 1968. Catalogue critique des types de poissons du Muséum national d'Histoire naturelle (Poissons à joues cuirassées). _Publs divers. Mus. natn. Hist. nat._, Paris, (23), 71 p.

--. 1972. Catalogue critique des types de poissons du Muséum national d'Histoire naturelle (Suite) (Mugiliformes et Polynémiformes). _Bull. Mus. natn. Hist. nat., Paris_, (3)(15) (Zool. 15):673-735.

--. 1973. Scorpaenidae, Triglidae, p. 579-590, _in_ Hureau, J.C & Th.

Monod (ed.), Check-list of the fishes of the north-eastern Atlantic and of the Mediterranean/Catalogue des poissons du nord-est Atlantique et de la Méditerranée. Paris, Unesco, 2 vol.

BLANC, M.; PAULIAN, P. 1957. Poissons des Iles Saint-Paul et Amsterdam. Mém. Inst. scient. Madagascar, (F)1:325-335.

BLANC, M.; CADENAT, J.; STAUCH, A. 1968. Contribution à l'étude de l'ich tyofaune de l'Ile Anno Bon. Bull. Inst. fond. Afr. noire, (A) 30(1): 238-256, 6 fig.

BLANCHARD, E. 1866. Les poissons des eaux douces de la France. Paris, xv + 656 p., fig.

BLAXTER, J. H. S.; WARDLE, C. S.; ROBERTS, B. L. 1971. Aspects of the circulatory physiology and muscle systems of deep-sea fish. J. mar. biol. Ass. U.K, 51(4):991-1006.

BLEEKER, P. 1849. Bijdrage tot de kennis der ichtyologische fauna van het eiland Bali met beschrijving van eenige nieuwe species. Verh. batav. Genoot. Kunst. Wet., 22:1-11.

--. 1849. Bijdrage tot de kennis der ichthyologische fauna van het eiland Madura, met beschrijving van eenige nieuwe species. Verh. batav. Genoot. Kunst. Wet., 22:1-16.

--. 1851. Over eenige nieuwe soorten van Pleuronecteoïden van den Indischen Archipel. Natuurk. Tijdschr. Ned.-Indië, 1:401-416.

--. 1851. Over eenige nieuwe soorten van Belone en Hemiramphus van Java. Natuurk. Tijdschr. Ned.-Indië, 1:93-95.

--. 1851. Derde bijdrage tot de kennis der ichthyologische fauna van Borneo met beschrijving van eenige nieuwe soorten van zoetwatervisschen. Natuurk. Tijdschr. Ned.-Indië, 2:57-70.

--. 1852. Bijdrage tot de kennis der Snoekachtige visschen van den Soenda-Molukschen Archipel. Verh. batv. Genoot. Kunst. Wet., 24:1-28.

--. 1852. Bijdrage tot de kennis der Balistini en Ostraciones van den Indischen Archipel. (Each group published under a separate sub-title: Balistini p. 1-27, Ostraciones p. 28-36). Verh. batav. Genoot. Kunst. Wet., 24(11):1-36, Aahangsel p. 37-38, 7 pl.

--. 1852. Nieuwe bijdrage tot de kennis der ichthyologische fauna van Ceram. Natuurk. Tijdschr. Ned.-Indië, 3:689-714.

--. 1853. Bijdrage tot de kennis der ichthyologische fauna van Solor. Natuurk. Tijdschr. Ned.-Indië, 5(2):67-96.

--. 1853. Nieuwe tientallen diagnostiche beschrijvingen van nieuwe of weinig bekende vischsoorten van Sumatra. Natuurk. Tijdschr. Ned.-Indië, 5:495-534.

--. 1853. Bijdrage tot de kennis der Muraenöiden en Symbranchöiden van den Indischen Archipel. Verh. batav. Genoot. Kunst. Wet., 25:76 p.

--. 1853. Nalezingen op de ichthyologie van Japan. Verh. batav. Genoot. Kunst. Wet., 25: 1-56.

--. 1854. Faunae ichthyologicae japonicae species novae. Natuurk. Tijdschr. ned.-Indië, 6:395-426.

--. 1854. Vijfde bijdrage tot de kennis der ichthyologische fauna van Celebes. Natuurk. Tijdschr. Ned.-Indië, 7:225-260.

--. 1855. Vijfde bijdrage tot de kennis der ichthyologische fauna van Ternate. Naturk. Tijdschr. Ned.-Indië, 8:295-304.

--. 1856. Beschrijvingen van nieuwe en weinig bekende vischsoorten van Amboina, verzameld op eene reis door den Molukschen Archipel, gedaan in het gevolg van Gouverneur-Generaal Duymaer van Twist. Act. Soc. Sci. Indo-Neerl., 1:1-76.

--. 1856. Bijdrage tot de kennis der ichthyologische fauna van het eiland Boero. Natuurk. Tijdschr. Ned.-Indië, 11:383-414.

--. 1857. Bijdrage tot de kennis der ichthyologische fauna van de Sangi-eilanden. Natuurk. Tijdschr. Ned.-Indië, 13:369-380.

--. 1857. Nieuwe nalezingen op de ichthyologie van Japan. Verh. batav.

Genoot. Kunst. Wet., 26:1-126, 6 pl.

--. 1857. Achtste bijdrage tot de kennis der vischfauna van Amboina. Act. Soc. Sci. Indo-Néerl., 2:1-102.

--. 1858. Tiende bijdrage tot de kennis der vischfauna van Celebes. Act. Soc. Sci. Indo-Néerl., 3:1-16.

--. 1859. Enumeratio specierum piscium hucusque in Archipelago Indico observatarum. Act. Soc. Sci. Indo-Neerl.,6:1-276.

--. 1860. Bijdrage tot de kennis der vischfauna van Nieuw-Guinea: Act. Soc. Sci. Indo-Néerl. (1859), 6(1):1-24.

--. 1860. Elfde Bijdrage tot de kennis der vischfauna van Amboina. Act. Soc. Sci. Indo-Néerl., 8(5):1-14.

--. 1860. Over eenige vischsoorten van de Kaap de Goede Hoop. Natuurk. Tijdschr. Ned.-Indië, 21:49-80.

--. 1861. Notice sur le genre Trachinus (Artedi) et ses espèces. Annls Sci. nat. (Zool.), (4)16:375-382.

--. 1861. Iets over de geslachten der Scaroïden en hunne Indisch-archi-pelagische soorten. Versl. Akad. Amst., 12:228-244.

--. 1861. Vischsoorten van Nieuw-Guinea, verzameld door F.G. Beckman. Naturrk. Tijdschr. Ned.-Indië, 22:98-100 (list).

--. 1862. Notices ichthyologiques. IV. Sur une nouvelle espèce de Cossy-phus oxycephalus). Versl. Akad. Amst., 14:128-130.

--. 1862. Sur quelques genres de la famille des Pleuronectoïdes. Versl. Akad. Amst., 13:422-429.

--. 1862-1878. Atlas ichthyologique des Indes Orientales Néerlandaises, publié sous les auspices du Gouvernement colonial néerlandais, Amsterdam. 1, Scaroïdes et Labroïdes, 1862: xxi + 168 p., pl. 1-48; 2, Siluroïdes, Chacoïdes et Hétérobranchoïdes, 1862:112 p., pl. 49-101; 3, Cyprins, 1863:150 p., pl. 102-144; 4, Murènes, Synbranches, Lepto-céphales, 1864:132 p., pl. 145-193; 5, Baudroies, Ostracions, Gymno-dontes, Balistes; 1865: 152 p., pl. 194-231; 6, Pleuronectes, Scom-brésoces, Clupées, Clupésoces, Chauliodontes, Saurides, 1866-72: 170 p., pl. 232-278; 7, Percoïdes I, 1873-76: 126 p., pl. 279-320; 8, Percoïdes II (Spariformes), Bogodoïdes, Cirrhitéoides, 1876-77: 156 p., pl. 321-354, 361-362; 9, Toxoteoidei, Chaetodontoidei, Nandoidei, 1876-1878: 80 p., pl. 355-360, 363-420.

--. 1863. Mémoire sur les poissons de la côte de Guinée. Natuurk. Verh. holland. Maatsch. Wet. Haarlem, (2)18:1-136, 28 pl.

--. 1863. Description de quelques espèces de poissons nouvelles ou peu connues de Chine envoyées au Musée de Leide par M. G. Schlegel. Ned. Tijdschr. Dierk., 1:135-150.

--. 1863. Sur une nouvelle espèce de Synaptura du Cap de Bonne Espéran-ce. Versl. Akad. Amst., 15:456-459.

--. 1864. Systema Muraenorum revisum. Ned. Tijdschr. Dierk., 2:113-122.

--. 1865. Description de quelques espèces inédites de poissons de l'ar-chipel des Moluques. Ned. Tijdschr. Dierk., 2:177-181.

--. 1865. Enumération des espèces de poissons actuellement connues de l'ile d'Amboine. Ned. Tijdschr. Dierk., 2:270-293.

--. 1865. Notice sur les Ostracions confondus sous le nom de O. quadri-cornis L. et description des Ostracion notacanthus et guineensis. Ned. Tijdschr. Dierk., 2:298-305, 1 pl.

--. 1866. Systema Balistidorum, Ostracionidorum, Gymnodontidorumque revisum. Ned. Tijdschr. Dierk., 3:8-19.

--. 1866. Synonyma Balistidorum, Ostracionidorum, Gymnodontidorumque Indo-archipelagicorum hucusque observatorum revisa, adjectis habita-tionibus citationibusque, ubi descriptiones figuraeque eorum recen-tiores reperiuntur. Ned. Tijdschr. Dierk., 3:20-40.

--. 1866. Sur les espèces d'Exocet de l'Inde archipélagique. Ned. Tijdschr. Dierk., 3:105-129.

--. 1866. Description d'une espèce inédite d'Exocet découverte par M. François Pollen. Ned. Tijdschr. Dierk., 3:130-133.

--. 1867. Description de quelques espèces nouvelles de Gobius de Madagascar. Archs neerl. Sci., 2: 403-420.

--. 1868. Notice sur la faune ichthyologique de l'île de Waigiou. Versl. Meded. K. Akad. Wet. Amst., (2)2:295-301.

--. 1868. Description de trois espèces inédites de poissons des îles d'Amboine et de Waigiou. Versl. Meded. K. Acad. Wet. Amst., (2)2: 331-335, 1 pl.

--. 1868. Description de deux espèces inédites d'Epinephelus rapportées de l'île de la Réunion par M.M. Pollen et van Dam. Versl. Meded K. Akad. Wet. Amst., (2)2:336-341.

--. 1868. Description et figure d'une nouvelle espèce de Trachypterus de l'île d'Amboine. Archs néerl. Sci., 3:279-280, pl. 12.

--. 1873. Sur le genre Parapristipoma et sur l'identité spécifique des Perca trilineata Thunb., Pristipoma japonicum Cuv. et Diagramma japonicum Bleeker. Archs néerl. Sci., 8:19-24.

--. 1873. Mémoire sur la faune ichthyologique de Chine. Ned. Tijdschr. Dierk., 4:113-154.

--. 1873. Troisième notice sur la faune ichthyologique des îles Arou. Versl. Meded. K. Akad. Wet. Amst., (2)7:35-39.

--. 1873. Description et figure d'une espèce insulindienne d'Orthagoriscus. Versl. Meded. K. Akad. Wet. Amst., (2)7:151-153., 1 pl.

--. 1874. Typi nonnuli generici piscium neglecti. Versl. Meded K. Akad. Wet. Amst., (2)8:367-371.

--. 1874. Esquisse d'un système naturel des Gobioides. Archs néerl. Sci., 9:289-331.

--. 1876. Description de quelques espèces insulindiennes inédites des genres Oxyurichthys, Paroxyurichthys et Cryptocentrus. Verh. K. Akad. Wet. Amst., (2)9:138-148.

--. 1876. Genera familiae Scorpaenoideorum conspectus analyticus. Versl. Meded. K. Akad. Wet. Amst., (2)9:294-300.

--. 1876. Systema Percarum revisum. Pars I et II. Archs néerl. Sci., 11:247-288, 289-340.

--. 1877. Notice sur les espèces nominales de Pomacentroïdes de l'Inde archipélagique. Archs néerl. Sci., 12:38-41.

--. 1877. Mémoire sur les Chromides marins ou Pomacentroïdes de l'Inde archipélagique. Natuurk. Verh. holland. Maatsch. Wet. Haarlem, (3)2:1-166.

--. 1878. Quatrième mémoire sur la faune ichthyologique de la Nouvelle-Guinée. Archs néerl. Sci., 13:35-66.

BLOCH, M. E. 1783. Oeconomische Naturgeschichte der Fische Deutschlands. 2, Berlin, 192 p., pl. 38-72.

--. 1785-1795. Naturgeschichte der ausländischen Fische. 9 vol. (= Allgemeine Naturgeschichte der Fische in 12 Teilen, vol. 4-12). Berlin. 1, 1785:viii + 136 p., pl. 109-144; 2, 1786:viii + 160 p., pl. 145-180; 3, 1787:xii + 146 p., pl. 181-216; 4, 1790:xii + 128 p., pl. 217-252; 5, 1791:vii + 152 p., pl. 253-288; 6, 1792:viii + 126 p., pl. 289-324; 7, 1793:xii + 144 p., pl. 325-360; 8, 1794:iv + 174 p., pl. 361-396; 9, 1795:ii + 192 p., pl. 397-432.

--. 1785-97. Ichthyologie ou Histoire naturelle générale et particulière des Poissons., Berlin. 1, 1785:5 + 206 + 1, pl. 1-37; 2, 1785:1 + 170 + 1, pl. 38-72; 3, 1786:1 + 160 + 1, pl. 73-107; 4, 1787:1 + 134 + 1, pl. 109-144; 5, 1787:1 + 130 + 1, pl. 145-180; 6, 1788:1 + 8 + 150 + 1, pl. 181-216; 7, 1797:1 + 8 + 104 + 1, pl. 217-252; 8, 1797:1 + 4 + 122 + 1, pl. 253-288; 9, 1797:1 + 110 + 1, pl. 289-324; 10, 1797:1 + 5 + 120 + 1, pl. 325-360; 11, 1797:2 + 136 + 1, pl. 361-396; 12, 1797:1 + 2 + 142 + 2, pl. 397-432.

--. 1788. Ueber zwey merkwürdige Fischarten (Notacanthus chemnitzii und Silurus militaris). Abh. Böhm. Ges. Wiss., (1787), 3:278-282, 2 pl.

--. 1789. Två utländska fiskar. K. svenska VetensAkad. nya Handl., 10:234-236, pl. 7.

--. 1792. Beschreibung zweyer neuen Fische. Schr. Ges. naturf. Freunde Berl., 10:422-424, pl.

--. 1792. Bemerkungen zu obiger Abhandlung des Herrn Abildgaard über den Ansauger (Myxine glutinosa Lin.). Schr. Ges. naturf. Freunde Berl., 10:244-251.

BLOCH, M. E.; SCHNEIDER, J. C. 1801. M. E. Blochii Systema Ichthyologiae iconibus cx illustratum. Post obitum auctoris opus inchoatum absolvit, correxit, interpolavit J. G. Schneider, Saxo. Berolini, lx + 584 p., 110 pl.

BLYTH, E. 1858. Report of the curator, zoological department, for may 1858. (p. 281-290: Fishes from Pegu, Calcutta, and elsewhere). J. Asiat. Soc. Beng., 27.

BOCOURT, F. M. 1868. Description de quelques acanthopthérygiens nouveaux appartenant aux genres Serranus et Mesoprion recueillis dans l'Amérique Centrale. Annls Sci. nat. (Zool.), (5)10:222-224.

BODDAERT, P. 1781. Beschreibung zweier merkwürdiger Fische (Sparus palpebratus and Muraena colubrina). Neue nord. Beitr. (Pallas), 2:55-57.

BOECK, B. van den. 1983. Information on Orange Roughy. Catch, Wellington, 10(11):9-10, 23.

BOEHLKE, J. E. 1953. A catalogue of the type specimens of recent fishes in the Natural History Museum of Stanford University. Stanford ichthyol. Bull., 5(1):1-168.

--. 1956. A synopsis of the eels of the family Xenocongridae (including the Chlopsidae and Chilorhinidae). Proc. Acad. nat. Sci. Philad., 108:61-95, 8 fig., pl. 7.

--. 1957. On the occurrence of garden eels in the western Atlantic, with a synopsis of the Heterocongrinae. Proc. Acad. nat. Sci. Philad., 109:59-79, pl. 4, fig. 1-7.

--. 1958. Substitute names for Nystactes Böhlke and Lucaya Böhlke, preoccupied. Copeia, 1958(1):59.

--. 1960. A new ophichthid eel of the genus Pseudomyrophis from the Gulf of Mexico. Notul. Nat., (329):1-8.

--. 1966. Order Lyomeri. In: Fishes of the western North Atlantic. Mem. Sears Fdn mar. Res., 1(5):603-628, 1 pl.

--. 1966. On the identity of the west Atlantic eel Pythonichthys sanguineus Poey. Notul. Nat., (392):5 p., 2 fig.

--. 1968. A new species of the ophichthid eel genus Verma from the West Atlantic, with comments on related species. Notul. Nat., (415):1-12.

--. 1981. Muraenidae, Ophichthidae, in Fischer, W., Bianchi G. & W.B. Scott (ed.), FAO species identification sheets for fishery purposes. Eastern Central Atlantic; fishing areas 34, 47 (in part). Canada Funds-in-Trust. Ottawa, Department of Fisheries and Oceans Canada, by arrangement with the Food and Agriculture Organization of the United Nations, vol. 1-7: pag. var.

BOEHLKE, J. E.; BOEHLKE, E. B. 1976. The chestnut moray, Enchelycore carychroa, a new species from the West Atlantic. Proc. Acad. nat. Sci. Philad., 127:137-146, 3 fig.

BOEHLKE, J. E.; CHAPLIN, C. C. G. 1968. Fishes of the Bahamas and adjacent tropical waters. Wynnewood, Pa., xxiii + 771 p., fig. 1-223, 511 fig. n. num., pl. 1-36, 7 pl. (plain) n. num.

BOEHLKE, J. E.; McCOSKER, J. E. 1975. The status of the ophichthid eel genera Caecula Vahl and Sphagebranchus Bloch, and the description of a new genus and species from fresh water in Brazil. Proc. Acad. nat. Sci. Philad., 127(1):1-11.

--. 1982. _Monopenchelys_, a new eel genus, and redescription of the type species, _Uropterygius acutus_ Parr (Pisces: Muraenidae). _Proc. Acad. nat. Sci. Philad._, 134:127-134, fig. 1-3.

BOEHLKE, J. E.; RANDALL, J. E. 1968. A key to the shallow-water west Atlantic cardinal fishes (Apogonidae), with descriptions of five new species. _Proc. Acad. nat. Sci. Philad._, 120(4):175-206, 6 fig., 4 tab.

BOEHLKE, J. E.; ROBINS, C. R. 1959. The characters and synonymy of the western Atlantic snake eel, _Ophichthus ophis_ Linnaeus. _Notul. Nat._, (320):1-9.

BOEHLKE, E. B.; ROSS, S. W. 1981. The occurrence of _Muraena robusta_ Osorio (Anguilliformes; Muraenidae) in the West Atlantic. _N.E. Gulf Sci._, 4(2):123-125.

BOELY, T. 1967. Etude préliminaire de quelques caractères de _Sardinella eba_ (C. et V.) des côtes du Sénégal. _Docum. scient. provis. Centre Rech. océanogr., Dakar-Thiaroye_, 3(5 p., mimeo).

--. 1967. Etude préliminaire de quelques caractères de _Sardinella eba_ (C. & V.) des côtes du Sénégal. _ICES_, C.M. 1967/J8. (mimeo).

--. 1971. La pêche industrielle de _Sardinella aurita_ dans les eaux sénégalaises de 1966 à 1970. _Docum. scient. provis. Centre Rech. océanogr., Dakar-Thiaroye_, (31) (15 p., mimeo).

--. 1979. Biologie des deux espèces de sardinelles (_Sardinella aurita_ Valenciennes, 1847 et _Sardinella maderensis_ Lowe, 1841) des côtes sénégalaises. Ph. D. thesis, Univ. Paris VI. ORSTOM _Trav. Doc. Micr._, (7), 219 p.

--. 1980. Etude du cycle sexuel de la sardinelle plate (_Sardinella maderensis_ Lowe, 1841) des côtes sénégalaises. _Cybium_ 4(1):77-88.

--. 1982. Etude du cycle sexuel de la sardinelle ronde (_Sardinella aurita_ Val. 1847) au Sénégal. _Océanogr. trop._, 17(1):3-13.

--. 1982. Les ressources en poissons pélagiques des côtes ouest africaines entre la Mauritanie et le fleuve Congo. _Rapp. P.-v. Réun. Cons. perm. int. Explor. Mer_, 180:423-431.

BOELY, T.; CHABANNE, J. 1975. Les poissons pélagiques côtiers au Sénégal. La pêche sardinière à Dakar; état actuel et perspectives. _Bull. Inst. fond. Afr. noire_, (A)37(4):859-886.

BOELY, T.; CHAMPAGNAT, C. 1968. Observations préliminaires sur _Sardinella aurita_ (C. et V.) des côtes sénégalaises. Symp. Living Resources Plateau Cont. Afr. Gibraltar to Cape Verde, Contr. N°53.

--. 1969. La pêche industrielle au Sénégal des poissons pélagiques côtiers en 1967 et 1968. _Docum. scient. provis. Centre Rech. océanogr., Dakar-Thiaroye_, (22)(9 p., mimeo).

--. 1970. Observations préliminaires sur _Sardinella aurita_ (C. et V.) des côtes sénégalaises. _Rapp. P.-v. Réun. Cons. perm. int. Explor. Mer_, 159:176-181.

BOELY, T.; DIEYE, I. 1971. La pêche sardinière au Sénégal en 1969. _Docum. scient. provis. Centre Rech. océanogr., Dakar-Thiaroye_, (28)(9 p. mimeo).

BOELY, T.; ELWERTOWSKI, J. 1970. Observations préliminaires sur la pêche de _Ethmalosa fimbriata_ (Bowdich) des eaux sénégalaises et son aspect biologique. _Rapp. P.-v. Réun. Cons. perm. int. Explor. Mer_, 159:182-183.

BOELY, T.; FREON, P. 1979. Les ressources pélagiques côtières. _In_: Troadec, J.P. & S. Garcia (ed.) Ressources halieutiques de l'Atlantique centre-est. _FAO Docums tech. Pêch._, (186.1):17-78, 12 fig. (Engl. version published 1980).

BOELY, T.; MARCHAL, E. 1977. Les pélagiques côtiers, p. 61-85. _In_: Le milieu marin de la Guinée Bissau et ses ressources vivantes. Le point des connaissances. ORSTOM, Paris.

BOELY, T.; ØSTVEDT, O. J. 1976. Les poissons pélagiques côtiers au Sénégal. Observations faites à bord du navire-usine *Astra* de la Mauritanie aux îles Bissagos. *Bull. Inst. fond. Afr. noire*, (A)38(3):677-702.

BOELY, T.; CHABANNE, J.; FREON P. 1979. Schémas migratoires, aires de concentrations et périodes de reproduction des principales espèces de poissons pélagiques côtières dans la zone sénégalo-mauritanienne. Annexe 4, p. 58-65. *In*: Anon., 1979.

BOELY, T.; CHAMPAGNAT, C.; CONAND, F. 1969. Reproduction et cycle biologique de *Sardinella aurita* (C. et V.) des côtes sénégalaises. *Docums scient provis. Centre Rech. océanogr., Dakar-Thiaroye* , (21) (13 p., mimeo).

BOELY, T.; FREON, P.; STEQUERT, B. 1982. La croissance de *Sardinella aurita* (Val. 1847) au Sénégal. *Océanogr. trop.*, 17:103-119, 11 fig.

BOELY, T.; CHABANNE, J.; FREON, P.; STEQUERT, B. 1978. Cycle sexuel et migrations de *Sardinella aurita* sur le plateau ouest-africain des îles Bissagos à la Mauritanie. Symposium Canaries Current : Upwelling and living resources (11-14 April 1978, Las Palmas), Paper N° 92 (12 p., mimeo).

--. 1982. Cycle sexuel et migrations de *Sardinella aurita* sur le plateau continental ouest-africain, des îles Bissagos à la Mauritanie. *Rapp. P.-v. Réunion. Cons. perm. int. Explor. Mer*, 180:350-355.

BOELY, T.; ØSTVEDT, O. J.; MYKLEVOLL, S.; SECK, M. 1974. Composition par espèce et par taille des captures du navire-usine *Astra*. *ICES*, C.M. 1974/J16 (8 p., mimeo).

BOESEMAN, M. 1947. Revision of the fishes collected by Burger and von Siebold in Japan. *Zoöl. Meded., Leiden*, 28:1-242, pl. 1-5.

--. 1963. An annotated list of fishes from the Niger Delta. *Zool. Verh., Leiden*, (61):1-48, 2 fig., 6 pl.

-- 1964. Scombroid types in the Leiden Museum Collection. *Proc. Symp. Scombroid Fishes, mar. biol. Ass. India*, 1(1):461-468, 5 pl.

--. 1984. Hexanchidae, *in* Whitehead, P.J.P., Bauchot, M.L., Hureau, J.C., Nielsen, J. & E. Tortonese, Fishes of the North-eastern Atlantic and the Mediterranean/Poissons de l'Atlantique du Nord-Est et de la Méditerranée, Paris, Unesco, 1:72-75, 3 fig., 3 maps.

--. 1984. Chlamydoselachidae, *in* Whitehead, P.J.P., Bauchot, M.L., Hureau, J.C., Nielsen, J. & E. Tortonese, Fishes of the North-eastern Atlantic and the Mediterranean/Poissons de l'Atlantique du Nord-Est et de la Méditerranée, Paris, Unesco, 1:76-77, 1 fig., 1 map.

BOLIN, R. L. 1939. A review of the myctophid fishes of the Pacific coast of the United States and of Lower California. *Stanford ichthyol. Bull.*, 1(4):89-156, fig. 1-29.

--. 1946. Lantern fishes from "Investigator" station 670, Indian Ocean. *Stanford ichthyol. Bull.*, 3(2):137-152, 5 fig.

--. 1959. Iniomi : Myctophidae. *Rep. scient. Results Michael Sars N. Atlant. deep Sea Exped.*, 4(2)(7):1-45.

BOLTACHEV, A.P. 1986. Some peculiarities on the distribution and behavior of *Diplospinus multistriatus* Maul (Gempylidae) in the Southeast Atlantic. *Vop. Ikhtiol.*, 26(5):715-719.

BOMBACE, G.; FROGLIA, C. 1973. Premières remarques sur les peuplements de l'étage bathyal de la Basse Adriatique. *Revue Trav. Inst. Pêch. marit.*, 37:159-161.

BONAPARTE, C. L. 1831. Saggio di una distribuzione metodica degli animali vertebrati. Prospetto del sistema d'Ittiologia generale. *G. Accad.*, 52:167-190.

--. 1832-41. Iconografia della fauna italica per le quattro classi degli animali vertebrati, tomo III, Pesci. Roma, without pagination, 75 puntate (in 30 fasc.), 78 pl. 1832: Fasc. I (Puntate 1-6), 2 pl.;

1833: II-V (7-28), 12 pl.; 1834: VI-XI (29-58), 12 pl.; 1835: XII-XIV (59-79), 12 pl.; 1836: XV-XVIII (80-93), 10 pl.; 1837: XIX-XXI (94-103 & 105-109), 5 pl. ; 1838: XXII-XXIII (104 & 110-120), 2 pl.; 1839: XXIV-XXVI (121-135), 8 pl.; 1840:XXVII-XXIX (136-154), 10 pl.; 1841:XXX (155-160), 5 pl.

--. 1838. Selachorum tabula analytica. _Nuovi Annali Sci. nat. Bologna_, 2:195-214.

--. 1845. Catalogo metodico dei pesci europei. _Atti 7ma Riun. scienzati ital., Napoli_, (2):1-95 (also publ. sep., Napoli, 1846, 97 p.).

BOND, G. W. 1981. Gonostomatidae, _in_ Fischer, W., Bianchi, G. & W.B. Scott (ed.), FAO species identification sheets for fishery purposes. Eastern Central Atlantic; fishing areas 34, 47 (in part). Canada Funds-in-Trust. Ottawa, Department of Fisheries and Oceans Canada, by arrangement with the Food and Agriculture Organization of the United Nations, vol. 1-7: pag. var.

BONDE, C. von, 1922. The Heterosomata (Flat fishes) collected by the S.S. "Pickle". _Rep. Fish. mar. biol. Surv. Un. S. Afr._ (1921), 2(spec. Rep. 1):1-29, 6 pl.

--. 1924. Shallow-water fishes procured by the S.S. "Pickle". _Rep. Fish. mar. biol. Surv. Un. S. Afr._ (1922), 3(Spec. Rep. 1):1-40, 9 pl.

--. 1928. Annexure "A". List of fishes, etc., procured by the S.S. "Pickle" during the period July 1, 1925, to May 25, 1927. _Rep. Fish. mar. biol. Surv. Un. S. Afr._ (1925-27), (5):16-85.

--. 1933. Annexure "A". Cape area survey: January-May 1932 - Table Bay-Saldanha Bay. List of fishes, etc., procured. _Rep. Fish. mar. biol. Surv. Un. S. Afr._ (1932), (10):32-84.

BONDE, C. von; SWART, D. B. 1923. The Platosomia (skates and rays) collected by the S.S. "Pickle". _Rep. Fish. mar. biol. Surv. Un. S. Afr._, 3 (Spec. Rep. 5):1-22, pl. 20-23.

BONE, Q. 1971. On the scabbard fish _Aphanopus carbo_. _J. mar. biol. Ass. U.K._, 51:519-225.

BONELLI, F. A. 1820. Description d'une nouvelle espèce de poisson de la Méditerranée appartenant au genre Trachyptère avec des observations sur les caractères de ce même genre. _Memorie Accad. Sci. Torino_, 24:485-494.

BONNATERRE, J. P. 1788. Ichthyologie. Tableau encyclopédique et méthodique des trois règnes de la nature. Paris, lvi + 215 p., 102 pl.

BORCEA, I. 1936. Note sur la biologie du Pomatome (Lufar) de la Mer Noire. _C.r. Séanc. Acad. Sci. Roum._, 1(3):222-223.

BORODATOV, V. A. 1958. Data on the biology of _Sardinella aurita_ in the Cape Verde area. _ICES_, C.M. 1958/Sardine Comm., Doc. 80 (6 p., mimeo).

--. 1959. (Results of the work of the South Atlantic Sardine Expedition). _Ryb. Khoz._, 35(9):10-16 (in Russian).

BORODATOV, V. A.; KARPECHENKO, J. L. 1958. (Prospects for the development of the fisheries in the eastern part of the Central Atlantic). _Ryb. Khoz._, 34(6):10-20 (in Russian).

BORODATOV, V. A.; KARPECHENKO, J. L.; PROBATOV, A. N; BIRYUKOV, N. P. 1960. Soviet investigations into the biology of _Sardinella aurita_ Valenciennes. FAO Proc. Wld scient. Meet. Biol. Sardines, 3:1221-1227.

BORODIN, N. A. 1928. Scientific results of the Yacht Ara Expedition during the years 1926 to 1928, while in command of William K. Vanderbilt. _Bull. Vanderbilt oceanogr. Mus._, 1(1):1-37, 5 pl. 2 map.

--. 1919. Some new deep-sea fishes. _Proc. New Engl. zool. Club_, 10:109-111.

--. 1930. Fishes (collected in 1929). _In_: Scient. Res. "Ara" Exped... 1926-1930... _Bull. Vanderbilt mar. Mus._, 1(2):39-64, pl. 1-2.

--. 1930. Some more new deep-sea fishes. Proc. New Engl. zool. Club, 11:87-92.

--. 1931. North-Atlantic deep-sea fishes. Bull. Mus. comp. Zool. Harv., 72(3):55-89; 1 fig., 5 pl.

--. 1934. Scientific results of the yacht "Ara". Mediterranean Cruise, 1933, in command of William K. Vanderbilt. Fishes. Bull. Vanderbilt mar. Mus., 1(4):103-123, pl. 1, 2.

BORODULINA, O. D. 1977. A new species of hatchetfish, Sternoptyx pseudo-diaphana (Sternoptychidae, Osteichthyes) from the waters of the southern hemisphere. Vop. Ikhtiol., 17:938-941, 2 fig. (in Russian).

--. 1978. Materials on the systematics and distribution on the oceanic hatchet-fishes genera Argyropelecus and Sternoptyx (Sternoptychidae, Osteichthyes). Trudy Inst. Okeanol. 111:28-60, fig. 1-16 (in Russian).

BORSIERI, C. 1904. Contribuzione alla conoscenza delle specie europee del genere Atherina. Annali Agric., (233):129-220, 3 fig., pl. 6-10 (sep. 92 p.).

BORY de SAINT-VINCENT, G. J. B. M. 1804. Voyages dans les quatre principales îles des mers d'Afrique. Paris, 3 vol. + atlas.

--. 1822. Dictionnaire classique d'Histoire naturelle, 1, Paris.

--. 1825. Dictionnaire classique d'Histoire naturelle, 8, Paris, 609 p.

BOSC, L. 1803. (Lophie unie). Nouv. Dict. Hist. nat., Paris, 13 ("Lophie variée"):313, pl. E30.

--. 1817. (Lophote, Lophotes). Dict. Sci. nat., Paris, 18:185-186.

BOUGIS, P. 1959. Atlas des Poissons. Poissons Marins, Paris, 1:201 p., 52 fig., 12 phot., 12 col. pl. (52 fig.); 2:234 p., 50 fig., 16 phot., 12 col. pl. (57 fig.).

BOUGIS, P.; RUIVO, M. 1957. Contribution à la connaissance de la morphologie de la thyroïde de Bathypterois dubius Vaillant, Résult. campagnes "Pr. Lacaze-Duthiers", II Algerie 1952 et Baléares 1953-1954. Vie Milieu, Suppl. 6:185-204.

BOUILLON, P.; TROADEC, J. P.; BARRO, M. 1969. Pêches au chalut sur les radiales de Jacqueville, Grand Lahou, Fresco et Sassandra (Côte d'Ivoire). Mars 1966-février 1967. Docums scient. provis. Centre Rech. océanogr. Abidjan, (36):1-13.

BOULENGER, G. A. 1889. Second account of the fishes obtained by Surgeon Major A.S.G. Jayakar at Muscat, east coast of Arabia. Proc. zool. Soc. Lond., 1889:236-246, 4 pl.

--. 1891. On young and eggs of Galeichthys feliceps. Proc. zool. Soc. Lond., 1891:148.

--. 1895. Catalogue of the perciform fishes in the British Museum (2ed). Vol. 1. London, xix + 394 p., 15 pl.

--. 1898. The flat fishes of Cape Colony. Mar. Invest. S. Afr., 1:1-4.

--. 1898. Descriptions of two new gobiiform fishes from the Cape of Good Hope. Mar. Invest. S. Afr., 1:8-9. (196-197).

--. 1899. Poissons nouveaux du Congo, (3rd part). Silures, Acanthopterygiens, Mastacembles, Plectognathes. Annls Mus. r. Congo belge, (Zool.), 1(3):39-58, 10 pl.

--. 1900. List of the fishes collected by Mr. J. S. Budgett in the River Gambia, with notes by J. S. Budgett, F.Z.S. Proc. zool. Soc. Lond., 1900:511-516.

--. 1901. Les poissons du bassin du Congo. Bruxelles, xii + 532 p., fig., 25 pl.

--. 1902. Notes on the classification of the teleostean fishes. Ann. Mag. nat. Hist., (7)9:197-204.

--. 1902. Contributions to the ichthyology of the Congo. II. On a collection of fishes from the Lindi river. Proc. zool. Soc. Lond., 1902:265-271, pl. 28-30.

--. 1903. Sur les affinités du genre Oreosoma. C. r. hebd. Séanc. Acad. Sci. Paris, 137:523-525.

--. 1905. Poissons de la Guinée espagnole. Mems R. Soc. esp. Hist. nat., 1:187-188.

--. 1905. A list of the freshwater fishes of Africa. Ann. Mag. nat. Hist., (7)16:36-60.

--. 1909. Descriptions of three new fishes from Portuguese Guinea. Ann. Mag. Nat. Hist., (8)4:429-431.

--. 1909-16. Catalogue of the fresh-water fishes of Africa in the British Museum (Natural History). London, 1, 1909: xi + 373 p., 270 fig.; 2, 1911: xii + 529 p., 382 fig.; 3, 1915: xii + 526 p., 351 fig.; 4, 1916: xxvii + 392 p., 194 fig.

--. 1910. On a large collection of fishes made by Dr W.J. Ansorge in the Quanza and Bengo rivers, Angola. Ann. Mag. nat. Hist., (8)6:537-561.

--. 1913. The luminous organs of Lamprotoxus flagellibarba. Scient. Invest. Fish. Brch Ire., (1912) (2):1-2, 1 pl.

BOULENGER, G. A.; BUDGETT, S. 1907. List of the fishes collected by Mr J.S. Budgett in the River Gambia, p. 95-99. In: The work of John Samuel Budgett (ed. J.G. Kerr). Cambridge.

BOUNHIOL, J. P. 1921. Sur la biologie de l'allache (Sardinella aurita Valenc.) des côtes d'Algérie. C. r. Ass. fr. Avanc. Sci., 44:279-281.

BOUNHIOL, J. P.; GAVARD. 1923. Une espèce nouvelle de Trachypterus Gouan: le Trachypterus gavardi Bounhiol. Bull. Inst. océanogr. Monaco, (432):1-4, fig.

BOURJOT, A. 1867-1868. Liste des Poissons que l'on rencontre le plus souvent au marché d'Alger ou Guide à la Pêcherie. Bull. Soc. Climatol., Alger, 1, 1867:418-433, 485-513, 556-582; 2, 1868:14-25, 134-138. (also reprint 1870, Alger, 132 p.).

BOUTIERE, H. 1958. Les Scorpaenidés des eaux marocaines. Trav. Inst. scient. chérif. (Zool.), (15): 83 p., 12 fig. (1 map), 5 pl., 8 tab.

--. 1959. Présence en Méditerranée de Scorpaena loppei Cadenat et de Scorpaena elongata Cadenat. Vie Milieu, 10(4):405-407.

BOUTIN, S. A. N. 1968. Ichthyoplancton, oeufs et larves de poissons téléostéens dans le Golfe de Gascogne en 1964. Rev. Trav. Inst. Pêch. marit., 32(4):413-476.

BOUTELLIER, P. 1918. La pêche maritime au Maroc (Casablanca, Fedala, Rabat). Casablanca, 95 p.

BOWDICH, S. 1825. Appendix: Zoology. p. 219-243. In: T.E. Bowdich, Excursions in Madeira and Porto Santo, during the autumn of 1823, while on his third voyage to Africa. London, xii + 278, 11 pl. (10 of them are not numb.), 57 fig. (fishes of Madeira, p. 121-125).

BRADBURY, M. G. 1981. Ogcocephalidae, in Fischer, W., Bianchi, G. & W.B. Scott (ed.), FAO species identification sheets for fishery purposes. Eastern Central Atlantic; fishing areas 34, 47 (in part). Canada Funds-in-Trust. Ottawa, Department of Fisheries and Oceans Canada, by arrangement with the Food and Agriculture Organization of the United Nations, vol. 1-7: pag. var.

BRADBURY, M. G. et al. 1971. Studies on the fauna associated with the deep scattering layers in the equatorial Indian Ocean, conducted on R/V Te Vega during October and November 1964. p. 409-452. In: Proc. int. Symp. biol. sound scattering in the ocean. (Farquhar, G. B., ed.) Washington, xi + 629 p.

BRANDT, J. F. 1854. Remarques sur le mémoire de M. Lowe, suivies d'une planche représentant le nouveau genre de poissons. Mém. Acad. imp. St. Pétersbourg Savants étrangers, 7:174-175.

BRANDES, C. H.; KOTTHAUS, A.; KREFFT, G. 1953. Rare fishes from the NW Area, Iceland. Annls biol. Copenh., (1952), 9:47-48.

BRANSTETTER, S. 1984. Carcharhinidae, in Whitehead, P. J.P., Bauchot,

M.L., Hureau, J.C., Nielsen J. & E. Tortonese, Fishes of the North-eastern Atlantic and the Mediterranean/Poissons de l'Atlantique du Nord-Est et de la Méditerranée, Paris, Unesco, 1:102-114, 14 fig., 14 maps.

--. 1984. Hemigaleidae, in Whitehead, P. J. P., Bauchot, M. L., Hureau, J.C., Nielsen, J. & E. Tortonese, Fishes of the North-eastern Atlantic and the Mediterranean/Poissons de l'Atlantique du Nord-Est et de la Méditerranée, Paris, Unesco, 1:115-116, 1 fig., 1 map.

--. 1984. Triakidae, in Whitehead, P.J.P., Bauchot, M.L., Hureau, J.C., Nielsen, J. & E. Tortonese, Fishes of the North-eastern Atlantic and the Mediterranean/Poissons de l'Atlantique du Nord-Est et de la Méditerranée, Paris, Unesco, 1:117-121, 5 fig., 5 maps.

BRAUER, A. 1901. Über einige von der Valdivia - Expedition gesammelte Tiefseefische und ihre Augen. Sber. Ges. Beförd. ges. Naturw. Marburg, 8:115-130.

--. 1902. Diagnosen von neuen Tiefseefischen, welche von der Valdivia-Expedition gesammelt sind. - Zool. Anz., 25(668):277-298.

--. 1904. Die Gattung Myctophum. Zool. Anz., 28:377-404.

--. 1906. Die Tiefseefische I. Systematischer Teil. Wiss. Ergebn. dt. Tiefsee-Exped. "Valdivia", 15(1):1-432, 176 fig., pl. 1-18.

--. 1908. Die Tiefseefische. II. Anatomischer Teil. Wiss. Ergebn. dt. Tiefsee-Exped. "Valdivia", 15(2):1-266, 11 fig., pl. 19-44.

BRAVO de LAGUNA, C. J. 1973. Elasmobranchii off Canary Islands. - ICES, C. M. 1973/J. 17 (5 p. mimeo).

--. 1975. Contribución al conocimiento de los peces condroictios del archipelago canario. Boln Inst. esp. Oceanogr., (201), 77 p., 140 fig.

BRAVO de LAGUNA, C. J.; SANTAELLA ALVAREZ, E. 1973. Observaciones biológico-pesqueras en el banco pesquero sahariano. Boln Inst. esp. Oceanogr., (171):1-79, 99 fig.

--. 1975. Observations on Raja miraletus (Pisces, Elasmobranchii, Rajidae) off West Africa. ICES, C. M. G12 (11 p., 1 fig., 10 tab., mimeo).

BRAVO de LAGUNA, J.; FERNANDEZ, M. A. R.; DELGADO, A. 1976. Problemática de investigaciónes sobre la sardina (Sardina pilchardus Walb.) de banco pesquero Sahariano. Investigacion pesq., 43(1):327-331.

BRAVO de LAGUNA, J.; FERNANDEZ, M. A. R.; SANTANA, J. C. 1976. The Spanish fishery on sardine (Sardina pilchardus, Walb.) off West Africa. ICES, C.M. J15 (7 p., mimeo).

BRAVO de LAGUNA, J; FERNANDEZ, M. A. R.; GONI, R.; DELGADO. 1979. Growth studies of Sardina pilchardus (Walb.), done by direct methods and scalimetry off West Africa. ICES, C.M. H52:1-7 (mimeo).

BREDER, C. M., Jr. 1925. Notes on fishes from three Panama localities, Gatun Spillway, Rio Tapica and Caledonia Bay. Zoologica, N.Y., 4(4):137-158, fig. 33-38.

--. 1927. Scientific results of the first oceanographic expedition of the "Pawnee" 1925. Fishes. Bull. Bingham oceanogr. Coll., 1(1):1-90, 36 fig.

--. 1928. Scientific results of the second oceanographic expedition of the "Pawnee" 1926: Nematognathi, Apodes, Isospondyli, Synentognathi, and Thoracostraci from Panama to Lower California. Bull. Bingham oceanogr. Coll., 2(2):1-25, 10 fig.

--. 1929. Field book of marine fishes of the Atlantic coast. New York, i-xxxvii, 1-332, fig.

--. 1932. On the habits and development of certain Atlantic Synentognathi. Publs Carnegie Instn, (435):1-35, 12 pl., 10 fig.

--. 1933. New locality records of Abudefduf analogus including one from the United States. Copeia, 1933(4):191-193, 2 fig.

--. 1938. A contribution to the life histories of Atlantic ocean flying-fishes. Bull. Bingham oceanogr. Coll., 6(5):1-126, 48 fig.

--. 1943. The eggs of Bathygobius soporator (Cuvier and Valenciennes) with a discussion of other non-spherical teleost eggs. Bull. Bingham oceanogr. Coll., 8(3):1-49.

--. 1944. The metamorphosis of Synodus foetens (Linnaeus). Zoologica. N.Y., 29(1):13-15.

--. 1946. An analysis of the deceptive resemblances of fishes to plant parts, with critical remarks on protective colouration, mimicry and adaptation. Bull. Bingham oceanogr. Coll., 10(2):1-49.

--. 1949. On the taxonomy and the post-larval of the surgeon fish, Acanthurus hepatus. Copeia, 1949(4):296.

--. 1951. Nocturnal and feeding behavior of the labrid fish Xyrichthys psittacus. Copeia, 1951:162-163.

BREDER, C. M., Jr; NICHOLS, J. T. 1930. West Indian forms of the flying fish, genus Cypselurus, with the description of a new species. Am. Mus. Novit., (417):1-9).

BREDER, C.; RASQUIN, P. 1952. The sloughing of the melanic area of the dorsal fin, an ontogenetic process in Tylosurus raphidoma. Bull. Am. Mus. nat. Hist., 99(1):1-24.

--. 1954. The nature of post-larval transformation in Tylosurus acus (Lacepède). Zoologica. N.Y., 39:17-30.

BREDER, C. M.; ROSEN, D. E. 1966. Modes of reproduction in fishes. New-York, xv + 941 p.

BRETHES, J. C. 1975. La bécasse de mer des eaux atlantiques marocaines. Premières observations sur la biométrie et la biologie. Trav. Docums Dev. Pêch. Maroc, 15:1-39, 18 fig.

BREVOORT, J. C. 1856. Notes on some figures of Japanese fish taken from recent specimens by the artist of the U. S. Japan Expedition. 253-288, pl. 3-12. In: Narrative of Commodore M. C. Perry's Expedition to Japan, 2. Washington, D.C.

BRIAN, A. 1924. Parasitologia mauritanica. Materiaux pour la faune parasitologique en Mauritanie. Bull. Com. Etud. hist. scient. Afr. occid. fr., 7(3):364-427.

BRICE, J. J. 1898. The fish and fisheries of the coastal waters of Florida. Rep. U.S. Commnr Fish. (1896), 22:263-342.

BRIGGS, J. C. 1952. Systematic notes on the oceanic fishes of the genus Lophotus. Copeia, 1952:206-207, fig.

--. 1955. A monograph of the Clingfishes (order Xenopterygii). Stanford ichthyol. Bull., 6: vi + 1-224, 114 fig., 15 maps.

--. 1958. A list of Florida fishes and their distribution. Bull. Fla St. Mus. biol. Sci., 2(8):223-318.

--. 1960. Fishes of worldwide (circumtropical) distribution. Copeia, 1960(3):171-180.

--. 1974. Marine Zoogeography. New York, 475 p. 65 fig.

--. 1986. Gobiesocidae, in P.J.P. Whitehead, M.L. Bauchot, J.C. Hureau, J. Nielsen & E. Tortonese. Fishes of the North-eastern Atlantic and the Mediterranean/Poissons de l'Atlantique du Nord-Est et de la Méditerranée, Paris, Unesco, 3:1351-1359, 7 + 1 fig.

BRIGHT, T. J.; PEQUEGNAT, W. E. 1969. Deep-sea hatchetfishes of the Gulf of Mexico. Q. Jl Fla Acad. Sci., 32(1):26-37.

BRISOUT de BARNEVILLE, C. N. F. 1846. Note sur les Diodoniens. Rev. Zool., 9:136-143.

BRITO HERNANDEZ, A. 1983. Tres especies nuevas para la fauna ictiológica de las Islas Canarias: Hoplostethus mediterraneus Cuvier, 1829, Sphoeroides cutaneus (Günther, 1870) y Blennius pilicornis Cuvier, 1829. Vieraea (1982), 12(1-2):17-26.

BRITO, A.; LOZANO, G. 1980. Primera cita para Canarias de Gobius niger

Linnaeus, 1758 (Pisces, Gobiidae) con algunos comentarios sobre la distribución de la fauna ictiológica de los fondos de la playa de las Teresitas (litoral NE de Tenerife). *Vieraea*, 10:171-180.

BROUSSONET, P. M. A., 1782. Ichthyologia, sistens piscium descriptiones et icones. Decas ichthyol. London, 4 p. without number + i-iv + 1-41, 11 pl.

BROWN, H. H. 1948. Note on the organisation of fishery research in British West Africa, p. 279-292. *In*: Conférence de la pêche maritime (16-22 January 1948, Dakar), Centre d'Etudes Sci. tech. Pêch. marit., Gouvernment Général Afr. Occid. Fr., Centre Nat. Inf. Econ., Paris, 292 p.

BROWNE, P. 1789. The civil and natural history of Jamaica... (2nd ed.), 503 + 20 ff. index, 49 pl., 1 map (inset.).

BROWNELL, C. L. 1978. Sur quelques collections de poissons littoraux de l'Atlantique marocain. *Bull. Inst. Pêch. marit. Maroc*, (23):111-133, 5 fig.

--. 1987. *Vanneaugobius dollfusi*, a new genus and species of a small gobiid with divided ventrals from Morocco (Pisces: Gobioidei). *Trans. R. Soc. S. Afr.*, 43:135-145.

BROWNELL, W. N.; RAINEY, W. E. 1971. Caribbean Research Institute Special Report. Research and development of deep water commercial and sport fisheries around the Virgin Islands Plateau. Contrib. Virgin Islands Ecol. Res. Stn, no 3, p. 1-88.

BRULHET, J. 1974. Résultats des travaux du N/O "Almoravide" effectués au cours du programme Joint 1/Cineca (mars-mai 1974). *Bull. Lab. Pêch. Nouadhibou*, 3:113-129, 8 fig.

--. 1975. Observations on the biology of *Mugil cephalus ashanteensis* and the possibility of its aquaculture on the Mauritanian coast. *Aquaculture*, 5:271-281.

BRULHET, J.; ABDALLAHI, A. O. 1974. Noms vernaculaires des principales espèces de poissons pêchées à Nouadhibou. *Bull. Lab. Pêch. Nouadhibou*, 3:153-156.

BRULHET, J.; MAIGRET, J. 1973. Résultats préliminaires de la campagne de chalutage expérimental (janvier-juin 1972). *Bull. Lab. Pêch. Nouadhibou*, 2:57-112, 14 fig., 24 maps.

BRULHET, J.; MAIGRET, J.; AROUNA, S. M. 1974. Résultats de la campagne de chalutage expérimental autour du Banc d'Arguin. *Bull. Lab. Pêch. Nouadhibou*, 3:77-112.

BRUNNICH, M. T. 1768. Ichthyologia Massiliensis, sistens piscium descriptiones eorumque apud incolas nomina. Accedunt Spolia Maris Adriatici. Hafniae et Lipsiae. 1st part: Pisces massilienses: xvi + 1-84; 2nd part: Spolia e Mari Adriatica reportata:85-110.

--. 1771. Collectio nova scriptorum Societatis Scientiarum Hafnensis, 3:418.

--. 1788. Om Sild-Tusten *Regalecus remipes*. *K. danske Vidensk. Selsk. Skr.* (n.s.), 3:414-418, pl. B (fig. 4-5).

--. 1788. Om en ny Fiskeart, den draabeplettede Pladefisk, fanget ved Helsingør i Nordsøen, 1786. *K. danske Vidensk. Selsk. Skr.*, (n.s.) 3:398-406, pl. A.

BRUUN, A. F. 1933. On the value of the number of vertebrae in the classification of the Exocoetidae. *Vidensk. Meddr dansk. naturh. Foren.*, 94:375-384.

--. 1934. Notes on the Linnean type-specimens of flying-fishes (Exocoeidae). *J. Linn. Soc. (Zool.)*, 39(263):133-135.

--. 1935. Flying-fishes (*Exocoetidae*) of the Atlantic, systematic and biological studies. *Dana Rep.*, (6):1-108, 30 fig., 7 pl.

--. 1936. Nordatlantens *Synaphobranchus* - Arter, Biologiog Udbredelse sammenlignet med Aals og Havaals. Nordiska (19. skandinaviska) naturforskarmötet i Helsingfors, 1936, 3 p.

--. 1936. Sur la distribution de quelques poissons océaniques d'après les expéditions danoises. Bull. Inst. océanogr. Monaco, (700):1-16, 6 fig.

--. 1937. Contributions to the life histories of the deep sea eels: Synaphobranchidae. Dana Rep., (9):31 p., 17 fig., 1 pl.

--. 1937. Notes sur les types des Exocets décrits par Cuvier et Valenciennes. Bull. Mus. natn. Hist. nat., Paris, (2)9:180-187.

--. 1937. Chascanopsetta in the Atlantic; a bathypelagic occurrence of a flat-fish, with remarks on distribution and development of certain other forms. Vidensk. Meddr dansk. naturh. Foren., 101:125-135, 1 fig., 3 tab., 1 pl.

--. 1950. General Report. List of Stations. Atlantide Rep., (1):7-47.

--. 1956. Animal life of the deep-sea bottom. In: Bruun et al. (ed.), The Galathea Deep-Sea Expedition, 1950-52. London, 296 p.

BUCKLAND, F. 1873. Familiar history of British fishes. London, 1-396 p., many fig.

--. 1881. Natural history of British fishes: their structure, economic uses, and capture by net and rod. Cultivation of fish ponds. Fish suited for acclimatization. Artificial breeding of salmon. London, 420 p., fig.

BUDKER, P. 1935. Sélaciens capturés dans la région de Dakar. Bull. Mus. natn. Hist. nat., Paris, (2)7(3):183-189.

BUEN, F. de. 1912. Peces de la costa mediterrànea de Marruecos. Boln R. Soc. esp. Hist. nat., 12:153-160.

--. 1917. Peces poco comunes de nuestras costas. Boln Pescas, 2:23-27, fig.

--. 1918. Los gobidos de la Peninsula Ibérica y Baleares. Nota II. Catalogo sistematico y ensayo de distribución geogràfica. Boln Pescas, 1918:291-337.

--. 1922. Correspondencia cientifica de los nombres vulgares empleados en las memorias estadísticas de Baleares y costas vascas. La pesca maritima en España en 1920. Boln Pescas, 7 (65-57):31.

--. 1923. Gobius de la Peninsula Ibérica y Baleares. Grupos Lesueurii, Colonianus, Affinis y Minutus. Mem. Inst. esp. Oceanogr., 3(3):121-266.

--. 1926. Catàlogo ictiológica del Mediterràneo español y de Marruecos, recopilando lo publicado sobre peces de la costas mediterrànea y próximas del Atlantico (Mar de España). Resultados Camp. int. Inst. esp. Oceanogr., 2:1-221, map.

--. 1928. Sobre dos especies del género Gobius (G. zebrus Risso y G. thori nov. nom.). Notas Resum. Inst. esp. Oceanogr., (2)(22):12 p.

--. 1928. El Gobius niger L. en aguas atlànticas y mediterràneas de Europa. Notas Resum. Inst. esp. Oceanogr., (2)(27):32 p.

--. 1928. Hallazgo del Makaira nigricans Lacepède en las costas españolas y consideraciones sobre los Xiphiformes en general. Notas Résum. Inst. esp. Oceanogr., (2)(28):1-12, 6 fig.

--. 1930. Sur une collection de Gobiinae provenant du Maroc. Essai de synopsis des espèces de l'Europe. Bull. Soc. Sci. nat. Maroc, 10:120-147.

--. 1930. Estados larvarios y juveniles de la Sarda sarda (Bloch). Trab. Inst. esp. Oceanogr., (3):1-32.

--. 1930. Ictiologia espanola. I. Scombriformes y Thunniformes. Boln Oceanogr. Pesc., 15(162):34-53.

--. 1930. Lebetus Winther 1877, Odondebuenia nov. gen. y Cabotia nov. gen. Trab. Inst. esp. Oceanogr., (5):30 p.

--. 1931. Clupéidés et leur pêche (4ème rapport). Rapp. P.-v. Réun. Commn int. Explor. scient. Mer Méditerr., 6:289-336.

--. 1932. Clupéidés et leur pêche (5ème rapport). Rapp. P.-v. Réun. Commn. int. Explor. scient. Mer Méditerr., 7:319-340, 6 fig.

--. 1932. Formas ontogénicas de peces (Nota primera). Notas Resum. Inst. esp. Oceanogr., (2)(57):1-38, 27 fig.

--. 1935-1936. Fauna ictiológica. Catálogo de los peces ibéricos: de la planicie continental, aguas dulces, pelágicos y los abismos próximos. Primera parte. Notas Resum. Inst. esp. Oceanogr., parte 1:(2)(88):1-89, pl. 1-20 (fig. 1-40), 1935; parte 2:(2)(89):91-149, pl. 21-53 (fig. 40-115), 1935; parte 3:(2)(94):151-173, 1936.

--. 1937. Diretmus argenteus. Faune Ichthyol. Atlant. N., Cahier 16:202.

--. 1940. Les Gobiides pélagiques ou vivant sur les fonds d'algues calcaires de l'Europe Occidentale. Bull. Inst. océanogr. Monaco, (790): 16 p.

--. 1950. Contribuciones a la ictiológia. III. La familia Istiophoridae y descripción de una especie uruguaya (Makaria perezi de Buen). Publnes cient. Serv. oceanogr. Pesca Montev., (5):163-178, 4 fig.

--. 1951. Contribuciones a la ictiologia. V-VI. Sobre algunas especies de Gobiidae de la colección del Laboratorio Arago (Banyuls-sur-Mer, Francia) y descripción de un nuevo género (Austrogobius) sudamericano. Bolm Inst. paul. Oceanogr., 2:55-69.

--. 1955. Pelagic fishes and oceanographic conditions along the northern and central coast of Chile. Proc. UNESCO Symp. phys. Oceanogr., Tokyo:153-155.

--. 1958. La familia Lepidotidae en Chile. Investnes zool. chil., 4:132-134.

--. 1958. Peces de la superfamilia Clupeoidae en aguas de Chile. Revta Biol. mar., 8:83-110.

--. 1959. Lampreas, tiburones, rayas y peces en la estación de biológia marina de Montemar, Chile. Revta Biol. mar., 9(1-3):3-200.

--. 1959. Notas sobre Ictiologia chilena con descripción de los especies nuevas. Revta Biol. mar., 9(1-3):257-270.

BUETTIKOFER, J. 1890. Reisebilder aus Liberia. 2 Thierwelt Fische:447-453. Leiden, 510 p., fig., pl.

BULLIS, H. R., Jr.; CARPENTER, J. S. 1966. Neoharriotta carri, a new species of Rhinochimaeridae from the southern Caribbean Sea. Copeia, 1966(3): 443-450, fig. 1-5.

BULLIS, H. R., Jr.; RIVAS, L. R. 1972. Zoogeography of Bathypterois in the western Atlantic. Proc. Symp. Invest. Resour. Caribbean Sea adj. Reg. (Curaçao, 18-26 Nov. 1965):369-373.

BULLIS, H. R., Jr; STRUHSAKER, P. J. 1970. Fish fauna of the western Caribbean upper slope. Q. Jl Fla Acad. Sci., 33:43-76, 5 fig., 3 tab.

BULLIS, H. R., Jr.; THOMPSON, J. R. 1965. Collection by the Exploratory Fishing Vessels Oregon, Silver Bay, Combat, and Pelican made during 1956-60 in the southwestern North Atlantic. Spec. scient. Rep. U.S. Fish Wildl. Serv. (Fish.), (510):1-130.

BURGESS, G. H.; LINK, G. W., Jr; ROSS, S. W. 1980. Additional marine fishes new or rare to Carolina waters. Northeast Gulf Sci., 3(2):74-87.

BURGESS, W. E. 1978. Butterflyfishes of the world. Neptune City, 832 p., many unnumbered photos, 11 fig. maps.

BURTON, M. 1933. Zool. Rec.; Pisces, 69(13):1-63.

BUSSING, W. A., 1965. Studies of the midwater fishes of the Peru-Chile Trench. In: Llano, G. A. (ed.) Biology of the Antarctic Seas II. Antarct. Res. Ser. (Am. geophys. U.), 5(1297):185-227, 16 fig.

BUSSING, W. A.; BUSSING, M. I. 1966. Antarctic Scopelarchidae: a new fish of the genus Benthalbella and the distribution of B. elongata (Norman). Bull. Sth. Calif. Acad. Sci., 65(1):53-61.

BUTLER, J. L. 1979. The nomeid genus Cubiceps (Pisces) with a description of a new species. Bull. mar. Sci., 29(2):226-241, 12 fig.

BUTLER, J. L.; AHLSTROM, E. H. 1976. Review of the deep-sea fish genus Scopelengys (Neoscopelidae) with a description of a new species S. clarkei from the Central Pacific. Fishery Bull., 74(1):142-150, 5 fig., 3 tab.

CABO, F. L. (see LOZANO CABO, F.)

CABRERA, J. B. L.; ALVAREZ, E. S. 1973. Observaciones biologico-pesqueras en el banco pesquero sahariano. Boln Inst. esp. Oceanogr., (171): 1-79.

CADENAT, J. 1935. Les Serranidés de la côte occidentale d'Afrique (du Cap Spartel au Cap Vert). Revue Trav. Off. (scient. tech.) Pêch. marit., 8(4):377-422, 30 fig.

--. 1936. Sur les stades jeunes de quelques poissons de chalut de la côte de Mauritanie. 1ère note. Revue Trav. Off. (scient. tech.) Pêch. marit., 9:293-322, 23 fig.

--. 1937. Recherches systématiques sur les poissons littoraux de la côte occidentale d'Afrique récoltés par le navire "Président Théodore Tissier" au cours de sa 5e croisière. Liste des poissons littoraux avec la diagnose préliminaire de 6 espèces nouvelles. Revue Trav. Off. (scient. tech.) Pêch. marit., 10(4):423-562, 61 fig.

--. 1938. Note sur deux poissons nouveaux de la côte occidentale d'Afrique (Neopercis Ledanoisi et Pterothrissa Belloci). Bull. Mus. natn. Hist. Nat., Paris, (2)10:361-368.

--. 1945. Les Scorpaenidae de l'Atlantique et de la Méditerranée. Première note. Le genre Scorpaena. Revue Trav. Off. (scient. tech.) Pêch. marit., 13:525-563, fig. 1-11.

--. 1946. Captures nouvelles de Paragaleus Budker et de Scorpaenodes africanus Pfaff. Bull. Mus. natn. Hist. nat., Paris, (2)18(4):319-323.

--. 1947. Noms vernaculaires des principales formes d'animaux marins des côtes d'Afrique occidentale française. Cat. Inst. fr. Afr. noire., 2:56 p.

--. 1948. Physionomie générale de la pêche maritime en Afrique occidentale française. Conf. Pêche marit., Dakar:25-60.

--. 1949. Description de quatre Téléostéens nouveaux de la côte occidentale d'Afrique. Bull. Mus. natn. Hist. nat., Paris, (2)21(6):663-671, 4 fig.

--. 1949. Notes sur une petite collection de poissons des Iles du Cap Vert. Bull. Inst. fr. Afr. noire, 11(3-4):332-339.

--. 1950. Poissons de mer du Sénégal. Init. afr., 3:1-345, fig. 1-241.

--. 1950. Note sur les merlus de la côte occidentale d'Afrique. Congr. Pêch. Pêcheries Un. fr. Outre-Mer, Marseille, 1950:128-130.

--. 1950. Rapport sur les Sélaciens des côtes du Sénégal et plus spécialement sur les Requins. Bull. Inst. fr. Afr. noire, (A)12:944-975.

--. 1950. Remarques au sujet de l'incubation buccale chez quelques espèces de poissons marins. Notes afr., (46):55.

--. 1950. Sur la présence de Microdesmus longipinnis Weymouth sur les côtes du Sénégal. C. r. prem. Conf. int. Afr. Ouest. Dakar, 1:190-191.

--. 1950. Notes sur quelques Scombridae des Iles du Cap-Vert. Congr. Pêch. Pêcheries Un. fr. Outre-Mer, Marseille, 1950:131-134.

--. 1951. Lista provisoria dos peixes observados nas ilhas de Cabo Verde, de 1 de Maio a 24 de Junho de 1950. Cabo Verde, 2(19):25-27.

--. 1952. Note au sujet des merlus de la région de Dakar. J. Cons. perm. int. Explor. Mer, 8(2):230-233.

--. 1953. Notes d'ichtyologie ouest-africaine. V [misprinted VI]. Poissons des campagnes du "Gérard-Tréca". <u>Bull. Inst. fr. Afr. noire</u>, (A)15(3):1051-1102, 43 fig.

--. 1954. Notes d'ichtyologie ouest-africaine. VI. Sur quelques espèces d'Apodes du genre <u>Caecula</u>. <u>Bull. Inst. fr. Afr. noire</u>, (A)16:238-244.

--. 1954. Notes d'ichtyologie ouest-africaine. VII. Biologie. Régime alimentaire. <u>Bull. Inst. fr. Afr. noire</u>, (A)16(2):564-583, 1 tab.

--. 1954. Notes d'ichtyologie ouest-africaine. VIII. Sur les mulets de la côte occidentale d'Afrique. <u>Bull. Inst. fr. Afr. noire</u>, (A)16: 584-591, 1 tab.

--. 1955. Remarques au sujet de <u>Merluccius senegalensis</u>. <u>Rapp. P.-v. Réun. Cons. perm. int. Explor. Mer.</u>, 137:47-48.

--. 1955. Sur les mulets de la côte occidentale d'Afrique. <u>Rapp. P-v. Réun. Cons. perm. int. Explor. Mer.</u>, 137:59-62.

--. 1956. Notes d'ichtyologie ouest-africaine. XII. Sur la présence du "Blue Marlin" <u>Makaira ampla</u> (Poey) sur les côtes d'Afrique occidentale (Dakar). <u>Bull. Inst. fr. Afr. noire</u>, (A)18:539-554.

--. 1956. Note d'ichtyologie ouest-africaine. XIII. Une espèce nouvelle d'apode Ophichthyidae des côtes de Sierra Leone: <u>Mystriophis creutzbergi</u>. <u>Bull. Inst. fr. Afr. noire</u>, (A)18:857-862.

--. 1956. Note d'ichtyologie ouest-africaine. XVI. Description d'une espèce nouvelle d'Ophichthyidae: <u>Caecula pantherina</u>. <u>Bull. Inst. fr. Afr. noire</u>, (A)18:1267-1271.

--. 1957. Notes d'ichtyologie ouest-africaine. XVII. Biologie. Régime alimentaire. <u>Bull. Inst. fr. Afr. noire</u>, (A)19(1):274-294.

--. 1958. Notes d'ichtyologie ouest-africaine. XVIII. Un "Diable de Mer" du Pacifique est, <u>Mobula lucasana</u> Beebe et Tee Van 1938, sur les côtes du Sénégal. <u>Bull. Inst. fr. Afr. noire</u>, (A)20(4):1427-1433.

--. 1958. Les Diables de Mer (Raies pélagiques de la famille des Mobulidae). <u>Notes afr.</u>, (80):116-120, 7 fig.

--. 1959. Notes d'ichtyologie ouest-africaine. XXIV. Molidae ouest-africains avec description d'une espèce nouvelle: <u>Pseudomola lassarati</u> de Côte d'Ivoire. <u>Bull. Inst. fr. Afr. noire</u>, (A)21(3):1112-1118, 11 fig.

--. 1959. Notes d'ichtyologie ouest-africaine. XXV. Description d'une <u>Mobula</u> de grande taille, à aiguillon caudal, de Côte d'Ivoire: <u>Mobula rancureli</u>, sp. nov. <u>Bull. Inst. fr. Afr. noire</u>, (A)21(4):1326-1331.

--. 1959. Notes d'ichtyologie ouest-africaine. XIX. Les Antennarius. <u>Bull. Inst. fr. Afr. noire</u>, (A)21(1):361-394, fig. 1-26.

--. 1960. Notes d'ichtyologie ouest-africaine. XXIX. Les Mobulidae de la côte occidentale d'Afrique. <u>Bull. Inst. fr. Afr. noire</u>, (A)22(3): 1053-1108, pl. 1-24.

--. 1960. Notes d'ichtyologie ouest-africaine. XXX. Poissons de mer ouest-africains observés du Sénégal au Cameroun et plus spécialement au large des côtes de Sierra-Leone et du Ghana. <u>Bull. Inst. fr. Afr. noire</u>, (A)22(4); 1358-1420, 3 pl.

--. 1960. Notes d'ichtyologie ouest-africaine. XXXI. Sur la présence d'un Ateleopidae, <u>Melanogloea ventralis</u> Barnard, 1941 (?) sur les côtes du Sénégal. <u>Bull. Inst. fr. Afr. noire</u>, (A)22(4):1424-1426, 1 pl. (2 fig.).

--. 1961. Notes d'ichtyologie ouest-africaine. XXXIV. Liste complémentaire des espèces de poissons de mer (provenant des côtes de l'Afrique occidentale) en collection à la section de biologie marine de l'IFAN à Gorée. <u>Bull. Inst. fr. Afr. noire</u>, (A)23(1):231-245.

--. 1962. Notes d'ichtyologie ouest-africaine. XXXVI. Sur le statut de quelques espèces de poissons de mer ouest-africains. <u>Bull. Inst. fr. Afr. noire</u>, (A)24(1):283-298, 6 pl.

--. 1962. Notes d'ichtyologie ouest-africaine. XXXVII. Sur quelques espèces nouvelles ou peu communes de la côte occidentale d'Afrique.

Bull. Inst. fr. Afr. noire, (A)24(1):305-312, 3 pl.

--. 1963. Notes d'ichtyologie ouest-africaine. XXXIX. Notes sur les requins de la famille des Carchariidae et formes apparentées de l'Atlantique ouest-africain (avec la description d'une espèce nouvelle: Pseudocarcharias pelagicus, classée dans un sous-genre nouveau). Bull. Inst. fr. Afr. noire, (A)25:526-543, 9 fig.

--. 1963 (?). Notes sur les Poissons marins des Iles du Cap Vert précédées de remarques sur les Poissons de cette région, de Madère et de Gambie, décrits et figurés par T.E. et S. Bowdich, Londres, 1825. 35 p. (manuscript deposited in MNHN Fish Department library no E7-461).

--. 1964. Notes d'ichtyologie ouest-africaine. XLI. Les Sphyraenidae de la côte occidentale d'Afrique. Bull. Inst. fr. Afr. noire, (A)26(2): 659-685.

--. 1964. Notes d'ichtyologie ouest-africaine. XLII. Les "Sars" des genres Puntazzo et Diplodus des eaux tropicales ouest-africaines. Bull. Inst. fr. Afr. noire (A)26(3):944-988, 26 fig.

CADENAT, J.; BLACHE, J. 1970. Description d'une espèce nouvelle, Tripterygion delaisi sp. nov., provenant de l'île de Gorée (Sénégal). Bull. Mus. natn. Hist. nat.. Paris, 41(5):1097-1105.

--. 1981. Requins de Méditerranée et de l'Atlantique (plus particulièrement de la côte occidentale d'Afrique). Faune trop., 21:1-330, 212 fig.

CADENAT, J.; MARCHAL, E. 1963. Résultats des campagnes océanographiques de la "Reine-Pokou" aux îles Sainte-Hélène et Ascension. Poissons. Bull. Inst. fr. Afr. noire, (A)25(4):1235-1315, 48 fig.

CADENAT, J.; MOAL, R. 1955. Note sur la sardine (S. pilchardus Walb.) dans la région du Cap Blanc. Rapp. P.-v. Réun. Cons. perm. int. Explor. Mer, 137:21-23.

CADENAT, J.; POSTEL, E. 1954. Acanthocybium solandri sur les côtes d'Afrique occidentale. Notes afr., (62):53-54.

CADENAT, J.; RANCUREL, P. 1960. Notes d'ichtyologie ouest-africaine. XXVI. Description d'une nouvelle espèce de Mobulidae de côte d'Ivoire: Mobula coilloti. Bull. Inst. fr. Afr. noire, (A)22(1):283-293.

CADENAT, J.; ROUX, C. 1964. Résultats scientifiques des campagnes de la "Calypso". Iles du Cap Vert III. Poissons téléostéens. Annls Inst. océanogr.. Monaco, 41(6):81-102.

CADENAT, J.; STAUCH, A. 1965. Sur la validité des genres Bathysolea (Roule, 1916) et Capartella (Chabanaud, 1950). Cah. ORSTOM Océanogr., 3(3):67-70.

CADENAT, J.; CAPAPE, C.; DESOUTTER, 1978. Description d'un Torpedinidae nouveau des côtes occidentales d'Afrique: Torpedo (Torpedo) bauchotae (Torpediniformes, Pisces). Cybium, 2(2):29-42, 9 fig., 1 tab.

CALDWELL, D. K. 1962. Western Atlantic fishes of the family Priacanthidae. Copeia, 1962(2):417-424, 4 fig.

--. 1962. Postlarvae of the blue marlin, Makaira nigricans, from off Jamaica. Contr. Sci., (53):1-11, 2 fig.

--. 1966. Marine and freshwater fishes of Jamaica. Bull. Inst. Jamaica (Sci. ser.), 17:1-120.

CALDWELL, D. K.; BRIGGS, J. C. 1957. Range extensions of western North Atlantic fishes with notes on some soles of the genus Gymnachirus. Bull. Fla St. Mus. biol. Sci., 2(1):1-11.

CALVARIO, J.; MARQUES, C.; POUSADA. 1980. Occurrence of Sphoeroides cutaneus (Günther, 1870) (Pisces, Tetrodontidae) off the Portuguese coast. Archos Mus. Bocage, (2)7(10):131-137.

CAMPBELL, W. D. 1879. On a new fish. Trans. Proc. N.Z. Inst., 11(27): 297-298.

CANESTRINI, G. 1861. Pleuronettidi del Golfo di Genova. Archo Zool. Anat. Fisiol., 1:1-44, 4 pl.

--. 1861. Intorno al sviluppo del _Dactylopterus_ _volitans_ C.V. ed al genere _Cephalacanthus_. Archo Zool. Anat. Fisiol., 1(1):45-51, pl. 4, fig. 4 + 4a-c & 5 + 5*.

--. 1861. I gobii del Golfo di Genova. Archo Zool. Anat. Fisiol., 1:121-157.

--. 1861. Sopra una nuova specie di _Tetrapturus_. Archo Zool. Anat. Fisiol., 1(1):259-261, 1 pl.

--. 1864. Note ittiologiche. Archo Zool. Anat. Fisiol., 3:100-112.

--. 1865. Sopra alcuni pesci poco noti o nuovi del Mediterraneo nota. Memorie Accad. Sci. Torino, (2)21:359-367, 2 pl.

--. 1866. Prospetto critico dei pesci d'acqua dolce d'Italia. Archo Zool. Anat. Fisiol., 4:47-187, pl.

--. 1872. Pesci d'Italia. Parte II. Pesci marini:37-208. In: E. Cornalia, 1870-74, Fauna d'Italia. Milano, 3:1-208.

CANTOR, T. E. 1849. Catalogue of Malayan fishes. J. Asiat. Soc. Beng., 18(2):983-1443, 14 pl.

CANTRAINE, F. J. 1835. Mémoire sur le _Serranus_ _tinca_. Bull. Acad. r. Belg. Cl. Sci., 2(4):207-208.

--. 1835. Sur un poisson nouveau trouvé dans le canal de Messine en janvier 1833. Bull. Acad. r. Belg. Cl. Sci., 2(1):23-24 (see also, ibid., 2(4):207-209).

--. 1837. Mémoire sur un poisson nouveau, trouvé dans le canal de Messine en janvier 1833. Nouv. Mém. Acad. r. Belg. Cl. Sci., (10):19 p., 2 pl.

--. 1838. Mémoire sur le _Serranus_ _tinca_. Nouv. Mém. Acad. r. Belg. Cl. Sci., (11):9 p., 1 pl. col.

CAPAPE, C. 1974. Premières données sur le cycle de la reproduction de _Dasyatis_ _centroura_ (Mitchill, 1815) et de _Gymnura_ _altavela_ (Linné, 1758) des côtes tunisiennes. Archs Inst. Pasteur Tunis, 51:345-356.

--. 1977. _Raja_ _africana_ n. sp., une nouvelle espèce pour les côtes ouest-africaines et tunisiennes. Bull. Soc. Sci. nat. Tunis., 12:69-78, fig. 1-6.

--. 1977. Deux nouvelles espèces pour les côtes ouest-africaines: _Raja_ _rouxi_ n. sp. et _Raja_ _dageti_ n. sp. (Pisces, Rajidae). Bull. Mus. natn. Hist. nat., Paris, (3)(482)(Zool. 339):1021-1038, fig. 1-6, 1977.

--. 1981. Nouvelle description de _Gymnura_ _altavela_ (Linnaeus, 1758)(Pisces, Rajiformes, Gymnuridae). Bull. Inst. natn. scient. tech. Oceanogr. Pêche Salammbo, 8:59-68, 6 fig.

CAPAPE, C.; DESOUTTER, M. 1978. Méthode d'étude, description critique et comparée des ptérygopodes de Rajidae: _Raja_ _radula_ Delaroche, 1809. Bull. Off. natn. Pêch. Tunis., 2(1-2):79-88, fig. 1-6.

--. 1978. Etude morphologique du neurocrâne, de la ceinture pelvienne et des ptérygopodes de deux Myliobatidae (Pisces, Rajiformes) des côtes tunisiennes, _Myliobatis_ _aquila_ (Linné, 1758) et _Pteromylaeus_ _bovinus_ (Geoffroy Saint Hilaire, 1817). Bull. Soc. Sci. nat. Tunis.,13:39-48, 8 fig.

--. 1979. Etude morphologique des ptérygopodes de _Torpedo_ _(Torpedo)_ _marmorata_ Risso, 1810 (Pisces, Torpedinidae). Neth. J. Zool., 29(3):443-449, 3 fig.

--. 1980. Nouvelle description de _Raja_ _asterias_ Delaroche, 1809 (Pisces, Rajiformes). Cybium, 4(4):29-40, fig. 1-5.

--. 1980. Nouvelles descriptions de _Torpedo_ _(Tetronarce)_ _nobiliana_ Bonaparte, 1835, et de _Torpedo_ _(Tetronarce)_ _mackayana_ Metzelaar, 1919 (Pisces, Torpedinidae). Bull. Mus. natn. Hist. nat., Paris, (4)2(A):325-342, 15 fig., 3 tab.

CAPART, A. 1953. Quelques copépodes parasites des poissons marins de la région de Dakar. Bull. Inst. fr. Afr. noire, (A)15(2):647-671.

CAPELLO, F. de B. 1867. Peixes novos de Portugal e da Africa occidental, e caractéres distinctivos d'outras especies já conhecidas. <u>Jorn. Sci. math. phys. nat.</u>, 1(2):154-169, fig.

--. 1867. Catálogo dos peixes de Portugal que existem no Museu de Lisboa. <u>Jorn. Sci. math. phys. nat.</u>, 1(3):233-264, fig., pl. 4, fig. 1-5; 307-313.

--. 1868. Catálogo dos peixes de Portugal que existem no Museu de Lisboa. <u>Jorn. Sci. math. phys. nat.</u>, 2(5):51-63.

--. 1869. Appendice ao catálogo dos peixes de Portugal que existem no Museu de Lisboa. <u>Jorn. Sci. math. phys. nat.</u>, 2(7):223-228.

--. 1871. Primeira lista dos peixes da Ilha da Madeira, Açores e das possessôes portuguezas d'Africa, que existem no Museu de Lisboa. <u>Jorn. Sci. math. phys. nat.</u>, 3(11):194-202; 3(12):280-282.

--. 1873. Primeira lista dos peixes da Ilha da Madeira, Açores, e das possessôes portuguezes d'Africa, que existem no Museu de Lisboa. (continuaçâo). <u>Jorn. Sci. math. phys. nat.</u>, 4(13):83-88.

CAPORIACCO, L. di. 1921. Revisione della specie mediterranee della famiglia Labridae. <u>Atti</u> <u>Soc.</u> <u>ital.</u> <u>Sci.</u> <u>nat.</u>, 60(1):49-101

CARAUSU, S. I. 1952. Tratat de Ichtiologie. Bucuresti, 802 p., 491 fig., 1 tab. (inset).

CARDENAS, S. 1976. Food habits of the Toad-fish, <u>Halobatrachus</u> <u>didactylus</u> (Schneider, 1801) in the Cadiz Bay (SW Spain). <u>ICES</u>, 13 p. (mimeo).

--. 1977. Régimen alimenticio del Sapo <u>Halobatrachus</u> <u>didactylus</u> (Schneider, 1801) - (Pisces Batrachoididae) en la Bahía de Cádiz (Espana sudoccidental). <u>Vie</u> <u>Milieu</u>, (A)27(1):111-129.

CARDOSO, J. 1895-1900. Notas africanas. I. Pescadores et pescarias no archipelago de Cabo Verdo. II. Plantas empregadas na pesca. III. Ichthyologia Cabo-Verdiana. <u>Anais</u> <u>Sci.</u> <u>nat.</u>. I. 1895, 2:171-174; 3: 93-96; 211-216. II. 1897, 4:69-76. III. 1900, 6:33-49.

CARTE, A. 1866. Notes on the genus <u>Chiasmodon</u>. <u>Proc.</u> <u>zool.</u> <u>Soc.</u> <u>Lond.</u>, 1866:35-39.

CARTER, H. J. 1983. <u>Apagesoma</u> <u>edentatum</u>, a new genus and species of ophidiid fish from the western North Atlantic. <u>Bull.</u> <u>mar.</u> <u>Sci</u>, 33(1): 94-101, 7 fig.

CARUS, J. V. 1893. Vertebrata. 1. class. Pisces: 498-711, In: Prodromus faunae Mediterraneae sive Descriptio Animalium Maris Mediterranei incolarum quam comparata silva rerum quatenus innotuit adiectis locis et nominibus vulgaribus... Stuttgart, 1889-93, 2: ix + 854 p.

CARUSO, J. H. 1981. Lophiidae, <u>in</u> Fischer, W., Bianchi, G. & W.B Scott (ed.), FAO species identification sheets for fishery purposes. Eastern Central Atlantic; fishing areas 34, 47 (in part). Canada Funds-in-Trust. Ottawa, Department of Fisheries and Oceans Canada, by arrangement with the Food and Agriculture Organization of the United Nations, vol. 1-7: pag. var.

--. 1986. Lophiidae, <u>in</u> P.J.P. Whitehead, M.L. Bauchot, J.C. Hureau, J. Nielsen & E. Tortonese. Fishes of the North-eastern Atlantic and the Mediterranean/Poissons de l'Atlantique du Nord-Est et de la Méditerranée, Paris, Unesco, 3:1362-1363, 2 fig.

CARUSO, J. H.; PIETSCH, T. W. 1986. Chaunacidae, <u>in</u> P.J.P. Whitehead, M.L. Bauchot, J.C. Hureau, J. Nielsen & E. Tortonese. Fishes of the North-eastern Altantic and the Mediterranean/Poissons de l'Atlantique du Nord-Est et de la Méditerranée, Paris, Unesco, 3:1369-1370, 1 fig.

CASINOS, A. 1978. The comparative feeding mechanism of Gadidae and Macrouridae. I. Functional morphology of the feeding apparatus. <u>Gegenbaurs</u> <u>morph.</u> <u>Jb.</u>, 124(3)5:434-449, 10 fig.

CASTELNAU, F. de. 1855. Animaux nouveaux ou rares recueillis pendant l'expédition dans les parties centrales de l'Amérique du Sud de Rio de Janeiro à Lima, et de Lima au Para. <u>In</u>: Expédition dans les parties centrales de l'Amérique du Sud pendant les années 1843 à 1847, 7e partie, Zoologie, 3, Poissons. Paris, xiv + 112 p., 50 pl.

--. 1861. Mémoire sur les poissons de l'Afrique australe. Paris, vii + 78 p.

--. 1872. Contribution to the ichthyology of Australia. Proc. Zool. Acclim. Soc. Victoria, 1:29-247.

--. 1873. Contributions to the ichthyology of Australia. Proc. Zool. Acclim. Soc. Victoria, 2:37-158.

--. 1875. Researches on the fishes of Australia. Philadelphia Centennial Expedition, p. 1-52.

--. 1878. Essay on the ichthyology of Port Jackson. Proc. Linn. Soc. N. S. W., 3:347-402.

CASTEX, M. N. 1967. Freshwater venomous rays, p.167-176. In: Animal Toxins, a collection of papers presented at the first international symposium on animal toxins (Atlantic City, April 1966). Pergamon Press, Oxford, New-York.

CASTILLO, R. & BRITO, A. 1982. Primera cita para las Islas Canarias de Gobius auratus Risso, 1810 (Pisces, Gobiidae). Investigación pesq., 46: 391-396.

CASTLE, P. H. J. 1961. Deep-water eels from Cook Strait, New Zealand. Zoology Publs Vict. Univ. Wellington, (27):1-30, 6 fig.

--. 1963. The systematics, development and distribution of two eels of the genus Gnathophis (Congridae) in Australasian waters. Zoology Publs Vict. Univ. Wellington, (34):15-47, 10 fig.

--. 1964. Eels and eel-larvae of the Tui Oceanographic Cruise, 1962, to the South Fiji Basin. Trans. R. Soc. N.Z., Zool., 5(7):71-84, 1 fig.

--. 1964. Deep-sea eels: family Synaphobranchidae. Galathea Rep., 7:29-42, 2 fig.

--. 1965. Leptocephali of the Nemichthyidae, Serrivomeridae, Synapho-branchidae, and Nettastomatidae in Australasian waters. Trans. R. Soc. N.Z., Zool., 5(11):131-146, 2 fig.

--. 1966. Die ichthyologische Ausbeute der ersten Westafrikafahrt des fischereitechnischen Forschungsschiffes "Walther Herwig". 3. The eel larvae (Leptocephali). Arch. FischWiss., 17(1):19-35, fig. 1-3.

--. 1969. An index and bibliography of eel larvae. Spec. Publs J.L.B. Smith Inst. Ichthyol. (7):121 p.

--. 1970. Ergebnisse der Forschungsreisen des FFS "Walther Herwig" nach Südamerika: XI. The Leptocephali. Arch. FischWiss, 21(1):1-21, 5 fig.

--. 1970. Distribution, larval growth and metamorphosis of the eel De-richthys serpentinus Gill, 1884 (Pisces: Derichthyidae). Copeia, 1970(3):444-452, 1 fig.

--. 1975. Fische des Indischen Ozeans. Ergebnisse... "Meteor"... A. Systematischer Teil, XV, Leptocephali (Anguilliformes). Meteor Forsch.-Ergebn., (D)(21):19-29, fig. 353.

CASTLE, P. H. J.; WILLIAMSON, G. R. 1975. Systematics and distribution of eels of the Muraenesox group (Anguilliformes, Muraenesocidae), a preliminary report and key. Spec. Publs J.L.B. Smith Inst. Ichthyol., (15):9 p., 4 fig.

CASTRO, J. L. 1978. Catálogo sistemático de los peces marinos que pene-tran a las aguas continentales de México con aspectos zoogeográficos y ecológicos. Ser. cient. Direccion gen. Inst. nac. Pesca. Mexico, (19):1-298, fig.

CASTRO, J. M. O. 1967. Nomenclatura portuguesa do pescado. Publcôes Gabin. Estud. Pescas, (39):1-288.

CATALANO, E. 1978. Ritrovamento di Hypleurochilus bananensis Poll nel Golfo di Palermo. Osservazioni su alcuni stadi larvali (Perciformes, Blenniidae). Naturalista sicil., (4)2(3-4):73-83.

CATESBY, M. 1743. The natural history of Carolina, Florida and the Bahama Islands: containing the figures of birds, beasts, serpents, insects and plants with their descriptions in English and French, etc. London, 2:1-100 + i-xliv + 6 p. n. num., 100 col. pl. (1 publ. in 1731) (2nd ed., 1754, revised by Mr. Edwards; 3rd ed., 1771, revised by G. Edwards).

CAVALIERE, A. 1955. Uovo ovarico e stadi giovanili di <u>Pomatomus</u> <u>telescopus</u> Risso. <u>Boll. Pesca Piscic. Idrobiol.</u>, 10 (n.s.) (1):99-103, 3 fig.

--. 1956. Su rari stadi larvali e giovanili di <u>Pomatomichthys</u> <u>constanciae</u> Gigl. e relativa differenziazione da stadi giovanili della specie affine <u>Pomatomus</u> <u>telescopium</u> Risso. <u>Boll. Pesca Piscic. Idrobiol.</u> 11(n.s.):121-127, 1 pl.

CAVALIERE, A.; BERDAR, A. 1976. Osservazioni biologiche su alcuni esemplari di <u>Bathophilus</u> <u>nigerrimus</u> pescati nello Stretto di Messina. <u>Memorie Biol. mar. Oceanogr.</u>, (n.s.)6(3):45-56.

CENTURIER-HARRIS, O. M. 1974. The appearance of lantern-fish in commercial catches. <u>S. Afr. shipp. News fishg Ind. Rev.</u>, 29(1):45.

CERVIGON, F. 1960. Peces recogidos en el curso de la campañas realizadas por las costas de Africa desde Cabo Espartel a Guinea Portuguesa. <u>Boln R. Soc. esp. Hist. nat.</u>, 58(2):245-252.

--. 1960. Peces recogidos en el curso de las campañas realizadas a bordo del "Costa Canaria" desde cabo Bojador a Guinea Portuguesa (Africa occidental) y consideraciones sobre su distribución. <u>Investigación pesq.</u>, 17:33-107, 56 fig.

--. 1961. Descripción de <u>Anodontus</u> <u>mauritanicus</u> nov. gen. nov. sp. (ordem Ateleopiformes) y <u>Cottunculus</u> <u>costae-canariae</u> nov. sp. (familia Cottidae) de la costas occidentales de Africa. <u>Investigación pesq.</u>, 19:119-128, fig. 1-7.

--. 1964. Neuvas citas de peces para Venezuela y datos sobre algunas especies poco conocidas. <u>Noved. cient. Mus. Hist. nat. La Salle</u>, Zool., (31):1-18.

--. 1966. Los Peces marinos de Venezuela. <u>Estac. Invest. mar. Margarita</u> Caracas, 1 (<u>Monogr.</u> 11):1-440, fig. 1-181, 1 map. 2(<u>Monogr.</u> 12): 441-951, fig. 182-385.

CETTI, F. 1777. Storia naturale di Sardegna, III. Anfibi e pesci di Sardegna, Sassari, 208 p., 5 pl.

CHABANAUD, P. 1925. Sur quelques Scombroides de la côte occidentale d'Afrique. <u>Bull. Soc. zool. Fr.</u>, 50:197-201.

--. 1925. <u>Monodichthys</u> <u>proboscideus</u> (gen. nov. et spec. nova) et remarques sur divers autres poissons Soléiformes. <u>Bull. Mus. natn. Hist. nat., Paris</u>, 31:356-361.

--. 1926. Sur divers poissons de mer de la côte occidentale d'Afrique. Description de deux espèces nouvelles. <u>Bull. Soc. zool. Fr.</u>, 51:8-16.

--. 1926. Sur les clupéidés du genre <u>Sardina</u> Antipa et de divers genres voisins. <u>Bull. Soc. zool. Fr.</u>, 51:156-163.

--. 1926. Sur un second exemplaire de <u>Monodichthys</u> <u>proboscideus</u> Chab., rectification de la diagnose générique et de la diagnose spécifique. <u>Bull. Mus. natn. Hist. nat., Paris</u>, 32:52-58.

--. 1926. Description d'une espèce nouvelle de Sole originaire de l'Atlantique oriental. <u>Bull. Mus. natn. Hist. nat., Paris</u>, 32:127-130.

--. 1926. Fréquence, symétrie et constance spécifique d'hyperostoses externes chez divers poissons de la familles des Sciénidés. <u>C. R. hebd. Séanc. Acad. Sci., Paris</u>, 182:1647-1649.

--. 1927. Les Soles de l'Atlantique oriental nord et des mers adjacentes. <u>Bull. Inst. océanogr. Monaco</u>, (488):1-67.

--. 1927. Description d'un poisson nouveau de la Baie du Cameroun, appartenant à la famille des Cerdalidae. <u>Bull. Mus. natn. Hist. nat., Paris</u>, 33:230-234.

--. 1927. Observations morphologiques et remarques sur la systématique des poissons hétérosomes Soléiformes. _Bull. Inst. océanogr. Monaco_, (500):1-15.

--. 1927. Sur quelques poissons hétérosomes de la Martinique. _Bull. Soc. zool. Fr._, 52:74-84, 7 fig.

--. 1927. Révision du genre _Heteromycteris_ Kaup (Pisces, Heterosomata, Soléiformes). _Ann. Mag. nat. Hist._, (9)20:523-527.

--. 1928. Description d'une nouvelle espèce d'Elasmobranches de la famille des Discobatidés, appartenant à la faune de l'Atlantique oriental tropical nord. _Bull. Soc. zool. Fr._, 53:419-424.

--. 1928. Remarques sur quelques genres de la famille des Soléidés. _Bull. Soc. zool. Fr._, 53:272-279.

--. 1928. Description d'un second exemplaire de _Leptocerdale aethiopicum_ Chab. _Bull. Soc. zool. Fr._, 53:280-285, fig. 1-4.

--. 1929. Observations sur la taxonomie, la morphologie et la bionomie des Soléidés du genre _Pegusa_. _Annls Inst. océanogr., Monaco_, 7(6): 215-260, 38 fig., 6 tab., 2 pl.

--. 1929. Sur _Platyrhina schoenleini_ Müller et Henle. _Bull. Soc. zool. Fr._, 54:558-562, fig. 1.

--. 1930. Les genres de poissons Hétérosomates de la sous-famille des _Soleinae_. _Bull. Inst. océanogr. Monaco_, (555):1-23.

--. 1930. Sur les _Rhinobatus_ du groupe de _cemiculus_ Geoffr. _Bull. Mus. natn. Hist. nat., Paris_, (2)2:86-88.

--. 1931. Notes ichthyologiques. I. Sur un grand spécimen de _Rhinobatus cemiculus_ Geoffr., originaire du Sénégal. _Bull. Soc. zool. Fr._, 56:112-118, fig. 1-4.

--. 1931. Sur divers poissons Soléiformes de la région indo-pacifique. _Bull. Soc. zool., Fr._, 56:291-305, 1 fig.

--. 1931. Les poissons pleuronectes de la Méditerranée (Pisces Heterosomata). _Mem. Ass. Naturalistes Nice Alpes-Maritimes_, 2:1-40 (Suppl. au Riviera scient., 18).

--. 1931. Sur la ceinture et quelques autres éléments morphologiques des poissons Hétérosomates, importance phylogénétique des caractères observés. _Bull. Soc. zool. Fr._, 56:386-398, 4 fig.

--. 1933. Poissons hétérosomes recueillis par M. le Professeur A. Gruvel et par MM. Ph. Dollfus et J. Liouville sur la côte atlantique du Maroc. _Mém. Soc. Sci. nat. phys. Maroc_, 35:1-111, 51 fig., 2 pl.

--. 1934. A propos de _Sardinella eba_ C.V., _maderensis_ Lowe, _aurita_ C.V. _Bull. Soc. zool. Fr._, 59:120-132.

--. 1937. Les Téléostéens dissymétriques du Mokattan inférieur de Tourah. 1ère partie. Classification des Téléostéens dissymétriques du Mokattan inférieur de Tourah. 2ème partie. Les Téléostéens dissymétriques du Mokattan inférieur de Tourah. _Mém. Inst. Egypte_, 32:xi, 1-121, 19 fig., 4 pl.

--. 1938. Chorologie des Soléidés de l'Atlantique nord et des mers adjacentes. _Bull. Stn biol. Arcachon_, 35:7-31, 3 fig.

--. 1938. Nouvelle description du genre _Microchirus_ (Pisces, Soleidae). _Bull. Soc. zool. Fr._, 63:316-322.

--. 1938. Contribution à la morphologie et à la systématique des Téléostéens dissymétriques. _Archs Mus. natn. Hist. nat., Paris_, (6)15:59-139, 75 fig., 9 pl.

--. 1939. Catalogue systématique et chorologique des Téléostéens dissymétriques du globe. _Bull. Inst. océanogr. Monaco_, (763):1-31.

--. 1940. Notules ichthyologiques. I. Diagnose préliminaire d'une espèce et d'une sous-espèce nouvelle de Pleuronectoidea. II. La 4ème fente branchiale des Achiridae. III. L'organe sensoriel pleurogrammique et la pigmentation zénithale des Achiridae (Note préliminaire). IV. Le squelette intermusculaire des Pleuronectoidea. _Bull. Mus. natn. Hist._

nat., Paris, (2)12:149-156.
--. 1941. Notules ichthyologiques. Troisième série. XII. A propos de l'organe pleurogrammique des Achiridae. XIII. Sur les différentes espèces dont se compose le genre Pegusa (Teleostei Soleidae). XIV. Addition à la synonymie d'un Téléostéen de la famille des Soleidae. XV. Présence possible de Solea ovata dans les eaux australiennes. Bull. Mus. natn. Hist. nat., Paris, (2)13:414-421.
--. 1941. Pluralité spécifique du genre Pegusa (Pleuronectoidea, Soléiformes). J. Wash. Acad. Sci., 31:109-114, 10 fig.
--. 1942. Contribution à l'étude des reliques ichthyologiques de la Téthys. C. r. somm. Séanc. Soc. Biogéogr., 1942:45-47.
--. 1943. Notules ichthyologiques (sixième série). XX. L'habitat du Soleidé Pegusa lascaris (Risso) ne serait-il pas circumafricain ? XXI. Le genre Dexillus Chabanaud. XXII. Nouveaux genres de la famille des Soleidae. Bull. Mus. natn. Hist. nat., Paris, (2)15:289-293.
--. 1944. Sarda chiliensis (C.V.). Remarques diverses. Bull. Inst. océanogr. Monaco, (860):1-6.
--. 1946. Notules ichthyologiques (suite). XXXI. Notation conventionelle de l'extension du maxillaire et de la position de l'apex du processus préoral chez les Cynoglossidés. XXXII. Description d'un nouveau Cynoglossus de la région du Sénégal. XXXIII. Définition d'un genre inédit appartenant à la famille des Cynoglossidae. Bull. Mus. natn. Hist. nat., Paris, (2)19:440-443.
--. 1948. Contribution à la morphologie des Citharidae. Bull. Soc. zool. Fr., 73:18-24, 4 fig.
--. 1948. Notules ichthyologiques. XXXVI. A propos de la famille des Citharidae: question de nomenclature. Bull. Mus. natn. Hist. nat. Paris, (2)20:150-153, 1 fig.
--. 1948. Description d'une nouvelle espèce de Soléidés, originaire de la côte occidentale de l'Afrique. Bull. Mus. natn. Hist. nat., Paris, (2)20:512-513.
--. 1949. Révision des Cynoglossidae (s. str.) de l'Atlantique oriental. Bull. Mus. natn. Hist. nat. Paris, (2)21:60-66, 202-209.
--. 1949. Les Cynoglossus de l'Atlantique. Bull. Mus. natn. Hist. nat. Paris, (2)21:516-521.
--. 1949. Notules ichthyologiques (suite). XXXIX. Présence de Bathysolea profundicola dans la Méditerranée. XL. Sur les Synaptura du groupe albomaculata. XLI. L'oeuf des Rhombosoleidae. Bull. Mus. natn. Hist. nat., Paris, (2)21(6):672-675.
--. 1949. Téléostéens dissymétriques (Heterosomata) In: Résultats scientifiques des croisières du Navire Ecole belge "Mercator". IV. Mém. Inst. r. Sci. nat. Belg., (2)33:3-102, 101 fig., 9 pl.
--. 1950. Sur deux espèces du genre Symphurus de l'Atlantique oriental. Ann. Mag. nat. Hist., (12)3:624-627.
--. 1950. Notules ichthyologiques (suite). XLII. Sur une apophyse du rachis d'un Cynoglossus (fig. 1). XLIII. Sur la musculature hypopharyngienne des Symphurinae. XLV. Nouvelle description de l'holotype d'un Cynoglossus de la mer Rouge. XLVI. Sur un Citharichthys stampflii de la côte du Gabon. Bull. Mus. natn. Hist. nat., Paris, (2) 22 (3): 336-340, 1 fig.
--. 1950. Définition et nomenclature des morphes pleurogrammiques des Cynoglossidae. Révision de quatre espèces du genre Cynoglossus. Bull. Mus. natn. Hist. nat., Paris, (2)22(6):713-716.
--. 1950. Description de nouveaux Soleidae capturés au cours de l'Expédition Océanographique Belge dans les eaux africaines de l'Atlantique sud (1948-49). Bull. Inst. r. Sci. nat. Belg., 26(55):1-19.
--. 1952. Description sommaire de deux Soléiformes de la côte atlantique de l'Afrique. Bull. Inst. r. Sci. nat. Belg., 28(69):1-5.

--. 1953. Sur <u>Citharichthys stampflii</u>, Paralychthidé de l'Atlantique o-
riental. <u>Rev. Zool. Bot. afr.</u>, **47**(3-4):390-399.

--. 1954. Révision des Soléidés du genre <u>Pegusa</u>. Description d'une espèce inédite. <u>Bull. Inst. fr. Afr. noire</u>, (A)16(1):243-282, 20 fig., 3
tab.

--. 1954. Nouvelle description de l'holotype de <u>Cynoglossus canariensis</u>
Steindachner. Synonymie de l'espèce. <u>Bull. Soc. zool. Fr.</u>, **79**:79-82.

--. 1955. Sur divers Pleuronectiformes appartenant au Naturhistorisches
Museum de Vienne. Nouvelle description de <u>Cynoglossus canariensis</u>.
<u>Öst. zool. Z.</u>, **5**(4):403-406.

--. 1955. Rectifications afférentes à la nomenclature et à la systématique des Pleuronectiformes du sous-ordre des Soleoidei. <u>Bull. Mus.
natn. Hist. nat., Paris</u>, (2)27(6):447-452.

--. 1955. Sur la répartition circum-africaine de diverses espèces de la
famille des Soleidae. <u>C. r. somm. seanc. Soc. Biogéogr.</u>, **32**:99-106, 2
fig., 2 maps.

CHABANAUD, P.; MONOD, Th. 1927. Les poissons de Port-Etienne. Contribution à la faune ichthyologique de la région du Cap Blanc (Mauritanie
Française). <u>Bull. Com. Etud. hist. scient. Afr. occid. fr.</u>, **9**(2):225-
287, 33 fig., 3 pl.

CHABANNE, J.; ELWERTOWSKI, J. 1973. Cartes des rendements de la pêche
des poissons pélagiques sur le plateau continental nord-ouest africain de 11° à 26°N. <u>Docum. scient. provis. Centre Rech. océanogr.,
Dakar-Thiaroye</u>, (49) (8 p., 88 maps, mimeo).

CHAINE, J. 1938. Recherches sur les otolithes des poissons. Etude descriptive et comparative de la sagitta des Téléostéens (suite). <u>Act.
Soc. linn. Bordeaux</u>, **90**:5-258, 18 pl.

--. 1942. Recherches sur les otolithes des poissons... (suite). <u>Act.
Soc. linn. Bordeaux</u>, **92**:3-135, 6 pl.

--. 1957. Recherches sur les otolithes des poissons. <u>Bull. Cent. Etud.
Rech. scient., Biarritz</u>, 1(4):465-551, 7 pl.

--. 1958. Recherches sur les otolithes de poissons. <u>Bull. Cent. Etud.
Rech. scient., Biarritz</u>, 2(2):159-241, 8 pl.

CHAINE, J.; DUVERGIER, J. 1935. Recherches sur les otolithes des poissons. Etude descriptive et comparative de la sagitta des Téléostéens.
<u>Act. Soc. linn. Bordeaux</u>, **87**:1-242, pl. 1-18.

CHAMPAGNAT, C. 1966. Indice relatif d'abondance saisonnière des sardinelles sur la Petite Côte du Sénégal. <u>Docum. scient. provis. Centre
Rech. océanogr., Dakar-Thiaroye</u>, (1) (11 p., mimeo).

--. 1967. La pêche industrielle des poissons pélagiques côtièrs au Sénégal en 1966. <u>Docum. scient. provis. Centre Rech. océanogr., Dakar-
Thiaroye</u>, (4) (5 p., mimeo).

CHAMPAGNAT, C.; DOMAIN, F. 1978. Migrations des poissons démersaux le
long des côtes ouest-africaines de 10° à 24° de latitude nord. <u>Cah.
ORSTOM, Océanogr.</u>, **16**(3-4):239-261, 13 fig.

CHAMPEAU, M.F. 1951. Des moeurs du "<u>Periophtalmus koelreuteri</u>" Gobiidé
de l'ouest africain. Trav. 2è Conf. int. Afr. occid. Bissau, (1947)3
(2):271-277, 3 pl.

CHAO, L. N. 1986. Sciaenidae, <u>in</u> P.J.P. Whitehead, M.L. Bauchot, J.C.
Hureau, J. Nielsen & E. Tortonese. Fishes of the North-eastern Atlantic and the Mediterranean/Poissons de l'Atlantique du Nord-Est et de
la Méditerranée, Paris, Unesco, 2:865-874, 3 + 7 fig.

CHAO, L. N.; MILLER, R. V. 1975. Two new species of sciaenid fishes.
(Tribe: Sciaenini) from the Caribbean sea and adjacent waters. <u>Bull.
mar. Sci.</u>, **25**:259-271.

CHAO, L. N.; TREWAVAS, E. 1981. Sciaenidae, <u>in</u> W. Fischer, G. Bianchi &
W.B. Scott (ed.), FAO species identification sheets for fishery purposes. Eastern Central Atlantic (Fishing Areas 34 and 47 in part).

Dept. Fish. Oceans Canada, Ottawa/FAO, Rome, vol. 1-7, pag. var.

CHAPMAN, W. N. 1919. Eleven new species and three new genera of oceanic fishes collected from the northwestern Pacific. Proc. U. S. natn. Mus., 86:501-542, fig. 58-70.

CHARLES-DOMINIQUE, E. 1983. Exposé synoptique des données biologiques sur l'ethmalose (Ethmalosa fimbriata S. Bowdich, 1825). Revue Hydrobiol. trop., 15(4):373-397.

CHARLES-DOMINIQUE, E.; ECOUTIN, J. M.; SAN GNANMILIN, A. 1980. La pêche artisanale en lagune Aby-Tendo-Ehy (Côte d'Ivoire). Premières estimations de la production. Docums scient. provis. Centre Rech. océanogr. Abidjan, 6(4):1-26 (mimeo).

CHAUVET, C. 1971. Répartition et abondance des Cynoglossidae (poissons plats) sur le plateau continental de la Côte d'Ivoire. Annls Univ. Abidjan, (E)4(1):287-299, 4 fig., 7 tab.

--. 1971. Données sur la vie sexuelle d'une espèce de poisson plat, Cynoglossus canariensis (Steind., 1882). Annls Univ. Abidjan, (E)4(1): 301-307, 5 fig.

--. 1971. Relation longueur-poids et variation saisonnière de la croissance chez une espèce de poisson-plat: Cynoglossus canariensis (Steind., 1882). Annls Univ. Abidjan, (E)4(1):309-315, 5 fig.

--. 1972. Croissance et détermination de l'âge par lecture d'écailles d'un poisson plat de Côte d'Ivoire; Cynoglossus canariensis. Docums scient. Centre Rech. océanogr. Abidjan, 3(1):1-18, fig., 6 tab.

CHAVEZ, E. A. 1972. Notas acerca de la ictiofauna del estuario del rio Tuxpan y sus relaciones con la temperatura y la salinidad. Mems IV Congr. Nac. Ocean., México:177-199.

CHEESEMAN, T. F. 1876. Notes on the swordfish (Xiphias gladius). Trans. Proc. N.Z. Inst., 8:210-220.

CHEN, J. T. F. 1951. Check-list of the species of fishes known from Taiwan (Formosa) (Continued). Q. Jl Taiwan Mus., 5(4):305-341.

CHEN, J. T. F.; LIU, M.; LEE, S. 1967. A review of the pediculate fishes of Taiwan. Biol. Bull. Tunghai Univ., (33)(= Ichthyol. 7):1-23.

CHIGIRINSKIY, A. I.; VOLKOV, A. F. 1971. Some morphological features and age-related changes in the feeding habits of the Atlantic saury (Scomberesox saurus Walb.). Vop. Ikhtiol., 11(5):868-876, 5 fig.

CHIRICHIGNO, F., N. 1962. Algunos peces nuevos y poco conocidos de la fauna marina del Perú. Ser. Divulgación cient. Serv. Pesq. Perú (17): 1-29, 24 fig.

--. 1969. Lista systemática de los peces marinos comunes para Ecuador-Perú-Chile. Lima, iii + 108 p., 160 fig.

--. 1974. Clave para identificar los peces marinos del Perú. Informe Inst. Mar. Perú, (44):1-387, 643 fig.

CHRISTENSEN, M. S. 1978. Trophic relationships in juveniles of three species of sparid fishes in the South African marine littoral. Fishery Bull., 76(2):389-401, 12 fig.

CHRISTENSEN, M. S.; WINTERBOTTOM, R. 1981. A correction factor for, and its application to, visual censuses of littoral fish. S. Afr. J. Zool., 16, 73-79.

CHRISTIANS, O. 1973. Der Degenfisch - ein neuer Nutzfisch ? Infn Fischw. Auslds, 20(6):184-185.

CHROMOV, N. S. 1960. (Plankton distribution and the food of sardinelles in the Dakar region). Ryb. Khoz., 36(1):7-12 (in Russian).

--. 1962. (Changes in the plankton distribution and the feeding of sardines in the commercially fished region on the western shores of Africa). Trudy vses. nauchno-issled. Inst. ryb. morsk. Khoz. Oceanogr., (46):214-235. (in Russian).

CHUNG, B. D. 1973. Biometrische Untersuchungen an Sardinella aurita Val. (Teleostei, Clupeida) in nordwestafrikanischen Gewässern. Fisch-Forsch., wiss. Schr. Reihe, 11(1):91-99.

CHURCH, R. 1971. Deepstar explores the ocean floor. Natn. geogr. Mag., 139(1):110-129.

CHYUNG, M. K. 1954. Korean fishes. Seoul, 517 + 13 p. (in Korean).

--. 1961. Illustrated encyclopedia. The fauna of Korea. 2. Fishes. Seoul, 861 p., 240 fig., 239 pl., 72 col. pl. (in Korean).

CIPRIA, G. 1939. Uova, sviluppo embrionale e post-embrionale di Scarus cretensis C.V. ottenute per fecondazione artificiale. Memorie R. Com. talassogr. ital., (261):1-5, 12 fig., 1 pl.

CISTERNAS, R. 1877. Ensayo descriptivo de los peces de agua dulce que habitan en la provincia de Valencia. An. Soc. esp. Hist. nat., 6:69-138.

CLARK, E. 1950. Notes on the behavior and morphology of some West Indian plectognath fishes. Zoologica, N.Y., 35(3):159-168, fig. 1-6.

CLARK, H. W. 1938. The Templeton Crocker Expedition of 1934-35. Additional new fishes. Proc. Calif. Acad. Sci., (4)22:179-185.

CLARK, J. R. 1973. Range and distribution of bluefish. p. 250-251. In: (A.L. Pacheco, ed.). Proc. workshop on egg, larval and juvenile stages of fish in Atlantic coast estuaries. Tech. Publ. n°1 natn. mar. Fish. Serv., Mid. Atl. coast. Fish. Cent., N.J.:338 p.

CLARK, R. S. 1913. Scottish National Antarctic Expedition. "Scotia" collection of fishes from St. Helena. Proc. R. phys. Soc. Edinb., 19 (3-4):47-53, 4 fig.

--. 1914. General report on the larval and postlarval teleosteans in Plymouth waters. J. mar. biol. Ass. U. K., 10(2):327-394, 11 fig.

--. 1915. Scottish national Antarctic Expedition: "Scotia" collection of Atlantic fishes. Rep. scient. Results Scott. natn. Antarct. Exped., 4(16):379-401, 4 fig., 1 photo.

CLARKE, F. E. 1878. On two new fishes. Trans. Proc. N.Z. Inst., 10:243-246, 1 pl.

CLARKE, T. A. 1973. Some aspects of the ecology of lanternfishes (Myctophidae) in the Pacific Ocean near Hawaii. Fishery Bull., 71(2):401-433.

--. 1974. Some aspects of the ecology of stomiatoid fishes in the Pacific ocean near Hawaii. Fishery Bull., 72(2):337-351.

CLARKE, T.A.; WAGNER, P.J. 1976. Vertical distribution and other aspects of the ecology of certain mesopelagic fishes taken near Hawaii. Fishery Bull., 74(3):635-645, 2 tab.

CLARKE, W. E. 1900. Occurrence of the long-finned tunny off the Orkney Islands. Ann. Scot. nat. Hist., 1900:248.

CLAUSEN, H. S. 1956. Biological and taxonomical notes on Nigerian freshwater Syngnathus (Linné, 1758) Kaup, 1856, with remarks on the taxonomic value of crista media trunci and c. superior caudae. Vidensk. Meddr. dansk naturh. Foren, 118:225-234, 2 pl.

CLEMENS, H. B.; NOWELL, J. C. 1963. Fishes collected in the eastern Pacific during tuna cruises, 1952 through 1959. Calif. Fish Game, 49 (4):240-269.

CLEMENS, W. A.; WILBY, G. V. 1961. Fishes of the Pacific Coast of Canada. Bull. Fish. Res. Bd Can., 68:1-443, fig. 1-281.

CLIFF, J. 1983. Early colonization by fish of an artificial reef in False Bay, South Africa. Trans. R. Soc. S. Africa, 45(1):63-71.

CLIFTON, H. E.; HUNTER, R. E. 1972. The sand tilefish, Malacanthus plumieri, and the distribution of coarse debris near West Indian coral reefs, p. 87-92, fig. 45-55. In: B.B. Collette & S.A. Earle (ed.), Results of the Tektite program: Ecology of coral reef fishes. Sci. Bull. nat. Hist. Mus., Los Angeles, 14:1-180, 69 fig., 13 tab.

CLIGNY, A. 1905. Poissons des côtes d'Espagne et de Portugal (Océan Atlantique). Première partie. Annls Stn aquic. Boulogne (n.s.), 1:1-

30.

--. 1905. Poissons des côtes d'Espagne et de Portugal (Océan Atlantique). Deuxième partie. Annls Stn aquic. Boulogne (n.s.), 1:69-92.

CLOQUET, H. 1816-45. Poissons. In: Dictionnaire des Sciences Naturelles, Paris, 60 vol. text + 2 vol. suppl., pl. Poissons: Atlas, 4, 85 col. pl.

--. [= H.C.] 1817. (Chironecte, Chironectes). Dict. Sci. nat., F.G. Levrault, Paris, 8: 597-599.

--. 1823. (Lophote, Lophotus). Dict. Sci. nat., Paris, 27:194-195.

COCCO, A. 1829. Su di alcuni nuovi pesci del mar di Messina. G. Sci. Lett. Arti Sicilia, 26(7 or 77):138-147.

--. 1833. Cenni sui genera Ruvettus e sui caratteri che lo distinguone. Ossni Peloritani, 13:18.

--. 1833. Su di alcuni pesci de'mari di Messina. G. Sci. Lett. Arti Sicilia, 42:9-21, pl. (also publ. in Maurolico, 2:236-244).

--. 1836. Su Paralepis hyalinus. Atti Accad. gioenia Sci. nat., 13:49-55, 1 pl.

--. 1838. Su di alcuni Salmonidi del mare di Messina; lettera al ch. Principe C.L. Bonaparte. Nuovi Annali Sci. nat.. Bologna, 1(2):161-194, pl. 5-8.

--. 1838. Osservazioni intorno taluni pesci del mare di Messina, lettera di Anastasio Cocco al Prof. Oronzio Gabrièle Costa di Napoli. G. Il Faro, 4, Anno 6:4.

--. 1839. Sopra una nuova specie di Trachurus G. Innominato, Messina, 3(7):56-59.

--. 1840. Su di alcuni nuovi pesci del mare di Messina. Maurolico G. Sci. Lett. Art. del nuovo Periodo. Messina, 4(5):236-244 (repeat of 1833).

--. 1844. Intorno ad alcuni nuovi pesci del mare di Messina. Lettera al sig. Augusto Krohn di Livonia. G. Gabin. Lett. Messina, 5:10 p., 1 pl.

--. 1846. Intorno ad uno Scaro nel mare di Messina. Atti sett. Adunanza Scient. Ital.. Napoli, p.748.

--. 1884-1886. Indice ittiologico del mare di Messina. Naturalista sicil., 3, 145-148, 176-179, 269-272, 328-332; 4, 25-29, 68-72, 85-88, 113-116, 177-180, 191-194, 228-232, 238-240, 291-294; 5, 11-16, 35-40, 62-67, (the last two notes in collab. with L. Facciola): 109-112, 143-149.

COCCO, A.; SCUDERI, L. 1835. Su di un nuove pesce del mare di Messina. G. Sci. Lett. Arti Sicilia, 51:264-270, 1 pl.

COCKERELL, T. D. A. 1913. The scales of the simenchelyid, ophidiid, brotulid and bregmacerotid fishes. Proc. biol. Soc. Wash., 26:75-77.

--. 1919. Some American Cretaceous fish scales, with notes on the classification and distribution of Cretaceous fishes. Bull. U.S. geol. Surv. (1918), 165-202.

COHEN, D. M. 1956. Neonesthes gnathoprora, a new species of astronesthid fish from the Atlantic Ocean. Zoologica. N. Y., 41(2)10:81-83.

--. 1958. Two new species of Bathylagus from the western North Atlantic with notes on other species. Breviora, (98):1-9, fig. 1-2.

--. 1958. A revision of the fishes of the sub-family Argentininae. Bull. Fla St. Mus. biol. Sci., 3:93-172.

--. 1960. Isospondyli 3. Argentinoidea (Argentinidae and Opisthoproctidae). Rep. scient. Results Michael Sars N. Atlant. deep Sea Exped., 4(2)(9):1-5, 1 fig.

--. 1961. A new genus and species of deepwater ophidioid fish from the Gulf of Mexico. Copeia, 1961:288-292.

--. 1963. A new genus and species of bathypelagic ophidioid fish from the western North Atlantic. Breviora, (196):1-8, 2 fig.

--. 1964. Suborder Argentinoidea. <u>Mem. Sears Fdn mar. Res.</u>, 1(4):1-70, fig. 1-20.

--. 1964. A review of the ophidioid fish genus <u>Luciobrotula</u> with the description of a new species from the western North Atlantic. <u>Bull. mar. Sci. Gulf Caribb.</u>, 14:387-398.

--. 1964. A review of the ophidioid fish genus <u>Oligopus</u> with the description of a new species from West Africa. <u>Proc. U.S. natn. Mus.</u>, 116:1-22, 5 pl.

--. 1973. Opisthoproctidae, p. 156-157, <u>in</u> Hureau, J.C & Th. Monod (ed.), 1973. Check-list of the Fishes of the north-eastern Atlantic and of the Mediterranean/Catalogue des poissons du nord-est Atlantique et de la Méditerranée. Paris, Unesco, 2 vol.

--. 1974. A review of the pelagic ophidioid fish genus <u>Brotulotaenia</u> with descriptions of two new species. <u>Zool. J. Linn. Soc.</u> , 55:119-149.

--. 1981. Argentinidae, Gadidae, Moridae, <u>in</u> Fischer, W., Bianchi, G. & W.B. Scott (ed.), FAO species identification sheets for fishery purposes. Eastern Central Atlantic; fishing areas 34, 47 (in part). Canada Funds-in-Trust. Ottawa, Department of Fisheries and Oceans Canada, by arrangement with the Food and Agriculture Organization of the United Nations, vol. 1-7: pag. var.

--. 1984. Argentinidae, <u>in</u> Whitehead, P.J.P., Bauchot, M.L., Hureau, J.C., Nielsen, J. & E. Tortonese, Fishes of the North-eastern Atlantic and the Mediterranean/Poissons de l'Atlantique du Nord-Est et de la Méditerranée, Paris, Unesco, 1:386-391, 6 + 6 fig. 6 map.

--. 1984. Bathylagidae, <u>in</u> Whitehead, P.J.P., Bauchot, M.L., Hureau, J.C., Nielsen, J. & E. Tortonese, Fishes of the North-eastern Atlantic and the Mediterranean/Poissons de l'Atlantique du Nord-Est et de la Méditerranée, Paris, Unesco, 1:392-394, 4 fig., 4 maps.

--. 1984. Opisthoproctidae, <u>in</u> Whitehead, P.J.P., Bauchot, M.L., Hureau, J.C., Nielsen, J. & E. Tortonese, Fishes of the North-eastern Atlantic and the Mediterranean/Poissons de l'Atlantique du Nord-Est et de la Méditerranée, Paris, Unesco, 1:395-398, 5 fig., 5 maps.

--. 1986. Bregmacerotidae, <u>in</u> P.J.P. Whitehead, M.L. Bauchot, J.C. Hureau, J. Nielsen & E. Tortonese. Fishes of the North-eastern Atlantic and the Mediterranean/Poissons de l'Atlantique du Nord-Est et de la Méditerranée, Paris, Unesco, 2:711-712, 2 fig.

--. 1986. Moridae, <u>in</u> P.J.P. Whitehead, M.L. Bauchot, J.C. Hureau, J. Nielsen & E. Tortonese. Fishes of the North-eastern Atlantic and the Mediterranean/Poissons de l'Atlantique du Nord-Est et de·la Méditerranée, Paris, Unesco, 2:713-723, 14 fig.

--. 1986. Melanonidae, <u>in</u> P.J.P. Whitehead, M.L. Bauchot, J.C. Hureau, J. Nielsen & E. Tortonese. Fishes of the North-eastern Atlantic and the Mediterranean/Poissons de l'Atlantique du Nord-Est et de la Méditerranée, Paris, Unesco, 2:724, 1 fig.

COHEN, D. M.; NIELSEN, J. G. 1978. Guide to the identification of genera of the fish order Ophidiiformes with a tentative classification of the order. <u>Circ. NOAA tech. Rep. NMFS</u>, (417):1-72, 103 fig.

COHEN, D. M.; PAWSON, D. L. 1977. Observations from the DSRV "Alvin" on populations of benthic fishes and selected larger invertebrates in and near DWD-106. NOAA Dumpsite Evaluation Rep., Rockville, Ma., 77-1 2:423-450.

COHEN, D. M.; RUSSO, J. L. 1979. Variation in the fourbeard rockling, <u>Enchelyopus cimbrius</u>, a North Atlantic gadid fish, with comments on the genera of rocklings. <u>Fishery Bull.</u>, 77(1):91-104, fig. 1-5.

COLES, R. J. 1915. Notes on the sharks and rays of Cape Lookout, N. C. <u>Proc. biol. Soc. Wash.</u>, 28:89-94.

COLIN, P. L. 1973. Burrowing behavior of the yellowhead jawfish, <u>Opisto-</u>

gnathus aurifrons. Copeia, 1973:84-90.
COLLETT, R. 1875. Norges Fiske med Bemaerkninger om deres Udbredelse. Forh. VidenskSelsk. Krist. (1874), 1875:1-240, pl. 1-2, 1 map (inset).
--. 1878. Fiske indsamlede under den norske Nordhavsexpeditions 2 Første togter, 1876 ag 1877. Forh. VidenskSelsk. Krist., (4):24 p.
--. 1879. Fiske fra Nordhavs-Expeditiones sidste togt, Sommeren 1878. Forh. VidenskSelsk. Krist. (1878) (14):1-106.
--. 1879. Meddelelser om Norges Fiske i Aarene 1875-1878. Forh. VidenskSelsk. Krist., (1):1-107.
--. 1879. Om den saakaldte Thynnus peregrinus, Coll. 1879. Forh. VidenskSelsk. Krist., (15):3 p.
--. 1879. On a new fish of the genus Lycodes from the Pacific. Proc. zool. Soc. London, 1879:381-382.
--. 1880. Den Norske Nordhavs-Expedition 1876-1878. Zoologi. Fiske. Christiania, 165 p., 3 fig., 5 pl., 1 map.
--. 1886. On a new pediculate fish from the sea off Madeira. Proc. zool. Soc. Lond., 1886 :138-142, pl. 15.
--. 1887. Aphanopus minor; en ny dybvandsfisk af Trichiuridernes familie fra Grønland. Forh. VidenskSelsk. Krist, (19):1-7.
--. 1889. Diagnoses de poissons nouveaux provenant des campagnes de l'"Hirondelle". I. Sur un genre nouveau de la famille des Muraenidae. Bull. Soc. zool. Fr., 14:122-132.
--. 1889. Diagnose de poissons nouveaux provenant des campagnes de l'"Hirondelle". III. Description d'une espèce nouvelle du genre Hoplostethus. Bull. Soc. zool. Fr., 14:306-308.
--. 1890. Diagnoses de poissons nouveaux provenant des campagnes de l'"Hirondelle". Bull. Soc. zool. Fr., 15:105-109.
--. 1896. Poissons provenant des campagnes du yacht l'"Hirondelle" (1885-88). Résult. Camp. scient. Prince Albert I, 10:viii + 198 p., pl. 1-6.
--. 1897. Om en Del for Norges Fauna nye Fiske, funde i 1880-96. Arch. Math. Naturv., 19:1-25.
--. 1897. Om Pterycombus brama Fries. Bergens Mus. Arb., 6:1-14, 2 pl.
--. 1904. Diagnoses of four hitherto undescribed fishes from the depths south of the Faroe Islands. Forh. VidenskSelsk. Krist., (9):1-7.
--. 1904. Meddelelser om Norges Fiske i Aarene 1884-1901. II. Forh. VidenskSelsk. Krist. 1903, (9):175 p.
--. 1905. Meddelelser on Norges Fiske i Aarene 1884-1901. III. Forh. VidenskSelsk. Krist., (7):46-481.
--. 1905. Fiske, indsamlede under "Michael Sars" Togter i Nordhavet 1900-1902. Rep. Norw. Fishery mar. Invest., 2(3):1-152, 2 pl.
--. 1905. On some fishes from the sea off the Azores. Zool. Anz., 28: 723-730.
COLLETTE, B. B. 1962. Hemiramphus bermudensis, a new halfbeak from Bermuda, with a survey of endemism in Bermudian shore fishes. Bull. mar. Sci. Gulf Caribb., 12(3):432-449.
--. 1965. Hemiramphidae from tropical West Africa. Atlantide Rep., (8):217-235, 9 fig., 5 tab.
--. 1966. Belonion, a new genus of fresh-water needlefishes from South America. Am. Mus. Novit., (2274):1-22, 7 fig.
--. 1966. Revue critique des types de Scombridae des collections du Muséum National d'Histoire Naturelle de Paris. Bull. Mus. natn. Hist. nat. Paris, (2)38(4):362-375.
--. 1981. Belonidae, Coryphaenidae, Hemiramphidae, Pomatomidae, Rachycentridae, Scombridae, Xiphiidae, in Fischer, W., Bianchi, G. & W.B. Scott (ed.), FAO species identification sheets for fishery purposes. Eastern Central Atlantic; fishing areas 34, 47 (in part). Canada

Funds-in-Trust. Ottawa, Department of Fisheries and Oceans Canada, by arrangement with the Food and Agriculture Organization of the United Nations, vol. 1-7: pag. var.

--. 1986. Coryphaenidae, in P.J.P. Whitehead, M.L. Bauchot, J.C. Hureau, J. Nielsen & E. Tortonese. Fishes of the North-eastern Atlantic and the Mediterranean/Poissons de l'Atlantique du Nord-Est et de la Méditerranée, Paris, Unesco, 2:845-846, 2 fig.

--. 1986. Scombridae, in P.J.P. Whitehead, M.L. Bauchot, J.C. Hureau, J. Nielsen & E. Tortonese. Fishes of the North-eastern Atlantic and the Mediterranean/Poissons de l'Atlantique du Nord-Est et de la Méditerranée, Paris, Unesco, 2:981-997, 1 + 15 fig.

COLLETTE, B. B.; CHAO, L. N. 1975. Systematics and morphology of the bonitos (Sarda) and their relatives (Scombridae, Sardini). Fishery Bull., 73(3):516-625.

COLLETTE, B. B.; NAUEN, C. E. 1983. FAO species catalogue. Vol. 2. Scombrids of the world. An annotated and illustrated catalogue of tunas, mackerels, bonitos and related species known to date. FAO Fish. Synop., (125) 2:137 p.

COLLETTE, B. B.; PARIN, N. V. 1970. Needlefishes (Belonidae) of the eastern Atlantic Ocean. Atlantide Rep., (11):7-60, 13 fig.

--. 1986. Belonidae, in P.J.P. Whitehead, M.L. Bauchot, J.C. Hureau, J. Nielsen & E. Tortonese. Fishes of the North-eastern Atlantic and the Mediterranean/Poissons de l'Atlantique du Nord-Est et de la Méditerranée, Paris, Unesco, 2:604-609, 5 + 2 fig.

--. 1986. Hemiramphidae, in P.J.P. Whitehead, M.L. Bauchot, J.C. Hureau, J. Nielsen & E. Tortonese. Fishes of the North-eastern Atlantic and the Mediterranean/Poissons de l'Atlantique du Nord-Est et de la Méditerranée, Paris, Unesco, 2:620-622, 2 + 3 fig.

COLLETTE, B. B.; RUSSO, J. L. 1979. An introduction to the Spanish mackerels, genus Scomberomorus. In: Proc. mackerel colloquium (E. L. Nakamura, ed.). Publ. Gulf St. mar. Fish. Comm., (4):3-16.

--. 1981. A revision of the scaly toadfishes, genus Batrachoides, with descriptions of two new species from the eastern Pacific. Bull. mar. Sci., 31(2):197-233, 14 tab., 10 fig.

COLLETTE, B. B.; UYENO, 1972. Pontinus niger, a synonym of the scorpionfish Ectreposebastes imus, with extension of its range to Japan. Jap. J. Ichthyol., 19(1):26-28, fig. 1.

COLLIGNON, J. 1957. Sea perch (Otolithus) fishery. Oceanography and sea fisheries on the West African coast. CCTA/CSA Symposium (21st-27th November 1957, Luanda). Paper N°57 (40 p. mimeo).

--. 1959. La systématique des Sciaenidés de l'Atlantique oriental. Bull. Inst. océanogr. Monaco, (1155):1-11.

--. 1960. Contribution à la connaissance des Otolithus des côtes d'Afrique equatoriale. Bull. Inst. Etud. centrafr. (n.s.), (19-20):55-94.

--. 1961. Le thazard ou melva dans l'Atlantique oriental. Bull. Inst. Pêch. marit. Maroc, (7):55-72.

--. 1970. Aspect du peuplement ichthyologique benthique du Cap Spartel au Cap Cantin. Rapp. P.-v. Réun. Cons. perm. int. Explor. Mer, 159: 202-209.

--. 1979. Sous-ordre des Scorpaenoides. In: Catalogue raisonné des poissons des mers marocaines (4ème partie). Bull. Inst. Pêch. marit. Maroc, (24):85-118, 19 fig., 7 photos.

COLLIGNON, J.; ALONCLE, H. 1960. Le régime alimentaire de quelques poissons benthiques des côtes marocaines. Bull. Inst. Pêch. marit. Maroc, (5):17-29.

--. 1972. Catalogue raisonné des poissons des mers marocaines. I. Cyclostomes, Sélaciens, Holocéphales. Bull. Inst. Pêch. marit. Maroc, (19):1-163, 53 fig., 35 phot.

COLLIGNON, J.; ROSSIGNOL, M.; ROUX, C. 1951. Mollusques, crustacés, poissons marins des côtes d'A.E.F. en collection au Centre d'Océanographie de l'Institut d'Etudes Centrafricaines de Pointe-Noire. Paris, 369 p., many fig.

COLLINS, B. L. 1954. Lista de peixes dos mares dos Açores. Açoreana, 5 (2):40 p.

COMPAGNO, L. J. V. 1973. Carcharhinidae, p. 23-31, in Hureau, J.C. & Th. Monod (ed.), 1973. Check-list of the fishes of the north-eastern Atlantic and of the Mediterranean/Catalogue des poissons du nord-est Atlantique et de la Méditerranée. Paris, Unesco, 2 vol.

--. 1978. Cetorhinidae, in Fischer, W. (ed.), FAO species identification sheets for fishery purposes. Western Central Atlantic. (Fishing Area 31). FAO, Rome, vol. 1-7, pag. var.

--. 1984. FAO species catalogue. Vol. 4. Sharks of the world. An annotated and illustrated catalogue of shark species known to date. Part 1. Hexanchiformes to Lamniformes. FAO Fish. Synop., (125) 4(1):1-249.

--. 1984. FAO species catalogue. Vol. 4. Sharks of the world. An annotated and illustrated catalogue of shark species known to date. Part 2. Carcharhiniformes. FAO Fish. Synop., (125) 4(2):251-655.

CONAND, F. 1970. Distribution et abondance des larves de quelques familles et espèces de poissons des côtes sénégambiennes en 1968. Docum. scient. provis. Centre Rech. océanogr.. Dakar-Thiaroye (26) (mimeo).

--. 1973. Estimation de l'abondance des larves de Sardinella aurita (C. et V.) sur la Petite Côte sénégalaise en 1971 et 1972. Docum. scient. provis. Centre Rech. océanogr.. Dakar-Thiaroye, (48) (13 p., mimeo).

--. 1976. Distribution et abondance des larves de Sardinella aurita (Val. 1847) dans la région du Sénégal en 1974. Cah. ORSTOM. Océanogr., 14(3):221-225.

--. 1978. Oeufs et larves de la sardinelle ronde (Sardinella aurita) au Sénégal: distribution, croissance, mortalité, variations d'abondance de 1971 à 1976. Cah. ORSTOM. Océanogr., 15(3):201-214.

--. 1978. Contribution à l'étude du cycle sexuel et de la fécondité de la sardinelle ronde, Sardinella aurita: pêche sardinière dakaroise en 1975 et premier semestre 1976. Cah. ORSTOM. Océanogr., 15(4):301-312.

--. 1979. Systématique des larves de clupéidés de l'Atlantique oriental entre 20°N et 15°S (eaux marines et saumâtres). Cah. ORSTOM. Océanogr., 16(1):3-8.

CONAND, F.; CREMOUX, J. L. 1972. Distribution et abondance des larves de sardinelles dans la région du Cap Vert de septembre 1970 à août 1971. Docum. scient. provis. Centre Rech. océanogr.. Dakar-Thiaroye, (36) (23 p., mimeo).

CONAND, F.; FAGETTI, E. 1971. Description et distribution saisonnière des larves de sardinelles des côtes du Sénégal et de la Gambie en 1968 et 1969. Cah. ORSTOM. Océanogr., 9(3):293-318 (also publ. in: Docum. scient. provis. Centre Rech. océanogr.. Dakar-Thiaroye, (29) (43 p., mimeo).

COOPER, J. G. 1863. On new genera and species of Californian fishes. Proc. Calif. Acad. Sci., 3(1):70-77, fig.

COPE, E. D. 1869. Supplement on some new species of American and African fishes. Trans. Am. phil. Soc., 13(2):400-407.

--. 1870. Observations on some fishes new to the American fauna, found at Newport, R.I. by Samuel Powell. Proc. Acad. nat. Sci. Philad., 22:118-121.

--. 1870. Partial synopsis of the fresh water fishes of North Carolina. Proc. Am. phil. Soc., 11:448-495, fig.

--. 1871. Contribution to the ichthyology of the Lesser Antilles. Trans. Am. phil. Soc., 14:445-483, 3 fig.

--. 1884. A carboniferous genus of sharks still living. Science. N.Y, 3:

275-276.

COPLEY, H. 1952. The game fishes of Africa. London, 276 p., 176 fig., 24 pl.

CORBIN, P. G. 1958. A new British fish (<u>Gobius forsteri</u>). <u>Nature, Lond.</u>, 181:1659.

CORNIDE, J. 1788. Ensayo de una historia de los peces y otras producciónes marinas de la costa de Galicia, arreglado al sistema del Caballero Carlos Linneo; con un tratado de las diversas pescas, y de las redes y aparejos con que se practican. Coruña, 263 p.

COSTA, A. 1862. Di un novello genere di pesce esocetidi. <u>Annuar. R. Mus. zool. R. Univ. Napoli</u>, 1:54-57.

COSTA, F. C. da; DA FRANCA, M. L. P. 1972. Contribuçao para o conhecimento da biologia de <u>Cantharus cantharus</u> (L.) (Pisces, Sparidae) de Angola. <u>Notas Cent. Biol. aquat. trop.</u>, (30):1-19.

--. 1973. Contribuçao para o conhecimento da biologia de <u>Lithognathus mormyrus</u> (L.) (Pisces, Sparidae) de Angola. <u>Notas Cent. Biol. aquat. trop.</u> (35):1-39.

COSTA, F. C. da; DA FRANCA, P.; DA FRANCA M. L. P. 1971. Nota sobre certos aspectos do comportamento de alguns Sciaenidae de Angola com possível interesse na sua pesca. <u>Notas Cent. Biol. aquat. trop.</u>, (23):1-27.

COSTA, M. J. 1979. Ateleopodidae, p. 353-354, <u>in</u> Tortonese, E. & J.C. Hureau (ed.), 1979. Supplément au Clofnam. <u>Cybium</u> 3(1):5-66 [=333-394].

--. 1980. <u>Phrynichthys wedli</u> Pietschmann, 1926 (Pisces, Diceratiidae), poisson nouveau pour les côtes du Portugal. <u>Cybium</u>, 4(1):89-90.

COSTA, M. J.; GEISTDOERFER, P. 1979. Nouvelle capture d'un <u>Guentherus altivela</u> Osorio, 1917 (Pisces, Ateleopodidae). <u>Cybium</u>, 3(3):101-103.

COSTA, M. J.; REINER, F. 1977. <u>Guntherus altivela</u> Osorio, 1917 (Pisces; Ateleopodidae) a new fish to the northeastern Atlantic. <u>Archos Mus. Bocage</u>, (2) (notas supl.) 32:1-6, fig. 1-2.

COSTA, O. G. 1829-57. Fauna del regno di Napoli, ossia enumerazione di tutti gli animali. Pesci. 3 pts, 70 fasc. Part 1, 1829-53:511 p., 60 pl.; part 2, 1842-53:148 p., 69 pl.; part 3, 1837-57:101 p., 17 pl. (for publication dates see Sherborn, 1937, <u>J. Soc. Biblphy nat. Hist.</u>, 1(2):35).

--. 1871. Illustrazione di un Centrolofo pescato nel Golfo di Napoli. <u>Annuar. R. Mus. zool. R. Univ. Napoli</u>, 6:84-87, pl. 1.

COUCH, J. 1832. Fishes new to the British fauna, contained in Couch's "History of the fishes of Cornwall". <u>Mag. nat. Hist.</u>, 5:15-24, fig. 2-10.

--. 1832. Further observations on some of the fishes of Cornwall. <u>Mag. nat. Hist.</u>, 5:311-316, 2 fig.

--. 1838. Description of a species of Ray-fish, not hitherto included in the British Fauna. <u>Ann. nat. Hist.</u> (n.s.), 2:71-73, 1 fig.

--. 1849. Description of <u>Scomber punctatus</u>, a species of mackerel not hitherto recognized by naturalists. <u>Zoologist</u>, 7: app.:29-31.

--. 1849. Description of <u>Brama pinna-squamata</u>, a supposed unrecognized British fish. <u>Zoologist</u>, 7: app. 26-28.

--. 1863. A history of the fishes of the British Islands. 2. London, vi + 1-265, fig., 63 pl.

--. 1865. A history of the fishes of the British Islands. 4. London, iv + 1-439, fig., 73 pl.

COURTENAY, W. R. 1967. Atlantic fishes of the genus <u>Rypticus</u> (Grammistidae). <u>Proc. Acad. nat. Sci. Philad.</u>, 119(6):241-293, 19 fig.

--. 1970. The R/V Pillsbury deep-sea Biological Expedition to the Gulf of Guinea, 1964-65. 14. Soap fishes of the genus <u>Rypticus</u> from Fernando Poo. <u>Stud. trop. Oceanogr.</u>, 4(2):276-280, 1 fig., 1 pl. 1 tab.

--. 1981. Grammistidae, in Fischer, W., Bianchi, G. & W. B. Scott (ed.), FAO species identification sheets for fishery purposes. Eastern Central Atlantic; fishing areas 34, 47 (in part). Canada Funds-in-Trust. Ottawa, Department of Fisheries and Oceans Canada, by arrangement with the Food and Agriculture Organisation of the United Nations, vol. 1-7: pag. var.

COWAN, C. F. 1969. Cuvier's Règne animal, first edition. J. Soc. Biblphy nat. Hist., 5(3):219.

CRADDOCK, J. E.; MEAD, G. W. 1970. Midwater fishes from the eastern South Pacific Ocean. Anton Bruun Rep., (3):1-46, 10 fig.

CRANE, J. M., Jr. 1966. Late tertiary radiation of viperfishes (Chauliodontidae) based on a comparison of recent and miocene species. Contr. Sci., (115):1-29, 13 fig.

--. 1968. Bioluminescence in the batfish Dibranchus atlanticus. Copeia, 1968:410-411.

CRESSEY, R. F.; COLLETTE, B. B. 1970. Copepods and needlefishes: a study in host-parasite relationships. Fishery Bull., 68(3):347-432.

CRESSEY, R. F.; COLLETTE, B. B.; RUSSO, J. L. 1983. Copepods and scombrid fishes: a study in host-parasite relationships. Fishery Bull., 81(2): 227-265.

CROSNIER, A; BERRIT, G. R. 1966. Fonds de pêche le long des côtes des Républiques du Dahomey et du Togo. Cah. ORSTOM. Océanogr., (Suppl.) 4(1):1-144, 2 map.

CROSNIER, A.; MARTEAU, J.; BERRIT, G. R.; STAUCH, A. 1965. Fonds de pêche le long des côtes de la République Fédérale du Cameroun. Cah. ORSTOM. Océanogr., (No. spéc., 1964), 1-132, fig., tab., map.

CUMMINGS, W. C. 1968. Reproductive habits of the sergeant major, Abudefduf saxatilis (Pisces, Pomacentridae) with comparative notes on four other damselfishes in the Bahama Islands. Unpubl. Doct. Diss., Univ. Miami.

CUNNINGHAM, J. T. 1910. On the marine fishes and invertebrates of St. Helena. Proc. zool. Soc. Lond., 1910:86-131, 4 fig., 4 pl.

CURTISS, A. 1938. A short zoology of Tahiti. Priv. publ. 193 p.

CUVIER, G. 1798. Tableau élémentaire de l'histoire naturelle des animaux. Paris, 1-710 p., 14 pl.

--. 1814. Observations et recherches critiques sur différents poissons de la Méditerranée, et à leur occasion sur des poissons d'autres mers, plus ou moins liés avec eux. Bull. Sci. (Soc. philomath.), Paris, 1814:80-92 (Report by A. D. = A. G. Desmarest).

--. 1814. Notice sur un poisson célèbre, et cependant presque inconnu des auteurs systématiques, appelé sur nos côtes de l'Océan, Aigle ou Maigre, et sur celles de la Méditerranée, Umbra, Fegaro et Poisson Royal; avec une description abrégée de sa vessie natatoire (lue en novembre 1813). Mém. Mus. Hist. nat., Paris, 1:1-21, 3 pl.

--. 1815. Observations et recherches critiques sur différents poissons de la Méditerranée, et à leur occasion sur des poissons d'autres mers, plus ou moins liés avec eux. Mém. Mus. Hist. nat., Paris, 1: 226-241, pl. 11; 312-330, pl. 16; 353-363; 451-466, pl. 23.

--. 1816. Poissons, p. 104-351. In: Le règne animal distribué d'après son organisation, pour servir de base à l'histoire naturelle des animaux et d'introduction à l'anatomie comparée. Poissons, 4 vol., Paris, 2: xviii + 532 p. (Dec.). (for dating see Whitehead, P.J.P. 1967: J. Soc. Biblphy nat. Hit., 4(6):300-301).

--. 1817. Article "Chironecte, Chironectes".In: Cloquet, H., Dict. Sci. nat., F.G. Levrault, Paris, 8:597-599.

--. 1817. Sur le genre Chironectes Cuv. (Antennarius Commers.). Mém. Mus. Hist. nat., Paris, 3: 418-435, pl. 16-18.

--. 1818. Sur les diodons, vulgairement Orbes épineux. Mém. Mus. Hist.

nat., Paris, 4:121-138.
--. 1829. Le règne animal distribué d'après son organisation, pour ser-
    vir de base à l'histoire naturelle des animaux et l'introduction à
    l'anatomie comparée. Poissons, (nouv. ed.) Paris, 2:122-406, pl. 10-
    13 (March, 1829).
--. 1832. Sur le poisson appelé Machaera. Nouv. Annls Mus. Hist. nat.,
    1:43-49, 1 pl.
CUVIER, G.; VALENCIENNES, A. 1828-49. Histoire naturelle des Poissons,
    Paris-Strasbourg, 22 vol. 11030 p., 621 pl. (numb. 1-650). 1, 1828:
    xvi + 573 p., pl. 1-8 (author: Cuvier); 2, 1828: xxi + 490 p., pl. 9-
    40 (Cuv.:1-238, 249-262, 387-490; Val.:238-249,262-386); 3, April
    1829: xxiii + 500 p., pl. 41-71 (Cuvier); 4, 1829: xxvi + 518 p., pl.
    72-99 (Cuvier); 5, 1830: xxviii + 499 p., pl. 100-140 (Cuvier); 6,
    1830: xxiv + 559 p., pl. 141-169 (Val.:1-425, 493-559; Cuv.:426-491);
    7, 1831: xxix + 531 p., pl. 170-208 (Cuv.:1-440; Val.:441-531); 8,
    1832: xix + 509 p., pl. 209-245 (Cuv.:1-470; Val.:471-509); 9, 1833:
    xxix + 512 p., pl. 246-279 (Cuv.:1-198, 330-359, 372-427; Val.:199-
    329, 359-371, 429-512); 10, 1835: xxiv + 482 p., pl. 280-306 (Val.);
    11, 1836: xx + 506 p., pl. 307-343 (Val.); 12, 1837: xxix + 507 p.,
    pl. 344-368 (Val.); 13, 1839: xix + 505 p., pl. 369-388 (Val.); 14,
    1840: xxii + 464 p., pl. 389-420 (Val.); 15, 1840: xxi + 540 p., pl.
    421-455 (Val.); 16, 1842: xx + 472 p., pl. 456-487 (Val.); 17, 1844:
    xxiii + 497 p., pl. 487-519 (Val.); 18, 1846: xix + 505 p., pl. 520-
    553 (Val.); 19, 1847: xix + 544 p., pl. 554-590 (Val.); 20, 1847:
    xviii + 472 p., pl. 591-606 (Val.); 21, 1848: xiv + 536 p., pl. 607-
    633 (Val.); 22, 1849: xx + 532 p., pl. 634-650 (Val.).

D'AUBREY, J. D. 1964. Preliminary guide to the sharks found off the east
    coast of Africa. Investl Rep. (S. Afr. Ass. mar. biol. Res.), Durban,
    (8):1-95.
DA COSTA, F. C. (see COSTA da).
DA FRANCA, P. (see FRANCA da).
DAGET, J. 1954. Les poissons du Niger Supérieur. Mém. Inst. fr. Afr.
    noire, 36:391 p., 141 fig.
--. 1960. La faune ichthyologique du bassin de la Gambie. Bull. Inst.
    fr. Afr. noire, (A)22:610-619, 2 fig.
--. 1961. Restes de poissons du Quaternaire saharien. Bull. Inst. fr.
    Afr. noire, (A)23:182-191, fig. 1-8.
--. 1979. Contributions à la faune de la République unie du Cameroun.
    Poissons de l'Ayina, du Dja et du Bas Sanaga. Cybium, 3(2):55-64.
DAGET, J.; DURAND, J. R. 1968; Etude du peuplement de poissons d'un
    milieu saumâtre tropical poikilohalin: la baie de Cocody en Côte
    d'Ivoire. Cah. ORSTOM, Hydrobiol. 2(2):91-111.
DAGET, J.; ILTIS, A. 1965. Poissons de Côte d'Ivoire (eaux douces et
    saumâtres). Mém. Inst. fr. Afr. noire, (74): xi + 1-385, 212 fig. 4
    pl.
DAGET, J.; STAUCH, A. 1913. Poissons de la partie camerounaise du Bassin
    de la Bénoué. Mém. Inst. fr. Afr. noire, (68):85-107, 4 fig.
--. 1968. Poissons d'eaux douces et saumâtres de la région côtière du
    Congo. Cah. ORSTOM, Hydrobiol., 2(2):21-50.
DAHLBERG, M. D. 1975. Guide to coastal fishes of Georgia and nearby
    States. Athens, xvi + 187 p., 187 fig.
DAMBECK, K. 1879. Die Verbreitung der Süss- und Brackwasser-Fische in
    Afrika. Jena. Z. Naturw., 13 (n. F.6):404-456.
DARDIGNAC, J. 1961. Les ombrines des côtes atlantiques du Maroc. Revue
    Trav. Inst. Pêch. marit., 25:263-279.
DARNELL, R. M. 1962. Fishes of the Rio Tamesi and related coastal la-

goons in East Central Mexico. Publs Inst. mar. Sci. Univ. Texas, 8:
299-365.

DAUDIN, F. M. 1816. "Antennaire". In: Dict. Sci. nat., F.G. Levrault,
Paris, 2:193.

DAVIES, D. H. 1956. The South African pilchard (Sardinops ocellata).
Migration, 1950-1955. Investl Rep. Div. Fish. Un. S. Afr., (24):1-52,
23 pl., 1 fig.

--. 1957. The biology of the South African pilchard (Sardinops ocellata)
Investl Rep. Div. Fish. Un. S. Afr., (32):1-11, 3 fig.

DAVIS, W. P. 1966. A review of the Dragonets (Pisces: Callionymidae) of
the western Atlantic. Bull. mar. Sci., 16(4):834-862.

DAVIS, W. P.; ROBINS, C. R. 1966. Pogonymus shango, a new callionymid
fish from Quartz Sand Beaches of Nigeria, with notes on related species. Stud. trop. Oceanogr., 4(1):106-116.

DAVY, B. 1972. A review of the lanternfish genus Taaningichthys (family
Myctophidae), with a description of a new species. Fishery Bull., 70
(1):67-78, 6 fig.

DAVY, J. 1834. Observations on the Torpedo, with an account of some
additional experiments on its electricity. Phil. Trans. R. Soc.,
1834(2): 531-550, pl. 22-24.

DAWSON, C. E. 1962. A new gobioid fish, Microdesmus lanceolatus, from
the Gulf of Mexico with notes on M. longipinnis (Weymouth). Copeia,
1962(2):330-336, fig. 1.

--. 1971. Occurrence and description of prejuvenile and early juvenile
Gulf of Mexico Cobia, Rachycentron canadum. Copeia, 1971(1):65-71,
fig. 1-2, tab. 1-3.

--. 1973. Microdesmus bahianus, a new western Atlantic wormfish (Pisces:
Mircrodesmidae). Proc. biol. Soc. Wash., 86(17):203-210, fig. 1-2.

--. 1974. A review of the Microdesmidae (Pisces: Gobioidea). 1. Cerdale
and Clarkichthys with descriptions of three new species. Copeia,
1974(2):409-448, fig. 1-24.

--. 1979. A new wormfish (Pisces: Microdesmidae) from the eastern tropical Atlantic. Copeia, 1979(2):203-205, fig. 1.

--. 1979. Review of the polytypic doryrhamphine pipefish Oostethus bra-
chyurus (Bleeker). Bull. mar. Sci., 29(4):464-480, 4 fig.

--. 1979. Notes on western Atlantic pipefishes with description of Syngnathus caribbaeus n. sp. and Cosmocampus n. gen. Proc. biol. Soc.
Wash., 92(4):671-676, 1 fig.

--. 1981. Note on west African pipefishes (Syngnathidae), with description of Enneacampus, n. gen. Proc. biol. Soc. Wash., 94(2):464-478.

--. 1982. Descriptions of Cosmocampus retropinnis sp. n., Minyichthys
sentus sp. n. and Amphelikturus sp. (Pisces, Syngnathidae) from the
eastern Atlantic region. Zool. Scripta, 11:135-140, 3 fig.

--. 1982. Fishes of the western North Atlantic. The pipefishes (subfamilies Doryrhamphinae and Syngnathinae). Mem. Sears Fdn mar. Res., 1
(8):4-172.

--. 1984. A new pipehorse (Syngnathidae) from western Australia, with
remarks on the subgenera of Acentronura. Jap. J. Ichthyol., 31(2):
156-180.

--. 1984. Revision of the genus Microphis Kaup (Pisces: Syngnathidae).
Bull. mar. Sci., 35(2):117-181.

--. 1986. Syngnathidae, in P.J.P. Whitehead, M.L. Bauchot, J.C. Hureau,
J. Nielsen & E. Tortonese. Fishes of the North-eastern Atlantic and
the Mediterranean/Poissons de l'Atlantique du Nord-Est et de la Méditerranée, Paris, Unesco, 2:628-639, 16 + 3 fig.

DAY, F. 1865. On the fishes of Cochin, on the Malabar coast of India.
Part. I. Acanthopterygii. Proc. zool. Soc. Lond., 1865:2-40.

--. 1869. Remarks on some of the fishes in the Calcutta Museum. Proc.

zool. Soc. Lond., 1869:511- 527.

--. 1875-1878, 1888. The fishes of India; being a natural history of the fishes known to inhabit the seas and fresh waters of India, Burma and Ceylon. 2 vol. London, xx + 778 p., suppl. (1888):779-816, 7 fig., 198 pl.

--. 1880-1884. The fishes of Great Britain and Ireland. London, Edinburgh, 1: cxii + 336 p., 5 + 7 fig., 92 pl.; 2:388 p., pl. 93-179.

--. 1883. On the occurrence of Paralepis coregonoides in Cornwall. Zoologist, (3)7:381-382, 506.

DAY, J. H. 1969. A guide to marine life on South African Shores. Cape Town.

DAY, J. H.; FIELD, J. G.; PENRITH, M. J. 1970. The benthic fauna and fishes of False Bay, South Africa. Trans. R. Soc. S. Afr., 39:1-108.

DEAN, B. 1904. Notes on the long snouted chimaeroid of Japan. J. Coll. Sci. imp. Univ., Tokyo, 19(4):1-20, 2 pl.

--. 1906. Chimaeroid fishes and their development. Publs Carnegie Instn, (32):195 p., 144 fig., pl.

DE BEAUFORT. (see BEAUFORT de).

DE BUEN. (see BUEN de).

DECAMPS, P. 1986. Luvaridae, in P.J.P. Whitehead, M.L. Bauchot, J.C. Hureau, J. Nielsen & E. Tortonese. Fishes of the North-eastern Atlantic and the Mediterranean/Poissons de l'Atlantique du Nord-Est et de la Méditerranée, Paris, Unesco, 2:998-999, 1 fig.

DE CIECHOMSKI, J. D. 1969. Nota sobre el hallozgo de larvas de un pez batipelagico, Idiacanthus sp., en el Atlantico sur frente a la Argentina. Physis, B. Aires, (B)28(77):239-246, fig. 1-6.

DE FILIPPI, F.; VERANY, G. B. 1859. Sopra alcuni pesci nuovi o poco noti del Mediterraneo. Memorie R. Accad. Sci. Torino, (2)18:187-199.

DEGENS, E. T.; DEUSEN, W. G.; HAEDRICH, R. L. 1969. Molecular structure and composition of fish otoliths. Mar. Biol., Berlin, 2(2):105-113, 11 fig., 3 tab.

DE GROOT, S. J.; NIJSSEN, H. 1971. Notes on the fishes collected by the R.V. "Tridens" on the north west African shelf, 19-25 January 1969. Bijdr. Dierk., 41(1):3-9, 4 fig.

DE JAGER. (see JAGER de).

DEKAY, J. E. 1842. Zoology of New York, or the New York Fauna, comprising detailed descriptions of all the animals hitherto observed within the State of New York, with brief notices of those occasionally found near its borders, and accompanied by appropriate illustrations. Part 4, Fishes. Nat. Hist. N. Y. Geol. Surv., Albany, 1(3-4):1-415 p., 79 pl.

DELAIS, M. 1951. Notes d'ichthyologie ouest-africaine. I. Note sur les Antennariides en collection au laboratoire de biologie marine de l'I.F.A.N. à Gorée. Bull. Inst. fr. Afr. noire, 13(1):145-150.

--. 1951. Notes d'ichthyologie ouest africaine. II. Les Gobiidae d'Afrique occidentale française en collection au Laboratoire de Biologie marine de l'I.F.A.N. à Gorée. Bull. Inst. fr. Afr. noire, 13:343-370.

--. 1952. Notes d'ichthyologie ouest-africaine. III. Notes sur les Lutjanidae. Bull. Inst. fr. Afr. noire, 14(4):1214-1227, 8 fig.

--. 1955. Note sur Liza dumerili Stein(d.) et L. hoefleri Stein(d.) Rapp. P.-v. Réun. Cons. perm. int. Explor. Mer, 137:63-65.

--. 1961. Comparison entre Mugil cephalus Linné de Méditerranée et sa sous-espèce de l'ouest-africain (Mugil cephalus Linné, s. sp. ashanteensis Bleeker). Rapp. P.-v. Réun. Commn. int. Explor. scient. Mer. Méditerr., 16:853-854.

DELAROCHE, F. E. 1809. Observations sur des poissons recueillis dans un voyage aux îles Baléares et Pythiuses.. Annls Mus. Hist. nat. Paris, 13:98-122.

--. 1809. Suite du mémoire sur les espèces de poissons observées à Ibiza. Observations sur quelques-uns des poissons indiqués dans le précédent tableau et descriptions des espèces nouvelles ou peu connues. ibid., 13:313-361, pl. 20-25. (both also separate, Paris:75 p., 6 pl.).

DE LEMA, T. (see LEMA, T. de)

DE LEO, G.; CATALANO, E.; PARRINELLO, N. 1976. Contributo alla conoscenza del Blennius cristatus L. 1758 (Perciformes Blenniidae). Mem. Biol. mar. Oceanogr., 6(6):209-228. Napoli.

DELFIN, F. T. 1901. Ictiologia Chilena. Catàlogo de los peces de Chile. (Republished from Revta chil. Hist. nat., 2-1898, 3-1899 and 4-1900). Valparaiso, 133 p.

--. 1903. Adición al "Catàlogo de los Peces de Chile", con descripción de una nueva especie. In Porter, C. E. Revta chil. Hist. nat., 7: 220-225.

DELSMAN, H. C. 1972. Fish eggs and larvae from the Java Sea. 11. The genus Trichiurus. Treubia, 9(4):338-351, fig.

DELSMAN, N. C. 1941. Résultats scientifiques des croisières du Navire-école Belge "Mercator". III. (3) Pisces. Mém. Mus. r. Hist. nat. Belg., (2)21:47-82, 12 fig.

DEMIDOV, V. F. 1961. Prespawning concentrations of Sardinella aurita in the Dakar and Takoradi areas. ICES, C.M. Sardine Committee, (116):1-7 (mimeo).

--. 1962. (Some features in the behaviour of the West African Sardinella aurita in the Dakar and Takoradi areas). Trudy azov.-chernomorsk. nauchno-issled. morsk. ryb. Khoz. Okeanogr., 20:25-44 (in Russian).

DEMIR, M. 1958. Marmara ve Kuzey-Dogu Ege'den üc derin deniz baligi nev'i. Hidrobiol. Mecm. Istanbul Univ., (A)4(3-4):134-150.

--. 1964. Distribution of meristic counts of common bonito (Sarda sarda Bl.) from Turkish waters. Proc. tech. Pap. gen. Fish. Coun. Mediterr., (7):455-457.

--. 1963. Synopsis of biological data on bonito Sarda sarda (Bloch) 1793. FAO Fish. Rep., (6)2:101-129, 13 fig.

DENTON, E. J. 1956. Recherches sur l'absorption de la lumière par le cristallin des poissons. Bull. Inst. océanogr. Monaco, (1071):1-10.

--. 1971. Reflectors in fishes. Scient. Am., 224(1):65-72.

DENTON, E. J.; NICOL, J. A. C. 1966. A survey of reflectivity in silvery teleots. J. mar. biol. Ass. U. K., 46:685-722.

DENTON, E. J.; LIDDICOAT, J. D.; TAYLOR, D. W. 1972. The permeability to gases of the swimbladder of the conger eel (Conger conger). J. mar. biol. Ass. U. K., 52:727-746.

DERANIYAGALA, P. E. P. 1931. Further notes on the anguilliform fishes of Ceylon. Spolia zeylan., 16(2):131-137, pl. 31-32.

--. 1933. Some larger Rhegnopteri of Ceylon. Ceylon J. Sci., 18(1):37-60, 6 fig. 4 pl.

--. 1951. The Istiophoridae and Xiphiidae of Ceylon. Spolia zeylan., 26 (2):137-142, 3 pl.

--. 1956. A new marlin, Makaira xantholineata, from Ceylon. Spolia zeylan., 28(1):23-24.

--. 1964. The beaked large scombroids of Ceylon. Proc. Symp. Scombroid Fishes, mar. biol. Ass. India, 1(1):441-442.

DESBROSSES, P. 1938. Sur les poissons-épées du genre Tetrapturus Rafin., 1810 rencontrés près des côtes d'Europe. Bull. Soc. zool. Fr., 63: 48-58, 1 fig.

--. 1938. Croissance et migration du requin griset, Hexanchus griseus (Bonnaterre, 1788) Rafinesque, 1810. Revue Trav. Off. (scient. tech.) Pêch. marit., 11:53-57, 1 fig.

DERSCHEID, J. M. 1924. Note sur certains Clupeidae de la côte W. de l'Afrique. _Revue zool. afr._, 12(2):278-282.

DESJARDINS, J. F. 1840. Description d'une nouvelle espèce de poisson de l'Ile Maurice appartenant à la famille des pectorales pédiculées et au genre chironecte. _Mag. Zool._, 1840:1-4.

DESMAREST, A. G. (=A.D.) 1814. _In_: G. Cuvier, Observations et recherches critiques sur différents poissons de la Méditerranée et, à leur occasion, sur des poissons des autres mers plus ou moins liés avec eux. _Bull. Soc. philomath. Paris_, 1814:80-92.

--. 1823. Première décade ichthyologique, ou description complète de dix espèces de poissons nouvelles ou imparfaitement connues, habitant la mer qui baigne les côtes de l'Ile de Cuba. _Mém. Soc. linn. Paris_, 2: 50 p., 7 pl. (also separate).

DESOUTTER, M. 1986. Acanthuridae, _in_ P.J.P. Whitehead, M.L. Bauchot, J.C. Hureau, J. Nielsen & E. Tortonese. Fishes of the North-eastern Atlantic and the Mediterranean/Poissons de l'Atlantique du Nord-Est et de la Méditerranée, Paris, Unesco, 2:962-963, 1 fig.

DESSIER, A. 1969. Note sur les stades larvaires et post-larvaires d'_Ilisha africana_ Bloch, 1795 (Pisces, Clupeidae). _Cah. ORSTOM, Océanogr._, 7(4):21-25.

DESVAUX, M. A. N. 1851. Essai d'ichthyologie des côtes océaniques et de l'intérieur de la France ou diagnose des poissons observés. _Mém. Soc. Agric. Sci. Angers_, (2)2:175 p., pl. I-III.

DE SYLVA, D. P. 1955. The osteology and phylogenetic relationships of the blackfin tuna, _Thunnus atlanticus_ (Lesson). _Bull. mar. Sci. Gulf Caribb._, 5(1):1-41.

--. 1976. Attacks by bluefish (_Pomatomus saltrix_) on humans in South Florida. _Copeia_, 1976(1):196-198.

--. 1981. Sphyraenidae, _in_ Fischer, W., G. Bianchi & W.B. Scott (ed.), FAO species identification sheets for fishery purposes. Eastern Central Atlantic (Fishing Areas 34 and 47 in part). Dept. Fish. Oceans Canada, Ottawa/FAO, Rome, vol. 1-7, pag. var.

--. 1963. Systematics and life history of the great barracuda, _Sphyraena barracuda_ (Walbaum). _Stud. trop. Oceanogr._, 1:1-179.

--. 1963. Postlarva of the white marlin, _Tetrapturus albidus_, from the Florida current off the Carolinas. _Bull. mar. Sci. Gulf Caribb._, 13 (1):123-132, 3 fig.

--. 1963. Preliminary report on the blue marlin sport fishery of Port Antonio, Jamaica. Inst. mar. Sci., Univ. Miami, Miami. 109 p., 5 fig. (mimeo).

--. 1973. Istiophoridae, p. 477-481, _in_ Hureau, J.C. & Th. Monod (ed.), Check-list of the Fishes of the north-eastern Atlantic and of the Mediterranean/Catalogue des poissons du nord-est Atlantique et de la Méditerranée. Paris, Unesco, 2 vol.

--. 1981. Sphyraenidae, _in_ Fischer, W., Bianchi, G. & W.B. Scott (ed.), FAO species identification sheets for fishery purposes. Eastern Central Atlantic; fishing areas 34, 47 (in part). Canada Funds-in-Trust. Ottawa, Department of Fisheries and Oceans Canada, by arrangement with the Food and Agriculture Organization of the United Nations, vol. 1-7:pag. var.

DE SYLVA, D. P.; RATHJEN, W. F. 1961. Life history notes on the little tuna, _Euthynnus alletteratus_, from the southeastern United States. _Bull. mar. Sci. Gulf Caribb._, 11(2):161-190.

DEVANY, T. 1969. Ecologic interpretation of distribution of the lantern-fishes (Myctophidae) in the Straits of Florida. PhD. thesis, University of Miami, Florida.

DEVINCENZI, G. J. 1924. Peces del Uruguay. _An. Mus. Hist. nat. Montevideo_, (2)1(5):139-290.

DE VIS, C. W. 1884. New fishes in the Queensland Museum, n° 3. _Proc._

<u>Linn. Soc. N.S.W.</u>, 9:537-547.

DEWITT, H. H.; GRECAY, P. A.; HACUNDA, J. S.; LINDSAY, B. P.; SHAW, R. F.; TOWNSEND, D. W. 1981. An addition to the fish fauna of the Gulf of Maine with records of rare species. <u>Proc. biol. Soc. Wash.</u>, 94: 669-674, 1 tab.

DIEUZEIDE, R. 1930. Sur quelques scombriniens des côtes algériennes. <u>Bull. Stn Agric. Pêche Castiglione</u>, (192)(2):127-167.

--. 1950. Sur un <u>Epigonus</u> nouveau de la Méditerranée (<u>Epigonus denticu-latus</u> sp. nov.). <u>Bull. Stn Aquic. Pêche Castiglione</u>, (n.s.)(2):89-105.

--. 1950. Le Serran impérial (<u>Serranus atricauda</u> Günther, 1874) en Médi-terranée. <u>Bull. Stn Aquic. Pêche Castiglione</u>, (n.s.)(2):159-169, 3 fig.

DIEUZEIDE, R.; NOVELLA, M.; (collab. ROLLAND, J.). 1953. Catalogue des poissons des côtes algériennes. I. <u>Bull. Stn Aquic. Pêche Castiglio-ne</u>, (1952) (n.s.)(4):1-135, 73 fig. n. num.

--. 1954-1955. Catalogue des poissons des côtes algériennes. <u>Bull. Stn Aquic. Pêche Castiglione</u>, part II (1953)1954, (n.s.)(5):1-258, 135 fig. n. num.; part III (1954)1955, (n.s.)(6):1-384, 202 fig. n. num.

DIEUZEIDE, R.; ROLAND, J. 1955. Sur un Stromateidae nouveau du genre <u>Cubiceps</u>. <u>Bull. Stn Aquic. Pêche Castiglione</u>, (n.s.)(7):341-368, 18 fig., 1 pl.

DODERLEIN, P. 1871. Descrizione di una notevole specie di sgomberoide (<u>Cybium Verany</u> Doderl.) presa di nelle acque di Sicilia. <u>G. Sci. nat. econ., Palermo</u>, 8:125-136.

--. 1878-79. Prodromo della fauna ittiologica della Sicilia ossia pros-petto metodico delle varie specie di pesci che vennero sin ora ris-contrate nei mari di Sicilia. <u>Atti Accad. Sci Lett., Palermo</u>, (n.s.) 6:1-63.

--. 1878-79. Prospetto metodico delle varie specie di pesci riscontrate sin'ora nelle acque marine e fluviali della Sicilia. <u>Atti Accad. Sci. Lett., Palermo</u>, (n.s.) 6:25-63.

--. 1879-1891. Manuale ittiologico del Mediterraneo, ossia senossi meto-dica delle varie specie di pesci riscontrate sin qui nel Mediterraneo ed in particolare nei mari di Sicilia. Palermo, 5 parts: 1, 1879: viii + 67 p; 2, 1881:1-120; 3, 1884:121-258; 4, 1889:1-188; 5, 1891: 189-320.

--. 1882. Una nota sopra lo <u>Scopelus doderleini</u> Facciolà. <u>Naturalista sicil.</u>, 1:258-263.

--. 1884. Rinvenimento una specie di pesce dell'esoticco genere <u>Pimelep-terus</u>, Lac. nell acque del Golfo di Palermo. <u>Naturalista sicil.</u>, 3: 81-86 and <u>G. Sci. nat. econ. Palermo</u>, 16:238-245.

DÖDERLEIN, L. 1882. Ein Stomiatide aus Japan. <u>Arch. Naturgesch.</u>, 48(1): 26-32.

DOLLFUS, R. P. 1955. Première contribution à l'établissement d'un fi-chier ichthyologique du Maroc Atlantique de Tanger à l'embouchure de l'Oued Dra. <u>Trav. Inst. scient. chérif.</u> (Zool.), (6):1-277, 1 map.

--. 1960. Exposé sommaire des récents progrès dans la connaissance ich-thyologique du Maroc atlantique. <u>Bull. Soc. Sci. nat. phys. Maroc</u>, 39:91-116, 1 pl.

--. 1961. Brèves additions à la connaissance ichthyologique du Maroc atlantique. <u>Bull. Soc. Sci. nat. phys. Maroc</u>, 41:97-100, pl. 5.

DOLLO, L. 1908. <u>Notolepis coatsi</u>. Poisson pélagique nouveau recueilli par l'Expédition Antarctique Nationale Ecossaise. Note préliminaire. <u>Proc. R. phys. Soc. Edinb.</u>, 28:58-65.

DOMAIN, F. 1970. Poissons démersaux du plateau continental sénégambien. Inventaires des chalutages effectués en 1969 à bord du "Laurent Ama-ro". <u>Docum. scient. provis. Centre Rech. océanogr. Dakar-Thiaroye</u>,

(23):1-300 (mimeo).

--. 1972. Poissons démersaux du plateau continental sénégambien. Application de l'analyse en composantes principales à l'étude d'une série de chalutages. Cah. ORSTOM, Océanogr., 10(2):111-123.

--. 1976. Les fonds de pêche du plateau continental ouest-africain entre 17°N et 12°N. Docum. scient. Centre Rech. océanogr. Dakar-Thiaroye, (61):23 p.

--. 1979. Les ressources démersales (Poissons). In: Les ressources halieutiques de l'Atlantique centre-est. FAO Doc. Tech. Pêches, (186.1):79-122, 3 fig.

DOMANIEVSKY, L. N. 1971. (On the problem of reactions to artificial light in various species of fishes on the western shores of Africa). Trudy atlant. nauchno-issled. Inst. ryb. Khoz. Okeanogr., (36):29-25 (in Russian).

--. 1971. La réaction à la lumière artificielle de quelques espèces de poissons de la côte occidentale de l'Afrique. UNDP (SF)/FAO, Project SEN/66/508, Survey and development of pelagic fish resources. Centre Rech. océanogr. Dakar-Thiaroye, (2):1-9 (mimeo, Engl. transl. of precedent reference).

--. 1974. On peculiarities in the biology of some little-known fish species from the north-west African coast. ICES, C.M. 1974/G4 (13 p., 15 fig., mimeo).

DOMANIEVSKY, L. N.; BARKOVA, N. A. 1979. Particularités de la répartition et état des stocks de sardine (Sardina pilchardus) au large de l'Afrique nord-occidentale. Annexe 8, p. 81-86. In: Anon., 1979.

DOMANIEVSKY, L. N.; STEPKINA, M. V.; TKACHENKO, V. A. 1970. (Some growth characteristics of Dentex macrophthalmus Bloch and Sardinella aurita Val. in the eastern Central Atlantic.). Vop. Ikhtiol., 10:861-869 (Engl. transl., J. Ichthyol., 10:652-658).

DONOVAN, E. 1802-1808. The natural history of British fishes including scientific and general descriptions of the most interesting species, and an extensive selection of accurately finished coloured plates, taken entirely from original drawings, purposely made from the specimens in a recent state, and for the most part whilst living. London, 1, 1802:(1-70), pl. I-XXIV; 2, 1803:(71-164), pl. XXV-XLVIII; 3, 1804:(165-300), pl. XLIX-LXXII; 4, 1806:(301-406), pl. LXXIII-XCVI; 5, 1808:(407-516), pl. XCVII-CXX.

DOOLEY, J. K. 1972. Fishes associated with the pelagic Sargassum complex, with a discussion of the sargassum community. Contr. mar. Sci., 16:1-32.

--. 1978. Systematics and biology of the tilefishes (Perciformes: Branchiostegidae and Malacanthidae), with descriptions of two new species. Circ. NOAA tech. Rep. NMFS, (411):5 + 78 p., 44 fig.

--. 1979. A rare and unusually large macrourid fish from the Canary Islands. Copeia, 1979, (3):541-542, 1 fig.

--. 1981. Branchiostegidae, Malacanthidae, in Fischer, W., Bianchi, G. & W.B. Scott (ed.), FAO species identification sheets for fishery purposes. Eastern Central Atlantic; fishing areas 34, 47 (in part). Canada Funds-in-Trust. Ottawa, Department of Fisheries and Oceans Canada, by arrangement with the Food and Agriculture Organization of the United Nations, vol. 1-7: pag. var.

DOOLEY, J. K.; TASSELL, J. V. van; BRITO, A. 1985. An annotated checklist of the shorefishes of the Canary Islands. Amer. Mus. Nov., (2824):1-49.

DORFMAN, D. 1978. The mudskipper, a piscatorial curiosity. Underwat. Natur., 11:25.

DOUMET, N. 1863. Description d'un nouveau genre de poissons de la Méditerranée. Revue Mag. Zool., (2)15:212-223, 1 pl.

DOUTRE, M. P. 1959. Composition chimique des poissons. Variations annuelles de la teneur en matières grasses de trois clupéidés du Sénégal (Ethmalosa fimbriata Bowdich, Sardinella eba C.V., Sardinella aurita C.V.). Comm. Sect. Tech. Pêch. marit., Dakar, (124):4 p. (mimeo).

--. 1959. Teneurs en matières protéiques, en phosphore et en calcium d'échantillons de farine fabriqués à partir de trois clupéidés du Sénégal. Revue Elev. Méd. vét. Pays trop., (n.s.), 12(3):305-312.

--. 1960. Les merlus du Sénégal. Mise en évidence d'une nouvelle espèce. Revue Trav. Inst. Pêch. marit., 24(4):513-536, 9 fig.

DRAGOVICH, A. 1969. Review of studies of tuna food in the Atlantic Ocean. Spec. scient. Rep. U.S. Fish Wildl. Serv. (Fish.), (593):1-21.

DRUZHININ, A. D. 1974. Rasprostraneniye, biologiya i promysel gorbylevykh ryb. (Distribution, biology and commercial use of croakers). Moscow 120 p., (in Russian).

--. 1974. On distribution and biology of drums (or croakers). Sciaenidae family - throughout the world ocean. Ichthyologia, 6:37-47. fig., tab.

DUARTE-BELLO, P. P. 1959. Catàlogo de peces cubanos. Monogr. Lab. Biol. mar., Marianao, 6:208 p.

DUARTE-BELLO, P. P.; BUESA, R. J. 1973. Catàlogo de peces cubanos (Primera revisiòn). I. Indice taxonòmico. Ciencias, La Habana, (Investnes mar.), (3):1-255.

DUBOIS, A. B.; OGILVY, C. S. 1978. Forces of the tail surface of swimming fish: thrust, drag and acceleration in bluefish (Pomatomus saltatrix). J. exp. Biol., 77(Dec):225-242.

DUBOIS, A. B.; CAVAGNA, G. A.; FOX, R. S. 1976. Locomotion of bluefish. J. exp. Zool., 195(2):223-236.

DU BUIT, M. H. 1972. A propos d'un Melanogloea ventralis Barnard conservé au Musée Océanographique de l'Odet (Pisces, Ateleopidae). Bull. Inst. fond. Afr. Noire, (A)34(2):482-484, fig. 1-2.

--. 1978. Alimentation de quelques poissons téléostéens de profondeur dans la zone du seuil de Wyville Thomson. Oceanol. Acta, 1:129-134.

DUCLERC, J. 1970. Répartition géographique et bathymétrique d'Helicolenus dactylopterus (Delaroche) et d'autres Scorpaenidés capturés par la "Thalassa" au Rio de Oro et en Mauritanie. Rapp. P.-v. Réun. Cons. perm. int. Explor. Mer., 159:210-214.

DUCROZ, J. 1957. La pêche du Bonga-fish au Cameroun. Oceanography and sea fisheries on the West African coast, CCTA/CSA. Symposium (20-27 November 1957, Luanda), Paper N°8 (5 p., mimeo).

--. 1962. Variations de la teneur en graisses des sardinelles de la région de Pointe Noire. Bull. Inst. Rech. sci. Congo, 1:101-107.

DUDNIK, Y. I. 1975. Some features of the geographical distribution of the Atlantic saury Scomberesox saurus in the South Atlantic in winter. Vop. Ikhtiol., 15(2):203-209.

--. 1975. Contribution to the ecology of the dwarf Atlantic saury Scomberesox sp. Parin (Pisces, Scomberesocidae). Okeanologii, 15(4):738-743. (in Russian, Engl. translation in: Oceanology, 1976, 15(4):503-504).

DÜBEN, M. W. von. 1845. Norriges Hafs-Fauna. K. svenska Vetensk Akad. Handl. (1844), 4(5)(14):110-116.

DÜBEN, M. W. von; KOREN, J. 1846. Ichthyologiska Bidrag. K. svenska Vetensk Akàd. Handl., (1844):27-120, pl. 2-3.

DUEL, D. G.; CLARK, J. R.; MANSUETI, A. J. 1966. Description of embryonic and early larval stages of bluefish, Pomatomus saltatrix. Trans. Am. Fish. Soc., 95(3):264-271.

DUHAMEL, G.; COMPAGNO, L.V. 1985. Lamnidae, Squalidae, p. 214-217. In: Fischer, W. & Hureau, J.C. (ed.). 1985. FAO Species identification sheets for fishery purposes. Southern Ocean (Fishing Areas 48, 58 and 88). Prepared and printed with the support of the Commission for the Conservation of Antarctic Marine Living Resources (CCAMLR), Rome, vol. 1.

DUHAMEL du MONCEAU, H. L. 1769-82. Traité général des pêches, et histoire des poissons qu'elles fournissent, tant pour la substance des hommes, que pour plusieurs autres usages qui ont rapport aux arts et au commerce. Paris, 4 vol. in-3, 250 pl.

DUMERIL, A. H. A. 1852. Monographie de la famille des Torpédiniens ou Poissons plagiostomes électriques, comprenant la description d'un genre nouveau, de trois espèces nouvelles, et de deux espèces nommées dans le Musée de Paris, mais non encore décrites. Revue Mag. Zool., 4:176-189, 227-244, 270-285.

--. 1855. Note sur un travail inédit de Bibron relatif aux poissons plectognathes gymnodontes (Diodons et Tetrodons). Revue Mag. Zool., 2(7):274-282.

--. 1861. Poissons de la côte occidentale d'Afrique. Archs Mus. Hist. nat., Paris, 10:137-268, pl. 13-23. (volume dated 1858-61, but livraisons 1-2 of 1858, livraisons 3-4 of 1861).

--. 1865-1870. Histoire naturelle des poissons ou ichthyologie générale. Paris, 2 vol., 1, 1865: Elasmobranches, Plagiostomes et Holocephales. 720 p. 2, 1870: Ganoides, Dipnés, Lophobranches. 624 p., Atlas: 14 pl.

DUMERIL, A. M. C. 1806. Zoologie analytique ou méthode naturelle de classification des animaux, rendue plus facile à l'aide de tableaux synoptiques. Paris, xxii + 344 p.

--. 1856. Ichthyologie analytique ou classification des poissons, suivant la méthode naturelle à l'aide de tableaux synoptiques. Mém. Acad. Sci. Paris, 27(1):511 p.

DUNBAR, M. J.; HILDEBRAND, H. H. 1952. Contribution to the study of the fishes of Ungava Bay. J. Fish. Res. Bd Can., 9(2):83-128, 1 fig.

DUNCKER, G. 1910. On some syngnathids ("pipe fish") from Ceylon. Spolia zeylan, 7:25-34, pl.

--. 1915. Revision der Syngnathidae. Mitt. naturh. Mus. Hamb., 32:9-120, 10 fig., 1 pl.

--. 1928. Teleostei Physoclisti 9. Gobiiformes. Tierwelt N.-u. Ostsee, (12g):121-148, 16 fig.

DUNCKER, G. et al., 1929. Pisces (Nachträge und Berichtigungen). In: G. Grimpe & E. Wagler. Tierwelt N.-u. Ostsee, 15(12): h141-h164.

DUNCKER, G.; LADIGES, W. 1960. Die Fische der Nordmark. Abh. Verh. naturw. Ver. Hamburg. (N.F.)3(Suppl.):1-432, 145 fig., 1 map.

DUNCKER, G.; MOHR, E. 1931. Die Fische der Südsee. Expedition der Hamburgischen wissenschalftlichen Stiftung 1908-1909. Mitt. Zool. St. Inst. Hamb., 44:57-84, 9 fig., 3 pl.

--. 1939. Revision der Ammodytidae. Mitt. zool. Mus., Berl., 24(1):8-31.

DUNNE, J. 1976. Littoral and benthic investigations on the west coast of Ireland. V. (Section A : faunistic and ecological studies). A contribution to the biology of the leopard-spotted goby, Thorogobius ephippiatus (Lowe) (Pisces: Teleostei: Gobiidae). Proc. R. Ir. Acad., (B)76:121-132.

DURAND, J. 1937. Note sur un Astronesthes niger Richardson 1845 pêché par un chalutier de la Rochelle. Annls Soc. Sci. nat. Charente-Infér., 3(1):1-7.

DURAND, J. R. 1967. Etude des poissons benthiques du plateau continental Congolais. Troisième partie. Etude de la répartition, de l'abondance et des variations saisonnières. Cah. ORSTOM. Océanogr., 5(2):3-68.

--. 1969. Etude de la similitude entre relevés d'un peuplement de poissons d'un milieu saumâtre tropical poïkilohalin. Verh. int. Verein

theor. angew. Limnol., 17:680-688.
DYKHUIZEN, P.; ZEI, M. 1971. Clupeid fisheries in the tropical eastern Atlantic., p.174-181. In: H. Kristjonsson (ed.), Modern fishing gear of the world, 3. Fishing News (Books), London, 537 p.

EBELING, A. W. 1962. Melamphaeidae I. Systematics and zoogeography of the species in the bathypelagic fish genus Melamphaes Günther. Dana Rep., (58):1-164, fig. 1-73.
--. 1986. Family Melamphaeidae. In Smith, M.M. & P.C. Heemstra (ed.). Smith's Sea Fishes. Johannesburg, Macmillan South Africa (Publ.) Ltd, xx + 1047 p., 144 pl.
EBELING, A. W.; WEED, W. H. 1963. Melamphaidae III. Systematics and distribution of the species in the bathypelagic fish genus Scopelogadus Vaillant. Dana Rep., (60):58 p., 23 fig.
--. 1973. Order Xenoberyces (Stephanoberyciformes). In: Fishes of the western North Atlantic. Mem. Sears Fdn mar. Res., 1(6):397-478, fig. 1-37, 4 tab.
EBELING, A. W.; IBARA, R. M.; LAVENBERG, R. J.; ROHLF, F. J. 1970. Ecological groups of deep-sea animals off southern California. Sci. Bull. nat. Hist. Mus. Los Angeles Cty, 6:1-43, fig. 1-9.
ECK, T. H. van (see VAN ECK, T. H.).
EDGERTON, H. E. 1955. Photographing the sea's dark underworld. Natn. geogr. Mag., 107(4):523-537.
EGE, V. 1918. Stomiatidae (Stomias). Rep. Dan. oceanogr. Exped. Mediterr., 2(A 4):1-28.
--. 1918. On the post-larval stages of the species of Paralepis inhabiting the north eastern part of the Atlantic, including the Mediterranean. Vidensk. Meddr dansk naturh. Foren., 69:243-246.
--. 1930. The North Atlantic and Mediterranean species of the genus Paralepis Cuv. A systematical and biological investigation. (Sudidae, Paralepis). Rep. Dan. oceanogr. Exped. Mediterr., 2(A 13):1-93, fig. 1-37.
--. 1933. On some new fishes of the families Sudidae and Stomiatidae. Preliminary note. Vidensk Meddr dansk naturh. Foren., 94:223-236.
--. 1934. The genus Stomias Cuv., taxonomy and biogeogeography. Dana Rep., (5):1-58, 12 fig., 1 pl.
--. 1939. A revision of the genus Anguilla Shaw: a systematic, phylogenetic, and geographical study. Dana Rep., (16):256 p., 53 fig., 6 pl.
--. 1948. Chauliodus Schn., bathypelagic genus of fishes. A systematic, phylogenetic and geographical study. Dana Rep., (31):1-148, 9 fig., 2 pl.
--. 1953. Paralepididae I (Paralepis and Lestidium). Taxonomy, ontogeny, phylogeny and distribution. Dana Rep., (40):1-184, 82 pl., fig. 1-33.
--. 1957. Paralepididae II (Macroparalepis). Taxonomy, ontogeny, phylogeny and distribution. Dana Rep., (43):1-101, fig. 1-24.
--. 1958. Omosudis Gthr., bathypelagic genus of fish. Dana Rep., (47):1-19, 3 fig.
EGGERT, B. 1935. Beitrag zur Systematik, Biologie und geographischen Verbreitung der Periophthalminae. Zool. Jahrb. (Syst.), 67, 29-116, 9 pl., 16 fig.
EHRENBAUM, E. 1894. Beiträge zur Naturgeschichte einiger Elbfische (Osmerus eperlanus L., Clupea finta Cuv., Acerina cernua L., Acipenser sturio L.). Wiss. Meersunters. (N.F.), 1(1):35-79, 3 pl.
--. 1901. Die Fische. Fauna arct., 2:65-168.
--. 1904. Eier und Larven von Fischen der deutschen Bucht. III. Fische mit festsitzenden Eiern. Wiss. Meeresunters. (Helgoland), 6(2): 127-200.

--. 1905-1909. Eier und Larven von Fischen des Nordischen Planktons. (Teil I), Lieferung 4, 1905: iv + i, 1-216 + iii, fig. 1-82; (Teil II), Lieferung 10, 1909: i + 217-413, fig. 83-148.

--. 1913-1914. Über Fische von Westafrika, besonders von Kamerun. Fischerbote, **5**,1913 (8):308-313, 1 fig.; (9):358-363, 4 fig.; (10):398-405, 3 fig.; (12):506-508, 1 fig.; **6**, 1914, (1):15-19; 2 fig.; (2): 53-57, 3 fig.; (3):106-112, 2 fig.; (5):193-200, 4 fig.; (6):254-259, 2 fig.; (7):289-296, 3 fig.; (8-10):337-346, 4 fig.; (11-12): 401-409, 4 fig. (published as book; Über Küstenfische..., 1915, Hamburg 1-85, 38 fig.).

--. 1924. Scombriformes. __Rep. Dan. oceanogr. Exped. Mediterr.__, **2**(A 11): 42 p.

--. 1928. Rare fishes in the North Sea. __Nature, Lond.__, **121**:709.

--. 1936. Naturgeschichte und wirtschaftliche Bedeutung der Seefische Nordeuropas. __Handb. Seefisch. Nordeur.__, **2**: x + 337 p., 276 fig., tab., 26 pl.

EHRENBAUM, E.; STRODTMAN, S. 1904. Eier und Jugendformen der Ostseefische. __Wiss. Meeresunters. (Kiel)__, (n.s.) **6**:51-126.

EHRICH, S. 1971. Fischfauna und Fischnahrung der Grossen Meteorbank. Diplomarbeit, Inst. Meereskunde, Univ. Kiel :106 p., 22 fig., 6 tab.

--. 1976. Zur Taxonomie, Ökologie und Wachstum von __Macroramphosus scolopax__ (Linnaeus, 1758) (Pisces, Syngnathiformes) aus dem subtropischen Nordostatlantik. __Ber. dt. wiss. Kommn Meeresforsch.__, **24**:251-266, 10 fig.

--. 1977. Die Fischfauna der Grossen Meteorbank. __"Meteor" Forsch.-Ergebn.__, (D)(25):1-23, fig. 1-3.

--. 1986. Macroramphosidae, in P.J.P. Whitehead, M.L. Bauchot, J.C. Hureau, J. Nielsen & E. Tortonese. Fishes of the North-eastern Atlantic and the Mediterranean/Poissons de l'Atantique du Nord-Est et de la Méditerranée, Paris, Unesco, **2**:627, 1 fig.

EHRICH, S.; JOHN, H.-C., 1973. Zur Biologie und Ökologie der Schnepfenfische (Gattung __Macrohamphosus__) vor Nordwestafrika und Überlegungen zum Altersaufbau der adulten Bestände der Grossen Meteorbank. __"Meteor" Forsch.-Ergebn.__, (D)(14):87-98, 5 fig.

EICHWALD, C. E., von. 1831. Pisces. 10-116. __In__: Zoologia specialis quam expositis animalibus tum vivis... adidit D. Eduardus Eichwald, Vilnae, pars post. 404 p., 2 pl.

--. 1838. Fauna Caspii maris primitiae. __Bull. Soc. Nat. Moscou__, **11**(1): 125-147.

EIGENMANN, C. H. 1885. A review of the genera and species of Diodontidae found in American seas. __Ann. N.Y. Acad. Sci.__, **3**:297-311.

--. 1891. New California fishes. __Am. Nat.__, **25**(290):153-156.

--. 1902. Description of a new oceanic fish found off southern New England. __Bull. U.S. Fish. Commn__ (1901), **21**:35-37.

EIGENMANN, C. H.; EIGENMANN, R. S. 1889. Notes from the San Diego biological laboratory. The fishes of the Cortez Banks. __W. Am. Scient.__, **6**:123-132.

--. 1890. Additions to the fauna of San Diego. __Proc. Calif. Acad. Sci.__, (2)**3**:1-24.

EIGENMANN, C. H.; KENNEDY, C. H. 1902. The leptocephalus of the American eel and other American leptocephali. __Bull. U. S. Fish Commn__, **21**:81-92, fig. 1-14.

EKSTRÖM, C. U. 1830-34. Fiskarne i Mörkö Skärgard. __K. svenska Vetensk Akad. Handl.__, 1830, 143-204; 1831:70-107, 270-321; 1834: pl. 1-74.

ELDRED, B. 1968. The larval development and taxonomy of the pygmy moray eel, __Anarchias yoshiae__ Kanazawa, 1952. __Leafl. Ser. Fla Dep. nat. Resour. mar. Res. Lab.__, **4**(1)(10):1-8, 3 fig.

--. 1970. Larva of the purplemouth moray _Gymnothorax_ _vicinus_ (Castelnau, 1855). _Leafl. Ser. Fla Dep. nat. Resour. mar. Res. Lab._, 4(1)(14):7 p. 2 fig.

--. 1970. Larva of the spotted moray, _Gymnothorax_ _moringa_ (Cuvier, 1829). _Leafl. Ser. Fla Dep. nat. Resour. mar. Res. Lab._, 4(1)(15):10 p., 2 fig.

ELST, R. van der. 1981. A guide to the common sea fishes of southern Africa. Ed. C. Smith, Cape Town, 367 p., illust.

ELWERTOWSKI, J.; BOELY, T. 1971. Répartition saisonnière des poissons pélagiques côtiers dans les eaux mauritaniennes et sénégalaises. _Docum. scient. provis. Centre Rech. océanogr.. Dakar-Thiaroye_, (32) (15 p., mimeo).

ELWERTOWSKI, J.; GONZALEZ ALBERDI, P.; CHABANNE, J.; BOELY, T. 1972. Première estimation des ressources pélagiques du plateau continental nord-ouest africain (zone de transition nord de l'Atlantique Centre-Est). _Docum. scient. provis. Centre Rech. océanogr. Dakar-Thiaroye_, (42)(34 p., mimeo).

EMERY, A. R. 1975. _Chromis_ Cuvier, 1814: the correct gender. _Copeia_ 1975(3):579-582.

--. 1978. Pomacentridae, _in_ Fischer, W. (ed.), FAO species identification sheets for fishery purposes. Western Central Atlantic. (Fishing Area 31). FAO, Rome, vol. 1-7, pag. var.

--. 1981. Pomacentridae, _in_ Fischer, W., Bianchi, G. & W.B. Scott (ed.), FAO species identification sheets for fishery purposes. Eastern Central Atlantic; fishing areas 34, 47 (in part). Canada Funds-in-Trust. Ottawa, Department of Fisheries and Oceans Canada, by arrangement with the Food and Agriculture Organization of the United Nations, vol. 1-7: pag. var.

EMERY, A. R.; BURGESS, W. E. 1974. A new species of Damselfish (_Eupomacentrus_) from the western Atlantic, with a key to known species of that area. _Copeia_, 1974(4):879-886, 4 fig.

EMERY, C. 1879. Contribuzioni all'ittiologia. I. Le metamorfosi del _Trachypterus_ _taenia_. _Atti Accad. naz. Lincei Memorie_, (3)3:390-397, 1 pl. (also publ. in _Mitt. zool. Stn Neapel_, 1:581-588, pl. 18, fig. 1-6).

--. 1883. Contribuzioni all'ittiologia. VII. Forme larvali di scopelidi. _Mitt. zool. Stn Neapel_, 4:408-419.

ENGELHARDT, R. 1912. Über einige neue Selachier-Formen. _Zool. Anz._, 39 (21/22):643-648.

--. 1913. Monographie der Selachier der Münchener Zoologischen Staatssammlung (mit besonderer Berücksichtigung der Haifauna Japans). _In_: Beiträge zur Naturgeschichte Ostasiens (F. Doflein ed.). _Abh. bayer. Akad. Wiss._, math.-phys. Kl., Suppl. 4(3):1-110, fig. 1-8, 1 map.

ERDMAN, D. S. 1956. Recent fish records from Puerto Rico. _Bull. mar. Sci. Gulf Caribb._, 6(4):315-340.

ESCHMEYER, W. N. 1965. Western Atlantic scorpionfishes of the genus _Scorpaena_, including four new species. _Bull. mar. Sci._, 15(1):84-164, 12 fig.

--. 1965. Three new scorpionfishes of the genera _Pontinus, Phenacoscorpius_ and _Idiastion_ from the western Atlantic Ocean. _Bull. mar. Sci._, 15(3):521-534, 5 fig.

--. 1969. A systematic review of the scorpionfishes of the Atlantic Ocean (Pisces, Scorpaenidae). _Occ. Pap. Calif. Acad. Sci._, (79):130 p., 13 fig., 15 tab.

--. 1969. A new scorpionfish of the genus _Scorpaenodes_ and _S. muciparus_ (Alcock) from the Indian Ocean, with comments on the limits of the genus. _Occ. Pap. Calif. Acad. Sci._, (76): 11 p., 3 fig.

--. 1971. Two new Atlantic scorpionfishes. _Proc. Calif. Acad. Sci._, (4)

37(17):501-508, 2 fig.

--.(in prep.). A new species of flying gurnard from the Indo-Pacific, with a synopsis of the family (Pisces: Dactylopteridae). _Proc. Calif. Acad. Sci._

--. 1981. Scorpaenidae, _in_ Fischer, W., Bianchi, G. & W.B. Scott (ed.), FAO species identifications sheets for fishery purposes. Eastern Central Atlantic; fishing areas 34, 47 (in part). Canada Funds-in-Trust. Ottawa, Department of Fisheries and Oceans Canada, by arrangement with the Food and Agriculture Organization of the United Nations vol. 1-7: pag. var.

ESCHMEYER, W. N.; BULLIS, H. R., Jr. 1968. Four advanced larval specimens of the blue marlin, _Makaira nigricans_, from the western Atlantic Ocean. _Copeia_, 1968(2):414-417.

ESCHMEYER, W. N.; COLLETTE, B. B. 1966. The scorpionfish subfamily Setarchinae, including the genus _Ectreposebastes. Bull. mar. Sci._, 16 (2):349-375, fig. 1-5.

ESCHMEYER, W. N.; HUREAU, J. C. 1971. _Sebastes mouchezi_, a senior synonym of _Helicolenus tristanensis_, with comments on _Sebastes capensis_ and zoogeographical considerations. _Copeia_, 1971(3):576-579, 1 fig.

ESCHMEYER, W. N.; RANDALL, J. E. 1975. The scorpaenid fishes of the Hawaiian Islands, including new species and new records (Pisces: Scorpaenidae). _Proc. Calif. Acad. Sci._, (4)40(11):265-334, fig. 1-25.

ESMARK, L. 1862. Beskrivelse over en ny Fiskeart, _Brama raschii_ Esm. _Forh. Vidensk. Selsk. Krist._, (1861):238-247, 1 pl.

--. 1871. Om tvende nye fiske-arter: _Argyropelecus elongatus, Maurolicus tripunctulatus. Forh. Vidensk. Selsk. Krist._, (1870):486-490.

ESTEVE, R. 1947. Révision des types de Myctophidés (Scopélidés) du Muséum. _Bull. Mus. natn. Hist. nat._, Paris, (2)19(1):67-69.

EUPHRASEN, B. A. 1788. Beskrifining på trenne fiskar. _K. svenska Vetensk. Akad. Handl._, 9:51-55.

--. 1791. _Scomber atun_ och _Echeneis tropica. K. svenska Vetensk. Akad. Handl._, 15:223-227, 1 pl.

EVANS, J. W. 1961. Normal stages of the early development of the flying fish _Hirundichthys affinis_ (Günther). _Bull. mar. Sci. Gulf. Caribb._, 11(4):483-502.

EVERMANN, B. W.; KENDALL, W. C. 1898. Descriptions of new or little known genera and species of fishes from the United States. _Bull. U.S. Fish Commn_, 17:125-133, 4 pl.

--. 1910. A comparison of the chub-mackerels of the Atlantic and Pacific Oceans. _Proc. U.S. natn. Mus._, 38(1748):327-328.

EVERMANN, B. W.; MARSH, M. C. 1899. Descriptions of new genera and species of fishes from Puerto-Rico. _Rep. U.S. Commnr Fish._, 24:351-362.

--. 1902. The fishes of Porto-Rico. _Bull. U.S. Fish Commn_ (1900), 20 (1): 51-350, pl. 1-49.

EVERMANN, B. W.; RADCLIFFE, L. 1917. The fishes of the West Coast of Peru and the Titicaca Basin. _Bull. U.S. natn. Mus._, 95:1-166, 14 pl.

EVERMANN, B. W.; SEALE, A. 1907. Fishes of the Philippine Islands. _Bull. Bur. Fish._, 26:49-110, 22 fig.

--. 1924. Report on the fishes collected by the Barbados-Antigua Expedition from the University of Iowa in 1918. _Stud. nat. Hist. Iowa Univ._, 10(4):25-40, 1 pl.

EYDOUX, J. F. T.; SOULEYET, F. L. A. 1841. Zoologie. Poissons: 155-215, pl. 1-10. _In_: Voyage autour du monde exécuté pendant les années 1836 et 1837 sur la corvette "La Bonite". Paris, 1:334 p., atlas. (for dates of publication consult M.L. Bauchot, P.J.P. Whitehead & Th. Monod, _Cybium_, 6(3):59-79).

EZIUZO, E. N. C. 1963. The identification of otoliths from West African demersal fish. _Bull. Inst. fr. Afr. noire_, (A)25(2):488-512, 21 fig.

--. 1965. Early development and metamorphosis of the reddish-brown eel, _Phyllogramma_ _regani_ Pellegrin. _Bull. Inst. fr. Afr. noire_, (A)27(1): 312-333, 18 fig.

FABRICIUS, J. C. 1798. Supplementum, J. C. Fabricii... Entomologia systematica emendata et aucta... adjectis synonymis, locis, observationibus, descriptionibus. Hafniae, 572 p.

FACCIOLA, L. 1881. Descrizione di due specie di _Blennius_ del mare di Messina. _Annuar. Soc. Naturalisti Modena_, (2)14:209-220. Modena.

--. 1882. Note sui pesci nuovi o poco noti dello stretto di Messina. _Naturalista sicil._, 1:166-168.

--. 1882. Note sui pesci dello stretto di Messina. _Naturalista sicil._, 1:184-189.

--. 1883. Note sui pesci dello stretto di Messina. _Naturalista sicil._, 2:145-148.

--. 1883. Descrizione di nuove specie di leptocephali dello Stretto di Messina. _Memorie Soc. tosc. Sci. nat._, 6(1):3-11, pl. 1.

--. 1884. Note sui pesci dello Stretto di Messina. _Naturalista sicil._, 3:51-54.

--. 1885. Su di alcuni rari Pleuronettidi del mar di Messina. Note preliminare. _Naturalista sicil._, 4:261-266.

--. 1887. Intorno a due Lepidogastrini ed un nuovo _Nettastoma_ del mare di Sicilia. _Naturalista sicil._, 6(9):163-167.

--. 1893. Le metamorfosi del _Conger_ _vulgaris_ e del _Conger_ _mystax_. _Naturalista sicil._, 13(3):56-60.

--. 1893. Sull'esistenza di forme di passagio da alcune specie di leptocefalidi agli adulti corrispondenti, 4: _Nettastoma_ _melanurum_ Raf. _Naturalista sicil._, 12:144-148.

--. 1897. Sunto di alcune ricerche su l'organizzazione e lo sviluppo dei Leptocefalidi. _Atti. Soc. Nat. Mat._, (3)14:122-145.

--. 1899. Sull'esistenza di due specie di _Nettastoma_. _Riv. ital. Sci. nat._, 19:29-32.

--. 1911. Generi dei vertebrati ittioidi del mare di Messina. _Bol. Soc. zool. ital._, (2)12:268-287.

--. 1914. Su di un nuovo tipo dei Nettastomidi. _Boll. Soc. zool. ital._, (3)3:39-47.

--. 1916. Labroidi del mare di Messina. _Monitore zool. ital._, 27(7): 140-152.

--. 1918. Specie della famiglia Maenidae viventi nel mare di Messina. _Monitore zool. ital._, 29(1):6-16.

FAGADE, S. O.; OLANIYAN, C. I. O. 1972. The biology of West African shad _Ethmalosa_ _fimbriata_ (Bowdich) in the Lagos Lagoon, Nigeria. _J. Fish Biol._, 4(4):519-533.

--. 1973. The food and feeding relationships of the fishes in the Lagos Lagoon. _J. Fish Biol._, 5(2):205-226.

--. 1974. Seasonal distribution of fish fauna of the Lagos Lagoon. _Bull. Inst. fond. Afr. noire_, (A)36(1):244-252.

FAGE, L. 1907. Essai sur la faune des poissons des îles Baléares et description de quelques espèces nouvelles. _Archs Zool. exp. gén._, (4)(7):69-93.

--. 1911. Sur une collection de poissons provenant de la côte méditerranéenne du Maroc. _C. r. Ass. fr. Avanc. Sci._, 40:578-582 (= _Bull. Soc. zool. France_, 1912, 36:215-220).

--. 1915. Sur quelques _Gobius_ méditerranéens (_G. Kneri_ Stndr., _G. elongatus_ Canestr., _G. niger_ L.). _Bull. Soc. zool. Fr._, 40:164-175.

--. 1918. Shore-Fishes. Rep. Dan. oceanogr. Exped. Mediterr., 2 (Biol. A 3):1-154, 114 fig., 16 map.

--. 1920. Engraulidae, Clupeidae. Rep. Dan. oceanogr. Exped. Mediterr., 2 (Biol. A 9):1-140, 50 fig.

FAGER, E. W.; LONGHURST, A. R. 1968. Recurrent group analysis of species assemblages of demersal fish in the Gulf of Guinea. J. Fish. Res. Bd Can., 25(1):1405-1421, 3 fig., 3 tab.

FAGETTI, E. 1970. Distribution and relative abundance of Clupeidae and Engraulidae larvae in the water of the continental shelf of Senegal and Gambia during 1968. UNDP (SF)/FAO, Project SEN/66/508. Survey and development of pelagic fish resources. Docum. scient. provis. Centre Rech. océanogr.. Dakar-Thiaroye, (1):26 p. (mimeo).

FAGETTI, E.; MARAK, R. 1972. Ichthyoplankton studies in West Africa - a review. FAO Fish. Circ., (137) (11 p., mimeo).

FAHAY, M. P. 1983. Guide to the early stages of marine fishes occurring in the western North Atlantic Ocean, Cape Hatteras to the southern Scotian Shelf. J. N.W. Atlant. Fishery Sci., 4, 423 p.

FAHAY, M. P.; OBENCHAIN, C. 1978. Leptocephali of the ophichthid genera Ahlia, Myrophis, Ophichthus, Pisodonophis, Callechelys, Letharchus, and Apterichtus on the Atlantic continental shelf of the United States. Bull. mar. Sci., 28(3):442-486.

FALK, U. 1965. Giftige Fische an der westafrikanischen Küste. Fisch. Forsch.. wiss. Schr. Reihe, 3:65-66, fig. 1-6.

--. 1965. Fischarten des nordwestafrikanischen Schelfs. Bildanhang. Fisch. Forsch. wiss. Schr. Reihe, 3:81-132, fig. 1-152.

FARRAN, G. P. 1924. Seventh report on the fishes of the Irish Atlantic slope. The macrurid fishes (Coryphaenoididae). Proc. R. Ir. Acad., 36(8):91-143, 11 fig., pl. 6-7.

FAST, T. N. 1957. The occurrence of the deep-sea Anglerfish, Cryptopsaras couesii in Monterey Bay, California. Copeia, 1957(3):237-240, 1 fig., 1 tab.

FEDOROV, V. V. 1986. Cottidae, in P.J.P. Whitehead, M.L. Bauchot, J.C. Hureau & E. Tortonese. Fishes of the North-eastern Atlantic and the Mediterranean/Poissons de l'Atlantique du Nord-Est et de la Méditerranée, Paris, Unesco, 3:1243-1260, 16 fig.

FEDOROV, V. V.; NELSON, J. S. 1986. Psychrolutidae, in P.J.P. Whitehead, M.L. Bauchot, J.C. Hureau, J. Nielsen & E. Tortonese. Fishes of the North-eastern Atlantic and the Mediterranean/Poissons de l'Atlantique du Nord-Est et de la Méditerranée, Paris, Unesco, 3:1261-1264, 3 fig.

FENAROLI, L. 1931. Notizie sulla pesca nel Mossamedes (Angola) e sui prodotti derivati, olio di pesce e guano. Boll. Pesca Piscic. Idrobiol., 7(3):448-456.

FERMOR, X. 1913. Wissenschaftliche Ergebnisse einer Reise von S. Awerinzew in die Tropen Afrikas. I. Einige Befunde zur Kenntnis von Ariodes polystaphylodon. Zool. Anz., 42(5):196-199, 3 fig.

FERNANDEZ, M. A. R. 1978. Contribución al estudio de la sardina (Sardina pilchardus Walb.) en aguas de Africa occidental. 2. Estimación de la abundancia de la población explotada por la española. Boln Inst. esp. Oceanogr., 4(4):43-61.

FERNANDEZ, M. A. R.; DELGADO, A.; BRAVO DE LAGUNA, J. 1978. Contribución al estudio de la sardina (Sardina pilchardus Walb.) en aguas de Africa occidental. 1. Relación entre la talla, la edad y el número de branchiospina. Boln Inst. esp. Oceanogr., 4(4):31-42.

FERNANDEZ-YEPEZ, A. 1948. Ichthyacus breederi nuevo género y especie de pez Synentognatho, de los Ríos de Sur América. Evencias, (4):4 p.

FERNHOLM, B. 1981. A new species of hagfish of the genus Myxine, with notes on other eastern Atlantic myxinids. J. Fish Biol., 19:73-82, 4 fig., 2 tab.

FERNHOLM, B.; VLADYKOV, V. D. 1984. Myxinidae, in Whitehead, P.J.P., Bauchot, M.L., Hureau, J.C., Nielsen, J. & E. Tortonese, Fishes of the North-eastern Atlantic and the Mediterranean/Poissons de l'Atlantique du Nord-Est et de la Méditerranée, Paris, Unesco, 1:68-69, 5 fig., 2 maps.

FERNHOLM, B.; WHEELER, A. C. 1983. The Linnaean fish specimens in the Swedish Museum of Natural History, Stockholm. Zool. J. Linn. Soc. Lond., 78:199-286.

FILHOL, H. 1884. Explorations sous-marines. Voyage du "Talisman". Nature. Paris, 22:119-122, 5 fig.; 134-138, 4 fig.; 147-151, 3 fig.; 161-164, 2 fig.; 182-186, 3 fig.; 198-202, 1 fig.; 230-234, 3 fig.; 278-282, 2 fig.; 326-330, 1 fig.; 391-394, 3 fig.

--. 1885. La vie au fond des mers (Bibliothèque de la nature). Paris, p. i-viii,1-303.

FILIPPI, F. de; VERANY, J. B. 1859. Sopra alcuni pesci nuovi o poco noti del Mediterraneo. Memorie Accad. Sci. Torino, (2)18:187-199, 1 pl. (also sep., Torino, 15 p., 1 pl.).

FISCHER, J. G. 1884. Über einige afrikanische Reptilien, Amphibien und Fische des naturhistorischen Museums. Jb. hamb. wiss. Anst., 1:1-39, 3 pl.

--. 1885. Ichthyologische und herpetologische Bemerkungen. Jb. hamb. wiss. Anst., 2:49-121, 4 pl.

FISCHER, W.; HUREAU, J.C. (ed.). 1985. FAO species identification sheets for fishery purposes. Southern Ocean (Fishing Areas 48, 58 and 88). Prepared and printed with the support of the Commission for the Conservation of Antarctic Marine Living Resources (CCAMLR). Rome, FAO, 1985, 1: xxiv + 1-232; 2: v + 233-472, num. fig.

FISCHER, W.; HUREAU, J.C. (ed.). 1987. Fiches FAO d'identification des espèces pour les besoins de la pêche. Océan Austral (Zones de pêche 48, 58 et 88) (Zone de la Convention CCAMLR). Publication préparée et publiée avec l'aide de la Commission pour la conservation de la faune et de la flore marines de l'Antarctique. Rome, FAO, 1987, 1:xxiv + 1-234; 2:v + 235-479, num. fig. (increased and corrected French edition of the 1985 English edition).

FISCHER, W.; HUREAU, J.C. (ed.). 1988. Fichas FAO de identificacion de especies para los fines de la pesca. Oceano Austral (Areas de pesca 48, 58 y 88). Publicacion preparada y publicada con el apoyo de la Comision para la conservacion de los recursos vivos marinos antarticos. Roma, FAO, 1988, 1:xxiv + 1-232; 2:vi +233-474.

FISCHER von WALDHEIM, G. 1813. Zoognosia tabulis synopticis illustrata in usum praelectionum Academiae Imperialis Medico-Chirurgicae Mosquenis. (3 ed.). Mosquae, 1: xii + 466 p.; 2: xxii + 605 p.

FISCHTAL, J. H.; THOMAS, J. D. 1973. Some hemiurid trematodes of marine fishes from Ghana. Proc. helminth. Soc. Wash., 38(2):181-189.

FISH, M. P. 1927. Contributions to the embryology of the American eel (Anguilla rostrata Lesueur). Zoologica. N. Y., 8(5):289-324, 14 fig.

FITCH, J. E. 1964. The ribbon fishes (family Trachipteridae) of the eastern Pacific Ocean, with a description of a new species. Calif. Fish Game, 50:228-240, fig.

--. 1979. The velvet whalefish, Barbourisia rufa, added to California's marine fauna, with notes on otoliths of whalefishes and possible related genera. Bull. Sth Calif. Acad. Sci., 78(1):61-67, 3 fig.

FITCH, J. E.; CRAIG, W. L. 1964. First records for the bigeye thresher (Alopias superciliosus) and slender tuna (Allothunnus fallai) from California, with notes on eastern Pacific scombrid otoliths. Calif. Fish Game, 50(3):195-206.

FITCH, J. E.; GOTSHALL, D. W. 1972. First record of the black scabbardfish, Aphanopus carbo, from the Pacific Ocean with notes on other

Californian trichiurid fishes. <u>Bull. Sth. Calif. Acad. Sci.</u>, 71:12-18.

--. 1968. Deep-water teleostean fishes of California. Berkeley, 155 p., 74 fig.

FITCH, J. E.; ROEDEL, P. M. 1963. A review of the frigate mackerels (genus <u>Auxis</u>) of the world. <u>FAO Fish. Rep.</u> (6)3:1329-1342, 2 fig.

FLEMING, J. 1822. The philosophy of zoology, or view of the structure, classification, etc., of animals. 2 vol. Edinburgh. Fishes, 2:305-397.

--. 1828. A history of British animals exhibiting the descriptive characters and systematical arrangements of the genera and species of quadrupeds, birds, reptiles, fishes, mollusca and radiata of the United Kingdom; including the indigenous, extirpated and extinct kinds, together with periodical and occasional visitants. Edinburg, London, xxiii + 565 p. (ed. 2, 1842, same pag.).

FONTAINE, M. 1941. Recherches sur quelques pigments sériques et dermiques de poissons marins. (Labridés et Cyclopteridés). <u>Bull. Inst. océanogr. Monaco</u>, (792):1-5.

FONTANA, A. 1969. Etude de la maturité sexuelle des sardinelles <u>Sardinella eba</u> (Val.) et <u>Sardinella aurita</u> C. et V. de la région de Pointe-Noire. <u>Cah. ORSTOM. Océanogr.</u>, 7:101-114.

FONTANA, A.; CHARDY, A. 1971. Note préliminaire sur les variations dans l'importance des pontes de <u>Sardinella aurita</u> dans la région de Pointe-Noire en fonction de certains facteurs hydrologiques et climatiques. <u>Docums scient. provis. Centre Rech. océanogr.. Pointe-Noire</u> (n.s.), (16) (7 p. mimeo).

FONTANA, A.; PIANET, R. 1973. Biologie des sardinelles <u>Sardinella eba</u> (Val.) et <u>Sardinella aurita</u> (Val.) des côtes du Congo et du Gabon. <u>Docums scient. provis. Centre Rech. océanogr.. Pointe-Noire</u> (n.s.), (31)(38 p., mimeo).

FORD, E. 1922. On the post-larvae of the wrasses occurring near Plymouth. <u>J. mar. biol. Ass. U.K.</u>, 12(4):693-699, 9 fig.

FOREY, P. L. 1973. Relationships of elopomorphs. p. 351-368, fig. 1-5. <u>In</u>: Greenwood, P.H., R.S. Miles & C. Patterson. Interrelationships of fishes. <u>Zool. J. Linn. Soc.</u>, 53 (Suppl. 1): xvi + 1 + 536, fig.

FORS[S]KAL, P. 1775. Descriptiones animalium; avium, amphibiorum, piscium, insectorum, vermium, quae in itinere orientali observavit. Havniae, 20 + xxxiv + 164 p., 1 map.

FORSTER, G. R. 1970. Line-fishing on the continental slope. 3. Mid-water fishing with vertical lines. <u>J. mar. biol. Ass. U.K.</u>, 51(1):73-77.

FORSTER, G. R.; BADCOCK, J. R.; LONGBOTTOM, M. R.; MERRETT, N. R.; THOMSON, K. S. 1970. Results of the Royal Society Indian Ocean Deep Slope Expedition, 1969. <u>Proc. R. Soc.</u>, (B)175:367-404.

FORSTER, J. G. A. 1777. A voyage around the world in His Britannic Majesty's sloop, Resolution, commanded by Capt. James Cook, during the years 1772, 3, 4 and 5. 2 vol. London, 1: xviii + 602 p.; 2:607 p.

FORSTER, J. R. 1788. Enchiridion historiae naturali inserviens, quo termini et delineationes ad avium, piscium, insectorum et plantarum adumbrationes intelligendas et concinnandas, secundum methodum systematis Linnaeani continentur. Halae, 248 p. (French transl. 1799).

FORTUNATOVA, K. R. 1940. Nutrition of <u>Scorpaena porcus</u> L. (on methods of quantitative study of nutrition dynamics in carnivorous marine fish). <u>Dokl. Akad. Nauk SSSR</u>, (2)29(3):242-247.

FOURMANOIR, P. 1957. Poissons téléostéens des eaux malgaches du canal de Mozambique. <u>Mém. Inst. scient. Madagascar</u>, (F)1:1-316.

--. 1969. Contenus stomacaux d'<u>Alepisaurus</u> (Poissons) dans le Sud-ouest Pacifique. <u>Cah. ORSTOM. Océanogr.</u>, 7(4):51-59.

--. 1970. Notes ichthyologiques (I). <u>Cah. ORSTOM. Océanogr.</u>, 8(2):19-33,

10 fig.

--. 1970. Notes ichthyologiques (II). <u>Cah. ORSTOM, Océanogr.</u>, **8**(3):35-46, 8 fig., 2 tab.

--. 1971. Notes ichthyologiques (III). <u>Cah. ORSTOM, Océanogr.</u>, **9**(2):267-278.

--. 1971. Notes ichthyologiques (IV). <u>Cah. ORSTOM, Océanogr.</u>, **9**(4):491-500, 8 fig.

--. 1976. Formes post-larvaires et juvéniles de poissons côtiers pris au chalut pélagique dans le Sud-ouest Pacifique. <u>Cah. Pacif.</u>, (19):47-88, 51 fig.

--. 1979. Découverte de très jeunes <u>Ruvettus pretiosus</u> Cocco dans un estomac de "Poissons-lancette", <u>Alepisaurus ferox</u> et de "Maquereau-frégate", <u>Auxis thazard</u>. Présence d'<u>Anotopterus</u> en eau tropicale. <u>Cah. Indo-Pacif.</u>, 1(4):445-446, 2 fig.

FOURMANOIR, P.; CROSNIER, A. 1964. Deuxième liste complémentaire des poissons du canal de Mozambique. Diagnoses préliminaires de 11 espèces nouvelles. <u>Trav. Centre océanogr. Pêch. Nosy-Bé</u>, (6):2-32, 15 fig., 5 pl.

FOWLER, H. W. 1901. Contributions to the ichthyology of the tropical Pacific. <u>Proc. Acad. nat. Sci. Philad.</u>, (1900):493-528, 4 pl.

--. 1901. Note on the Odontostomidae. <u>Proc. Acad. nat. Sci. Philad.</u>, 53:211-212.

--. 1901. <u>Myctophum phengodes</u> in the North Atlantic. <u>Proc. Acad. nat. Sci. Philad.</u>, 53:620-621.

--. 1902. Description of a new hemirhamphid. <u>Proc. Acad. nat. Sci. Philad.</u>, (1901):293-294, fig.

--. 1903. New or little known Mugilidae and Sphyraenidae. <u>Proc. Acad. nat. Sci. Philad.</u>, **55**(3):743-752.

--. 1904. Description of a new lantern fish. <u>Proc. Acad. nat. Sci. Philad.</u>, **55**:754-755.

--. 1904. A collection of fishes from Sumatra. <u>J. Acad. nat. Sci. Philad.</u>, (2)**12**:495-560, 21 pl.

--. 1905. New, rare or little known scombroids. N° 1. <u>Proc. Acad. Nat. Sci. Philad.</u>, **56**:757-771.

--. 1905. New, rare or little known scombroids. N° 2. <u>Proc. Acad. Nat. Sci. Philad.</u>, **57**:56-88.

--. 1905. Some fishes from Borneo. <u>Proc. Acad. nat. Sci. Philad.</u>, **57**:455-523, fig. 1-16.

--. 1906-07. New, rare or little known scombroids. N° 3. <u>Proc. Acad. nat. Sci. Philad.</u>, **58**:114-122, 3 fig.

--. 1907. Notes on Serranidae. <u>Proc. Acad. nat. Sci. Philad.</u>, **58**:77-113, fig. 1-13, pl. 3-4.

--. 1907. A supplementary account of the fishes of New Jersey. <u>A. Rep. New Jersey State Mus.</u>, (1906):251-350, 1 pl.

--. 1910. Notes on batoïd fishes. <u>Proc. Acad. nat. Sci. Philad.</u>, **62**:468-475, 2 fig.

--. 1911. A description of the fossil remains of the Cretaceous, Eocene and Miocene formations of New Jersey. <u>Bull. geol. Surv. New Jers.</u>, 4:22-192, fig. 1-108, pl. 1-10.

--. 1911. Notes on clupeoid fishes. <u>Proc. Acad. nat. Sci. Philad.</u>, **63**:204-221.

--. 1911. Notes on salmonid and related fishes (<u>Stomias, Synodus</u>). <u>Proc. Acad. nat. Sci. Philad.</u>, **63**:551-571.

--. 1912. Records of fishes for the middle Atlantic states and Virginia. <u>Proc. Acad. nat. Sci., Philad.</u>, **64**:34-59, 2 fig.

--. 1912. Description of nine new eels, with notes on other species. <u>Proc. Acad. nat. Sci. Philad.</u>, **64**:8-33, 9 fig.

--. 1915. The genus <u>Cryptotomus</u> Cope. <u>Copeia</u>, 1915 (14):[3].

--. 1917. A few reptiles and fishes from Batanga, West Africa. <u>Copeia</u>, 1917(45):53.

--. 1918. New and little-known fishes from the Philippines. <u>Proc. Acad. nat. Sci. Philad.</u>, 70:2-71.

--. 1919. The fishes of the United States "Eclipse" Expedition to West Africa. <u>Proc. U. S. natn. Mus.</u>, 56(2294):195-292, 13 fig.

--. 1920. Notes on synentognathous fishes. <u>Proc. Acad. nat. Sci. Philad.</u>, 71:2-15.

--. 1920. Notes on tropical American fishes. <u>Proc. Acad. nat. Sci. Philad.</u>, 71:128-155.

--. 1922. Notes on hemibranchiate and lophobranchiate fishes. <u>Proc. Acad. nat. Sci. Philad.</u>, 73(3):437-448, 2 fig.

--. 1923. New fishes obtained by the American Museum Congo Expedition 1909-1915. <u>Am. Mus. Novit.</u>, (103):1-6.

--. 1923. Fishes from Madeira, Syria, Madagascar and Victoria, Australia. <u>Proc. Acad. nat. Sci. Philad.</u>, 75:33-45.

--. 1923. New or little known Hawaiian fishes. <u>Occ. Pap. Bernice P. Bishop Mus.</u>, 8(7):376-378.

--. 1925. New taxonomic names of West African marine fishes. <u>Am. Mus. Novit.</u>, (162):1-5.

--. 1928. The fishes of Oceania. <u>Mem. Bernice P. Bishop Mus.</u>, 10:1-540, 82 fig., 49 pl.

--. 1928. Note on <u>Clupea fimbriata</u> Bowdich. <u>Copeia</u>, (167):38-39.

--. 1929. Fishes from Florida and the West Indies. <u>Proc. Acad. nat. Sci. Philad.</u>, 80:151-173.

--. 1931. The fishes of Oceania. Supplement 1. <u>Mem. Bernice P. Bishop Mus.</u>, 11(5):313-381, 7 fig.

--. 1931. The fishes of the families Pseudochromidae, Lobotidae... and Teraponidae, collected by the United States Bureau of Fisheries Steamer "Albatross", chiefly in Philippine Seas and adjacent waters. <u>Bull. U.S. natn. Mus.</u>, Bull. 100, 11: ix + 388 p., 29 fig.

--. 1931. The fishes obtained by Mr. James Bond at Grenada, British West Indies, in 1929. <u>Proc. Acad. nat. Sci Philad.</u>, 82:269-277, fig. 1-2.

--. 1932. The fishes obtained by the Pinchot South Seas Expedition of 1929, with descriptions of one new genus and three new species. <u>Proc. U.S. natn. Mus.</u>, 80(2906):1-16.

--. 1933. Description of a new long-finned tuna (<u>Semathunnus guildi</u>) from Tahiti. <u>Proc. Acad. nat. Sci. Philad.</u>, 85:163-164, pl. 12.

--. 1934. The fishes of Oceania. Supplement 2. <u>Mem. Bernice P. Bishop Mus.</u>, 11(6):385-466, 4 fig.

--. 1934. Description of new fishes obtained 1907 to 1910, chiefly in the Philippine Islands and adjacent seas. <u>Proc. Acad. nat. Sci. Philad.</u>, 85:233-367, fig. 1-117, 12 pl.

--. 1934. Zoological results of the third de Schauensee Siamese Expedition, Part I - Fishes. <u>Proc. Acad. nat. Sci. Philad.</u>, 86:67-163, fig. 1-129, pl. 12.

--. 1934. Fishes obtained by Mr. H.W. Bell-Marley chiefly in Natal and Zululand in 1929 to 1932. <u>Proc. Acad. nat. Sci. Philad.</u>, 86:405-514, fig. 1-53.

--. 1935. Description of a new scorpaenid fish (<u>Neomerinthe hemingwayi</u>) from off New Jersey. <u>Proc. Acad. nat. Sci. Philad.</u>, 87:41-43, 1 fig.

--. 1935. South African fishes received from Mr. H.W. Bell-Marley in 1935. <u>Proc. Acad. nat. Sci. Philad.</u>, 87:361-408, 39 fig.

--. 1936. The marine fishes of West Africa, based on the collection of the American Museum Congo Expedition 1909-15. <u>Bull. Am. Mus. nat. Hist.</u>, 70(1), Jan. 21, 1936: vii + 606 p., fig. 1-275; (2), Nov. 18, 1936:607-1493, fig. 276-567.

--. 1937. Notes on fishes from the Gulf Stream and the New Jersey coast.

Proc. Acad. nat. Sci. Philad., **89**:297-308, 1 fig.

--. 1938. Descriptions of new fishes obtained by the United States Bureau of Fisheries Steamer "Albatross", chiefly in Philippine seas and adjacent waters. Proc. U.S. natn. Mus., **85**(3032):31-135, fig. 6-61.

--. 1938. The fishes of the George Vanderbilt South Pacific Expedition, 1937. Monogr. Acad. nat. Sci. Philad., 2: i-viii + 1-349, 11 pl.

--. 1938. Studies of Hong-Kong Fishes. Hongkong Nat., Suppl., (6):1-52,5 fig.

--. 1940. The fishes obtained by the Wilkes Expedition, 1838-1842. Proc. Am. phil. Soc., **82**(5):733-800, 76 fig.

--. 1941. The George Vanderbilt Oahu Survey. The fishes. Proc. Acad. nat. Sci. Philad., **93**:247-279.

--. 1941. A list of the fishes known from the coast of Brazil. Archos Zool. Est. S. Paulo, 3(6):115-184.

--. 1941. Contributions to the biology of the Philippine Archipelago and adjacent regions. The fishes of the groups Elasmobranchii, Holocephali, Isospondyli, and Ostariophysi obtained by the United States Bureau of Fisheries Steamer "Albatross" in 1907 to 1910, chiefly in the Philippine Islands and adjacent seas. Bull. U.S. natn. Mus., Bull. 100, 13: I-X, 1-879, fig. 1-30.

--. 1943. Descriptions and figures of new fishes obtained in Philippine seas and adjacent waters by the United States Bureau of Fisheries Steamer "Albatross". Bull. U.S. natn. Mus., Bull. Bull. 100, 14(2): 51-91, 25 fig.

--. 1944. Fishes of Chile. Systematic catalog, pt. 2. Revta chil. Hist. nat., **46-48**:15-343.

--. 1944. Fishes obtained in the New Hebrides by Dr Edward L. Jackson. Proc. Acad. nat. Sci. Philad., **96**:155-199.

--. 1944. The fishes. Results of the fifth George Vanderbilt Expedition, 1941. (Bahamas, Caribbean sea, Panama, Galapagos Archipelago and Mexican Pacific islands). Monogr. Acad. nat. Sci. Philad., 6:57-529, fig. 1-268, pl. 1-20.

--. 1947. The bull-eye (Cookeolus boops) from off the New Jersey coast. Notul. Nat., (189):1-3, fig. 1.

--. 1949. Fishes of Oceania. Supplement 3. Mem. Bernice P. Bishop Mus., 12(2):37-186.

--. 1952. A list of the fishes of New Jersey, with off-shore species. Proc. Acad. nat. Sci. Philad., 104:89-151, 1 fig.

--. 1956. Fishes of the Red Sea and southern Arabia. I. Branchiostomida to Polynemida. Jerusalem, 1-240, fig. 1-117.

--. 1958. Some new taxonomic names of fish-like vertebrates. Notul. Nat., (310):1-16.

--. 1972. A catalogue of world fishes (XVIII). Q. Jl Taiwan Mus., 26(1-2):1-111.

FOWLER, H. W.; BEAN, B. A. 1923. Descriptions of eighteen new species of fishes from the Wilkes Exploring Expedition, preserved in the United States National Museum. Proc. U.S. natn. Mus., 63:1-27.

--. 1928. The fishes of the families Pomacentridae... collected by... "Albatross"... Philippine seas and adjacent waters. Bull. U.S. natn. Mus., Bull. 100, 7: i-viii, 1-525, 49 pl.

--. 1930. Contribution to the biology of the Philippine Archipelago and adjacent regions. The fishes of the families Amiidae, Chandidae, Duleidae and Serranidae obtained by the United States Bureau of Fisheries Steamer "Albatross" in 1907 to 1910 chiefly in the Philippine Islands and adjacent seas. Bull. U.S. natn Mus. Bull. 100, 10: i-xi, 1-334, 27 fig.

FOWLER, H. W; PHILLIPS, R. J. 1910. A new fish of the genus Paralepis from New Jersey. Proc. Acad. nat. Sci. Philad., 62:403-406, 1 fig.

FOWLER, H. W.; STEINITZ, H. 1956. Fishes from Cyprus, Iran, Iraq, Israel and Oman. <u>Bull. Res. Coun. Israel</u>, (B)5(3-4):260-292.

FRADE, F. 1929. Sur quelques thons peu connus de l'Atlantique. <u>Bull. Soc. port. Sci. nat.</u>, 10(20):227-243.

--. 1931. Données biométriques sur trois espèces de thon de l'Atlantique oriental. <u>Rapp. P.-v. Réun. Cons. perm. int. Explor. Mer</u>, 70:117-126.

--. 1931. Sur le nombre de rayons des nageoires et des pinnules branchiales chez le thon rouge Atlantique. <u>Bull. Soc. port Sci. nat.</u>, 11(10):139-144.

-. 1932. Sur les caractères ostéologiques à utiliser pour la détermination des thonidés de l'Atlantique oriental et de la Méditerranée. <u>Rapp. P.-v. Réun. Cons. perm. int. Explor. Mer</u>, 7 (n. s.):79-90.

--. 1960. Le thon patudo <u>Parathunnus obesus</u> (Lowe) et sa pêche. <u>Estudos Ensaios Docum. Jta Invest. Ultramar</u>, (69):1-74.

FRADE, F.; POSTEL, E. 1955. Contribution à l'étude de la reproduction des Scombridés et Thonidés de l'Atlantique tropical. <u>Rapp. P.-v. Réun. Cons. perm. int. Explor. Mer</u>, 137:33-35.

FRADE, F.; VILELA, H. 1962. Le thon et le germon, <u>Thunnus thynnus</u> (L.) et <u>Germo alalunga</u> (Bonn.). Morphologie, biologie et pêche. <u>Estudos Ensaios Docum. Jta Invest. Ultramar</u>, (98):1-92.

FRADE, F.; BACELAR, A.; GONZALES, B. 1946. Trabalhos da Missâo Zoologies da Guiné. <u>Anais Jta Miss. geogr.</u>, 1:261-415, fig. 1-50.

FRANCA, M. L. P. da. 1964. Peixes emalhados nas redes lagosteiras em aguas costeiras de Angola. <u>Notas mimeogr. Cent. Biol. pisc.</u>, (40):1-13, 1 fig., 3 maps, 19 tab.

FRANCA , M. L. P. da; COSTA, F. C. da 1972. Contribiuçâo para o conhecimento da biologia de <u>Boops boops</u> (L.) (Pisces, Sparidae) de Angola. <u>Notas Cent. Biol. aquat.trop.</u> (29):1-20.

--. 1972. Nota sobre a relaçâo entre o rendimento de pesca e a hora do dia em alguns Sciaenidae de Angola. <u>Notas Cent. Biol. aquat. trop.</u>, (28):1-17.

--. 1973. Contrbuçâo para o conhecimento da biologia de <u>Pagrus pagrus</u>, <u>Pagrus ehrenbergi</u> Val. e <u>Pagrus auriga</u> Val. (Pisces, Sparidae) de Angola. <u>Notas Cent. Biol. aquat. trop.</u>, (34):1-109, 34 fig.

--. 1973. Myctophidae colhidos nas águas do Arquipélago de Cabo Verde pelo navio de estudos "Walter Herwig" em 1970. <u>Notas Cent. Biol. aquat. trop.</u>, (37):1-16, 1 fig.

FRANCA, M. L. P. da ; VASCONCELOS, M. de S. 1962. Peixes do Arquipélago de Cabo Verde. <u>Notas mimeogr. Cent. Biol. pisc.</u>, (28):1-85.

--. 1962. Contribution à l'étude de <u>Pseudolepidaplois pfaffi</u> Bauchot and Blanc (1961)(Teleostei Labroidei). <u>Mems Jta Invest. Ultramar</u>, (2)36:105-122, 5 pl.

FRANCA, M. L. P. da; COSTA, F. C. da; FRANCA, P. da. 1970. Contribuiçâo para o conhecimento da biologia dos Zeidae de Angola. <u>Notas Cent. Biol. aquat. trop.</u>, (22):1-35, 16 diagrams, 18 tab.

FRANCA, P. da. 1957. Contribuçâo para o conhecimento dos Stromateidae de Angola. <u>Anais Jta Invest. Ultramar</u>, 12(2):7-26, 2 pl.

--. 1958. Contribuiçâo para o conhecimento dos Sciaenidae de Angola. <u>Mems Jta Invest. Ultramar.</u>, (4)(2):7-51, 11 fig.

--. 1969. Sobre a distribuiçâo dos Trichiuridae (Pisces, Perciformes) que ocorrem na costa de Angola. <u>Notas Cent. Biol. aquat. trop.</u>, (16):1-19.

--. 1971. Hipótese acerca da provável ocorrência de <u>Merlucius merlucius paradoxus</u> Franca 1960 em águas Angolanas. <u>Notas Cent. Biol. aquat. trop.</u>, (26):1-18, 1 fig.

FRANCA, P. da; FERREIRA, M. L. 1967. Contribuiçâo para o conhecimento dos Ateleopidae do Atlântico oriental Africano. <u>Notas mimeogr. Cent. Biol. aquat. trop.</u>, (5):1-24, fig. 1-13.

FRANCA, P. da; FRANCA, M. L. P. da. 1968. Ocorrência de _Kyphosus inci-sor_ (Cuvier) (Pisces, Perciformes, Kyphosidae) na costa de Angola. _Notas Cent. Biol. aquat. trop._, (9):1-16, 4 fig.

--. 1969. Ocorrência de una espécie de _Cottunculus_ (Pisces, Percifor-mes, Cottunculidae) em aguas Angolanas. _Notas Cent. Biol. aquat. trop._, (13):1-15, 6 fig.

FRANCA, P. da; PICCIOCHI, P. 1958. Contribuçâo para o conhecimento dos Pomadasyidae de Angola. _Mems Jta Invest. Ultramar_, (2)4:145-195, 7 fig.

FRANCA, P. da; COSTA, F. C. da; FRANCA, M. L. P., da. 1970. Contribuiçâo para o conhecimento da biologia dos Sciaenidae de Angola. _Notas Cent. Biol. aquat. trop._, (20):1-137.

FRANZ, V. 1910. Die japanischen Knochenfische der Sammlungen Haberer und Doflein. _Abh. bayer. Akad. Wiss._ (math.-phys. Kl.), Suppl. 4(1):1-135, 7 fig., 11 pl.

FRASER, J. H. 1956. Scottish plankton investigations. _Annls biol. Co-penh._ (1954), 11:26-27.

FRASER, T. H. 1972. Comparative osteology of the shallow water cardinal fishes (Perciformes: Apogonidae) with reference to the systematics and evolution of the family. _Ichthyol. Bull. Rhodes Univ._, (34):1-105, 4 pl.

--. 1973. The fish _Elops machnata_ in South Africa. _Spec. Publ. J.L.B. Smith Inst. Ichthyol._ (11):1-6.

FRASER, T. H; MAYER, G. F. 1978. Apogonidae, _in_ Fischer, W. (ed.), FAO species identification sheets for fishery purposes. Western Central Atlantic (Fishing Area 31). FAO, Rome, vol. 1-7, pag. var.

FRASER, T. H.; ROBINS, C. R. 1970. The R/V Pillsbury deep sea biological expedition to the Gulf of Guinea, 1964-65. 18. A new Atlantic genus of Cardinal fishes with comments on some species from the Gulf of Guinea. _Stud. trop. Oceanogr._, 4(2):302-315, 5 fig., tab.

FRASER-BRUNNER, A. 1931. Some interesting West African fishes, with descriptions of a new genus and two new species. _Ann. Mag. nat. Hist._, (10)8:217-225, fig. 1-4.

--. 1933. A revision of the chaetodont fishes of the subfamily Pomacan-thinae. _Proc. zool. Soc. Lond._, 1933:543-599, fig. 1-29, pl. 1 .

--. 1935. Notes on the plectognath fishes. I. A synopsis of the genera of the family Balistidae. _Ann. Mag. nat. Hist._, (10)15:658-663, 2 fig.

--. 1935. New or rare fishes from the Irish Atlantic slope. _Proc. R. Ir. Acad._, 42(B-9):319-326, fig. 1-5.

--. 1938. Notes on the classification of certain British fishes. _Ann. Mag. nat. Hist._, (11)2(11):410-416.

--. 1943. Notes on the plectognath fishes. VIII. The classification of the suborder Tetraodontoidea, with a synopsis of the genera. _Ann. Mag. nat. Hist._, (11)10:1-18, 4 fig.

--. 1949. Note on the electric rays of the genus _Torpedo_. _Ann. Mag. nat. Hist._, (12)2:943-947, 1 fig.

--. 1949. On the fishes of the genus _Euthynnus_. _Ann. Mag. nat. Hist._, (12)2:622-627.

--. 1949. A classification of the fishes of the family Myctophidae. _Proc. zool. Soc. Lond._, 118:1019-1106, 167 fig., pl. 1.

--. 1950. The fishes of the family Scombridae. _Ann. Mag. nat. Hist._, (12)3:131-163.

--. 1951. The ocean sunfishes (fam. Molidae). _Bull. Br. Mus. nat. Hist. (Zool.)_, 1, (6):89-121, 18 fig.

FRASER-BRUNNER, A.; PALMER, G. 1951. The gadid fishes of the genus _Mol-va. Ann. Mag. nat. Hist._, (12)4:188-192.

FREDERIKSEN, K. 1957. Report to the Government of Liberia on the development of marine fisheries. FAO Rep., (715):25 p. (mimeo).

FREDERIKSEN, R. D. 1976. Retinal tapetum containing discrete reflectors and photoreceptors in the bathypelagic teleosts Omosudis lowei. Vidensk. Meddr dansk naturh. Foren., 139:109-146, fig. 1-30.

FREMINVILLE, C. P. P. 1813. Description de quelques nouvelles espèces de poissons de l'ordre des Branchiostèges. Nouv. Bull. Sci. (Soc. philomath.) Paris, 3(67):249-255.

FREON, P.; STEQUERT, B. 1978. Les poissons pélagiques côtiers au Sénégal: structure démographique des captures des sardinières dakarois en 1976. Centre Rech. océanogr. Dakar-Thiaroye, Archive 57, (mimeo).

--. 1978. Note sur la présence de Sardina pilchardus (Walb.) au Sénégal. Symposium Canaries Current: Upwelling and living resources (11-14 April 1978, Las Palmas): Docum. scient. provis. Centre Rech. océanogr. Dakar-Thiaroye, (67): 12 p., 7 fig., 8 tab. (mimeo).

--. 1979. Note sur la présence de Sardina pilchardus (Walb.) au Sénégal: étude de la biométrie et interprétation. Cybium, 3(2):65-90.

--. 1982. Note sur la présence de Sardina pilchardus (Walb.) au Sénégal: étude de la biométrie et interprétation. Rapp. P.-v. Réun. Cons. perm. int. Explor. Mer, 180:345-349.

FREON, P.; BOELY, T.; STEQUERT, B. 1979. Les poissons pélagiques côtiers au Sénégal: relations taille/poids des principales espèces d'intérêt commercial. Annex 13, p. 114-121. In: Anon., 1979.

FREON, P.; STEQUERT, B.; BOELY, T. 1978. La pêche des poissons pélagiques côtiers des Iles Bissagos au nord de la Mauritanie: descriptions et interactions des pêcheries. Symposium Canaries Current: Upwelling and living resources (11-14 April 1978, Las Palmas), Paper no 93 (mimeo).

--. 1979. Les pêches sénégalaises. Description et analyse des captures et des rendements des principales espèces pélagiques côtières. Annex 3, p. 22-57. In: Anon., 1979.

--. 1980. La pêche des poissons pélagiques côtiers de l'ouest des Iles Bissagos au nord de la Mauritanie: description des types d'exploitation. Cah. ORSTOM, Océanogr., 16(3):209-227.

FREON, P.; STEQUERT, B.; SOW, I. 1978. Les poissons pélagiques côtiers au Sénégal: recueil des statistiques de la pêche des sardiniers dakarois en 1977. Centre Rech. océanogr. Dakar-Thiaroye, Archive 58: 32 p. (mimeo).

FRICKE, R. 1981. Revision of the genus Synchiropus (Teleostei: Callionymidae). Theses zoologicae, Braunschweig, 194 p.

--. 1983. A new species of the genus Callionymus from St Helena (Teleostei: Callionymidae). Annali Mus. civ. Stor. natur. Giacomo Doria, 84: 393-399.

--. 1985. Protogrammus, a new genus of callionymid fishes, with a redescription of P. sousai from the eastern Atlantic. Jap. J. Ichthyol., 32(3):294-298, 3 fig.

--. 1985. Polytypy of Synchiropus agassizi (Goode & Bean, 1888) with the description of a new subspecies from the western Atlantic (Teleostei: Callionymidae). Annali Mus. civ. Stor. nat. Giacomo Doria, 85:235-249, 3 fig., 3 tab.

--. 1986. Callionymidae, in P.J.P. Whitehead, M.L. Bauchot, J.C. Hureau, J. Nielsen & E. Tortonese, Fishes of the north-eastern Atlantic and the Mediterranean/Poissons de l'Atlantique du Nord-Est et de la Méditerranée, Paris, Unesco, 3:1086-1093, 9 + 13 fig.

--. 1986. Draconettidae, in P.J.P. Whitehead, M.L. Bauchot, J.C. Hureau, J. Nielsen & E. Tortonese, Fishes of the North-eastern Atlantic and the Mediterranean/Poissons de l'Atlantique du Nord-Est et de la Méditerranée, Paris, Unesco, 3:1094-1095, 1 fig.

FRICKE, R.; SMITH-VANIZ, W. F. 1981. Callionymidae, in Fischer, W.,

Bianchi, G. & W.B. Scott (ed.), FAO species identification sheets for fishery purposes. Eastern Central Atlantic; fishing areas 34, 47 (in part). Canada Funds-in-Trust. Ottawa, Department of Fisheries and Oceans Canada, by arrangement with the Food and Agriculture Organization of the United Nations, vol. 1-7: pag. var.

FRIES, B. F. 1837. Pterycombus, ett nytt fisk-slägte från Ishafvet. K. svenska VetenskAkad. Handl., 1837:14-22, pl. 2.

FRITSCH, G. 1886. Ergebnisse der Vergleichungen an den elektrischen Organen der Torpedineen. Arch. Anat. Physiol. (Physiol.), 1886:358-370, 1 fig.

FRITZSCHE, R. A. 1978. Scombridae - mackerels and tunas, p. 61-167. In: Development of fishes of the Mid-Atlantic Bight. An atlas of egg, larval and juvenile stages. 5. Chaetodontidae through Ophidiidae. U.S. Fish Wildl. Serv.. biol. Serv. Program, FWS/OBS-78/12. 340 p., 178 fig.

--. 1981. Aulostomidae, Fistulariidae, in Fischer, W, Bianchi, G. & W.B. Scott (ed.), FAO species identification sheets for fishery purposes. Eastern Central Atlantic; fishing areas 34, 47 (in part). Canada Funds-in-Trust. Ottawa, Department of Fisheries and Oceans Canada, by arrangement with the Food and Agriculture Organization of the United Nations, vol. 1-7: pag. var.

FRØILAND, Ø. 1972. Funn av liten laksetobis ved Bjørnøya. Fauna, Oslo, 25:183-185, 1 fig.

FROST, G. A. 1926. A comparative study of the otoliths of the neopterygian fishes (continued). Orders Haplomi, Heteromi, Iniomi, Lyomeri, Hypostomides, Salmopercae, Synentognathi, Microcyprini, Solenichthyes. Ann. Mag. nat. Hist., (9)18:465-483, pl. 20-22.

--. 1927. A comparative study of the otoliths of the neopterygian fishes. Pt VIII. Ann. Mag. nat. Hist., (9)20:298-315, pl. 5.

--. 1928. A comparative study of the otoliths of the neopterygian fishes (continued). Pt IX. Ann. Mag. nat. Hist., (10)1(4):451-456, pl. 17.

--. 1928. A comparative study of the otoliths of the neopterygian fishes. Pt. X. Ann. Mag. nat. Hist., (10)2:328-331, 1 pl.

--. 1929. A comparative study of the otoliths of the neopterygian fishes. Ann. Mag. nat. Hist., (10)4(19):120-130, pl. 1.

--. 1930. A comparative study of the otoliths of the neopterygian fishes (concluded). Pt XIV. Ann. Mag. nat. Hist., (10)5:621-627, pl. 23.

FULTON, T. W. 1901. Ichthyological notes. Rep. Fishery Bd Scotl., (19): 290.

FURNESTIN, J. 1943. Contribution à l'étude biologique de la sardine atlantique. (Sardina pilchardus Walb.). Revue Trav. Off. (scient. tech.) Pêch. marit., 13:221-234.

--. 1947. Premières observations sur la biologie de la sardine marocaine (Sardina pilchardus Walb.) C.r. Séanc. mens. Soc. Sci. nat. phys. Maroc, 13:88-89.

--. 1950. Etude comparative de quelques caractères métriques des sardines du Golfe de Gascogne et du Maroc. Rapp. P.-v. Réun. Cons. perm. int. Explor. Mer., 126:37-42 (also Mém. Off. scient. tech. Pêch. marit., (14):37-42).

--. 1950. Premières observations sur la biologie de la sardine marocaine. Rap. P.-v. Cons. perm. int. Explor. Mer, 126:57-61 (also Mém. Off. scient. tech. Pêch. marit. (spec. ser.),(14):57-61).

--. 1950. Les pêches maritimes et les industries du poisson au Maroc. In: Congrès pêches et des pêcheries dans l'Union française d'outre-mer. (October 1950, Marseille), Inst. Colonial de Marseille:90-103.

--. 1952. L'influence des variations de la température sur les migrations des sardines du Maroc. Annls biol. Copenh., 8:82.

--. 1953. Ultra-sons et pêche à la sardine au Maroc, ch. II. Observations diverses sur le comportement de la sardine en fonction de certains facteurs du milieu. <u>Bull. Inst. Pêch. marit. Maroc</u>, (1):38-49.

--. 1955. Remarques sur quelques caractères de la sardine de Mauritanie (<u>Sardina pilchardus</u> Walb.). <u>Rapp. P.-v. Cons. perm. int. Explor. Mer.</u>, 137:24-25.

--. 1955. La ponte de la sardine et de l'anchois dans les eaux marocaines au cours des années 1951-2. <u>Rapp. P.-v. Cons. perm. int. Explor. Mer</u>, 137:25-28.

FURNESTIN, J.; COUPE, R. 1950. Les caractéristiques morphologiques des anchois (<u>Engraulis encrasicholus</u> Linné) du Maroc. <u>J. Cons. perm. int. Explor. Mer</u>, 16:182-184.

FURNESTIN, J.; FURNESTIN, M. L. 1959. La reproduction de la sardine et de l'anchois des côtes atlantiques du Maroc (saison et aires de ponte). <u>Revue Trav. Inst. Pêch. marit.</u>, 23(1):70-104.

--. 1970. La sardine marocaine et sa pêche. Migrations trophiques et génétiques en relation avec l'hydrologie et le plancton. <u>Rapp. P.-v. Réun. cons. perm. int. Explor. Mer.</u>, 159:165-175.

FURNESTIN, J.; VINCENT, A. 1955. Persistance de la phase migratoire chez l'alose marocaine adaptée aux eaux douces. <u>C. r. hebd. Séanc. Acad. Sci. Paris</u>, 240:355-357.

FURNESTIN, J.; DARDIGNAC, J.; MAURIN, C.; VINCENT, A.; COUPE, R.; BOUTIERE, H. 1958. Données nouvelles sur les poissons du Maroc Atlantique. <u>Revue Trav. Inst. Pêch. marit.</u>, 22(4):379-493, 75 fig.

FURSA, T. J.; MOVCHAN, Y. U. 1978. The fish fauna of the south-west coast of India. <u>Vop. Ikhtiol.</u>, 18:387-398, 3 fig., 2 tab. (Engl. transl. <u>J. Ichthyol.</u>, 18:351-361).

GAETANI, D. de. 1937. Contributo alla conoscenza dello sviluppo postembrionale in <u>Apogon imberbis</u>. <u>Memorie R. Com. talassogr. ital.</u>, (243): 1-14, 10 fig.

GAIMARD, J. P. 1842-1856. Voyages en Scandinavie, en Laponie, au Spitzberg et aux Féroé sur la corvette "Recherche". Poissons, Section VII Zoologie, Plates. Paris. [Poissons: pl. 1-16 + 18-22].

GALL, J. (see LE GALL, J.).

GANKOV, A. A. 1959. Aimed fishing for sardines. <u>Ryb. Khoz.</u>, 35(1):37-42 (in Russian).

GARCIA CABRERA, R. C.; NAVARRO, F. de P. 1955. Premières données sur la sardine de Ténérife (Iles Canaries). <u>Notes Rapp. Inst. scient. tech. Pêch. marit.</u>, (n.s.)11:37-56.

GARMAN, S. 1884. New sharks: <u>Chlamydoselachus anguineus</u> and <u>Heptranchias pectorosus</u>. <u>Bull. Essex Inst.</u>, 16:3-15, 1 pl.

--. 1885. The generic name of the Pastinacas or "Sting-Rays". <u>Proc. U.S. Nat. Mus.</u>, 8(14):221-224.

--. 1888. On the lateral line system of the Selachia and Holocephala. <u>Bull. Mus. comp. Zool. Harv.</u>, 17:57-120, pl. 1-53.

--. 1896. Report on the fishes collected by the Bahama Expedition of the State University of Iowa, under Professor C.C. Nutting in 1883. <u>Bull. Labs nat. Hist. St. Univ. Ia</u>, 4(1):76-93, pl. 1-4.

--. 1899. Reports on an exploration off the west coasts of Mexico, Central and South America, and off the Galapagos Islands in charge of Alexander Agassiz, by the U.S. Fish Commission Steamer "Albatross" during 1891, Lieut. Commander Z.L. Tanner, U.S.A. commanding. XXVI. The Fishes. <u>Mem. Mus. comp. Zool. Harv.</u>, 24:1-431, pl. 1-85 + A-M, 1 map.

--. 1901. Genera and families of the chimaeroids. <u>Proc. New Engl. zool. Club</u>, 2:75-77.

--. 1903. Some fishes from Australia. <u>Bull. Mus. comp. Zool. Harv.</u>, 39

(8):229-241, pl. 1-5.

--. 1904. The Chimaeroids (Chismopnea Raf., 1815; Holocephala Müll., 1834), especially Rhinochimaera and its allies. Bull. Mus. comp. Zool. Harv., 41(2):243-272, pl. 1-15.

--. 1906. New Plagiostomia. Bull. Mus. comp. Zool. Harv., 46(11):203-208.

--. 1908. New Plagiostomia and Chismopnea. Bull. Mus. comp. Zool. Harv., 51(9):251-256.

--. 1911. The Chismopnea (Chimaeroids). Mem. Mus. comp. Zool. Harv., 40: 79-101.

--. 1913. The Plagiostomia (Sharks, Skates and Rays). Mem. Mus. comp. Zool. Harv., 36: xiii + 528, 77 pl.

GARNAUD, J. 1950. La reproduction et l'incubation branchiale chez Apogon imberbis. Bull. Inst. océanogr. Monaco, (977):10 p., 7 fig.

--. 1960. La ponte, l'éclosion, la larve du Balistes capriscus Linné, 1758. Bull. Inst. océanogr. Monaco, (1169):6 p., 2 fig.

GARRICK, J. A. F. 1971. Harriotta raleighana, a long-nosed Chimaera (Family Rhinochimaeridae), in New Zealand waters. Jl R. Soc. N.Z., 1(3/4):203-213, fig. 1.

GARRICK, J. A. F.; INADA, T. 1975. Dimensions of long-nosed Chimaera Harriotta raleighana from New Zealand. N.Z. J. mar. freshwat. Res., 9(2):159-167, fig. 1-3.

GARRICK, J. A. F.; MORELAND, J. M. 1968. Notes on a bramble shark, Echinorhinus cookei, from Cook Strait, New Zealand. Rec. Dom. Mus.. Wellington, 6:133-139.

GATTI, M. A. 1903. Ricerche sugli organi luminosi dei pesci. Annali Agric., 233:7-126,1 pl.

GAUDECHOUX, J. P.; RICHER DE FORGES, B. 1983. Inventaire ichthyologique des eaux mauritaniennes. Docum. Cent. natn. Rech. océanogr. Pêch.. Nouadhibou, (3):1-22.

GAY, C. 1848. Historia física y política de Chile. Zoología, 2. Paris, Santiago, 372 p. (Atlas separate).

GEHRINGER, J. W. 1956. Observations on the development of the Atlantic sailfish Istiophorus americanus (Cuvier), with notes on an unidentified species of istiophorid. Fishery Bull. Fish. Wildl. Serv. U.S., 57(110):139-171, 40 fig.

--. 1959. Early development and metamorphosis of the ten-pounder Elops saurus Linnaeus. Fishery Bull. Fish. Wildl. Serv.. U.S., 59(155): 615-647.

--. 1970. Young of the Atlantic sailfish, Istiophorus platypterus. Fishery Bull., 68(2):177-190, 5 fig.

GEISTDOERFER, P. 1972. Dentition pharyngienne et tubes digestifs de quelques Macrouridae (Téléostéens, Gadiformes). Bull. Mus. natn. Hist. nat.. Paris, (3)(79)(Zool. 58):901-916.

--. 1973. Régime alimentaire des Macrouridae (Téléostéens, Gadiformes) atlantiques et méditerranéens, en relation avec la morphologie du tube digestif. Bull. Mus. natn. Hist. nat.. Paris, (3)(161) (Ecol. gén. 17):285-295.

--. 1973. Sur quelques particularités histologiques de l'intestin de Chalinura mediterranea (Macrouridae, Gadiformes). C. r. hebd. Séanc. Acad. Sci.. Paris, (D)276:331-333, 1 pl.

--. 1974. Histologie du tube digestif de Coelorhynchus coelorhynchus et de Chalinura mediterranea (Macrouridae, Gadiformes, Pisces). Bull. Mus. natn. Hist. nat.. Paris, (3)(222)(Zool. 150):641-660, 3 pl.

--. 1975. Ecologie alimentaire des Macrouridae (Téléostéens, Gadiformes). Alimentation. Morphologie et histologie de l'appareil digestif. Place des Macrouridae dans la chaine alimentaire profonde. Thèse Doct. Etat, Univ. Paris VI, 1975, CNRS no A.O. 11.826, 315 p., 39 fig., 8 pl. Arch. Docum. micro-édition, Mus. nat. Hist. nat., Paris. (Sci. nat.) SN 75 601 226, 1979.

--. 1977. Etude biomécanique du mouvement de fermeture des mâchoires chez Ventrifossa occidentalis et Coelorhynchus coelorhynchus (Macrouridae, Gadiformes). Bull. Mus. natn. Hist. nat., Paris, (3)(481)(Zool. 338):993-1020, 8 fig., 1 pl.

--. 1978. Ecologie alimentaire des Macrouridae. Revue Trav. Inst. Pêch. marit., 42(3):177-260, 21 fig., 5 pl.

--. 1980. Connaissances nouvelles sur la biologie de la famille des Macrouridae (Téléostéens, Gadiformes). Vie Milieu 1978-1979, (AB) 28-29(2):323-351.

--. 1986. Macrouridae, in P.J.P. Whitehead, M.L. Bauchot, J.C. Hureau, J. Nielsen & E. Tortonese. Fishes of the North-eastern Atlantic and the Mediterranean/Poissons de l'Atlantique du Nord-Est et de la Méditerranée, Paris, Unesco, 2:644-676, 34 fig.

GEISTDOERFER, P.; HUREAU, J.-C. 1985. Moridae, p. 306-309, in: Fischer, W. & Hureau, J.C. (ed.). FAO Species identification sheets for fishery purposes. Southern Ocean (Fishing Areas 48, 58 and 88). Prepared and printed with the support of the Commission for the Conservation of Antarctic Marine Living Resources (CCAMLR). Rome, Vol. 2.

GEISTDOERFER, P.; RANNOU, M. 1971. A propos des Chalinura méditerranéens (Téléostéens, Macrouridae). Bull. Mus. natn. hist. nat., Paris, (2)42(5):1009-1018, 1 fig.

--. 1972. Poissons benthiques récoltés en Méditerranée occidentale par le N.O. Jean Charcot. Bull. Mus. natn. Hist. natn., Paris, (3)(25)(Zool. 19):101-110, 3 fig.

GEISTDOERFER, P.; HUREAU, J.-C.; RANNOU, M. 1970. Deux poissons abyssaux nouveaux capturés dans l'Atlantique Nord et Est: Bathytyphlops azorensis n.sp. (Ipnopidae) et Lycenchelys labradorensis n. sp. (Zoarcidae). Bull. Mus. natn. Hist. nat. Paris, (2)42(3):452-459, 3 fig.

--. 1971. Liste préliminaire des espèces de poissons de profondeur récoltées au cours de la campagne "Noratlante" du N.O. Jean Charcot en Atlantique Nord (août-octobre 1969). Bull. Mus. natn. Hist. nat., Paris, (2)42(6):1177-1185.

--. 1971. Liste préliminaire des poissons:363-368. In: Résultats campagne Noratlante N.O. Jean Charcot (3 août-2 novembre 1969). Publication CNEXO, no 01-1971:1-385.

GEOFFROY SAINT-HILAIRE, E. 1817. Poissons de la Mer Rouge et de la Méditerranée. In: Description de l'Egypte... Planches Histoire Naturelle, 1: pl. 18-27. Paris.

GEOFFROY SAINT-HILAIRE, I. 1827. Histoire naturelle des poissons du Nil (suite). p. 265-310. Suivi de: Histoire naturelle des poissons de la Mer Rouge et de la Méditerranée p. 311-350. In: Description de l'Egypte... Histoire naturelle, 1, Paris.

GEORGE, C. J.; ATHANASSIOU, V. 1966. Additions to the check-list of the fishes of the coastal waters of Lebanon. Misc. Pap. nat. Sci. Am., Univ. Beirut, 5:6-8.

GEORGE, C. J.; ATHANASSIOU, V.; TORTONESE, E. 1970. The presence of a third species of the genus Sphyraena (Pisces) in the marine waters of Lebanon. Annali Mus. civ. Stor. nat. Giacomo Doria, 78:256-363.

GEORGHIEV, J. M. 1966. Composition en espèces et caractéristique des Gobiidés (Pisces) en Bulgarie. Izv. nauchnoizsled. Inst. Rib. Stop. Okeanogr. (Varna), 7:159-226, 57 fig., tab.

GERLOTTO, F. 1976. Biologie de Ethmalosa fimbriata (Bowdich) en Côte d'Ivoire. II. Etude de la croissance en lagune par la méthode Petersen. Docums scient. Centre Rech. océanogr. Abidjan, 7(2):1-27.

--. 1977. Biologie de Ethmalosa fimbriata (Bowdich) en Côte d'Ivoire. III. Etude des migrations en lagune Ebrié. Docums scient. Centre

Rech. océanogr. Abidjan, 10(2):3-41.

GERO, J. B. de. 1961. Composition biochimique de la sardine (Sardina pilchardus) des côtes atlantiques du Maroc. Bull. Inst. Pêch. marit. Maroc, (6):69-75.

GHAZEL A.L. 1975. Faune du Maroc: poissons d'eau douce du Maroc. Institut d'études et de recherches pour l'arabisation. Série albums didactiques. 61 p., fig. [in Arabic and French].

GHENO, Y. 1968. Détermination de l'âge et croissance de Sardinella aurita Val. de la région de Pointe-Noire. Docums scient. provis. Centre Rech. océanogr. Pointe-Noire, (430):21 p. (mimeo).

--. 1970. Note sur les sardinelles immatures de l'estuaire du Gabon. Docums scient. provis. Centre Rech. océanogr. Pointe-Noire, (n.s.) (6):8 p. (mimeo).

--. 1971. La pêche des poissons côtiers de surface à Pointe-Noire en 1969. Docums scient. provis. Centre Rech. océanogr. Pointe-Noire, (n.s.)(13):10 p. (mimeo).

--. 1973. Première estimation de la mortalité des sardinelles des côtes congolaises. Docums scient. provis. Centre Rech. océanogr. Pointe-Noire, (n.s.)(32):16 p. (mimeo).

--. 1975. Nouvelle étude sur la détermination de l'âge et de la croissance de Sardinella aurita Val. dans la région de Pointe-Noire. Cah. ORSTOM. Océanogr., 13(3):251-262.

GHENO, Y.; CAMPOS ROSADO, F. 1972. Distributions de fréquence de longueur des sardinelles Sardinella aurita Val. et Sardinella eba Val. débarquées à Pointe-Noire et à St Paul de Loanda (juin 1969 - octobre 1970). Docums scient. provis. Centre Rech. océanogr. Pointe-Noire, (26):14 p. (mimeo).

GHENO, Y.; FONTANA, A. 1973. La pêche des sardinelles à Pointe-Noire en 1970, 1971 et 1972. Docums scient. provis. Centre Rech. océanogr. Pointe-Noire, (33):9 p. (mimeo).

GHENO, Y.; LE GUEN, J. C. 1968. Détermination de l'âge et croissance de Sardinella eba (Val.) dans la région de Pointe-Noire. Cah. ORSTOM. Océanogr., 6(2):69-82.

GHENO, Y.; MARCILLE, J. 1971. La pêche de Sardinella aurita par les thoniers à l'appât vivant sur les côtes du Congo et du Gabon. Docums scient. provis. Centre Rech. oceanogr. Pointe-Noire, (n.s.)(12):20 p. (mimeo).

GHENO, Y.; POINSARD, F. 1968. Observations sur les jeunes sardinelles de la baie de Pointe-Noire (Congo). Cah. ORSTOM. Océanogr., 6(2):53-67.

--. 1969. La pêche des sardinelles (Sardinella aurita Val. et Sardinella eba Val.) à Pointe-Noire de 1964 à 1968. Cah. ORSTOM. Océanogr., 7(3):69-93.

GHENO, Y.; RIBEIRO, F. 1968. Note sur trois échantillons de Sardinella eba (Val.) en provenance de St-Paul de Loanda. Docums scient. provis. Centre Rech. océanogr. Pointe-Noire, (436):9 p. (mimeo).

GIBBS, R. H. Jr. 1957. A taxonomic analysis of Myctophum affine and M. nitidulum, two lantern-fishes previously synonymised, in the western north Atlantic. Deep Sea Res., 4:230-237.

--. 1959. A synopsis of the post larvae of western Atlantic lizard-fishes (Synodontidae). Copeia, 1959(3):232-236.

--. 1960. The stomiatoid fish genera Eustomias and Melanostomias in the Pacific, with descriptions of two new species. Copeia, 1960 (3):200-203.

--. 1960. Alepisaurus brevirostris, a new species of lancetfish from the western North Atlantic, Breviora, (123):1-14, fig. 1-3.

--. 1964. Fam. Astronesthidae. In: Fishes of the western North Atlantic. Mem. Sears Fdn mar. Res., 1(4):311-350, fig. 77-91.

--. 1964. Fam. Idiacanthidae. In: Fishes of the western North Atlantic.

Mem. Sears Fdn mar. Res., 1(4):512-522, 4 fig.

--. 1968. Photonectes munificus, a new species of melanostomiatid fish from the South Pacific Subtropical Convergence, with remarks on the convergence fauna. Contr. Sci., (149):1-6, 1 fig.

--. 1969. Taxonomy, sexual dimorphism, vertical distribution and evolutionary zoogeography of the bathypelagic fish genus Stomias (family Stomiatidae). Smithson. Contr. Zool., (31):1-25.

--. 1971. Notes on fishes of the genus Eustomias (Stomiatoidei, Melanostomiatidae) in Bermuda waters, with the description of a new species. Proc. biol. Soc. Wash., 84(29):235-244.

--. 1978. Exocoetidae, in Fischer, W. (ed.), FAO species identification sheets for fishery purposes. Western Central Atlantic (Fishing Area 31). Rome, FAO, vol. 1-7, pag. var.

--. 1981. Alepisauridae, Exocoetidae, in Fischer, W., Bianchi, G. & W.B. Scott (ed.), FAO species identification sheets for fishery purposes. Eastern Central Atlantic; fishing areas 34, 47 (in part). Canada Funds-in-Trust. Ottawa, Department of Fisheries and Oceans Canada, by arrangement with the Food and Agriculture Organization of the United Nations, vol. 1-7: pag. var.

--. 1984. Astronesthidae, in Whitehead, P.J.P., Bauchot, M.L., Hureau, J.C., Nielsen, J. & E. Tortonese, Fishes of the North-eastern Atlantic and the Mediterranean/Poissons de l'Atlantique du Nord-Est et de la Méditerranée, Paris, Unesco, 1:325-335, 13 + 5 fig., 13 maps.

--. 1984. Chauliodontidae, in Whitehead, P.J.P., Bauchot, M.L., Hureau, J.C., Nielsen, J. & E. Tortonese, Fishes of the North-eastern Atlantic and the Mediterranean/Poissons de l'Atlantique du Nord-Est et de la Méditerranée, Paris, Unesco, 1:336-337, 2 fig., 2 maps.

--. 1984. Stomiidae, in Whitehead, P.J.P., Bauchot, M.L., Hureau, J.C., Nielsen, J. & E. Tortonese, Fishes of the North-eastern Atlantic and the Mediterranean/Poissons de l'Atlantique du Nord-Est et de la Méditerranée, Paris, Unesco, 1:338-340, 3 + 5 fig., 4 maps.

--. 1984. Melanostomiidae, in Whitehead, P.J.P., Bauchot, M.L., Hureau, J.C., Nielsen, J. & E. Tortonese, Fishes of the North-eastern Atlantic and the Mediterranean/Poissons de l'Atlantique du Nord-Est et de la Méditerranée, Paris, Unesco, 1:341-365, 30 + 37 fig., 30 maps.

--. 1984. Malacosteidae, in Whitehead, P.J.P., Bauchot, M.L., Hureau, J.C., Nielsen, J. & E. Tortonese, Fishes of the North-eastern Atlantic and the Mediterranean/Poissons de l'Atlantique du Nord-Est et de la Méditerranée, Paris, Unesco, 1:366-370, 5 + 1 fig., 7 maps.

--. 1984. Idiacanthidae, in Whitehead, P.J.P., Bauchot, M.L., Hureau, J.C., Nielsen, J. & E. Tortonese, Fishes of the North-eastern Atlantic and the Mediterranean/Poissons de l'Atlantique du Nord-Est et de la Méditerranée, Paris, Unesco, 1:371-372, 1 fig., 1 map.

GIBBS, R. H., Jr.; COLLETTE, B. B. 1959. On the identification, distribution and biology of the dolphins, Coryphaena hippurus and C. equiselis. Bull. mar. Sci. Gulf. Caribb., 9(2):117-152, 18 fig.

--. 1967. Comparative anatomy and systematics of the tunas, genus Thunnus. Fishery Bull., 66(1):65-130.

GIBBS, R. H., Jr.; HURWITZ, B. A. 1967. Systematics and zoogeography of the stomiatoid fishes, Chauliodus pammelas and C. sloani, of the Indian Ocean. Copeia, 1967(4):798-805, 3 fig.

GIBBS, R. H., Jr.; MORROW, J. E. 1973. Astronesthidae, p. 126-129, in Hureau, J.C. & Th. Monod (ed.), Check-list of the fishes of the north-eastern Atlantic and of the Mediterranean/Catalogue des poissons du nord-est Atlantique et de la Méditerranée. Paris, Unesco, 2 vol.

GIBBS, R. H., Jr.; STAIGER, J. C. 1970. Eastern tropical Atlantic flyingfishes of the genus <u>Cypselurus</u> (Exocoetidae). <u>Stud. trop. Ocea-</u> <u>nogr.</u>, 4(2):432-466, 9 fig.

GIBBS, R. H., Jr.; WILIMOVSKY, N.J. 1966. Fam. Alepisauridae. <u>Mem. Sears</u> <u>Fdn mar. Res.</u>, 1(5):482-497, fig. 173-176.

GIBBS, R. H., Jr.; GOODYEAR, R. H.; KEENE, M. J.; BROWN, D. W. 1971. Biological studies of the Bermuda Ocean Acre. II. Vertical distribution and ecology of the lanternfishes (family Myctophidae). Rep. U.S. Navy Underwater Systems Center, Smithson. Instn Washington,:1-141 p., 71 fig., 85 tab.

GIBERT, A. M. 1913. Fauna Ictiológica de Catalunya. Catalech rahonat dels peixos observats en el litoral y en les aygues dolces catalanes. J. Bartra Labor, Barcelona, 96 p.

GIBSON, R. N. 1970. Observations on the biology of the giant goby <u>Gobius</u> <u>cobitis</u> Pallas. <u>J. Fish Biol.</u>, 2:281-288.

--. 1972. The vertical distribution and feeding relationships of inter-tidal fish on the Atlantic coast of France. <u>J. Anim. Ecol.</u>, 41:189-207.

GIBSON, R. N.; EZZI, I. A. 1978. The biology of a Scottish population of Fries' goby, <u>Lesueurigobius friesii</u>. <u>J. Fish Biol.</u>, 12:371-389.

GIGLIOLI, E. H. 1880. Elenco dei Mammiferi..., e Catalogo degli Anfibi e dei Pesci italiani. <u>In</u>: Catalogo... Sezione italiana, Esposizione intern. di Pesca. Berlino, 1880, (11):63-177 (also sep., Firenze, 1880:18-55).

--. 1882. New deep-sea fish from the Mediterranean. <u>Nature. Lond.</u>, (27): 198-199.

--. 1883. Intorno a due nuovi pesci del Golfo di Napoli. <u>Zool. Anz.</u>, 6: 397-400.

--. 1884. Esplorazione talassografica del Mediterraneo eseguita sotto gli auspici del Governo italiano, p. 199-291. <u>In</u>: Giglioli, E.H & R. Issel Pelagos. Saggi sulla vita e sui prodotti del mare. Genoa, 436 p., 32 fig., 1 map.

--. 1893. Di una nuova specie di Macruride appartenente alla fauna abis-sale del Mediterraneo. <u>Zool. Anz.</u>, 16(428):343-345.

GILBERT, C.H. 1890. A preliminary report on the fishes collected by the steamer "Albatross" on the Pacific coast of North America during the year 1889, with descriptions of twelve new genera and ninety-two new species. <u>Proc. U.S. natn. Mus.</u>, 13:49-126.

--. 1891. Scientific results of explorations by the U.S. Fish Commission steamer "Albatross" XXI. Descriptions of apodal fishes from the tro-pical Pacific. <u>Proc. U.S. natn. Mus.</u>, 14:347-352.

--. 1891. Scientific results of the explorations by the U.S. Fish Com-mission steamer "Albatross". XXII. Descriptions of thirty-four new species of fishes collected in 1888 and 1889, principally among the Santa Barbara Islands and in the Gulf of California. <u>Proc. U.S. natn. Mus.</u>, 14:539-566.

--. 1891. Description of a new species of eel (<u>Sphagebranchus kendalli</u>). <u>Bull. U.S. Fish Commn</u> (1889), 9:310.

--. 1895. The ichthyological collections of the steamer "Albatross" during the year 1890 and 1891. <u>Rep. U.S. Commnr Fish.</u>, 19(6):393-476.

--. 1905. The deep-sea fishes of the Hawaiian Islands. <u>Bull. U.S. Fish Commn</u> (1903), 23(2):577-713, fig. 230-276, pl. 66-101.

--. 1906. Certain scopelids in the collection of the Museum of Compara-tive Zoology. <u>Bull. Mus. comp. Zool. Harv.</u>, 46(14):255-263.

--. 1908. Reports on the scientific results of the U.S. Fish Commission steamer "Albatross", lantern fishes. <u>Mem. Mus. comp. Zool. Harv.</u>, 26(6):217-237, 6 pl.

--. 1911. Notes on lantern fishes from southern seas, collected by J.T. Nichols in 1906. <u>Bull. Am. Mus. nat. Hist.</u>, 30(2):13-19.

--. 1913. The lantern fishes of Japan. <u>Mem. Carneg. Mus.</u>, 6(2):67-107, pl. 11-14.

--. 1915. Fishes collected by the United States fisheries steamer "Albatross" in Southern California in 1904. <u>Proc. U.S. natn. Mus.</u>, 48(2075):305-380, pl. 14-22.

GILBERT, C. H.; BURKE, C. V. 1912. Fishes from Bering Sea and Kamchatka. <u>Bull. Bur. Fish., Wash.</u>, 30:31-96, 37 fig.

GILBERT, C. H.; CRAMER, F. 1897. Report on the fishes dredged in deep water near the Hawaiian Islands, with descriptions and figures of twenty-three new species. <u>Proc. U.S. natn. Mus.</u>, 19:403-435, pl. 36-48.

GILBERT, C. H.; HUBBS, C. L. 1916. Report on the Japanese macrouroid fishes collected by the United States fisheries steamer "Albatross" in 1906, with a synopsis of the genera. <u>Proc. U.S. natn. Mus.</u>, 51(2149):135-214, pl. 8.

--. 1920. The macrouroid fishes of the Philippine islands and the east Indies. <u>Bull. U.S. natn. Mus.</u>, Bull. 100, 1(7):369-588, 40 fig.

GILBERT, C. R. 1967. A revision of the hammer-head sharks. <u>Proc. U.S. natn. Mus.</u>, 119(3539):1-88, pl. 1-10.

--. 1973. Sphyrnidae, p. 32-34, <u>in</u>: Hureau, J.C. & Th. Monod (ed.), Check-list of the fishes of the north-eastern Atlantic and of the Mediterranean/Catalogue des poissons du nord-est Atlantique et de la Méditerranée. Paris, Unesco, 2 vol.

GILCHRIST, J. D. F. 1902. Catalogue of fishes recorded from South Africa. <u>Mar. Invest. S. Afr.</u>, 1(6):97-179.

--. 1902. South African fishes. <u>Mar. Invest. S. Afr.</u>, 2:101-113, pl. 5-10.

--. 1903. Descriptions of new South African fishes. <u>Mar. Invest. S. Afr.</u>, 2:203-211, pl. 13-18.

--. 1904. The development of South African fishes. Part II. <u>Mar. Invest. S. Afr.</u>, 3:131-152, pl. 5-11.

--. 1906. Description of fifteen new South African fishes, with notes on other species. <u>Mar. Invest. S. Afr.</u>, 4:143-171, pl. 37-51.

--. 1913. Review of the South African Clupeidae (herrings) and allied families of fishes. <u>Mar. biol. Rep., Cape Tn</u>, (1):46-66.

--. 1914. Observation on the habits of some South African fishes. <u>Mar. biol. Rep. Cape Tn</u>, (2):90-115.

--. 1914. The snoek and allied fishes in South Africa. <u>Mar. biol. Rep., Cape Tn</u>, (2):116-125.

--. 1921. The reproduction of deep-sea fishes. <u>Ann. Mag. nat. Hist.</u>, (9)7:173-177, 1 pl.

--. 1921. Fisheries and marine biological survey. Report No 1 for the year 1920. <u>Rep. Fish. mar. biol. Surv. Un. S. Afr.</u>, (1): iv + 1-111, 9 pl., 4 map.

--. 1922. Deep-sea fishes procured by the S.S. "Pickle" (Part I). <u>Rep. Fish. mar. biol. Surv. Un. S. Afr.</u> (1921), 2 (Spec. Rep. 3):41-79, pl. 7-12.

--. 1922. Report. Annexure "A". List of Fishes, etc., procured. Annexure "B". Liste of stations of S.S. "Pickle". <u>Rep. Fish. mar. biol. Surv. Un. S. Afr.</u> (1921), 2:7-59.

GILCHRIST, J. D. F.; BONDE, C. von. 1924. The Stromateidae (Butter Fishes) collected by the S.S. "Pickle". <u>Rep. Fish. mar. biol. Surv. Un S. Afr.</u> (1922), 3(Spec. Rep. 4):1-12, pl. 17-19.

--. 1924. Deep-sea fishes procured by the S.S. "Pickle" (Part II). <u>Rep. Fish. mar. biol. Surv. Un. S. Afr.</u> (1922), 3(Spec. Rep. 7):1-24, 6 pl.

GILCHRIST, J. D. F.; THOMPSON, W. W. 1909. Descriptions of fishes from the coast of Natal. Pt. II. <u>Ann. S. Afr. Mus.</u>, 6(3):213-279.

--. 1911. Descriptions of fishes from the coast of Natal. Pt. III. Ann. S. Afr. Mus., 11:29-58.

--. 1914. Description of fishes from the Coast of Natal. Pt. IV. Ann. S. Afr. Mus., 13(3):65-95.

--. 1916. Descriptions of four new South African fishes. Mar. biol. Rep., Cape Tn, 3:56-61, 4 unnumbered fig.

--. 1917. A catalogue of the sea fishes recorded from Natal, part I. Ann. Durban Mus., 1:255-290; part II: ibid.:291-431.

GILL, T. N. 1858. Synopsis of the fresh water fishes of the western portion of the Island of Trinidad, W.I. Ann. Lyceum nat. Hist. N.Y., 6:363-430.

--. 1859. Description of a new generic form of Gobiinae from the Amazon river. Ann. Lyceum nat. Hist. N.Y., 7:45-48.

--. 1859. Description of Hyporhamphus, a new genus of fishes allied to Hemirhamphus Cuv. Proc. Acad. nat. Sci. Philad., 11:131.

--. 1859. Description of a third genus of Hemirhamphidae. Proc. Acad. nat. Sci. Philad., 11:155-157.

--. 1859. Description of a new genus of Salarianae from the West Indies. Proc. Acad. nat. Sci. Philad., 11:168.

--. 1860. Notes on the nomenclature of North American fishes. Proc. Acad. nat. Sci. Philad., 11:20-21.

--. 1860. Monograph of the genus Labrisomus Swainson. Proc. Acad. nat. Sci. Philad., 12, 102-108.

--. 1860. On the genus of Callionymus of authors. Proc. Acad. nat. Sci. Philad., (1859):128-130.

--. 1861. Monograph of the genus Labrax of Cuvier. Proc. Acad. nat. Sci. Philad., (1860):108-119.

--. 1862. Synopsis of the sub-family Clupeinae, with descriptions of new genera. Proc. Acad. nat. Sci. Philad., (1861):33-38.

--. 1862. Synopsis generum Rhyptici et affinium. Proc. Acad. nat. Sci. Philad., (1861):52-54.

--. 1862. Catalogue of the fishes of the eastern coast of North America, from Greenland to Georgia. Proc. Acad. nat. Sci. Philad., 13:1-63.

--. 1862. On several new generic types of fishes. Proc. Acad. nat. Sci. Philad., 13:77-78.

--. 1862. Revision of the genera of North America Sciaenidae. Proc. Acad. nat. Sci. Philad., 13:79-89.

--. 1862. Notes on some genera of fishes of the western coast of North America. Proc. Acad. nat. Sci. Philad., 13:164-168.

--. 1862. Analytical synopsis of the order Squali and revision of the nomenclature of the genera. Ann. Lyceum nat. Hist. N.Y., 7:367-413.

--. 1863. Note on the sciaenoids of California. Proc. Acad. nat. Sci. Philad., 14:16-18.

--. 1863. Synopsis of the family of cirrhitoids. Proc. Acad. nat. Sci. Philad., 14:102-122.

--. 1863. Description of a new species of Cirrhitus. Proc. Acad. nat. Sci. Philad., 14:122-124.

--. 1863. On the limits and arrangement of the family of scombroids. Proc. Acad. nat. Sci. Philad., 14:124-127.

--. 1863. Catalogue of the fishes of Lower California in the Smithsonian Institution, collected by Mr. J. Xanthus. Part I, Proc. Acad. nat. Sci. Philad., 14:140-151.

--. 1863. On a new genus of fishes allied to Aulorhynchus and on the affinities of the family Aulorhynchoidae, to which it belongs. Proc. Acad. nat. Sci. Philad., 14:233-235.

--. 1863. Remarks on the relations of the genera and other groups of Cuban fishes. Proc. Acad. nat. Sci. Philad., 14:235-242.

--. 1863. Catalogue of fishes of Lower California in the Smithsonian

Institution collected by Mr. J. Xanthus. Part II. Proc. Acad. Nat. Sci. Philad., 14:242-246.

--. 1863. Catalogue of the fishes of Lower California, in the Smithsonian Institution, collected by Mr. J. Xanthus. Part III, Proc. Acad. nat. Sci. Philad., 14:249-262.

--. 1863. Notice of a collection of the fishes of California presented to the Smithsonian Institution by Mr. Samuel Hubbard. Proc. Acad. nat. Sci. Philad., 14(6):274-282.

--. 1863. Note on some genera of fishes of western North America. Proc. Acad. nat. Sci. Philad., 14:329-333.

--. 1864. Catalogue of the fishes of Lower California, in the Smithsonian Institution, collected by Mr. J. Xanthus. Part IV, Proc. Acad. nat. Sci. Philad., 15(2):80-88.

--. 1864. Descriptions of some new species of Pediculati and on the classification of the group. Proc. Acad. nat. Sci. Philad., 15:88-92.

--. 1864. Descriptive enumeration of a collection of fishes from the western coast of Central America, presented to the Smithsonian Institution by Captain John M. Dow. Proc. Acad. nat. Sci. Philad., (1863):162-180.

--. 1864. On an unnamed generic type allied to Sebastes. Proc. Acad. nat. Sci. Philad., 15(5):207-209.

--. 1864. Descriptions of the gobioid genera of the western coast of temperate North America. Proc. Acad. nat. Sci. Philad., (1863):252-267.

--. 1864. On the gobioids of the eastern coast of the United States. Proc. Acad. nat. Sci. Philad., 15:267-271.

--. 1864. Note on the genera of Hemirhamphinae. Proc. Acad. nat. Sci. Philad., (1863):272-273.

--. 1864. Second contribution to the selachology of California. Proc. Acad. nat. Sci. Philad., 16:147-151.

--. 1864. Note on the paralepidoids and microstomatoids, and on some pecularities of arctic ichthyology. Proc. Acad. nat. Sci. Philad., 16:187-189.

--. 1864. Synopsis of the pleuronectoids of California and North Western America. Proc. Acad. nat. Sci. Philad., 16:194-198.

--. 1864. Description of a new generic type of pleuronectoids in the collection of the Geological Survey of California. Proc. Acad. nat. Sci. Philad., 16:198-199.

--. 1864. Notes on the labroids of the western coast of North America. Proc. Acad. nat. Sci. Philad., 15:221-224.

--. 1865. On a new family of fishes related to the blennioids. Ann. Lyceum nat. Hist. N.Y., 8:141-144.

--. 1877. The genus Sebastapistes, proposed for the reception of Scorpaena guttata Girard, Scorpaena strongia Cuv. & Val., and Sebastichthys cyanostigma Bleeker. In: T.H. Streets, Contributions to the natural history of the Hawaiian and Fanning Islands. Bull. U.S. natn. Mus., (7):62.

--. 1878. Synopsis of the pediculate fishes of the eastern coast of extratropical North America. Proc. U.S. natn. Mus., 1:215-224.

--. 1878. Note on the Ceratidae. Proc. U.S. natn. Mus., 1:228.

--. 1879. On the proper specific name of the common pelagic antennariid Pterophryne. Proc. U.S. natn. Mus., 1:223-226.

--. 1879. Note on the Maltheidae. Proc. U.S. natn. Mus., 1(1878):231-232.

--. 1883. Cryptopsaras couesii. Forest Streams, 1883:284.

--. 1884. Diagnosis of new genera and species of deep-sea fish-like vertebrates. Proc. U.S. natn. Mus., 6:253-260.

--. 1884. New families of fishes (Stephanoberycidae and Derichthyidae) recently added to the deep-sea fauna. Am. Nat., 18:433.

--. 1884. Synopsis of the plectognath fishes. Proc. U.S. natn. Mus., 7:411-427.

--. 1889. Gleanings among the pleuronectids and observations on the name Pleuronectes. Proc. U.S. natn. Mus., 11:593-606.

--. 1890. The characteristics of the Dactylopteroidea. Proc. U.S. natn. Mus., 13(818):243-248, pl. 19.

--. 1890. The osteological characteristics of the family Simenchelyidae. Proc. U.S. natn. Mus., 13:239-242.

--. 1892. Notes on the Tetraodontoidea. Proc. U.S. natn. Mus., 14:705-720.

--. 1893. A comparison of antipodal faunas. Mem. natn. Acad. Sci., 6:91-124.

--. 1896. On the application of the name Teuthis to a genus of fishes. Proc. U.S. natn. Mus., 18:179-189.

--. 1896. Lipophrys a substitute for Pholis. Am. Nat., 30:498.

--. 1905. Flying fishes and their habits. Rep. Smithson. Instn, (1904):495-515, 2 fig., 4 pl.

--. 1905. The tarpon and lady-fish and their relatives. Smithson. misc. Collns, 48(3):31-46, fig. 1-7, pl. 17-21.

--. 1909. Angler Fish: their kinds and ways. Rep. Smithson. Instn, (1908):565-615, fig. 1-49.

GILL, T. N.; RYDER, J. A. 1883. Diagnoses of new genera of nemichthyoid eels. Proc. U.S. natn. Mus., 6:260-262.

--. 1884. On the anatomy and relations of the Eurypharyngidae. Proc. U.S. natn. Mus., 6(17):262-273.

GILL, T. N.; SMITH, H. M. 1905. A new family of jugular acanthopterygians. Proc. biol. Soc. Wash., 18:249-250.

GILL, T. N.; TOWNSEND, C. H. 1897. Diagnoses of new species of fishes found in the Bering Sea. Proc. biol. Soc. Wash., 11:231-234.

GILTAY, L. 1928. Notes ichthyologiques. II. Une espèce nouvelle de Rhinobatus du Congo belge (Rhinobatus congolensis, nov. sp.). Annls Soc. r. zool. Belg., 59:21-27, fig. 1.

--. 1935. Notes sur quelques poissons marins du Congo Belge. Bull. Mus. r. Hist. nat. Belg., 11(36):1-15, 3 fig.

GINSBURG, I. 1932. A revision of the genus Gobionellus (family Gobiidae). Bull. Bingham oceanogr. Coll., 4:3-51, 7 fig.

--. 1933. Descriptions of new and imperfectly known species and genera of gobioid and pleuronectid fishes in the United States National Museum. Proc. U.S. natn. Mus., 82:9-23.

--. 1951. The eels of the northern Gulf coast of the United States and some related species. Tex. J. Sci., 3:431-485, fig. 1-16.

--. 1953. The taxonomic status and nomenclature of some Atlantic and Pacific populations of yellowfin and bluefin tunas. Copeia, 1953(1):1-10.

--. 1953. Western Atlantic scorpion fishes. Smithson. misc. Collns, 121(8):103 p., 6 fig.

--. 1955. Fishes of the family Percophididae from the coast of eastern United States and the West Indies, with descriptions of four new species. Proc. U.S. natn. Mus., 104(3347):623-639.

--. 1954. Four new fishes and on little-known species from the east coast of the United States and the Gulf of Mexico. J. Wash. Acad. Sci., 44(8):256-264, 6 fig.

GIORNA, M. E. 1809. Suite et conclusion du mémoire sur des poissons d'espèces nouvelles et de genres nouveaux. Memorie Accad. Sci. Torino (1805-08), 16:177-180.

GIOVENE, G. M., 1829. Di alcuni pesci del mare di Puglia. Mem. Soc. Ital., 20:21-42.

GIRARD, C. F. 1858. Notes upon various new genera and new species of fishes, in the museum of the Smithsonian Institution and collected in connection with the U.S. and Mexican Boundary Survey: Major William Emory, Commissioner. Proc. Acad. nat. Sci. Philad., 10:167-171.

--. 1859. Ichthyological notices. Proc. Acad. nat. Sci. Philad., 11:56-68 (separate, 1859).

--. 1859. Ichthyology of the boundary. In: United States and Mexican Boundary Survey, 2(2):1-85. Washington, 2 vol.

GIRGENSOHN, O. G. L. 1846. Anatomie und Physiologie des Fischnervensystems. Mém. prés. Acad. imp. Sci. St. Pétersb., 5:275-589, 15 pl.

GISTEL, J. 1848. Naturgeschichte des Tierreichs für höhere Schulen. Stuttgart, xvi + 216 p., fig., 32 pl. col.

--. 1850. Gonichthys, ein Fisch aus der Bai von Madeira. Isis, (München), (5):71-72.

GJØSAETER, J. 1973. Age, growth and mortality of the myctophid fish, Benthosema glaciale (Reinhardt), from western Norway. Sarsia, 52:1-14.

--. 1978 Aspects of the distribution and ecology of the Myctophidae from the western and northern Arabian Sea. Dev. Rep. Indian Ocean Programme, (43)2:62-108.

GJØSAETER, J.; BLINDHEIM, J. 1982. Observations on mesopelagic fish off North-west Africa between 16° and 27°N. Rapp. P.-v. Réun. Cons. int. Explor. Mer, 180:381-398.

GLOERFELT-TARP, T.; KAILOLA, P. 1984. Trawled fishes of southern Indonesia and northwestern Australia. Singapore, xvi + 406 p., 3 pl., many fig.

GMELIN, J. F. 1789. Pisces, p. 1126-1516. In: Caroli a Linné, Systema Naturae per regna tria naturae, secundum classes, ordines, genera, species; cum characteribus, differentiis, synonymis, locis (ed. XIII). Lipsiae, 1(3).

GODSIL, H. C.; BYERS, R. D. 1944. A systematic study of the Pacific tunas. Calif. Fish. Game, (60):1-131, 74 fig.

GOFFIN, A. 1909. Les pêcheries et les poissons du Congo. Bruxelles, 231 p., num. fig.

GOHAR, H. A. F.; HAMDY, A. R. 1963. The jaws and teeth of Aetobatus narinari. Publs mar. biol. Stn Ghardaqa, 12:178-190.

GOHAR, H. A. F.; MAZHAR, F. M. 1964. The elasmobranchs of the northwestern Red Sea. Publs mar. biol. Stn Ghardaqa, 13:1-144, pl. 1-16.

--. 1964. The internal anatomy of Selachii from the northwestern Red Sea. Publs mar. biol. Stn Ghardaqa, 13:145-239, pl. 1-11.

GOLDFUSS, G. A. 1820. Handbuch der Zoologie, 2te Abt. Nürnberg, xxiv + 510 p., pl. 3-4.

GOLOVAN, G. A. 1974. Preliminary data on the composition and distribution of the bathyal ichthyofauna (in the Cap Blanc area). Okeanologija, 14(2):355-358, 1 fig., 1 tab. (Engl. transl. Oceanology, 14:288-290).

--. 1976. Rare and firstly recorded chondrostean and teleostean fishes of the continental slope of West Africa. Trudy Inst. Okeanol., 104:277-317, fig. 1-11. (in Russian).

--. 1978. Composition and distribution of the ichthyofauna of the continental slope of north-western Africa. Trudy Inst. Okeanol., 111:195-258, fig. 1-7. (in Russian).

GOLOVAN, G. A.; PAKHORUKOV, N. P. 1975. Some data on the morphology and ecology of Alepocephalus bairdi (Alepocephalidae) of the central and eastern Atlantic. J. Ichthyol. 15(1):44-50, 3 fig.

--. 1979. The widespread fish, Narcetes stomias, from the central part of the eastern Atlantic. Biol. morja, Vladivostok, 1979(3):86-88.

--. 1980. New data on the ecology and morphometry of Alepocephalus ros-

tratus Risso (Alepocephalidae). Vop. Ikhtiol., 20(3):463-472.

---. 1986. New records of rare species of cartilaginous fishes. Vop. Ikhtiol., 26(1):153-154. (in Russian).

GOLVAN, Y. 1956. Parasites de poissons de mer ouest-africains récoltés par J. Cadenat, VIII - Acanthocéphales. Bull. Inst. fr. Afr. noire, (2)18(2):467-485.

GOMON, M. F. 1981. Labridae, in Fischer, W., Bianchi, G. & W.B. Scott (ed.), FAO species identification sheets for fishery purposes. Eastern Central Atlantic; fishing areas 34, 47 (in part). Canada Funds-in-Trust. Ottawa, Department of Fisheries and Oceans Canada, by arrangement with the Food and Agriculture Organization of the United Nations, vol. 1-7: pag. var.

GOMON, M. F.; LUBBOCK, R. 1980. A new hogfish of the genus Bodianus (Teleostei: Labridae) from islands of the Mid-Atlantic Ridge. Northeast Gulf Science, 3(2):104-111, 3 fig., 1 tab.

GONCALVES, B. C. 1955. Peixes coligidos pela Missâo Zoológica da Guiné (1945-1946). Anais Jta Invest. Ultramar, 10(4)(1):115-163.

GONZALES, A. P. 1972. A contribution to the identification of west African species of Pomadasys (Osteichthyes, Perciformes, Pomadasyidae). Bull. Inst. fond. Afr. noire, (A)34(1):160-168, fig.

GOODE, G. B. 1876. Catalogue of the fishes of the Bermudas, based chiefly on the collections of the United States National Museum. Bull. U.S. natn. Mus., (5):82 p. (also Smithson. misc. Collns, 13(5)).

---. 1880. A preliminary catalogue of the fishes of the St. John's River and the east coast of Florida, with descriptions of a new genus and three new species. Proc. U. S. natn. Mus., 2:108-121.

---. 1880. A study of the trunk-fishes (Ostraciontidae), with notes upon the American species of the family. Proc. U.S. natn. Mus., 2:261-283.

---. 1880. Descriptions of seven new species of fishes from deep soundings on the southern New England coast, with diagnoses of two undescribed genera of flounders and a genus related to Merlucius. Proc. U.S. natn. Mus., 3:337-350.

---. 1881. Fishes from the deep water on the south coast of New England obtained by the United States Fish Commission in the summer of 1880. Proc. U.S. natn. Mus., 3:467-486.

---. 1882. The taxonomic relations and geographical distribution of the members of the swordfish family, Xiphiidae. Proc. U.S. natn. Mus., 4:415-433.

---. 1883. Materials for a history of the swordfishes. Rep. U.S. Commnr Fish., 8:287-394, 24 pl.

---. 1884. The generic names Amitra and Thyris replaced. Proc. U.S. natn. Mus., 6:109.

---. 1884. The swordfish family. Xiphiidae, p. 336-359. In: G.B. Goode et al.: Fishes, p. 163-682. In: The fisheries and fishery industries of the United States. Section I. Washington, xxxiv + 895 p.

---. 1886. The beginnings of natural history in America. Proc. biol. Soc. Wash., 3:35-105.

GOODE, G. B.; BEAN, T. H. 1879. The identity of Rhinonemus caudacuta (Storer) Gill, with Gadus cimbrius Linn. Bull. U.S. natn. Mus., 1:348-349.

---. 1879. A catalogue of the fishes of Essex County, Massachusetts, including the fauna of Massachusetts Bay and the contiguous deep waters. Bull. Essex Inst., 11:1-38.

---. 1879. Description of Alepocephalus bairdii, a new species of fish from the deep-sea fauna of the western Atlantic. Proc. U.S. natn. Mus., 2:55-57.

---. 1880. Catalogue of a collection of fishes sent from Pensacola, Florida, and vicinity, by Mr. Silas Stearns, with descriptions of six new species. <u>Proc. U.S. natn. Mus.</u>, 2:121-156.

--. 1881. Description of a new species of fish, <u>Apogon pandionis</u>, from deep water off the mouth of Chesapeake Bay. <u>Proc. U.S. natn. Mus.</u>, 4:160-161.

--. 1882. <u>Benthodesmus</u>, a new genus of deep-sea fishes allied to <u>Lepidopus</u>. <u>Proc. U.S. natn. Mus.</u>, 4:379-383.

--. 1882. Descriptions of twenty-five new species of fish from the southern United States, and three new genera, <u>Letharcus, Ioglossus</u>, and <u>Chriodorus</u>. <u>Proc. U.S. natn. Mus.</u>, 5:412-437.

--. 1883. A list of the species of fishes recorded as occurring in the Gulf of Mexico. <u>Proc. U.S. natn. Mus.</u>, 5:234-240.

--. 1883. Reports on the results of dredging under the supervision of Alexander Agassiz, on the east coast of the United States, during the summer of 1880, by the U.S. Coast Survey Steamer "Blake", Commander J.R. Bartlett, U.S.N., commanding. XIX. Report on the fishes. <u>Bull. Mus. comp. Zool. Harv.</u>, 10(5):183-226.

--. 1885. On the American fishes in the Linnean collection. <u>Proc. U.S. natn. Mus.</u>, 8:193-209.

--. 1886. Description of <u>Leptophidium cervinum</u> and <u>L. marmoratum</u>, new species from deep water off the Atlantic and Gulf coasts. <u>Proc. U.S. natn. Mus.</u>, 1885, 8:422-424.

--. 1886. Descriptions of new fishes obtained by the United States Fish Commission mainly from deep water off the Atlantic and Gulf coasts. <u>Proc. U.S. natn. Mus.</u>, 1885, 8:589-605.

--. 1886. Reports on the results of dredging etc. XXVIII. Description of thirteen species and two genera of fishes from the "Blake" collection. <u>Bull. Mus. comp. Zool. Harv.</u>, 12(5):153-170.

--. 1895. On Cetomimidae and Rondeletiidae, two new families of bathybial fishes from the northwestern Atlantic. <u>Proc. U.S. natn. Mus.</u>, 17(1012):451-454, 3 fig.

--. 1895. A revision of the order Heteromi, with a description of the new generic types <u>Macdonaldia</u> and <u>Lipogenys</u>. Scientific results of explorations by the U.S. Fish Commission Steamer "Albatross". No 29. <u>Proc. U.S. natn. Mus.</u>, 17(1013):455-470.

--. 1895. On <u>Harriotta</u>, a new type of chimaeroid fish from the deeper waters of the northwestern Atlantic. Scientific results of explorations by the U.S. Commission Steamer "Albatross". No 30. <u>Proc. U.S. natn. Mus.</u>, 17:471-473, pl. 19.

--. 1896. Oceanic ichthyology, a treatise on the deep-sea and pelagic fishes of the world, based chiefly upon the collections made by steamers Blake, Albatross and Fish Hawk in the northwestern Atlantic. <u>Mem. Mus. comp. Zool. Harv.</u>, 1 (text): xxxv + 1-553; 2(atlas): xxiii + 1-26, 123 pl., 417 fig.

GOODYEAR, R. H., 1969. Records of the alepocephalid fish <u>Photostylus pycnopterus</u> in the Indian and Pacific Oceans. <u>Copeia</u>, 1969(2):398-400.

--. 1973. Malacosteidae, p. 142-143, <u>in</u> Hureau, J. C. & Th. Monod (ed.), Check-list of the fishes of the north-eastern Atlantic and of the Mediterranean/Catalogue des poissons du nord-est Atlantique et de la Méditerranée. Paris, Unesco, 2 vol.

GOODYEAR, R. H.; GIBBS, R. H., Jr. 1969. Ergebnisse der Forschungsreisen des FFS "Walther Herwig" nach Südamerika. X. Systematics and zoogeography of stomiatoid fishes of the <u>Astronesthes cyaneus</u> species group (family Astronesthidae), with descriptions of three new species. <u>Arch. FischWiss.</u>, 20 (2/3):107-131.

GOODYEAR, R. H.; ZAHARUNEC, B. J.; PUGH, W. L.; GIBBS, R. H., Jr. 1972. Ecology and vertical distribution of Mediterranean midwater fishes. p. 91-229. <u>In</u>: Goodyear, R. H., R. H. Gibbs, Jr., C. F. E. Roper, R.

C. Kleckner, M. J. Sweeney, B. J. Zahuranec & W. L. Pugh, Mediterranean Biological Studies. 2 vol., 1 - reports; 2 - appendices. Final Rep. to the Office of Naval Research Contract Nr. NOO 14-67-A-0399-0007.

GOPINATH, K. 1946. Notes on the larval and post-larval stages of fish found along the Trivandrum coast. Proc. natn. Inst. Sci. India, (B)12:7-21.

GORBUNOVA, N. N. 1965. Seasons and conditions of spawning of the scombroid fishes (Pisces, Scombroidei). Trudy Inst. Okeanol., 80:36-51 (in Russian).

--. 1972. Systematics, distribution and biology of the fishes of the genus Vinciguerria (Pisces, Gonostomatidae). Trudy Inst. Okeanol., 93:70-109, 18 fig. (in Russian).

--. 1977. Larvae of some species of trichiuroid fishes (Pisces, Trichiuroidae: Gempylidae, Trichiuridae). Trudy Inst. Okeanol., 109:133-148. (in Russian).

--. 1977. Distribution of some gempylid species larvae (Pisces, Trichiuridae). Okeanologija, 17(6):1107-1112. (in Russian).

--. 1981. Larvae of the genus Vinciguerria (Gonostomatidae) with a key for their identification. Vop. Ikhtiol., 21(4):742-745.

--. 1982. Larvae of trichiuroid fishes from collection of the International Mexican Biological Centre. Trudy Inst. Okeanol., 118:85-106.

GORELOVA, T. A. 1978. The feeding of lanternfishes, Ceratoscopelus warmingi and Bolinichthys longipes, of the family Myctophidae in the western part of the Pacific Ocean. J. Ichthyol., 18(4):588-598.

--. 1980. Feeding of deep-sea fishes of the genus Cyclothone (Gonostomatidae, Pisces). Okeanologija, 20(2):321-328.

GOREN, M. 1978. Comparative study of Bathygobius fuscus (Rüppell) and related species of the Red Sea, including B. fishelsoni n. sp. Senckenberg. biol., 58:267-273.

GOSLINE, W. A. 1959. Four new species, a new genus, and a new suborder of Hawaiian fishes. Pacif. Sci., 13(1):67-77, 6 fig.

--. 1961. Some osteological features of modern lower teleostean fishes. Smithson. misc. Collns, 142(3):1-42.

--. 1968. The suborders of perciform fishes. Proc. U.S. natn. Mus., 124(3647):1-78, 12 fig.

--. 1980. The evolution of some structural systems with references to the interrelationships of modern lower teleostean fish groups. Jap. J. Ichthyol., 27(1):1-28.

GOSLINE, W. A.; BROCK, V. E. 1960. Handbook of Hawaiian fishes. Honolulu, IX + 372 p., fig. 1-277.

GOSSE, P. H. 1851. A naturalist's sojourn in Jamaica. London, 508 p.

GOUAN, A. 1770. Historia piscium, sistens ipsorum anatomen externam, internam, atque genera in classes et ordines redacta. Histoire des poissons, contenant la description anatomique de leurs parties externes et internes et le caratère des divers genres rangés par classes et par ordres. Strasbourg, xviii (x2) + 1-228 (x2)(lat. + fr.) + 229-252, pl. 1-4.

GRAAE, M. J. F. 1967. Lestidium bigelowi, a new species of paralepidid fish with photophores. Breviora, (277):1-10, fig. 1-3.

GRANDPERRIN, R.; RIVATON, J. 1966. "Coriolis": Croisière "Alizé". Individualization de plusieurs ichtyofaunes le long de l'équateur. Cah. ORSTOM. Océanogr., 4(4):35-49.

GRANT, C. J.; COWPER, T. R.; RIED, D. D. 1978. Age and growth of snoek, Leionura atun (Euphrasen), in south-eastern Australian waters. Austr. J. mar. Freshwat. Res., 29(4):435-444.

GRANT, E. M. 1972. Guide to fishes. Brisbane, xxiv + 472 p.

GRAS, R. 1958. La pêche des ethmaloses dans le lac Nokové. Notes Docums

Pêch. Piscic. Centre tech. Forest. trop., (ser. D.R.)(4):11 p., (mimeo).

--. 1961. Liste des poissons du Bas-Dahomey, faisant partie de la collection du Laboratoire d'Hydrobiologie du Service des Eaux, Forêts et Chasses du Dahomey. Bull. Inst. fr. Afr. noire, (A)23(2):572-586.

GRASSE, P. P. (ed.). 1958. Traité de Zoologie: anatomie, systématique, biologie. Tome 13. Agnathes et poissons: anatomie, éthologie, systématique. 3:1813-2758, fig. 1308-1889. Paris.

GRASSET, G. 1971. La pêche expérimentale effectuée avec une senne tournante et coulissante par la pirogue du projet en mai-juin 1971. UNDP (SF)/FAO, Project SEN/66/508. Survey Devel. pelag. Fish Resources. Docum. scient. provis. Centre Rech. océanogr. Dakar-Thiaroye, (Tech. Rept. 7/1971): 16 p., mimeo.

GRASSI, G. B. 1896. The reproduction and metamorphosis of the common eel (Anguilla vulgaris). Q. Jl microsc. Sci., 39(3):371-385, 4 fig.

--. 1897. The reproduction and metamorphosis of the common eel (Anguilla vulgaris). Proc. R. Soc., 60:260-271.

--. 1912. Nuova contribuzione alla storia dello sviluppo dei murenoidi. Atti Accad. naz. Lincei Rc., (5)21(1):15-20.

--. 1913. Metamorfosi dei murenoidi: ricerche sistematiche ed ecologiche. Jena, x + 211 p., 15 pl.

--. 1915. Contributo alla conoscenza delle uova e delle larve dei murenoidi (Aggiunta II alla monografia sulla metamorfosi dei murenoidi). Atti Accad. naz. Lincei Rc., (5)10(16):693-711.

--. 1917. Contributo alla conoscenza delle uova e delle larve dei murenoidi (Aggiunta III alla monografia sulla metamorfosi dei murenoidi). Memorie R. Com. talassogr. ital., (45):1-32, 2 pl.

GRASSI, G. B.; CALANDRUCCIO, S. 1892. Le leptocefalide e la loro transformazione in murenide. Atti. Accad. naz. Lincei Rc., (5)1(2): 375-379.

--. 1896. Sullo sviluppo dei murenoidi. Atti. Accad. naz. Lincei Rc., (5)5(1):348-349.

GRASSLE, J. F.; SANDERS, H. L.; HESSLER, R. R.; ROWE, G. T.; McLELLAN, T. 1975. Pattern and zonation: a study of the bathyal megafauna using the research submersible "Alvin". Deep Sea Res., 22:457-481.

GRAY, J. E. 1830-1834. Illustrations of Indian Zoology: chiefly selected from the collection of Major-General Hardwicke. 2 vol. London, 1:100 pl., 2:102 pl. (for dating, see W. R. Dawson, 1946, J. Biblphy nat. Hist., 2(3):55-69).

--. 1838. Description of a new species of Tetrapturus from the Cape of Goode Hope. Ann. nat. Hist., 1(36):313.

--. 1851. List of the specimens of fish in the collection of the British Museum. Part I. Chondropterygii. London, x + 160 p., 2 pl.

--. 1854. Catalogue of fish collected and described by Laurence Theodore Gronow, now in the British Museum. London, vii + 196 p.

--. 1864. Notice of a portion of a new form of animal (Myriosteon higginsii), probably indicating a new group of Echinodermata. Proc. zool. Soc. Lond., 1864:163-166, 5 fig.

GREENFIELD, D. W. 1981. Holocentridae, in Fischer, W., Bianchi, G. & W.B. Scott (ed.), FAO species identification sheets for fishery purposes. Eastern Central Atlantic; fishing areas 34, 47 (in part). Canada Funds-in-Trust. Ottawa, Department of Fisheries and Oceans Canada, by arrangement with the Food and Agriculture Organization of the United Nations, vol. 1-7: pag. var.

GREENWOOD, P. H. 1970. Skull and swimbladder connections in fishes of the family Megalopidae. Bull. Br. Mus. nat. Hist., (Zool), 19:121-135, 3 pl.

GREENWOOD, P. H.; ROSEN, D. E. 1971. Notes on the structure and relationships of the alepocephalid fishes. Am. Mus. Novit., (2473):1-41, 25 fig.

GREENWOOD, P. H.; ROSEN, D. E.; WEITZMANN, S. H.; MYERS, G. S. 1966. Phyletic studies of teleostean fishes, with a provisional classification of living forms. Bull. Am. Mus. nat. Hist., 131(4):339-455, fig. 1-9, pl. 21-23.

GREGORY, W. K. 1933. Fish skulls: a study of the evolution of natural mechanism. Trans. Am. phil. Soc., 23(2):i-vii, 75-481, fig. 1-302.

GREGORY, W. K.; CONRAD, G. M. 1937. The comparative osteology of the swordfish (Xiphias) and the sailfish (Istiophorus). Am. Mus. Novit., (952):1-25, 12 fig.

--. 1939. Body-forms of the black marlin (Makaira nigricans marlina) and striped marlin (Makaira mitsukurii) of New Zealand and Australia. Bull. Am. Mus. nat. Hist., 76(8):443-456, 2 fig., pl. 3-4.

--. 1943. The osteology of Luvarus imperialis, a scombroid fish; a study in adaptative evolution. Bull. Am. Mus. nat. Hist., 81(2):225-283, 38 fig.

GREY, M. 1953. Fishes of the family Gempylidae, with records of Nesiarchus and Epinnula from the western Atlantic and description of two new subspecies of Epinnula orientalis. Copeia, 1953(3):135-141.

--. 1955. Notes on a collection of Bermuda deep-sea fishes. Fieldiania, Zool., 37:265-302, fig. 45-56.

--. 1955. The fishes of the genus Tetragonurus Risso. Dana Rep., (41):1-75, 16 fig.

--. 1956. The distribution of fishes found below a depth of 2000 meters. Fieldiana, Zool., 36(2):75-337.

--. 1956. New records of deep-sea fishes, including a new species, Oneirodes bradburyae, from the Gulf of Mexico. Copeia, 1956(4):242-246, fig. 1-2.

--. 1958. Descriptions of abyssal benthic fishes from the Gulf of Mexico. Fieldiana, Zool., 39(16):149-183, fig. 22-28.

--. 1959. Deep-sea fishes from the Gulf of Mexico with the description of a new species, Squalogadus intermedius (Macrouroididae). Fieldiana, Zool., 39(29):323-346, fig. 53-57.

--. 1959. Three new genera and one new species of the family Gonostomatidae. Bull. Mus. comp. Zool. Harv., 121(4):167-184, fig. 1-3.

--. 1960. Description of a western Atlantic specimen of Scombrolabrax heterolepis Roule and notes on fishes of the family Gempylidae. Copeia, 1960(3):210-215, 3 fig.

--. 1964. Gonostomatidae. In: Fishes of the western North Atlantic. Mem. Sears. Fdn mar. Res., 1(4):78-240, fig. 21-60.

GREY, Z. 1928. Big game fishing in New Zealand seas. Nat. Hist., N.Y., 28(1):46-52, 6 fig.

GRIEG, J. A. 1912. Ichthyologiske notiser. Bergens Mus. Arb., 1911 (6): 1-37, 2 pl.

GRIFFIN, T. L. 1927. Addition to the fish fauna of New Zealand. Trans. Proc. N.Z. Inst., 58:136-150.

--. 1936. Revision of the eels of New Zealand. Trans. R. Soc. N.Z., 66(1):12-26, 6 fig.

GRIFFINI, A. 1903. Ittiologia italiana. Descrizione dei pesci di mare e d'acqua dolce. Milano, xii + 475, 244 fig.

GRIFFITH, E.; SMITH, C. H. 1834. The class Pisces arranged by the Baron Cuvier, with supplementary additions. London, viii + 680 p.

GRINDLEY, J. R.; PENRITH, M. J. 1965. Notes on the bathypelagic fauna of the seas around South Africa. Zool. afr., 1(2):275-295.

GRONOVIUS, L. T. 1754. Museum ichthyologicum, sistens piscium indigenorum & quorundam exoticorum, qui in Museo Laurentii Theodori Gronovii.

Leiden, 1:70 p., 4 pl.

--. 1763. Zoophylacii Gronoviani fasciculus primus exhibens animalia quadrupeda, amphibia atque pisces, quae in museo suo adservat, rite examinavit, systematice disposuit, descripsit atque iconibus illustravit Laur. Theod. Gronovius, J.U.D.... Lugduni Batavorum, 136 p., 14 pl.

--. 1772. Animalium rariorum fasciculus. Pisces. Acta helvetica, 7:43-52, 2 pl.

--. 1854. Catalogue of fishes collected and described by Theodore Gronow, now in the British Museum. Edited from the manuscript, by J. E. Gray. London, vii + 196 pp.

GROOT (see DE GROOT).

GROS, P. 1980. Description mathématique de la croissance de Lesueurigobius friesii (Teleostei, Gobiidae). Essai de synthèse. Oceanologica Acta, 3:161-168.

GRUVEL, A. 1908. Les pêcheries des côtes du Sénégal et des rivières du sud. Paris, 245 p.

GRUVEL, A.; BESNARD, W. 1937. Atlas de poche des principaux produits marins rencontrés sur les marchés du Maroc. Paris, 217 p., 127 fig.

GRUVEL, A.; BOUYAT, A. 1906. Les pêcheries de la côte occidentale d'Afrique. Paris, 280 p.

GUBANOV, Ye. P. 1972. On the biology of the thresher shark (Alopias vulpinus Bonnaterre) in the northwest Indian Ocean. J. Ichthyol., 12: 591-600.

GUDGER, E. W. 1910. Notes on some Beaufort fishes, 1909. Am. Nat., 44: 395-403.

--. 1912. Natural history notes on some Beaufort, N. C. fishes, 1910. 1. Elasmobranchs, with special reference to utero-gestation. Proc. biol. Soc. Wash., 25:141-156.

--. 1913. Natural history notes on some Beaufort, N. C. fishes, 1912. Proc. biol. Soc. Wash., 26:97-109.

--. 1924. Notes on the behavior of the gray snapper, a common West Indian Fish. Nat. Hist., N.Y., 24(2):252-253.

--. 1929. On the morphology, coloration and behavior of seventy teleostean fishes of Tortugas, Florida. Pap. Tortugas Lab., 26:149-204, pl. 1-4 (also publ. Publs Carnegie Instn, (391)).

--. 1930. The opah or moonfish, Lampris luna, on the eastern coast of North America. Am. Nat., 64:168-178, fig.

--. 1931. The triple-tail, Lobotes surinamensis, its names, occurrence on our coasts and its natural history. Am. Nat., 65:49-69.

--. 1937. The natural history and geographical distribution of the pointed-tailed Ocean Sunfish (Masturus lanceolatus) with notes on the shape of the tail. Proc. zool. Soc. Lond., (A)107(3):353-396, 5 pl.

--. 1937. An albino tarpon, Tarpon atlanticus, the only known specimen. Am. Mus. Novit., (944):1-4, 2 text fig.

--. 1945. The frogfish, Antennarius scaber, uses its lure in fishing. Copeia, 1945(2):111-112.

GÜNTHER, A. 1859. Catalogue of the Acanthopterygian fishes in the collection of the British Museum. 1. Gasterosteidae, Berycidae, Percidae, Aphredoderidae, Pristipomatidae, Mullidae, Sparidae. London, xxxi + 524 p.

--. 1860. Catalogue of the Acanthopterygian fishes in the collection of the British Museum. 2. Squamipinnes, Cirrhitidae, Triglidae, Trachinidae, Sciaenidae, Polynemidae, Sphyraenidae, Trichiuridae, Scombridae, Carangidae, Xiphiidae. London, xxi + 548 p.

--. 1860. Alepidosaurus, ein Meerwels. Arch. Naturgesch., 26(1):121-123.

--. 1861. A preliminary synopsis of the labroid genera. Ann. Mag. nat. Hist., (3)8:382-389.

--. 1861. Catalogue of the Acanthopterygian fishes in the collection of the British Museum. 3. Gobiidae, Discoboli, Oxudercidae, Batrachidae, Pediculati, Blenniidae, Acanthoclinidae, Comephoridae, Trachypteridae, Lophotidae, Teuthididae, Acronuridae, Hoplognathidae, Malacanthidae, Nandidae, Polycentridae, Labyrinthici, Luciocephalidae, Atherinidae, Mugilidae, Ophiocephalidae, Trichonotidae, Cepolidae, Gobiesocidae, Psychrolutidae, Centriscidae, Fistularidae, Mastacembelidae, Notacanthi. London, xxv + 586 p.

--. 1862. Catalogue of the fishes in the British Museum. 4. Catalogue of the Acanthopterygii Pharyngognathi and Anacanthini in the collection of the British Museum. London, xxi + 534 p.

--. 1862. Descriptions of new species of reptiles and fishes in the collection of the British Museum. Proc. zool. Soc. Lond., 1862(2): 188-194.

--. 1863. On the European species of the genus Labrax. Ann. Mag. nat. Hist, (3)12:174-175.

--. 1864. On some new species of Central-American fishes. Proc. zool. Soc. Lond., 1864:23-27, pl. 3, 4.

--. 1864. Report of a collection of fishes made by Messrs. Dow, Godman and Salvin in Guatemala. Proc. zool. Soc. Lond., 1864:144-154.

--. 1864. On a new genus of pediculate fish from the sea of Madeira. Proc. zool. Soc. Lond., 1864:301-303, pl. 25.

--. 1864. Catalogue of the fishes in the British Museum. 5. Catalogue of the Physostomi, containing the families Siluridae, Characinidae, Haplochitonidae, Sternoptychidae, Scopelidae, Stomiatidae in the collection of the British Museum. London, xxii + 455 p., fig.

--. 1866. Catalogue of the fishes in the British Museum. 6. Catalogue of the Physostomi, containing the families Salmonidae, Percopsidae, Galaxidae, Mormyridae, Gymnarchidae, Esocidae, Umbridae, Scombresocidae, Cyprinodontidae in the collection of the British Museum. London, xv + 368 p., fig.

--. 1866. Note on west-African species of Hemirhamphus. Ann. Mag. nat. Hist., (3)18:427.

--. 1867. Additions to the knowledge of Australian reptiles and fishes. Ann. Mag. nat. Hist., (3)20:45-68.

--. 1867. New fishes from the Gaboon and Gold Coast. Ann. Mag. nat. Hist., (3)20:110-117.

--. 1868. Catalogue of the fishes in the British Museum. 7. Catalogue of the Physostomi, containing the families Heteropygii, Cyprinidae, Gonorhynchidae, Hyodontidae, Osteoglossidae, Clupeidae, Chirocentridae, Alepocephalidae, Notopteridae, Halosauridae, in the collection of the British Museum. London, xx + 512 p.

--. 1868. Report on a collection of fishes made at St Helena by J.C. Melliss, Esq.. Proc. zool. Soc. Lond., 1868:225-228, 2 pl.

--. 1869. Report on a second collection of fishes made at St. Helena by J.C. Melliss, Esq. Proc. zool. Soc. Lond., 1869:238-239, pl. 1.

--. 1869. The fishes of the Nile. p. 197-268. In: J. & B.H. Petherick: Travels in Central Africa and explorations of the western Nile tributaries. 2: Appendix C. London.

--. 1870. Catalogue of the fishes in the British Museum. 8. Catalogue of the Physostomi, containing the families Gymnotidae, Symbranchidae, Muraenidae, Pegasidae and of the Lophobranchii, Plectognathi, Dipnoi, Ganoidei, Chondropterygii, Cyclostomata, Leptocardii, in the collection of the British Museum. London, xxv + 549 p.

--. 1871. Report on several collections of fishes recently obtained for the British Museum. Proc. zool. Soc. Lond., 1871:652-675.

--. 1872. On some new species of reptiles and fishes collected by J. Brenchley, Esq. Ann. Mag. nat. Hist., (4)10:418-426.

--. 1873. Erster ichthyologischer Beitrag nach Exemplaren aus dem Museum Godeffroy. J. Mus. Godeffroy, 1(2):97-103, 6 fig. n.num.

--. 1873. Zweiter ichthyologischer Beitrag nach Exemplaren aus dem Museum Godeffroy. J. Mus. Godeffroy, 1(4):89-92.

--. 1873-75. Andrew Garrett's Fische der Südsee, beschrieben und redigirt von A.C.L.G. Günther. 1:iv + 1-128, 5 fig., col. pl. 1-83. Heft 1, 1873, J. Mus., Godeffroy, 2(3):1-24, pl. 1-20; Heft 2, 1874, J. Mus. Godeffroy, 2(5):25-56, pl. 21-40; Heft 3, 1874, J. Mus. Godeffroy, 2(7):57-96, pl. 41-60; Heft 4, 1875, J. Mus. Godeffroy, 2(9): 97-128, pl. 61-83.

--. 1874. Notice of some new species of fishes from Morocco. Ann. Mag. nat. Hist., (4)13:230-232, pl. 13-14.

--. 1876. Remarks on fishes, with descriptions of new species in the British Museum, chiefly from southern seas. Ann. Mag. nat. Hist., (4)17:389-402.

--. 1876-1881. Andrew Garrett's Fische der Südsee, beschrieben und redigirt von A.C.L.G. Günther.2. Heft 5, 1876, J. Mus. Godeffroy, 4(11): 129-168, 1 fig., pl. 84-100; Heft 6, 1877, J. Mus. Godeffroy, 4(13): 169-216, 2 fig., pl. 101-120; Heft 7, 1881, J. Mus. Godeffroy, 4 (15):217-256, pl. 121-140.

--. 1877. Preliminary notes on new fishes collected in Japan during the expedition of H.M.S. "Challenger". Ann. Mag. nat. Hist., (4)20:433-446.

--. 1878. Preliminary notices of deep-sea fishes collected during the voyage of H.M.S. "Challenger". Ann. Mag. nat. Hist., (5)2:17-28, 179-187, 248-251.

--. 1880. An introduction to the study of fishes. Edinburgh, 720 p., 320 fig.

--. 1880. Report on the shore fishes. Rep. scient. Results Voyage Challenger, Zool., 1(6):1-82, pl. 1-32.

--. 1881. Report on a collection (of fishes) made by Mr. T. Conry in Ascension Island. Ann. Mag. nat. Hist., (5)8:430-440.

--. 1881. Account of the zoological collection made during the survey of H.M.S. "Albert" in the Straits of Magellan. Proc. zool. Soc. Lond., 1881:1-141, 11 pl.

--. 1882. Report on the fishes. p. 677-680. In: R.N. Tizards & J. Murray: Exploration of the Faeröe Channel, during the summer of 1880, in H.M.S. ship "Knight Errant." Proc. R. Soc. Edinb., 11.

--. 1887. Report of the deep-sea fishes collected by H.M.S. Challenger during the years 1873-76. Rep. scient. Results Voyage Challenger, Zool., 22: lxv, 1-268, fig. 1-7, pl. 1-66.

--. 1889. Fishes. In: W.S. Green. Report of a deep-sea trawling cruise on the S.W. coast of Ireland. Ann. Mag. nat. Hist., (6)4:415-420.

--. 1889. Report on the pelagic fishes collected by H.M.S. Challenger, during the years 1873-76. Rep. scient. Results Voyage Challenger, Zool., 31:1-47, pl. 1-6.

GUICHENOT, A. 1848. Peces de Chile. p. 137-372, 11 pl. In: C. Gay: Historia fisica y politica de Chile. 2. Paris, Santiago, 372 p. (plates in separate Atlas).

--. 1850. Histoire naturelle des reptiles et des poissons. Explor. scient., Algérie 1840-42. Sci. Phys., Zool., 5:1-148, pl. 1-4 (Rept.) + 1-8 (Poiss.).

--. 1853. Poissons. In: Sagra, R. de la. Histoire physique, politique et naturelle de l'île de Cuba. Paris, 2:206 p.

--. 1865. Catalogue des Scaridés de la collection du Musée de Paris. Mém. Soc. Sci. nat. math. Cherbourg, 11:1-75.

--. 1866. Catalogue des poissons de Madagascar de la collection du Musée de Paris, avec la description de plusieurs espèces nouvelles. Mém.

Soc. Sci. nat. math. Cherbourg, (2)12:129-148.

--. 1867. Notice sur le Salarichthys, nouveau genre de poissons de la famille des Blennioides, et description de l'espèce type. Mém. Soc. Sci. nat. math. Cherbourg, 13:96-100.

--. 1868. Index generum ac specierum Anthiadidorum hucusque in Museo Parisiensi observatorum. Annls Soc. linn. Dep. Maine Loire, 10:80-87.

GUIMARAES, A.; PEREIRA, R. 1881. Description d'un nouveau poisson du Portugal. Jorn. Sci. math. phys. nat., 8:222-224, fig.

--. 1882. Lista dos peixes da Ilha da Madeira, Açores e das possessôes portuguezas d'Africa, que existem no Museu de Lisboa. Supplemento. Jorn. Sci. math. phys. nat., 9(33):30-39.

--. 1884. Lista dos peixes da Ilha da Madeira, Açores e das possessôes portuguezas d'Africa, que existem no Museu de Lisboa (Segundo supplemento). Jorn. Sci. math. phys. nat., 10(37):11-28.

GUNNERUS, J. E. 1765. Efterretning om Berglaxen, en rar Norsk fisk, som kunde kaldes: Coryphaenoides rupestris. Trondhiemske Selsk. Skr., 3 (4):50-58.

--. 1768. Laxe-störjen, Scombro pelagico Linn. K. norske Vidensk. Selsk. Skr., 4:92-94, pl. 12, fig. 1.

GUNTER, G. 1942. A new Scorpaena from the Texas coast, with notes on Scorpaena mystax Jordan and Starks. Copeia, 1942(2):105-111, 1 fig.

--. 1942. A list of the fishes of the mainland of North and Middle America recorded from both fresh and sea water. Am. Midl. Nat., 28(2): 305-326.

--. 1956. A revised list of euryhalin fishes of North and Middle America. Am. Midl. Nat., 56(2):345-354.

GUSHCHIN, A.; KUKUEV, E. J. 1981. On composition of ichthyofauna of the northern part of the Middle-Atlantic Ridge. p. 36-40, 1 tab. In: Fishes of the open ocean. (N.V. Parin, ed.). Moscow, 119 p.

GUTHERZ, E. J. 1981. Bothidae, in Fischer, W., Bianchi, G. & W.B. Scott (ed.), FAO species identification sheets for fishery purposes. Eastern, Central Atlantic; fishing areas 34, 47, (in part). Canada Funds-in-Trust. Ottawa, Department of Fisheries and Oceans Canada, by arrangement with the Food and Agriculture Organization of the United Nations, vol. 1-7: pag. var.

GUTIERREZ, M.; ESTABLIER, R.; ARIAS, A. 1978. Acumulación y efectos histopatológicos del cadmio y el mercurio en el sapo (Halobatrachus didactylus). Investigación pesq., 42(1):141-154, fig.

HAEDRICH, R. L. 1964. Food habits and young stages of North Atlantic Alepisaurus (Pisces, Iniomi). Breviora, (201):1-15.

--. 1965. Identification of deep-sea mooring-cable biter. Deep Sea Res.. 12: 773-776, fig. 1-3.

--. 1966. The stromateoid fish genus Icichthys: notes and a new species. Vidensk. Meddr dansk naturh. Foren., 129:199-213.

--. 1967. The stromateoid fishes: systematics and a classification. Bull. Mus. comp. Zool. Harv., 135(2):31-139, 56 fig., 2 tab.

--. 1981. Ariommidae, Bramidae, Centrolophidae, Nomeidae, Stromateidae, in Fischer, W., Bianchi, G., & W.B. Scott (ed.), FAO species identification sheets for fishery purposes. Eastern Central Atlantic; fishing areas 34, 47 (in part). Canada Funds-in-Trust. Ottawa, Department of Fisheries and Oceans Canada, by arrangement with the Food and Agriculture Organization of the United Nations, vol. 1-7: pag. var.

--. 1986. Bramidae, in P.J.P. Whitehead, M.L. Bauchot, J.C. Hureau, J. Nielsen & E. Tortonese. Fishes of the North-eastern Atlantic and the Mediterranean/Poissons de l'Atlantique du Nord-Est et de la Méditerranée, Paris, Unesco, 2:847-853, 6 fig.

--. Centrolophidae, in P.J.P. Whitehead, M.L. Bauchot, J.C. Hureau, J. Nielsen & E. Tortonese. Fishes of the North-eastern Atlantic and the Mediterranean/Poissons de l'Atlantique du Nord-Est et de la Méditerranée, Paris, Unesco, 3:1177-1182, 4 fig.

--. 1986. Nomeidae, in P.J.P. Whitehead, M.L. Bauchot, J.C. Hureau, J. Nielsen & E. Tortonese. Fishes of the North-eastern Atlantic and the Mediterranean/Poissons de l'Atlantique du Nord-Est et de la Méditerranée, Paris, Unesco, 3:1183-1188, 4 + 1 fig.

--. 1986. Tetragonuridae, in P.J.P. Whitehead, M.L. Bauchot, J.C. Hureau, J. Nielsen & E. Tortonese. Fishes of the North-eastern Atlantic and the Mediterranean/Poissons de l'Atlantique du Nord-Est et de la Méditerranée, Paris, Unesco, 3:1189-1191, 2 fig.

--. 1986. Stromateidae, in P.J.P. Whitehead, M.L. Bauchot, J.C. Hureau, J. Nielsen & E. Tortonese. Fishes of the North-eastern Atlantic and the Mediterranean/Poissons de l'Atlantique du Nord-Est et de la Méditerranée, Paris, Unesco, 3:1192-1193, 1 fig.

HAEDRICH, R. L.; CERVIGON, F. 1969. Distribution of the centrolophid fish _Schedophilus pemarco_, with notes on its biology. _Breviora_, (340):1-9.

HAEDRICH, R. L.; CRADDOCK, J. E. 1968. Distribution and biology of the opisthoproctid fish _Winteria telescopa_ Brauer 1901. _Breviora_, (294):1-11, 4 fig.

HAEDRICH, R. L.; HENDERSON, N. R. 1974. Pelagic food of _Coryphaenoides armatus_, a deep benthic rattail. _Deep Sea Res._, 21:739-744, 1 fig.

HAEDRICH, R. L.; HORN, M. H. 1969. A key to the stromateoid fishes. Technical Report. Woods Hole Oceanographic Institution Ref. No 69-70, Sept. 1969, 46 p. (Unpublished manuscript). (Second ed., 1972, Ref. No 72-15).

HAEDRICH, R. L.; NIELSEN, J. 1966. Fishes eaten by _Alepisaurus_ (Pisces, Iniomi) in the southeastern Pacific Ocean. _Deep Sea Res._, 13:909-919.

HAEDRICH, R. L.; POLLONI, P. T. 1974. Rarely seen fishes captured in Hudson Submarine Canyon. _J. Fish. Res. Bd Can._, 31:231-234.

--. 1976. A contribution to the life history of a small rattail fish, _Coryphaenoides carapinus_. _Bull. Sth. Calif. Acad. Sci._, 75(2):203-211, 3 fig.

HAEDRICH, R. L.; ROWE, G. T.; POLLONI, P. T. 1975. Zonal and faunal composition of epibenthic populations on the continental slope south of New England. _J. mar. Res._, 33:191-212.

HAEDRICH, R. L.; WITTENBERG, J.; NELSON, G. J. 1973. A septum in the eye of osteoglossoid fishes. _Copeia_, 1973(3):594-595.

HAFFNER, R. E. 1952. Zoogeography of the bathypelagic fish, _Chauliodus_. _Syst. Zool._, 1:112-133.

HAIGH, E. H. 1972. Larval development of three species of economically important South African fishes. _Ann. S. Afr. Mus._, 59(3):47-70, fig. 1-11.

HALLIDAY, R. G. 1970. Growth and vertical distribution of the glacier lanternfish, _Benthosema glaciale_, in the northwestern Atlantic. _J. Fish. Res. Bd Can._, 27:105-116.

HALLIDAY, R. G.; SCOTT, W. B. 1969. Records of mesopelagic and other fishes from the Canadian Atlantic with notes on their distribution. _J. Fish. Res. Bd Can._, 26(10):2691-2702.

HALY, A. 1875. Descriptions of new species of fish in the collection of the British Museum. _Ann. Mag. nat. Hist._, (4)15:268-270.

HAMDY, A. R. 1973. Studies on the batoid features in the skull of _Pteroplatea altavela_. The neurocranium. _Proc. Egypt. Acad. Sci._, 24:11-14.

--. 1973. Studies on the batoid features in the skull of _Pteroplatea altavela_. The visceralcranium. _Proc. Egypt. Acad. Sci._, 24:15-18.

HAMDY, A. R.; KHALIL, M. S. 1963. The hyoid arch in Batoidei. _Proc._

Egypt. Acad. Sci., 17:70-73, fig. 1-4.

--. 1963. The connections and relations of the neurocranium and viscero-
cranium of Rhinobatus granulatus, Raia miraletus, Pteroplatea altave-
la, Aetomylus milvus, and Stoasodon narinari. Proc. Egypt. Acad.
Sci., 17:60-69, fig. 1-5.

HAMILTON, H. 1916. Notes on the occurrence of the genus Trachipterus in
New Zealand. Trans. Proc. N.Z. Inst., 48:370-382, fig.

HAMILTON-BUCHANAN, F. 1822. An account of the fishes found in the River
Ganges and its branches. Edinburgh, London, 405 p., 39 pl.

HAMMOND, D. L.; CUPKA, D. M. 1975. A sportsman's field guide to the
billfishes, mackerels, little tunas and tunas of South Carolina.
Educ. Rep. S. Carol. Wildl. mar. Res. Dep., (3):1-32, 27 fig.

HANEDA, Y. 1938. Über den Leuchtfisch Malacocephalus laevis (Lowe). Jap.
J. med. Sci. Trans. Abstr., (Biophys.), 5(3):355-366.

--. 1951. The luminescence of some deep-sea fishes of the families Gadidae and Macrouridae. Pacif. Sci., 5:372-378, 4 fig.

HARDER, W. 1960. Zur Anatomie des Darmtraktes einiger Gadiformes und
Macruriformes (Osteichthyes, Pisces). Kurze Mitt. Inst. FischBiol.
Univ. Hamb., (10):1-41, 27 fig.

HARDY, A. C. 1956. The open sea. Its natural history: the world of
plankton. London, 335 p.

HARDY, J. D. Jr. 1978. Development of fishes of the Mid-Atlantic Bight.
An atlas of egg, larval and juvenile stages. 2. Anguillidae through
Syngnathidae. U.S. Fish. Wildl. Serv., biol. Serv. Program, FWS/OBS
78/12. 458 p., 245 fig.

--. 1978. Development of fishes of the Mid-Atlantic Bight. An atlas of
egg, larval and juvenile stages. 3. Aphredoderidae through Rachycentridae. U.S. Fish. Wildl. Serv., biol. Serv. Program, FWS/OBS 78/12.
394 p., 230 fig.

HARRIS, V. A. 1960. On the locomotion of the mud-skipper Periophthalmus
koelreuteri (Pallas)(Gobiidae). Proc. zool. Soc. Lond., 134:107-135,
fig. 1-16.

HARRISSON, C. M. H. 1967. On methods for sampling mesopelagic fishes.
Symp. zool. Soc. Lond., (19):71-126.

HARRISSON, C. M. H.; PALMER, G. 1968. On the neotype of Radiicephalus
elongatus Osorio with remarks on its biology. Bull. Br. Mus. nat.
Hist., (Zool.), 16(5):185-211, 6 fig.

HARRY, R. R. 1948. A new name for Paralepis danae, a species of fish
from near Cape Verde, Africa. Copeia, 1948(3):221.

--. 1951. Deep-sea fishes of the Bermuda Oceanographic Expeditions.
Family Paralepididae. Zoologica, N.Y., 36(1):17-35, fig. 1-9.

--. 1952. Deep-sea fishes of the Bermuda Oceanographic Expeditions.
Families Cetomimidae and Rondeletiidae. Zoologica, N.Y., 37(1):55-72,
1 pl, 4 fig.

--. 1953. Studies on the bathypelagic fishes of the family Paralepididae. 1. Survey of the genera. Pacif. Sci., 7(2):219-249, fig. 1-22.

--. 1953. Studies on the bathypelagic fishes of the family Paralepididae
(Order Iniomi). 2. A revision of the North Pacific species. Proc.
Acad. nat. Sci. Philad., 105:169-230, fig. 1-28.

HART, J. L. 1973. Pacific fishes of Canada. Bull. Fish. Res. Bd Can.,
180: ix, 1-740, fig., pl.

HART, T. J.; MARSHALL, N. B. 1951. Breeding ground of the pilchards off
the coast of South-West Africa. Nature, Lond., 168:272.

HARTMANN, A. R.; CLARKE, T. A. 1975. The distribution of myctophid fishes across the central equatorial Pacific. Fishery Bull., 73(3):633-
641.

HARTMANN, J. 1970. Juvenile saury pike (Scomberesox saurus Walb.), an
example of ichthyoneuston. J. Cons. perm. int. Explor. Mer, 33(2):

245-255, 9 fig.

--. 1970. Verteilung und Nahrung des Ichthyoneuston im subtropischen Nordostatlantik. "Meteor" Forsch. Ergebn., (D)(8):1-60.

HARTSUIJKER, L. 1973. Premiers résultats commerciaux de l'introduction de la seine tournante dans la pêche piroguière de la Petite Côte du Sénégal. UNDP (SF)/FAO, Project SEN/66/508, Survey Devel. pelag. Fish Resources. Docum. scient. provis. Centre Rech. océanogr. Dakar-Thiaroye, (Tech. Rept. 1/1973): 29 p. (mimeo).

HARVEY, E.N. 1931. Stimulation by adrenalin of the luminescence of deep-sea fish. Zoologica, N.Y., 12(6):67-69.

HARWOOD, I. 1827. On a newly discovered genus of serpentiform fishes. Phil. Trans. R. Soc., 117:49-57, fig. 1-4, pl. 7.

HASAN, C.; SERBAN, A. 1959. Contribution à l'étude des écailles de quelques poissons téléostéens de la Mer Noire. Lucr. Sta. zool. marit. Agigea, (vol. festiv.):173-194, fig. 1-23.

HASS, H. 1969. Unterscheidungsmerkmale dottersackloser Larven von Clupeiformes aus der Unterelbe. Arch. FischWiss., 20(1):22-25.

HASSELT, J. C., van. 1824. Extrait d'une seconde lettre sur les poissons de Java, écrite par M. van Hasselt à M. C.-J. Temminck, datée de Tjecande, résidence de Bantam, 29 décembre 1822. Bull. Sci. nat. Géol. (de Férussac), 2(2):374-377.

HAÜY, R. J. 1787. Encyclopédie méthodique. Histoire naturelle. 3. Poissons. Paris, Liège, 435 p.

HEBIG, W. 1978. Neubeschreibungen von Fischen in der Zeitschrift "Aquarien Terrarien", nebst Bemerkungen zu einer neuen Farbvariante von Girardinus metallicus Poey, 1854. Zool. Abh. staat. Mus. Tierk, Dresden, 35(7):113-128.

HECHT, T.; HECHT, A. 1978. A descriptive systematic study of the otoliths of the neopterygian marine fishes of South Africa. Part II. The delimitation of teleost orders, some systematic notes and a provisional new phyletic order sequence. Trans. R. Soc. S. Afr., 43(2):199-218, fig. 1-37.

HECKEL, J. J. 1840. Ichthyologische Beiträge zu den Familien der Cottoiden, Scorpaenoiden, Gobioiden und Cyprinoiden. Annln naturh. Mus. Wien, 2:145-164, pl. 8-9.

HECKEL, J. J.; KNER, R. 1858. Die Süsswasserfische der Österreichischen Monarchie mit Rücksicht auf die angrenzenden Länder. Leipzig, xii + 388 p., 203 fig.

HECTOR, J. 1875. Notes on New Zealand ichthyology. Trans. Proc. N. Z. Inst., 7:239-250.

--. 1877. Notes on New Zealand ichthyology. Trans. Proc. N. Z. Inst., 9 (62):465-469, pl.

HEEMSTRA, P. C. 1980. A revision of the zeid fishes (Zeiformes: Zeidae) of South Africa. Ichthyol. Bull. Rhodes Univ., (41):1-18, 5 fig., 2 pl.

--. 1981. Centracanthidae, Emmelichthyidae, in Fischer, W., Bianchi, G. & W.B. Scott (ed.), FAO species identification sheets for fishery purposes. Eastern Central Atlantic; fishing areas 34, 47 (in part). Canada Funds-in-Trust. Ottawa, Department of Fisheries and Oceans Canada, by arrangement with the Food and Agriculture Organization of the United Nations, vol. 1-7: pag. var.

--. 1986. Families Trachichthyidae and Stephanoberycidae. In: Smith, M.M. & P.C. Heemstra, (ed.). Smith's Sea Fishes. Johannesburg, Macmillan South Africa (Publ.) Ltd., 1047 p., 144 pl.

HEEMSTRA, P. C.; KANNEMEYER, S. X. 1984. The families Trachipteridae and Radiicephalidae (Pisces, Lampriformes) and a new species of Zu from South Africa. Ann. S. Afr. Mus., 94(2):13-39, 11 fig., pl., 2 tab.

HEEMSTRA, P. C.; RANDALL, J. E. 1977. A revision of the Emmelichthyidae

(Pisces: Perciformes). _Aust. J. mar. Freshwat. Res._, **28**:361-396, fig. 5.

HEEZEN, B.C.; HOLLISTER, C.D. 1971. The face of the deep. London, 659 p.

HEFFORD, A. E. 1910. Notes on teleostean ova and larvae observed at Plymouth in spring and summer 1909. _J. mar. biol. Ass. U. K._, **9**(1): 1-58, 2 pl., 16 fig.

HELDEN, L. V.; WIRTZ, P. 1985. A comparison of _Blennius ocellaris_ L. 1758, _B. riodourensis_ Metzelaar 1919, and _B. normani_ Poll, 1949 (Pisces, Blenniidae). _Spixiana_, München, **8**(2):197-217.

HELLER, E.; SNODGRASS, R. E. 1903. Papers from the Hopkins Stanford Galapagos Expedition, 1898-1899. XV. New fishes. _Proc. Wash. Acad. Sci._, **5**:189-229.

HEM, S. 1976. Etat des stocks pélagiques côtiers en Côte d'Ivoire. _Docums scient. Centre Rech. océanogr. Abibjan_, **7**(2):1-20.

HEMMING, F. (ed.). 1954. Opinion 225. Validation, under the plenary powers, of _Belone_ Cuvier, 1817 (Class Pisces) and _Raphistoma_ Hall, 1847 (Class Gastropoda). _Opin. Decl. int. Commn zool. Nom._, **4**(15): 161-176.

HEMPEL, G.; WEIKERT, H. 1972. The neuston of the subtropical and boreal north-eastern Atlantic Ocean. A review. _Mar. Biol._, Berlin, **13**(13): 70-88, 16 fig., 6 tab.

HENLE, F. G. J. 1834. Ueber _Narcine_, eine neue Gattung electrischer Rochen nebst einer Synopsis der electrischen Rochen. Berlin, 44 p., 4 pl.

HENTSCHEL, E. 1936. Allgemeine Biologie des südatlantischen Ozeans. _Wiss. Ergebn. dt. atlant. Exped. "Meteor"_, **11**(2):i-xii + 1-344 p.

HERALD, E. S. 1953. Family Syngnathidae: Pipefishes. _In_: Schultz, L. P., _et al._, Fishes of the Marshall and Marianas Islands. _Bull. U. S. nat. Mus._, Bull. 202, 1:231-278, fig. 36-44.

HERALD, E. S. 1961. Living fishes of the world. London, 303 p., 145 pl.

HERALD, E. S.; RANDALL, J. E. 1972. Five new Indo-Pacific pipefishes. _Proc. Calif. Acad. Sci._, (4)**39**(11):121-140, 6 fig.

HERMANN, J. 1781. Ueber ein neues amerikanisches Fischgeschlecht, _Sternoptyx diaphana_, der durchsichtige Brustfaltenfisch. _Naturforscher_, **16**:33-34.

--. 1804. Observationes zoologicae quibus novae complures aliaeque animalium species discribuntur et illustrantur: opus posthumum edidit F. L. Hammer. Pars prior. Parisiis, Argentorati, 8 + 332 p.

HERRE, A. W. C. T. 1923. A review of the eels of the Philippine Archipelago. _Philipp. J. Sci._, **23**(2):123-236, 11 pl., 14 fig.

--. 1926. Four new Philippine fishes. _Philipp. J. Sci._, **31**:533-541, 3 pl.

--. 1927. Gobies of the Philippines and the China Sea. _Monogr. Philipp. Bur. Sci._, (23):352 pp.

--. 1930. _Busuanga_ Herre, new genus. _Science, N.Y._, **71**:132.

--. 1931. A check list of the fishes from the Solomon Islands. _J. panPacif. Res. Instn_, **6**(4):4-9.

--. 1936. Fishes of the Crane Pacific Expedition. _Publs Field Mus. nat. Hist._, (353)(Zool. 21):1-472, fig. 1-50.

--. 1942. Notes on a collection of fishes from Antigua and Barbados, British West Indies. _Stanford Univ. Publs_ (biol. Sci.), **7**(2):285-305.

--. 1945. Additions to the fish fauna of the Philippine Islands. _Copeia_, 1945(3):146-149.

--. 1946. New genera of Eleotridae and Gobiidae and one new species from West Africa. _Proc. biol. Soc. Wash._, **59**:121-126.

--. 1950. A new name for _Hanno_, a genus of African gobies. _Stanford ichthyol. Bull._, 3:198.

--. 1953. Check list of Philippine Fishes. _Res. Rep. U.S.. Fish. Wildl._

Serv., 20:1-977.

HERRING, P. J. 1972. Bioluminescence of searsid fishes. J. mar. biol. Ass. U. K., 52:879-887, fig. 1-2.

--. 1976. Carotenoid pigmentation of whalefishes. Deep Sea Res., 23:235-238.

--. 1977. Bioluminescence in an evermannellid fish. J. Zool., 181:297-307.

HERSEY, J. B. 1967. Deep sea photography. Johns Hopkins oceanogr. Stud., 3:310 p.

HICKLING, C. F. 1925. A new type of luminescence in fishes. J. mar. biol. Ass. U.K. (n.s.), 13:914-929.

--. 1926. A new type of luminescence in fishes II. J. mar. biol. Ass. U. K., (n.s.)14:495-507.

--. 1931. A new type of luminescence in fishes III. The gland in Coelorhynchus coelorhynchus Risso. J. mar. biol. Ass. U.K., 17:853-867.

--. 1951. Report of the Fisheries Adviser on his visit to the Gambia. Govt. Printer, Bathurst, spec. Publ. No 9/51, 7 p.

HIKITA, T. 1957. A preliminary list of fishes from the northern Pacific Ocean. J. Fac. Sci. Hokkaido Univ., (6)13:46-48.

HILDEBRAND, H. H. 1958. Estudios biológicos preliminares sobre la laguna Madre de Tamaulipas. Ciencia Méx., 17(79):151-173.

HILDEBRAND, S. F. 1943. Notes on the affinity, anatomy and development of Elops saurus Linnaeus. J. Wash. Acad. Sci., 33:90-94.

--. 1946. A descriptive catalogue of the shore fishes of Peru. Bull. U.S. natn. Mus., 189: xi + 530 p., 95 fig.

--. 1948. A new genus and five new species of American fishes. Smithson. Misc. Collns, 110(9):1-15.

--. 1963. Family Elopidae. In: Fishes of the western North Atlantic. Mem. Sears Fdn mar. Res., 1(3):111-131, fig. 19-21.

--. 1964. Family Engraulidae, p. 152-249 and Family Clupeidae (excluding Harengula and Dorosoma), p. 257-454. Fishes of the western North Atlantic. Mem. Sears Fdn mar. Res., 1(3):1-630.

HILGENDORF, F. M. 1878. Ueber das Vorkommen einer Brama-Art und einer neuen Fischgattung Centropholis aus der Nachbarschaft des Genus Brama in den japanischen Meeren. Sber. Ges. naturf. Freunde Berl., 1878:1-2.

--. 1879. Beiträge zur Ichthyologie Japan's. Sber. Ges. naturf. Freunde Berl., 1879:78-81.

--. 1888. Die Fische der Azoren. Arch. Naturgesch., 54:205-213.

--. 1889. Ueber eine Fischsammlung von Haiti, welche zwei neue Arten, Poecilia tridens und Eleotris maltzani, enthält. Sber. Ges. naturf. Freunde Berl., 1889:51-55.

HILTON, T. 1961. The coastal fisheries of Ghana. Bull. Ghana geogr. Ass., 9:34-50.

HIRASAKA, K.; NAKAMURA, H. 1947. On the Formosan spearfishes. Bull. oceanogr. Inst. Taiwan., (3):9-24, 1 fig., 3 pl.

HIYAMA, Y. 1940. Description of two new species of fish, Raja tobitukai and Chlorophthalmus acutifrons. Jap. J. Zool., 9(1):169-170, fig. 1-2.

HJORT, J. 1912. Fishes from the sea-bottom: 387-456, fig. 253-314. Pelagic animal life: 561-659, fig. 389-489. In: Murray, J. & J. Hjort. The Depths of the Ocean. London, xx + 821 p., 575 fig., 9 pl., 4 maps.

HOEK, P. P. C. 1888. Aanteekeningen omtrent larven en jongen voornaamste in het verslag besproken vischsoorten. Tijdschr. ned. dierk. Vereen., (Suppl. Pt 2):274-319, 4 pl.

HOESE, D. F. 1986. Family No. 240: Gobiidae. Family No. 241: Eleotridae. In: Smith, M.M. & P.C. Heemstra (ed.), Smiths' Sea Fishes, 774-811.

Johannesburg, Macmillan South Africa (Publ.), 1047 p., 144 pl.
HOESE, D. H.; MOORE, R. H. 1977. Fishes of the Gulf of Mexico, Texas, Louisiana, and adjacent waters. Texas, xv + 327 p., fig.
HOESE, D. R.; WINTERBOTTOM, R. 1979. A new species of <u>Lioteres</u> (Pisces, Gobiidae) from Kwazulu, with a revised checklist of South African gobies and comments on the generic relationships and endemism of western Indian Ocean gobioid. <u>Life Sci. occ. Pap. r. Ontario Mus.</u>, (31):1-13, 2 fig.
HOESLANDT, H. 1958. Reproduction d'alose atlantique (<u>Alosa alosa</u> Linné) et transfert au Bassin Méditerranéen. <u>Verh. int. Verein. theor. angew. Limnol.</u>, 13:736-742, 1 fig., pl. 4-5.
HOEVEN, J. van der. 1858. Handbook of Zoology (transl. by Rev. W. Clarke). Cambridge, 2 vol., 1856-58.
HOLLARD, H. L. G. M. 1853. Monographie de la famille des balistides "Introduction" and "Première partie. Des Balistides en général". <u>Annls Sci. nat. (Zool.)</u>, (3)(20):71-114, pl. 1-3.
--. 1855. Monographie de la famille des balistides. Deuxième partie (1). <u>Annls Sci. nat. (Zool.)</u>, (4)(1):39-72, 303-339, 3 pl.
--. 1855. Monographie de la famille des balistides. <u>Annls Sci. nat. (Zool.)</u>, (4)(4):5-27, pl. 1 (5 fig.).
--. 1857. Monographie de la famille des ostracionides. <u>Annls Sci. nat. (Zool.)</u>, (4)(7):121-170, pl. 13 (9 fig.).
HOLLISTER, G. 1937. Caudal skeleton of the Bermuda shallow water fishes III. Order Iniomi: Synodontidae. <u>Zoologica, N.Y.</u>, 22(4):385-399.
HOLLY, M. 1930. Synopsis der Süsswasserfische Kameruns. <u>Sber. Akad. Wiss. Wien</u>, (1)139(3-4):195-281, fig., 2 pl.
HOLM, A. 1957. Specimina Linnaeana i Uppsala bevarade zoologiska samlingar från Linnés tid. <u>Uppsala Univ. Arsskr</u>, 6:1-68.
HOLT, E. W. L. 1890. On the ova of <u>Gobius</u>. <u>Ann. Mag. nat. Hist.</u>, (6)6: 34-40.
--. 1891. Survey of fishing grounds, west coast of Ireland, 1890. I: On the eggs and larvae of Teleosteans. <u>Scient. Trans. R. Dubl. Soc.</u>, (2)4(7):435-474, pl. 47-52.
--. 1891. Survey of fishing grounds, west coast of Ireland, 1890. Preliminary note on the fish obtained during the cruise of the SS. "Fingal", 1890. <u>Scient. Proc. R. Dubl. Soc.</u>, 7:121-123.
--. 1893. Survey of fishing grounds, west coast of Ireland, 1890-91: On the eggs and larval and post-larval stages of Teleosteans. <u>Scient. Trans. R. Dubl. Soc.</u>, (2)5(2):5-120, 15 pl.
--. 1898. La Girelle royale et la Girelle de Giofredi doivent-elles toutes deux être rapportées à l'espèce dimorphique <u>Coris julis</u> (Linné)? <u>Bull. Mus. Hist. nat. Marseille</u>, 1:151-162, 1 fig., 1 tab.
--. 1898. Contributions to our knowledge of the plankton of the Faeroe Channel. No 5. Report on a collection of very young fishes obtained by Dr. G.H. Fowler. <u>Proc. zool. Soc. Lond.</u>, 1898:550-566, pl. 46-47.
--. 1899. Recherches sur la reproduction des poissons osseux principalement dans le Golfe de Marseille. <u>Annls Mus. Hist. nat. Marseille</u>, 5 (2):1-128, pl. 1-9 (106 fig.).
HOLT, E. W. L.; BYRNE, L. W. 1898. An observation of the colour-changes of a wrasse, <u>Labrus maculatus</u>, Donovan. <u>J. mar. biol. Ass. U.K.</u>, (2)5 :193.
--. 1898. Notes on the reproduction of teleostean fishes in the southwestern district. <u>J. mar. biol. Ass. U. K.</u>, 5:333-340.
--. 1906. The marine fauna of the coast of Ireland. Part VIII. First report on the fishes of the Irish Atlantic slope. <u>Rep. Sea inld Fish. Ire.</u>, (2)(1905)(2):29-54, 2 fig., 1 pl. (also as: <u>Scient. Invest. Fish. Brch Ire.</u>, (1905)(2):1-28, 2 fig., 1 pl.).
--. 1907. Biscayan plankton. Pt X. The fishes. <u>Trans. Linn. Soc. Lond.</u>,

(2)**10**(7):189-201, fig. 1-5.
--. 1908. Second report on the fishes of the Irish Atlantic slope. <u>Rep. Sea inld Fish. Ire.</u>, (2)(1906)(5):141-201, 3 fig., 5 pl. (also as: <u>Scient. Invest. Fish. Brch Ire.</u>, (1906)(5):1-63, 3 fig., 5 pl.).
--. 1908. New deep-sea fishes from the south-west coast of Ireland. <u>Ann. Mag. nat. Hist.</u>, (8)1:86-95, fig., pl. 3.
--. 1909. Preliminary note on some fishes from the Irish Atlantic slope. <u>Ann. Mag. nat. Hist.</u>, (8)3:279-280.
--. 1910. Third report on the fishes of the Irish Atlantic slope. The Holocephali or Chimaeras. <u>Scient. Invest. Fish. Brch Ire.</u>, (1908)(4): 1-26, 3 fig., 4 pl.
--. 1911. Fifth report on the fishes of the Irish Atlantic slope. The fishes of the genus <u>Scopelus</u>. <u>Scient. Invest. Fish. Brch Ire.</u>, (1910) (6):1-33, 8 fig., 1 pl.
--. 1913. Sixth report on the fishes of the Irish Atlantic slope. The families Stomiatidae, Sternoptychidae and Salmonidae. <u>Scient. Invest. Fish. Brch Ire.</u>, (1912)(1):1-28, 11 fig., 2 pl.
HOLT, E. W. L.; CALDERWOOD, W. L. 1895. Report on the rarer fishes. <u>In</u>: Survey of fishing-grounds, west coast of Ireland, 1890-91. <u>Scient. Trans. R. Dubl. Soc.</u>, 5(2):360-524, pl. 39-44.
HOLTEN, H. S. 1802. <u>Trichiurus gladius</u>, en ny fisk fra Portugal. <u>Skr. naturh. Selsk. Kjøbenhavn</u>, 5:19-26, fig.
HOLTHUIS, L. B. 1966. Comments on the type-species of <u>Sciaena</u> Linnaeus, 1758. <u>Bull. zool. Nom.</u>, 23:2.
HOLZLÖHNER, S. 1975. On the recent stock development of <u>Sardina pilchardus</u> Walbaum off Spanish Sahara. <u>ICES</u>, C.M. 1975/5 (13 p., mimeo).
HORN, M. H. 1972. Systematic status and aspects of the ecology of the elongate ariommid fishes (Suborder Stromateoidei) in the Atlantic. <u>Bull. mar. Sci.</u>, 22(3):537-558.
--. 1973. Systematic comparison of the stromateid fishes <u>Stromateus brasiliensis</u> Fowler and <u>Stromateus stellatus</u> Cuvier from coastal South America with a review of the genus. <u>Bull. Br. Mus. nat. Hist.</u> (Zool.), 24(7):317-339, 9 fig.
HORN, M. H.; TEAL, J. M.; BACKUS, R. H. 1970. Petroleum lumps on the surface of the sea. <u>Science, N.Y.</u>, 168(3928):245-246.
HORNELL, J. 1928. Report on the fishery resources of Sierra Leone. Govt. Printing Office, Freetown, 51 p.
--. 1929. The principal fishes of economic value in Sierra Leone. <u>Sierra Leone Stud.</u>, 14:3-9.
--. 1935. Report on the fisheries of Palestine. London, 106 p., 14 fig.
HORSBURGH, D. B. 1935. A revision of two species of <u>Vinciguerria</u>, a genus of deep-sea fishes. <u>Proc. Calif. Acad. Sci.</u>, (4)21(19):225-232.
HOUOT, G. S.; WILLM, P. H. 1955. Two thousand fathoms down. New York, 192 p.
HOUTTUYN, M. 1764. Natuurlijke historie of uitvoerige Beschrijving der dieren, planten en mineraalen, volgens het samenstel van den heer Linnaeus. 1 Deel, 7. Amsterdam, 446 p., pl. 57-62.
--. 1782. Beschrijving van eenige Japansche visschen, en andere zeeschepselen. <u>Verh. Holl. Maatsch. Weet. Haarlem</u>, 20(2):311-350.
HOWARD, J. K.; STARCK II, W. A. 1975. Distribution and relative abundance of billfishes (Istiophoridae) of the Indian Ocean. <u>Stud. trop. Oceanogr.</u>, (13): i-viii + 1-31, 38 maps in atlas.
HOWARD, J. K.; UEYANAGI, S. 1965. Distribution and relative abundance of billfishes (Istiophoridae) of the Pacific Ocean. <u>Stud. trop. Oceanogr.</u>, (2): i-x + 1-134, 37 fig., 38 maps in atlas.
HOWARD, K. T.; MARSHALL, N. B.; WIMPENNY, R. S. 1955. A voyage to the black hake grounds off the west African coast. <u>Rapp. P-v. Réun. Cons. perm. int. Explor. Mer</u>, 137:53-56.

HOWE, K. M.; STEIN, D. L.; BOND, C. E. 1979. First records off Oregon of the pelagic fishes *Paralepis atlantica*, *Gonostoma atlanticum* and *Aphanopus carbo*. *Fishery Bull.*, 77(3):700-703.

HOWELL-RIVERO, L. 1932. The apodal fishes of Cuba. *Proc. New Engl. zool. Club*, 13:3-26.

--. 1935. Some new and rare Cuban eels. *Mems Soc. cub. Hist. nat. "Felipe Poey"*, 8(6):339-344, fig. 1-3.

HUBBS, C. L. 1916. Notes on the marine fishes of California. *Univ. Calif. Publs Zool.*, 16(13):153-169, pl. 18-20.

--. 1929. *Oostethus*: a new generic name for a doryrhamphine pipefish. *Occ. Pap. Mus. Zool. Univ. Mich.*, 199:1-4.

--. 1929. The generic relationships and nomenclature of the Californian sardine. *Proc. Calif. Acad. Sci.*, (4)18(11):261-265.

--. 1938. Fishes from the caves of Yucatan. *In*: A.S. Pearse (ed.). Fauna of the caves of Yucatan. *Publs Carnegie Instn*, (491):261-295, 4 pl.

--. 1944. Species of the circumtropical fish genus *Brotula*. *Copeia*, 1944:162-178.

--. 1945. Phylogenetic position of the Citharidae, a family of flatfishes. *Misc. Publs Mus. Zool. Univ. Mich.*, (63):1-38, 1 fig.

--. 1958. *Dikellorhynchus* and *Kanazawaichthys*: nominal fish genera interpreted as based on prejuveniles of *Malacanthus* and *Antennarius*, respectively. *Copeia*, 1958:282-285.

--. 1966. Comments on the type-species of *Sciaena* Linnaeus, Z.N.(S.) 850. *Bull. zool. Nom.*, 23:2-5.

HUBBS, C. L.; FOLLETT, W. I. 1978. Anatomical notes on an adult male of the deep-sea ophidiid fish *Parabassogigas grandis* from off California. *Proc. Calif. Acad. Sci.*, 41:388-399.

HUBBS, C. L.; GIOVANNOLI, L. 1931. Records of the rare Sunfish *Masturus lanceolatus* for Japan and Florida. *Copeia*, 1931(3):135-137.

HUBBS, C. L.; KAMPA, E. M. 1946. The early stages (egg, prolarva, and juvenile) and the classification of the California flyingfish. *Copeia*, 1946(4):188-218, 5 fig.

HUBBS, C. L.; RECHNITZER, A. B. 1958. A new fish, *Chaetodon falcifer*, from Guadalupe Island, Baja California, with notes on related species. *Proc. Calif. Acad. Sci.*, (4)29(8):273-313, 1 tab., 3 pl., 1 map.

HUBBS, C. L.; WISNER, R. L. 1964. *Parvilux*, a new genus of myctophid fishes from the north-eastern Pacific, with two new species. *Zoöl. Meded., Leiden*, 39:445-463.

--. 1980. Revision of the sauries (Pisces, Scomberesocidae) with descriptions of two new genera and one new species. *Fishery Bull.*, 77 (3): 521-66, 17 fig.

HUBBS, C. L.; FOLLETT, W. I.; DEMPSTER, L. J. 1979. List of the fishes of California. *Occ. Pap. Calif. Acad. Sci.*, (133):1-51.

HUBBS, C. L.; MEAD, G. W.; WILIMOVSKI, N. J. 1953. The widespread, probably antitropical distribution and the relationship of the bathypelagic iniomous fish *Anotopterus pharao*. *Bull. Scripps Instn Oceanogr.*, 6(5):173-197, 1 fig., pl. 2-6.

HUBRECHT, A. A. W. 1881. On a collection of fishes from the St. Paul's river, Liberia, with description of three new species. *Notes Leyden Mus.*, 3:66-71.

HUGHES, G. M.; IWAI, T. 1978. A morphometric study of the gills in some Pacific deep-sea fishes. *J. Zool.*, 184:155-170, 4 fig., 2 pl.

HULLEY, P. A. 1966. The validity of *Raja rhizacanthus* Regan and *Raja pullopunctata* Smith, based on a study of the clasper. *Ann. S. Afr. Mus.*, 48(20):497-514, fig. 1-8.

--. 1970. An investigation of the Rajidae of the west and south coasts of southern Africa. *Ann. S. Afr. Mus.*, 55(4):151-220, fig. 1-21, pl.

1-13.

--. 1972. The origin, interrelationship and distribution of southern African Rajidae. Ann. S. Afr. Mus., 60(1):1-103, 59 fig.

--. 1972. A report on the mesopelagic fishes collected during the deep-sea cruises of R.S. "Africana II", 1961-1966. Ann. S. Afr. Mus., 60(6):197-236, fig. 1-3.

--. 1972. Mesopelagic fishes from Vema Seamount (IK Station 52). Ann. S. Afr. Mus., 60(7):237-244, fig. 1-2.

--. 1984. Neoscopelidae, in Whitehead, P.J.P., Bauchot, M.L., Hureau, J.C., Nielsen, J. & E. Tortonese, Fishes of the North-eastern Atlantic and the Mediterranean/Poissons de l'Atlantique du Nord-Est et de la Méditerranée, Paris, Unesco, 1:426-428, 3 + 2 fig., 3 maps.

--. 1984. Myctophidae, in Whitehead, P.J.P., Bauchot, M.L., Hureau, J.C., Nielsen, J. & E. Tortonese, Fishes of the North-eastern Atlantic and the Mediterranean/Poissons de l'Atlantique du Nord-Est et de la Méditerranée, Paris, Unesco, 1:429-483, 59 + 69 fig., 59 maps.

--. 1985. Myctophidae, p. 320-326. In: Fischer, W. & Hureau, J.C. (ed.). FAO Species identification sheets for fishery purposes. Southern Ocean (Fishing Areas 48, 58 and 88). Prepared and printed with the support of the Commission for the Conservation of Antarctic Marine Living Resources (CCAMLR). Rome, vol. 2.

HULLEY, P. A.; STEHMANN, M. 1977. The validity of Malacoraja Stehmann 1970 (Chondrichthyes, Batoidei, Rajidae) and its phylogenetic significance. Ann. S. Afr. Mus., 72(12):227-237, fig. 1-5.

HUREAU, J.-C. 1973. Pomatomidae, p. 369-370, in: Hureau, J.C. & Th. Monod (ed.). Check-list of the fishes of the north-eastern Atlantic and of the Mediterranean/Catalogue des poissons du nord-est Atlantique et de la Méditerranée (Clofnam). Paris, Unesco, 2 vol.

--. 1978. Statut actuel des espèces de poissons décrites par A. Risso, p. 89-92. In: Monod, Th. & J.C. Hureau, Antoine Risso, 1777-1845, volume publié à l'occasion du bicentenaire de sa naissance. Annls Mus. Hist. nat. Nice (1977), 5.

--.1981. Polymixiidae, in: Fischer, W., Bianchi, G. & W.B. Scott (ed.), FAO species identification sheets for fishery purposes. Eastern Central Atlantic (Fishing Areas 34 and 47 in part). Dept. Fish. Oceans Canada, Ottawa/FAO, Rome vol. 1-7, pag. var.

--. 1984. Megalopidae, in Whitehead, P.J.P., M.L. Bauchot, J.C. Hureau, J. Nielsen & E. Tortonese, Fishes of the North-eastern Atlantic and the Mediterranean/Poissons de l'Atlantique du Nord-Est et de la Méditerranée, Paris, Unesco, 1:226-227, 1 map.

--. 1985. Bothidae, p. 260-264. In: Fischer, W.& Hureau, J.C. (ed.). FAO Species identification sheets for fishery purposes. Southern Ocean (Fishing Areas 48, 58 and 88). Prepared and printed with the support of the Commission for the Conservation of Antarctic Marine Living Resources (CCAMLR). Rome, vol. 2.

--. 1986. Polymixiidae, in P.J.P. Whitehead, M.L. Bauchot, J.C. Hureau, J. Nielsen & E. Tortonese. Fishes of the North-eastern Atlantic and the Mediterranean/Poissons de l'Atlantique du Nord-Est et de la Méditerranée, Paris, Unesco, 2:738-739, fig.

--. 1986. Priacanthidae, in P.J.P. Whitehead, M.L. Bauchot, J.C. Hureau, J. Nielsen & E. Tortonese. Fishes of the North-eastern Atlantic and the Mediterranean/Poissons de l'Atlantique du Nord-Est et de la Méditerranée, Paris, Unesco, 2:800-802, 2 fig.

--. 1986. Mullidae, in P.J.P. Whitehead, M.L. Bauchot, J.C. Hureau, J. Nielsen & E. Tortonese. Fishes of the North-eastern Atlantic and the Mediterranean/Poissons de l'Atlantique du Nord-Est et de la Méditerranée, Paris, Unesco, 2:877-882, 5 fig.

--. 1986. Uranoscopidae, in P.J.P. Whitehead, M.L. Bauchot, J.C. Hureau,

J. Nielsen & E. Tortonese. Fishes of the North-eastern Atlantic and the Mediterranean/Poissons de l'Atlantique du Nord-Est et de la Méditerranée, Paris, Unesco, 2:955-956, 1 fig.

--. 1986. Polynemidae, _in_ P.J.P. Whitehead, M.L. Bauchot, J.C. Hureau, J. Nielsen & E. Tortonese. Fishes of the North-eastern Atlantic and the Mediterranean/Poissons de l'Atlantique du Nord-Est et de la Méditerranée, Paris, Unesco, 3:1205-1206, 1 fig.

--. 1986. Triglidae, _in_ P.J.P. Whitehead, M.L. Bauchot, J.C. Hureau, J. Nielsen & E. Tortonese. Fishes of the North-eastern Atlantic and the Mediterranean/Poissons de l'Atlantique du Nord-Est et de la Méditerranée, Paris, Unesco, 3:1230-1238, 8 + 11 fig.

--. 1986. Peristediidae, _in_ P.J.P. Whitehead, M.L. Bauchot, J.C. Hureau, J. Nielsen & E. Tortonese. Fishes of the North-eastern Atlantic and the Mediterranean/Poissons de l'Atlantique du Nord-Est et de la Méditerranée, Paris, Unesco, 3:1239-1240, 1 + 1 fig.

--. 1986. Platycephalidae, _in_ P.J.P. Whitehead, M.L. Bauchot, J.C. Hureau, J. Nielsen & E. Tortonese. Fishes of the North-eastern Atlantic and the Mediterranean/Poissons de l'Atlantique du Nord-Est et de la Méditerranée, Paris, Unesco, 3:1241-1242, 1 fig.

HUREAU, J.-C.; FISCHER, W. 1985. Myxinidae, Petromyzonidae, p. 210. _In_: Fischer, W. & Hureau, J.C. (ed.). FAO Species identification sheets for fishery purposes. Southern Ocean (Fishing Areas 48, 58 and 88). Prepared and printed with the support of the Commission for the Conservation of Antarctic Marine Living Resources (CCAMLR). Rome, vol.2.

HUREAU, J.-C.; LITVINENKO, N. I. 1986. Scorpaenidae, _in_ P.J.P. Whitehead, M.L. Bauchot, J.C. Hureau, J. Nielsen & E. Tortonese. Fishes of the North-eastern Atlantic and the Mediterranean/Poissons de l'Atlantique du Nord-Est et de la Méditerranée, Paris, Unesco, 3:1211-1229, 18 + 6 fig.

HUREAU, J.-C.; MONOD, Th. (ed.). 1973. Check-list of the fishes of the north-eastern Atlantic and of the Mediterranean/Catalogue des poissons du nord-est Atlantique et de la Méditerranée (Clofnam). Paris, 1: xxii + 683 p.; 2: 331 p. [Second edition in 1980 with a Supplement edited by Tortonese, E. & J.C. Hureau, first published in _Cybium_ 1979, 3(1):5-66 (= p. 333-394 of the second edition)].

HUREAU, J.-C.; NIELSEN, J. G. 1981. Les poissons Ophidiiformes des campagnes du N.O. "Jean Charcot" dans l'Atlantique et la Méditerranée. _Cybium_, 5(3):3-28, 23 fig.

HUREAU, J.-C.; GEISTDOERFER, P.; RANNOU, M. 1979. The ecology of deep-sea benthic fishes. _Sarsia_, **64**:103-108.

HUREAU, J.-C.; STAIGER, J. C.; NIELSEN, J. G. 1979. A new species of deep-sea brotulid fish, _Typhlonus delosommatus_, from the tropical Atlantic Ocean. _Bull. mar. Sci._, **29**(2):272-277.

HUTTON, F. W. 1873. Contributions to the ichthyology of New Zealand. _Trans. Proc. N.Z. Inst._, 5:259-272, pl. 7-12 and 15 [15 inserted before 12].

--. 1875. Descriptions of new species of New Zealand fish. _Ann. Mag. nat. Hist._, (4)**16**:313-314.

--. 1891. List of New Zealand fishes. _Trans. Proc. N.Z. Inst._, 22:275-285.

HUTTON, F. W.; (HECTOR, J.). 1872. Fishes of New Zealand. Catalogue with diagnoses of the species; with notes on the edible species by James Hector. Wellington, 133 p., 12 pl.

IBANEZ, M. 1971. Variaciones morfométricas en el desarrollo postlarvario y juvenil de un pez batibéntico: _Trachyrhynchus scabrus_ (Raf.). San Sebastian, 99 p. (mimeo.).

IDA, H.; TOMINAGA, Y. 1971. On a _Macristium_ specimen from the Indian Ocean. _Jap. J. Ichthyol._, **18**:103.

IDYLL, C. P. 1964. Abyss. New York, 396 p., fig.

IHERING, H. von. 1897. Os peixes da costa do mar no estado do Rio Grande do Sul. _Revta Mus paul._, 2:25-63.

ILJIN, B. S. 1930. Le sytème des Gobiides. _Trab. Inst. esp. Oceanogr._, (2):63 p.

ILLIGER, J. K. W. 1811. Prodromus systematis mammalium et avium additis terminis zoogeographicis utriusque classis, eorumque versione germanica. Berolini, 301 p.

IMAI, S. 1954. On two flying-fishes of the genus _Hirundichthys_ Breder and their juveniles from Japan. _Mem. Fac. Fish. Kagoshima Univ._, 3 (2):62-72, fig. 1-8 (in Japanese).

--. 1957. On the Stomiatoidea of Suruga Bay and Sagami Bay. Suisangaku-shusei, p. 553-563. Tokyo Univ. Press, (in Japanese).

--. 1959. Studies on the life histories of the flying-fishes found in the adjacent waters of Japan. 1. _Mem. Fac. Fish. Kagoshima Univ._, 7: 1-85, pl. 1-41 (in Japanese).

INADA, T. 1981. Merlucciddae, _in_ Fischer, W., Bianchi, G. & W.B. Scott (ed.), FAO species identification sheets for fishery purposes. Eastern Central Atlantic; fishing areas 34, 47 (in part). Canada Funds-in-Trust. Ottawa, Department of Fisheries and Oceans Canada, by arrangement with the Food and Agriculture Organization of the United Nations, vol. 1-7: pag. var.

IRVINE, F. R. 1931. Salt water fish of the Gold coast. Nature Study Leafl., Accra, (D2):1-28, fig.

--. 1947. The fishes and fisheries of the Gold Coast. London, 352 p., 217 fig.

IRVINE, F. R.; BROWN, A. P. 1931. Salt water fish of the Gold Coast, pt. 2. Nature Study Leafl., Accra, (D2), 28 p., 99 fig.

ISAREV, A. T. 1971. Results of the fishes research studies at the Valdivia Bank. _Trudy Atlant. nauchno-issled. Inst. ryb. Khoz. Oceanogr._, (41):78-81. (in Russian).

ISHIYAMA, R. 1958. Studies on the rajid fishes (Rajidae) found in the waters around Japan. _J. Shimonoseki Coll. Fish._, 7(2/3):193-394, fig. 1-86, pl. 1-3.

ISHIYAMA, R.; HUBBS, C. L. 1968. _Bathyraja_, a genus of Pacific skates (Rajidae) regarded as phyletically distinct from the Atlantic genus _Breviraja_. _Copeia_, 1968(2):407-410, fig. 1-2.

IVERSEN, E. S.; YOSHIDA, H. O. 1957. Notes on the biology of the wahoo in the Line Islands. _Pacif. Sci._, 11:370-379.

IWAI, T. 1960. Luminous organs of the deep-sea squaloid shark, _Centroscyllium ritteri_ Jordan and Fowler. _Pacif. Sci._, 14:51-54.

IWAI, T.; NAKAMURA, J.; MATSUBARA, K. 1965. Taxonomic study of the tunas. _Spec. Rep. Misaki mar. biol. Inst._ (2):1-51 (in Japanese, English transl. _Bur. comml Fish. Ichthyol. Lab_. No 38).

IWAMOTO, T. 1966. The macrourid fish _Cetonurus globiceps_ in the Gulf of Mexico. _Copeia_, 1966(3):439-442, 3 fig.

--. 1970. Macrourid fishes of the Gulf of Guinea. _Stud. trop. Oceanogr._, (4)(2):316-431, 27 fig.

--. 1973. Macrourids (Gadiformes: Pisces) collected off Angola by the R/V Undaunted, with the description of a new species. _Proc. biol. Soc. Wash._, **86**(31):373-384, 3 fig.

--. 1981. Macrouridae, _in_ Fischer, W., Bianchi, G. & W.B. Scott (ed.), FAO species identification sheets for fisheries purposes. Eastern Central Atlantic; fishing areas 34, 47 (in part). Canada Funds-in-Trust. Ottawa, Department of Fisheries and Oceans Canada, by arrangement with the Food and Agriculture Organization of the United Nations, vol. 1-7: pag. var.

IWAMOTO, T.; GEISTDOERFER, P. 1985. Macrouridae, p. 292-305. In: Fischer, W. & Hureau, J.C. (ed.). FAO Species identification sheets for fishery purposes. Southern Ocean (Fishing Areas 48, 58 and 88). Prepared and printed with the support of the Commission for the Conservation of Antarctic Marine Living Resources (CCAMLR). Rome, vol. 2.

IWAMOTO, T.; STEIN, D. L. 1974. A systematic review of the rattail fishes (Macrouridae: Gadiformes) from Oregon and adjacent waters. Occ. Pap. Calif. Acad. Sci., (111):1-79, 25 fig.

IWAMOTO, T.; McCOSKER, J. E.; BARTON, O. 1976. Alepocephalid fishes of the genera Herwigia and Bathylaco, with the first Pacific record of H. kreffti. Jap. J. Ichthyol., 23(1):55-59, 1 fig.

JAGER, B. v. D. de. 1955. The development of the snoek (Thyrsites atun), a fish predator of the pilchard. Investl Rep. Div. Fish. Un. S. Afr., (19):1-16.

--. 1960. Synopsis on the biology of the South African pilchard Sardinops ocellata (Pappe). FAO Proc. Wld scient. Meet. Biol. Sardines, 2:97-114.

JAHN, A. E.; BACKUS, R. H. 1976. On the mesopelagic fish faunas of slope water, Gulf Stream and northern Sargasso Sea. Deep Sea Res., 23:223-234.

JAQUET, ML. 1907. Considérations sur les Scorpénides de la mer de Nice. Bull. Inst. océanogr. Monaco, (109):1-48, 33 fig.

--. 1920. Contribution à l'anatomie du Simenchelys parasiticus Gill. Résult. Camp. scient. Prince Albert I, 56:1-76, 5 pl.

JAMES, P.S.B.R. 1967. The ribbon-fishes of the family Trichiuridae of India. Mem. mar. biol. Ass. India, 1:1-228.

JENKINS, O. P. 1904. Report on collections of fishes made in the Hawaiian islands, with descriptions of new species. Bull. U.S. Fish Commn (1902), 22:417-511, 4 pl., 56 fig.

JENSEN, A. C. 1966. Life history of the spiny dogfish. Fishery Bull., 65: 527-554.

--. 1967. Observations on pelagic fishes off the west coast of Africa. Bull. mar. Sci., 17(1):42-51.

JENSEN, A. S. 1902. Ichthyologiske studier. Vidensk. Meddr dansk naturh. Foren., 63:191-215.

--. 1904. The fishes of East-Greenland. Meddr Grønland, 29:211-278, pl. 11-12.

--. 1926. Investigations of the "Dana" in West Greenland waters, 1925. Rapp. P.-v. Réun. Cons. perm. int. Explor. Mer, 39:85-100.

--. 1941. A new deep sea berycoid fish from Davis Strait Caristius groenlandicus n. sp. Vidensk. Meddr dansk naturh. Foren., 105:49-53, fig. 1, pl. 1.

--. 1942. Contribution to the ichthyofauna of Greenland 1-3. Spolia zool. Mus. haun., 2:7-44, 8 pl.

--. 1948. Contribution to the ichthyofauna of Greenland 8-24. Spolia zool. Mus. haun., 9:1-182, 26 fig., 4 pl.

JENYNS (post. Blomefield), L. 1835. A manual of British vertebrate animals. Cambridge, London, xxxii + 559 p.

--. 1842. The zoology of the voyage of H.M.S. Beagle, under the command of Captain Fitzroy, R.R,... during the years 1832 to 1836. (ed. C.W. Darwin) Pt. IV. Fish. London. 172 p., pl. 1-29

JESPERSEN, P. 1915. Sternoptychidae. Rep. Dan. oceanogr. Exped. Mediterr., (Biol. A2):1-41.

--.1916. Saccopharynx ampullaceus Harwood. Vidensk. Meddr dansk naturh. Foren., 67:145-154, 1 pl.

JESPERSEN, P.; TÅNING, A. V. 1919. Some Mediterranean Sternoptychidae. Vidensk. Meddr dansk naturh. Foren., 70:215-226, pl. 17.

--. 1926. Mediterranean Sternoptychidae. Rep. Dan. oceanogr. Exped. Mediterr., 2 (Biol., A 12):1-52, fig. 1-30.

JOHANESSON, K. A.; LOSSE, G. F. 1971. School parameters and abundance estimates of anchovy recorded off Takoradi, Ghana, in September 1971. UNDP (SF)/FAO, Results Regional Fisheries Survey West Africa, Abidjan, (Draft Rep. 16):40 p. (mimeo).

JOHN, H. -C. 1973. Oberflächennahes Ichthyoplankton der Kanarenstrom-Region. "Meteor" Forsch.-Ergebn., (D)(15):36-50.

--. 1975. Untersuchungen am oberflächennahen Ichthyoplankton des mittleren und südlichen Atlantischen Ozeans. Dissert., Fachber. math.-nat., Univ. Kiel, 1-186.

--. 1975. Fischereibiologische Untersuchungen in der argentinischen Seehechtfischerei. Ber. dt. wiss. Kommn Meeresforsch., 24:23-45.

--. 1976. Die Häufigkeit des Ichthyoplanktons an der Oberfläche des mittleren und südatlantischen Ozeans. Ber. dt. wiss. Kommn Meeresforsch., 25(1-2):23-36.

--. 1977. Zur Kenntnis des Ichthyoneustons im nordwest-afrikanischen Küstenwasser während des Winters. "Meteor" Forsch.-Ergebn., (D)(27): 1-4.

--. 1978. Solar and lunar rhythms in the occurrence of Astronesthes niger (Pisces) at the sea surface. J. inter-disciplinary Cycle Res., 9(3):169-177.

--. 1982. Horizontal and vertical distribution of sardine and other fish larvae. Rapp. P.-v. Réun. Cons. perm. int. Explor. Mer, 180:359-364.

JOHNELS, A. G. 1954. Notes on fishes from the Gambia River. Ark. Zool., (2)6:327-411, 19 fig.

JOHNSEN, S. 1927. On some bathypelagic stages of macrurid fishes. Nyt Mag. Naturvid., 65:221-242, 1 pl.

JOHNSON, F. R. 1957. The preparation of edible fishmeal from Sardinella (Ghana). Oceanography and sea fisheries on the West African coast, CCTA/CSA Symposium (20-27 November 1957, Luanda). (mimeo).

JOHNSON, J. Y. 1862. Descriptions of some new genera and species of fishes obtained at Madeira. Proc. zool. Soc. Lond., 1862(2):167-180, 1 fig., pl. 22-23.

--. 1862. Notes on rare and little known fishes taken at Madeira. Ann. Mag. nat. Hist., (3)10:161-172, 274-287; 11:237-248.

--. 1863. Descriptions of five new species of fishes obtained at Madeira. Proc. zool. Soc. Lond., 1863(1):36-46, pl. 7.

--. 1864. Descriptions of three new genera of marine fishes obtained at Madeira. Proc. zool. Soc. Lond., 1863(3):403-410.

--. 1864. Descriptions of three new genera of marine fishes obtained at Madeira. Ann. Mag. nat. Hist., (3)14:70-78.

--. 1865. Description of a new genus of trichiuroid fishes obtained at Madeira (Nealotus tripes), with remarks on the genus Dicrotus Günther, and on some allied genera of Trichiuridae. Proc. zool. Soc. Lond., 1865(2):434-437.

--. 1866. Description of Trachichthys darwini, a new species of berycoid fish from Madeira. Proc. zool. Soc. Lond., 1866(2):311-315, pl. 32.

--. 1890. On some new species of fishes from Madeira. Proc. zool. Soc. Lond., 1890(3):452-459.

JOHNSON, R. K. 1969. A review of the fish genus Kali (Perciformes: Chiasmodontidae). Copeia, 1969(3):386-391.

--. 1974. A Macristium larva from the Gulf of Mexico with additional evidence for the synonymy of Macristium with Bathysaurus (Myctophiformes: Bathysauridae). Copeia, 1974(4):973-977.

--. 1974. Five new species and a new genus of alepisauroid fishes of the

Scopelarchidae (Pisces: Myctophiformes). _Copeia_, 1974(2):449-457.

--. 1974. A revision of the alepisauroid family Scopelarchidae (Pisces, Myctophiformes). _Fieldiana: Zool._, 66:1-249, 60 fig., 10 tab.

--. 1975. A new myctophid fish, _Bolinichthys_ _distofax_, from the western and central North Pacific Ocean, with notes on other species of _Bolinichthys._ _Copeia_, 1975(1):53-60.

--. 1982. Fishes of the families Evermannellidae and Scopelarchidae: systematics, morphology, interrelationships and zoogeography. _Fieldiana: Zool._, (n.s.)(12):1-252, 74 fig.

--. 1984. Scopelarchidae, _in_ P.J.P. Whitehead, M.L. Bauchot, J.C. Hureau, J. Nielsen & E. Tortonese, Fishes of the North-eastern Atlantic and the Mediterranean/Poissons de l'Atlantique du Nord-Est et de la Méditerranée, Paris, Unesco, 1:484-488, 4 fig., 4 maps.

--. 1984. Evermannellidae, _in_ P.J.P. Whitehead, M.L. Bauchot, J.C. Hureau, J. Nielsen & E. Tortonese, Fishes of the North-eastern Atlantic and the Mediterranean/Poissons de l'Atlantique du Nord-Est et de la Méditerranée, Paris, Unesco, 1:489-493, 4 + 3 fig., 4 maps.

JOHNSON, R. K.; BARNETT, M. A. 1972. Geographic meristic variation in _Diplophos_ _taenia_ Günther (Salmoniformes: Gonostomatidae). _Deep Sea Res._, 19:813-821, fig. 1-4.

JOHNSON, R. K.; BERMAN, L. U. 1986. Stylephoridae, _in_ P.J.P. Whitehead, M.L. Bauchot, J.C. Hureau, J. Nielsen & E. Tortonese, Fishes of the North-eastern Atlantic and the Mediterranean/Poissons de l'Atlantique du Nord-Est et de la Méditerranée, Paris, Unesco, 2:736-737, 1 + 1 fig.

JOHNSON, R. K.; COHEN, D. M. 1974. Results of the research cruises of FRV "Walther Herwig" to South America. XXX. Revision of the chiasmodontid fish genera _Dysalotus_ and _Kali_, with descriptions of two new species. _Arch._ _FischWiss._, 25(1/2):13-46.

JOHNSON, R. K.; GLODEK, G. S. 1975. Two new species of _Evermannella_ from the Pacific Ocean, with notes on other midwater species endemic to the Pacific Central or the Pacific Equatorial Water Masses. _Copeia_, 1975(4):716-730.

JOHNSON, R. K.; KEENE, M. J. 1986. Chiasmodontidae, _in_ P.J.P. Whitehead, M.L. Bauchot, J.C. Hureau, J. Nielsen & E. Tortonese, Fishes of the North-eastern Atlantic and the Mediterranean/Poissons de l'Atlantique du Nord-Est et de la Méditerranée, Paris, Unesco, 2:957-961, 6 fig.

JOHNSTON, G. 1834. A list of the fishes of Berwickshire, exclusive of the salmons. _Proc._ _Hist._ _Berwicksh._ _nat._ _Club_, 1:170-176.

JONES, P. W.; MARTIN, F. D.; HARDY, J. D. 1978. Development of fishes of the Mid-Atlantic Bight. 1. Acipenseridae through Ictaluridae. U.S. Fish Wildl Serv., biol. Serv. Program, FWS/OBS-78/12. 366 p., 199 fig.

JONES, S. 1958. Notes on eggs, larvae and juveniles of fishes from Indian waters. I. _Xiphias_ _gladius_ Linnaeus. _Indian_ _J._ _Fish._, 5(2):357-361, 4 fig.

JONES, S.; KUMARAN, M. 1964. Eggs, larvae and juveniles of Indian scombroid fishes. _Proc._ _Symp._ _Scombroid_ _Fishes,_ _mar._ _biol._ _Ass._ _India_, 1(1): 343-378, 108 fig.

--. 1964. Distribution of larval billfishes (Xiphiidae and Istiophoridae) in the Indo-Pacific with special reference to the collections made by the Danish Dana Expedition. _Proc._ _Symp._ _Scombroid_ _Fishes,_ _mar._ _biol._ _Ass._ _India_, 1(1): 483-498, 7 fig.

JONES, S.; SILAS, E. G. 1961. Indian Tunas. A preliminary review, with a key for their identification. _Indian_ _J._ _Fish._, 7(2):369-393.

--. 1963. Tuna and tuna-like fishes from the Indian seas. _FAO_ _Fish._ _Rep._, (6)3:1775-1796.

--. 1964. A systematic review of the scombroid fishes of India. _Proc._

symp. Scombroid Fishes, mar. biol. Ass. India, 1(1):1-105, 15 fig., 9 pl.

JONSSON, G. 1974. Sjaldsédir fiskar arid 1973. Aegir, 67(6):111-112, fig. 1-4.

--. 1975. Skrá um islenska fiska ásamt lýsingu á nokkrum peirra. Hafrannsoknir, 7:1-76, 50 fig.

JONSSON, G.; MAGNUSSON, J.; MAGNUSSON, J. V. 1978. Icelandic observations of rare fishes in 1976. Annls biol., Copenh. (1976),33:180-183.

JONSSON, J. 1957. Nýjar tegundir vid Island sidan 1926. p. 536-550. In: B. Saemundsson. Fiskarnir (2nd ed.). Reykjavik, 583 p., 266 + 21 fig.

JORDAN, D. S. 1884. Notes on a collection of fishes from Pensacola, Florida, obtained by Silas Stearns, with descriptions of two new species etc. Proc. U.S natn. Mus., 7:33-40.

--. 1884. The fishes of the Florida Keys. Bull. U.S. Fish Commn, 4:11-86.

--. 1885. List of fishes collected at Key West, Florida, with notes and descriptions. Proc. U.S. natn. Mus., 7:103-150.

--. 1885. A catalogue of the fishes known to inhabit the waters of North America, north of the Tropic of Cancer, with notes on the species discovered in 1883 and 1884. Rep. U.S. Commnr Fish., (1884):1-185.

--. 1885. List of fishes from Egmont Key, Florida, in the Museum of Yale College, with descriptions of two new species. Proc. Acad. nat. Sci. Philad., (1884):42-46.

--. 1885. A list of the fishes known from the Pacific coast of tropical America, from the Tropic of Cancer to Panama. Proc. U.S. natn. Mus., 8:361-394.

--. 1886. List of fishes collected at Havana, Cuba, in December, 1883, with notes and descriptions. Proc. U.S. natn. Mus., 9:31-55.

--. 1886. On the generic name of the tunny. Proc. Acad. nat. Sci. Philad., 40:180.

--. 1886. Fishes of Havana, Cuba. Proc. U.S. natn. Mus., 9:50-51.

--. 1887. A catalogue of the fishes known to inhabit the waters of North America north of the Tropic of Cancer, with notes on the species discovered in 1883 and 1884. Rep. U.S. Commnr Fish. (1885), 13:789-974.

--. 1887. A preliminary list of the fishes of the West Indies. Proc. U.S. natn. Mus., 9:554-608.

--. 1888. Description of a new species of Xyrichthys (X. jessiae) from the Gulf of Mexico. Proc. U.S. natn. Mus., 10:698.

--. 1890. Scientific results of explorations by the U.S. Fish Commission Steamer Albatross. No IX. Catalogue of fishes collected at Port Castries, St. Lucia, by the Steamer Albatross, November, 1888. Proc. U.S. natn. Mus., 12:645-652.

--. 1891. A review of the labroid fishes of America and Europe. Rep. U.S. Commnr Fish. (1890), 15:599-699, 11 pl.

--. 1891. A list of fishes collected in the harbor of Bahia, Brazil, by the U.S. Fish Commission steamer "Albatross". Proc. U.S. natn. Mus., 13(829):313-336.

--. 1895. The fishes of Sinaloa. Proc. Calif. Acad. Sci., (2) 5:377-513, 29 pl.

--. 1900. Notes on recent fish literature. Am. Nat., 34:897-898.

--. 1904. Notes on fishes collected in the Tortugas Archipelago. Bull. U.S. Fish Commn, 22:539-544.

--. 1905. A guide to the study of fishes. New York, 1: xxvi + 624 p., 393 fig., 38 portr., col. front; 2: xxvii + 599 p., 506 fig., col. front.

--. 1917. Notes on Glossamia and related genera of cardinal fishes. Copeia, (44):46-47.

--. 1919. New genera of fishes. Proc. Acad. nat. Sci. Philad., 70:341-344.

--. 1919. On Elephenor, a genus of fishes from Japan. Ann. Carneg. Mus., 12:329-344, pl. 54-58.

--. 1917-1920. The genera of fishes, from Linnaeus to... 1920,... with the accepted type of each. A contribution to the stability of scientific nomenclature. Leland Stanford jr Univ. Publs Univ. Ser., part I, 1917(27):1-161; part II, 1919(36):ix + 163-284 + xiii; part III, 1919(39):285-410 + xv; part IV, 1920(43):441-576 + xviii (reprint 1968).

--. 1921. Description of deep-sea fishes from the coast of Hawaii, killed by a lava flow from Mauna Loa. Proc. U.S. natn. Mus., 59:643-656, 8 fig.

--. 1923. A classification of fishes including families and genera as far as known. Stanford Univ. Publs. (Biol. Sci.), 3(2):77-243 + x.

--. 1925. The fossil fishes of the Miocene of southern California. Stanford Univ. Publs (Biol. Sci.), 4:1-51, 22 pl.

JORDAN, D. S.; BOLLMAN, C. H. 1889. List of fishes collected at Green Turtle Cay, in the Bahamas, by Charles L. Edwards, with descriptions of three new species. Proc. U.S. natn. Mus., 11:549-553.

--. 1890. Descriptions of new species of fishes collected at the Galapagos Islands and along the coast of United States of Colombia, 1887-1888. Proc. U.S. nat. Mus., 12:149-183.

--. 1892. A preliminary review of the apodal fishes or eels inhabiting the waters of America and Europe. Rep. U.S. Commnr Fish., 16:581-677, 8 pl.

JORDAN, D. S.; DICKERSON, M. C. 1908. Notes on a collection of fishes from the Gulf of Mexico at Veracruz and Tampico. Proc. U.S. nat. Mus., 34:11-12.

JORDAN, D.S.; EDWARDS, C. L. 1886. Review of the American species of Tetraodontidae. Proc. U.S. natn. Mus., 11:549-553.

JORDAN, D. S.; EIGENMANN, C. H. 1889. A review of the Sciaenidae of America and Europe. Rep. U.S. Commnr Fish.(1886), 14:343-451, 4 pl.

--. 1890. A review of the genera and species of Serranidae found in the waters of America and Europe. Bull. U.S. Fish Commn, 8:329-433.

JORDAN, D. S.; EVERMANN, B. W. 1887. Description of six new species of fishes from the Gulf of Mexico, with notes on other species. Proc. U.S. natn. Mus., 9:466-476.

--. 1896. A check-list of the fishes and fish-like vertebrates of North and Middle America. Rep. U.S. Commnr Fish., 21:207-584.

--. 1896-1900. The fishes of North and Middle America: a descriptive catalogue of the species of fish-like vertebrates found in the waters of North America, north of the Isthmus of Panama. Bull. U.S. natn. Mus., 47. 1, 1896: lx + 1-1240; 2, 1898: xxx + 1241-2183; 3, 1898: xxiv + 2183a-3136; 4, 1900: cl + 3137-3313; 392 pl. (958 fig.).

--. 1902. American food and game fishes. A popular account of all the species found in America north of the equator, with keys for ready identification, life histories and methods of capture. New York, 1 + 573 p., many fig.

--. 1903. Descriptions of new genera and species of fishes from the Hawaiian Islands. Bull. U.S. Fish Commn (1902), 22:161-208.

--. 1905. The aquatic resources of the Hawaiian Islands. I. The shore fishes of the Hawaiian Islands, with a general account of the fish fauna. Bull. U.S. Fish Commn (1903), 23(1): i-xxviii, 1-574, 229 fig., 65 pl., 73 color pl.

--. 1926. A review of the giant mackerel-like fishes, tunnies, spearfishes and swordfish. Occ. Pap. Calif. Acad. Sci., 12:1-113, 1 fig., 20 pl.

--. 1927. New genera and species of North American fishes. Proc. Calif. Acad. Sci., (4)(16):501-507.

JORDAN, D. S.; FESLER, B. 1893. A review of the sparoid fishes of America and Europe. Rep. U.S. Commnr Fish. (1889-91), 17:421-544, 55 pl.

JORDAN, D. S.; FORDICE, M. W. 1887. A review of the American species of Belonidae. Proc. U.S. natn. Mus., 9(575):339-361.

JORDAN, D. S.; FOWLER, H. W. 1903. Review of the elasmobranchiate fishes of Japan. Proc. U.S. natn. Mus., 26(1324):593-674, 10 fig., pl. 26-27.

JORDAN, D. S.; GILBERT, C. H. 1881. Description of a new species of Nemichthys (Nemichthys avocetta), from Puget Sound. Proc. U.S. natn. Mus., 3:409-410.

--. 1881. Description of a new species of Paralepis (Paralepis coruscans), from the Straits of Juan de Fuca. Proc. U.S. natn. Mus., 3:411-413.

--. 1882. Descriptions of thirty-three new species of fishes from Mazatlan, Mexico. Proc. U.S. natn. Mus., 4:338-365.

--. 1882. Notes on the fishes of the Pacific coast of the United States. Proc. U.S. natn. Mus., 4:29-70.

--. 1882. Notes on fishes observed about Pensacola, Florida, and Galveston, Texas, with descriptions of new species. Proc. U.S. natn. Mus., 5:241-307.

--. 1882. Synopsis of the fishes of North America. Bull. U.S. natn. Mus., 16: lvi + 1-1018.

--. 1883. Notes on a collection of fishes from Charleston, South Carolina, with descriptions of three new species. Proc. U.S. natn. Mus., 5:580-620.

--. 1883. Description of two new species of fishes (Myrophis vafer and Chloroscombrus orqueta) from Panama. Proc. U.S. natn. Mus., (1882)5:645-647.

--. 1884. Descriptions of ten new species of fishes from Key West, Florida. Proc. U.S. natn. Mus., 7:24-32.

JORDAN, D. S.; GOSS, D. K. 1889. A review of the flounders and soles (Pleuronectidae) of America and Europe. Rep. U.S. Commnr Fish. (1886), 14:225-342, 9 pl., 23 fig.

JORDAN, D. S.; GUNN, J. A. 1899. List of fishes collected at the Canary Islands by O.F. Cook, with description of four new species. Proc. Acad. nat. Sci. Philad., 50:339-348.

JORDAN, D. S.; HUBBS, C. L. 1917. Notes on a collection of fishes from Port Said, Egypt. Ann. Carneg. Mus., 11:461-468.

--. 1919. A monographic review of the family Atherinidae or Silversides. Leland Stanford jr Univ. Publs Univ. Ser., 40:87 p., 12 pl.

--. 1925. Record of fishes obtained by David Starr Jordan in Japan, 1922. Mem. Carneg. Mus. 10(2):93-346, fig. 1, pl. 5-12.

JORDAN, D. S.; JORDAN, E. K. 1922. A list of the fishes of Hawaii, with notes and descriptions of new species. Mem. Carneg. Mus., 10(1):1-92, fig. 1-7, pl. 1-4.

JORDAN, D. S.; MEEK, S. E. 1885. A review of the American species of flying fishes (Exocoetus). Proc. U.S. natn. Mus., 8:44-67.

JORDAN, D. S.; METZ, C. W. 1913. A catalogue of the fishes known from the waters of Korea. Mem. Carneg. Mus., 6(1):1-65, 10 pl., 67 fig.

JORDAN, D. S.; RICHARDSON, R. E. 1908. Fishes of the islands of the Philippine archipelago. Bull. Bur. Fish.. Wash. (1907), 27:233-287.

JORDAN, D. S.; RUTTER, C. 1897. A collection of fishes made by Joseph Seed Roberts in Kingston, Jamaica. Proc. Acad. Nat. Sci. Philad., 49:91-133.

JORDAN, D. S.; SEALE, A. 1905. List of fishes collected in 1882-83 by Pierre Louis Jouy at Shanghai and Hong Kong, China. Proc. U.S. natn.

Mus., 29:517-529.
--. 1906. The fishes of Samoa. Description of the species found in the Archipelago, with a provisional check-list of the fishes of Oceania. Bull. Bur. Fish., Wash. (1905), 25:173-455, fig. 1-111, col. pl. 33-53.
--. 1907. Fishes of the islands of Luzon and Panay. Bull. Bur. Fish., Wash. (1906), 26:1-48.
JORDAN, D. S.; SINDO, M. 1902. A review of the pediculate fishes or anglers of Japan. Proc. U.S. natn. Mus., 24(1261):361-381, 7 fig.
JORDAN, D. S.; SNYDER, J. O. 1900. A list of fishes collected in Japan by Keinosuke Otaki, and by the United States Steamer "Albatross", with description of fourteen new species. Proc. U.S. natn. Mus., 23 (1213):335-380, pl. 9-20.
--. 1901. A review of the apodal fishes or eels of Japan, with descriptions of nineteen new species. Proc. U.S. natn. Mus., 23:837-890.
--. 1901. A preliminary check list of the fishes of Japan. Annotnes zool. jap., 3(2-3):31-159.
--. 1901. Description of nine new species of fishes contained in museums of Japan. J. Coll. Sci. imp. Univ. Tokyo, 15:301-311, pl. 15-17.
--. 1902. A review of the labroid fishes and related forms found in the waters of Japan. Proc. U.S. natn. Mus., 24:595-662, 10 fig.
--. 1904. On a collection of fishes made by Mr. Alan Owston in the deep waters of Japan. Smithson. misc. Collns, 45:230-240, pl. 58-63.
JORDAN, D. S.; STARKS, E. C. 1895. The fishes of Puget Sound. Proc. Calif. Acad. Sci., (2)5:785-855, 38 pl.
--. 1904. A review of the scorpaenoid fishes of Japan. Proc. U.S. natn. Mus., 27; 91-175, 21 fig., pl. I-II.
--. 1904. List of fishes dredged by the steamer "Albatross" off the coast of Japan in the summer of 1900, with descriptions of new species and a review of the Japanese Macrouridae. Bull. U.S. Fish Commn (1902), 22:577-628, 49 fig., pl. 1-8.
--. 1906. A review of the flounders and soles of Japan. Proc. U.S. natn. Mus., 31(1484):161-246, 27 fig.
JORDAN, D. S.; SWAIN, J. 1884. Descriptions of scaroid fishes from Havana and Key West, including five new species. Proc. U.S. natn. Mus., 7:81-102.
--. 1885. A review of the American species of marine Mugilidae. Proc. U.S. natn. Mus., 7:261-275.
--. 1885. A review of the American Species of Epinephelus and related genera. Proc. U.S. natn. Mus., 7:358-410.
--. 1885. Description of three new species of fishes (Prionotus stearnsi, Prionotus ophryas and Anthias vivanus) collected at Pensacola, Florida, by Mr. Silas Stearns. Proc. U.S. natn. Mus., 7:541-545.
JORDAN, D. S.; THOMPSON, W. F. 1914. Record of the fishes obtained in Japan in 1911. Mem. Carneg. Mus., 6:205-313, 87 fig., 19 pl.
JORDAN, D. S.; EVERMANN, B. W.; CLARK, H. W. 1930. Checklist of the fishes and fishlike vertebrates of North and Middle America north of the northern boundary of Venezuela and Colombia. Rep. U.S. Commnr Fish. (1928), (part II):1-670.
JORDAN, D. S.; EVERMANN, B. W.; TANAKA, S. 1927. Notes on new or rare fishes from Hawaii. Proc. Calif. Acad. Sci., (4)16(20):649-680, 3 pl.
JORDAN, D. S.; TANAKA, S.; SNYDER, J. O. 1913. A catalogue of the fishes of Japan. J. Coll. Sci. imp. Univ. Tokyo, 33(1):1-497, fig. 1-396.
JORDANO, D.; MURUVE, M. 1959. Ocho peces tropicales en mercados espanoles y cuatro primera ataciones para las pesquerias canario-africanas. Archos Zootecnia, 8(30):103-129, fig. 1-8.
JORDAS, I. 1978. Povodom ulova rjede jadranske vrste ribe- Lepidopus caudatus. Morsko Ribarst., 30(4):188-190, 4 fig.

JØRENSEN, J. M.; MUNK, O. 1979. Photophores and presumably luminous chin barbel and pectoral fin ray elements of Thysanactis dentex (Pisces: Stomiatoidea). Acta Zool., Stockh. 60:33-42, fig.
JOUBIN, L. 1905. Cours d'Océanographie.Bull. Inst. océanogr. Monaco, (45): 185 p., 177 fig., 3 pl.
--. 1929-38. Faune ichthyologique de l'Atlantique nord. Fiches faunisti-ques, 418 pl. (With systematic and alphabetic index by H. Blegvad, 1938). Copenhagen.
JOUBIN, L.;ROULE, L. 1918. Observations sur la nourriture des thons de l'Atlantique. Bull. Inst. océanogr. Monaco, (348):1-7.

KADER, D. A. el. 1972. Etude de la nutrition de certains clupeidés (Poissons téléostéens) de la Côte d'Ivoire. UNDP (SF)/FAO, Results Regional Fisheries Survey West Africa, Abidjan, (Sci. Rept. II)(33 p., mimeo).
KÄHSBAUER, P. 1962. Beitrag zur Kenntnis der Fischfauna von Nigeria. Annln naturh. Mus. Wien, (1961)65:135-165.
--. 1968. Über einige westafrikanische Syngnathiformes. Atlantide Rep., (10):263-273.
KAMOHARA, T. 1936. Supplementary note on the fishes collected in the vicinity of Kochi (VIII). Zool., Mag., Tokyo, 48:17-22, 5 fig.
--. 1936. On the Zeomorphi of Japan. Botany Zool., Tokyo, 4(2):357-362, 5 fig.
--. 1938. Gempylidae of Japan. Annotnes zool. jap., 17(1):45-50, 1 pl.
--. 1938. On the offshore bottom-fishes of Prov. Tosa, Shihoku, Japan. Tokyo, 86 p., 43 fig.
--. 1940. Scombroidei (Exclusive of Carangiformes). Section Percomorphi Subclass Teleostomi. Fauna Nipponica, 15-2(5). Tokyo, viii + 225 p. 102 fig. (in Japanese).
--. 1952. Revised descriptions of the offshore bottom-fishes of Prov. Tosa, Shikohu, Japan. Rep. Kôchi Univ. nat. Sci., (3):1-122, 100 fig.
--. 1957. Notes on twenty additions to the marine fish fauna of Prov. Tosa, Japan, including one new genus (family Peristediidae). Res. Rep. Kôchi Univ., 6(5):1-6, 1 fig.
--. 1960. Coloured illustrations of the fishes of Japan (in Japanese). Osaki, (1): xi + 1-135, 76 fig., 64 pl.
--. 1964. Revised catalogue of fishes of Kochi Prefecture, Japan. Rep. Usa mar. biol. Stn Kochi Univ., 11(1):1-99, 63 fig.
--. 1967. Fishes of Japan in color. Osaka, xv + 135 p., 64 pl.
KANAZAWA, R. H. 1952. More new species and new records of fishes from Bermuda. Fieldiana, Zool., 34(7):71-100, fig. 9-13.
--. 1958. A revision of the eels of the genus Conger with descriptions of four new species. Proc. U.S. natn. Mus., 108:219-267, 7 fig., 4 pl.
--. 1961. A new eel, Coloconger cadenati and a redescription of the heterocongrid eel Taenioconger longissimus, (Günther) both from the coast of Senegal. Bull. Inst. fr. Afr. noire, (A)23(1):108-115, 2 pl.
--. 1961. Paraconger, a new genus with three new species of eels (family Congridae). Proc. U.S. natn Mus., 113:1-14, 3 fig., 2 pl.
KANAZAWA, R. H.; MAUL, G. E. 1967. Description of a new genus and spe-cies of the eel family Nemichthyidae from the eastern Atlantic. Boca-giana, (12):1-6, 3 fig.
KARLOVAC, J. 1953. Sternoptychidae, Stomiatidae and Scopelidae in the Adriatic Sea. Izv. vses. nauchno-issled. Inst. ozern. rechn. ryb. Khoz., 5(2B):1-46, 5 map.
KARMOVSKAYA, E. 1977. Studies on taxonomy and distribution of the genus Borodinula (Nemichthyidae, Osteichthyes) with description of a new

species. Trudy Inst. Okeanol., 109:186-210, 8 fig.
--. 1979. Leptocephales of the Anguilliformes in the Indo-West-Pacific and Australian-New Zealand areas (on materials collected during the 57th cruise of the R/V "Vityaz" and 16th cruise of the R/V "Dmitry Mendeleew". Trudy Inst. Okeanol., 106:97-124, 14 fig., 2 tab. (in Russian, English summary).
KARNELLA, C.; GIBBS, R. H., Jr. 1977. The lanternfish Lobianchia dofleini: an example of the importance of life-history information in prediction of oceanic sound scattering, p. 361-379. In: ANDERSEN, N.R. & ZAHURANEC B.J. (ed). Ocean Sound Scattering Prediction. New York.
KARPECHENKO, J.L. 1960. (West African sardines). Ryb. Khoz., (36)(2):16-25 (in Russian).
KARRER, C. 1968. Über Erstnachweise und seltene Arten von Fischen aus dem Südatlantik (argentinisch-südbrasilianische Küste). Zool. Jb. (Syst.), 95:542-570, 9 fig., 6 tab., 1 map.
--. 1971. Die Otolithen der Moridae (Teleostei, Gadiformes) und ihre systematische Bedeutung. Zool. Jb. (Syst.), 98:153-204, fig. 1-20.
--. 1972. Die Gattung Harriotta Goode & Bean, 1895 (Chondrichthyes, Chimaeriformes, Rhinochimaeridae). Mit Beschreibung einer neuen Art aus dem Nordatlantik. Mitt. zool. Mus. Berl., 48(1):203-221, fig. 1-5.
--. 1973. Über das Vorkommen von Fischarten im Nordwestatlantik (Neufundland - Baffinland). Fisch.-Forsch., wiss. Schr. Reihe, 11(1):73-90, fig. 1-32.
--. 1973. Über Fische aus dem Südostatlantik. Mitt. zool. Mus. Berl., 49:191-257, 29 fig.
--. 1975. Über Fische aus dem Südostatlantik (Teil 2). Mitt. zool. Mus. Berl., 51(1):63-82, 1 pl.
--. 1976. Über drei mesopelagische Fischarten aus dem Golf von Guinea. Mitt. zool. Mus. Berl., 52(1):177-182, 1 pl.
--. 1982. Anguilliformes du Canal de Mozambique (Pisces, Teleostei). Faune trop., 23:1-116, fig. 1-31.
--. 1986. Oreosomatidae, in P.J.P. Whitehead, M.L. Bauchot, J.C. Hureau, J. Nielsen & E. Tortonese. Fishes of the North-eastern Atlantic and the Mediterranean/Poissons de l'Atlantique du Nord-Est et de la Méditerranée, Paris, Unesco, 2:775-776, 1 fig.
KARTAS, F.; BAUCHOT, M. L. 1974. Redescription de Lepidotrigla dieuzeidei. Comparaison avec les espèces méditerranéenes et est-atlantiques du genre Lepidrotrigla. Bull. Mus. natn. Hist. nat., Paris, (3)(268) (Zool. 190):1785-1807.
KASKIN, N. J. 1968. (Plankton distribution and the food of the sardine off the Atlantic coast of Africa). Ryb. Khoz., 1968:1-40 (in Russian).
--. 1975. Finds of Photostylus pycnopterus (Pisces, Alepocephalidae) in the northeastern Atlantic, the Gulf of Mexico and the Caribbean. J. Ichthyol., 15(5):817-820. 1 fig.
--. 1977. Gill structure of Scopelarchus guentheri family Scopelarchidae) in different parts of the range. J. Ichthyol., 17(4):668-673.
KAUP, J. J. 1826. Beiträge zur Amphibiologie und Ichthyologie. Isis (Oken), 19: column. 87-89.
--. 1827. Vermischte Bemerkungen. Isis (Oken), 20: column. 624-625.
--. 1853. Uebersicht der Lophobranchier. Arch. Naturgesch., 19:226-234.
--. 1855. Uebersicht über die Species einiger Familien der Sclerodermen. Arch. Naturgesch., 21(1):215-233.
--. 1856. Uebersicht der Aale. Arch. Naturgesch., 22(1):41-77.
--. 1856. Catalogue of the apodal fish in the British Museum. London, 163 p., 19 pl.
--. 1856. Catalogue of lophobranchiate fish in the collection of the

British Museum. London, 76 p., 4 pl.
--. 1858. Uebersicht der Soleinae, der vierten Subfamilie der Pleuronectidae. <u>Arch</u>. <u>Naturgesch</u>., **47**:94-104.
--. 1858. Uebersicht der Plagusinae, der fünften Subfamilie der Pleuronectidae. <u>Arch</u>. <u>Naturgesch</u>., **47**:105-110.
--. 1860. On some new genera and species of fishes collected by Drs Keferstein and Heckel at Messina. <u>Ann</u>. <u>Mag</u>. <u>nat</u>. <u>Hist</u>., (3)**6**:270-273.
--. 1860. Neue aalähnliche Fische des Hamburger Museums. <u>Abh</u>. <u>naturw</u>. <u>Ver</u>. <u>Hamb</u>., **4**(2):1-35, 5 pl.
--. 1873. Über die Familie Triglidae nebst einigen Worten über die Classification. <u>Arch</u>. <u>Naturgesch</u>., **39**:(1):71-94.
KAWAGUCHI, K.; AIOI, K. 1972. Myctophid fishes of the genus <u>Myctophum</u> (Myctophidae) in the Pacific and Indian Oceans. <u>J</u>. <u>oceanogr</u>. <u>Soc</u>. <u>Japan</u>, **28**:161-175.
KAWAGUCHI, K.; SHIMIZU, H. 1978. Taxonomy and distribution of the lanternfishes, genus <u>Diaphus</u> (Pisces, Myctophidae) in the western Pacific, eastern Indian Oceans and the Southeast Asian Seas. <u>Bull</u>. <u>Ocean Res</u>. <u>Inst</u>. <u>Univ</u>. <u>Tokyo</u>, **10**:1-145.
KAWAGUCHI, K.; IKEDA, H.; TAMURA, M.; UEYANAGI, S. 1972. Geographical distribution of surface-migrating myctophid fishes (Genus <u>Myctophum</u>) in the tropical and subtropical Pacific and Indian Oceans. <u>Bull</u>. <u>Far Seas Fish</u>. <u>Res</u>. <u>Lab</u>., **6**:23-37.
KEENE, M. J.; TIGHE, K. A. 1984. Beryciformes: developmental and relationships. <u>In</u>: Ontogeny and systematics of fishes based on an international symposium dedicated to the memory of Elbert Halvor Ahlstrom. <u>Spec</u>. <u>Publ</u>. <u>Am</u>. <u>Soc</u>. <u>Ichthol</u>. <u>Herpet</u>., (1):383-392.
KEMP, N. R. 1977. Detailed comparisons of the dentitions of extant hexanchid sharks and Tertiary hexanchid teeth from South Australia and Victoria, Australia (Selachii: Hexanchidae). <u>Mem</u>. <u>natn</u>. <u>Mus</u>. <u>Vict</u>., **39**:61-83.
KENDALL, A. W. Jr.; WALFORD, L. A. 1979. Sources and distribution of bluefish, <u>Pomatomus</u> <u>saltatrix</u> larvae and juvenile off the east coast of the United States. <u>Fishery</u> <u>Bull</u>., **12**(1):229-242.
KENDALL, W. C.; GOLDSBOROUGH, E. L. 1911. Reports on the scientific results of the expedition to the tropical Pacific, in charge of Alexander Agassiz, by the U.S. Fish Commission Steamer "Albatross", from August 1899 to March 1900. The shore fishes. <u>Mem</u>. <u>Mus</u>. <u>comp</u>. <u>Zool</u>. <u>Harv</u>., **26**(7):241-344, pl. 1-7.
KENSLEY, B.; PENRITH, M.-L. 1973. The constitution of the intertidal fauna of rocky shores of Moçamedes, southern Angola. <u>Cimbebasia</u>, (A)1 (9):113-123.
KERSTAN, M.; SAHRHAGE, D. 1980. Biological investigations on fish stocks in the waters off New Zealand. <u>Mitt</u>. <u>Inst</u>. <u>Seefisch</u>., (29):1-168 + 5, 160 fig.
KESSLER, K. 1874. Opisanie ryb, prinadlezhash chikh k semeinstvam, obshchim chernomu i kaspiiskomu moryam. <u>Trudy</u> <u>petrogr</u>. <u>Obshch</u>. <u>Estest</u>., 5:191-224.
KHROMOV, N. S. (see CHROMOV, N. S.).
KIENER, A.; SPILLMANN, C. J. 1969. Contributions à l'étude systématique et écologique des Athérines des côtes françaises. <u>Mém</u>. <u>Mus</u>. <u>natn</u>. <u>Hist</u>. <u>nat</u>., <u>Paris</u>, (A)60(2):33-74, 8 pl.
--. 1972. Note complémentaire à l'étude systématique et écologique d'<u>Atherina</u> <u>boyeri</u> Risso (Poissons, "Cyprinidae ?") dans sa zone de dispersion actuelle. <u>Bull</u>. <u>Mus</u>. <u>natn</u>. <u>Hist</u>. <u>nat</u>., <u>Paris</u>, (3)(55)(Zool. 41):563-580.
KIKAWA, S. 1975. Synopsis on the biology of the shortbill spearfish, <u>Tetrapturus</u> <u>angustirostris</u> Tanaka, 1914 in the Indo-Pacific areas. <u>NOAA</u> <u>tech</u>. <u>Rep</u>. <u>NMFS</u> <u>SSRF</u>, (675)(Part 3):39-54, 9 fig.

KING, D. P. F. 1977. Distribution and relative abundance of eggs of the South West African pilchard _Sardinops ocellata_ and anchovy _Engraulis capensis_, 1971/72. _Fish. Bull. Un. S. Afr._, 9:23-31.

KING, J. E.; IVERSON, R. T. B. 1962. Midwater trawling for forage organisms in the Central Pacific, 1951-56. _Fishery Bull. Fish Wildl. Serv. U.S._, 62(210):271-321.

KISHINOUYE, K. 1915. A study of the mackerels, cybiids and tunas. _Suisan Gakkai Ho_, 1(1):1-24 (transl. by W. F. van Campen, U.S. Dept. Interior, _Spec. Scient. Rep. Fish._, 1950, (24):1-24).

--. 1923. Contributions to the comparative study of the so-called scombroid fishes. _J. Coll. Agric. imp. Univ. Tokyo_, 8(3):293-475.

--. 1926. A new aberrant form of the _Cybium_ from Japan. _J. Coll. Agric. imp. Univ. Tokyo_, 7(4):377-382.

KITAHARA, T. 1897. Scombridae of Japan. _J. imp. Fish. Bur., Tokyo_, 6:15 p. (in Japanese with 3 p. English summary).

KLAUSEWITZ, W. 1971. _Nystactichthys halis_ (Böhlke), ein Neunachweis für die Kanarischen Inseln und den Ost-Atlantik (Pisces: Congridae: Heterocongrinae). _Senckenberg. biol._, 52:31-34, 1 fig.

KLAUSEWITZ, W.; EIBL-EIBESFELDT, J. 1959. Neue Röhrenaale von den Maldiven und Nikobaren (Pisces, Apodes, Heterocongridae). _Senckenberg. biol._, 40:135-153, 17 fig.

KLAWE, W. L. 1977. What is tuna ? _Mar. Fish Rev._, Wash., 39(1268):1-5, 2 fig.

--. 1978. World catches of tunas and tuna-like fishes in 1975. _Internal Rep. Inter Am. trop. Tuna Commn_, 11:ii + 1-191, 6 fig.

KLAWE, W. L.; SHIMADA, B. M. 1959. Young scombroid fishes from the Gulf of Mexico. _Bull. mar. Sci. Gulf Caribb._, 9(1):100-115.

KLEIN, J. T. 1775-1881. Neuer Schauplatz der Natur, nach den Richtigsten Beobachtungen und Versuchen, in Alphabetischer Ordnung, vorgestellt durch eine Gesellschaft von Gelehrten. Leipzig. (= Gesellschaft Schauplatz), 1, 1775: xiv + 1044 p.; 2, 1776:842 p.; 3, 1776:836 p.; 4, 1777:874 p.; 5, 1777:840 p.; 6, 1778:5 + 782 p.; 7, 1779:820 p.; 8, 1779:824 p.; 9, 1780:832 p.; 10, 181:604 p.

KLIMAJ, A. 1970. Fishing yield and species composition of trawl catches of M.T. "Ramada" on the fishing grounds of the shelf off Mauritania and Senegal in 1967. _Rapp. P.-v. Réun. Cons. perm. int. Explor. Mer_, 156:254-260.

--. 1978. Fishery atlas of the Southwest African shelf, bottom and pelagic catches, 3:173 p. Transl. from Polish: Atlas rybacki szelfu Afryki południowo - zachodniej, Połowy denne i pelagizne. Gdynia, 3,1976: 206 p., unnumb. fig.

KLUNZINGER, C. B. 1870. Synopsis der Fische des Roten Meeres I. Theil. Percoiden, Mugiloiden. _Verh. zool.-bot. Ges. Wien_, 20:669-834.

--. 1871. Synopsis der Fische des Rothen Meeres. II. Theil. _Verh. zool.-bot. Ges. Wien_, 21:441-688.

--. 1884. Die Fische des Rothen Meeres. Eine kritische Revision mit Bestimmungs-Tabellen. Teil I. Acanthopteri veri Owen. Stuttgart, 133 p., fig., 13 pl.

KNAPP, L. W. 1981. Percophidae, Platycephalidae, _in_ Fischer, W., Bianchi, G. & W.B. Scott (ed.), FAO species identification sheets for fishery purposes. Eastern Central Atlantic; fishing areas 34, 47 (in part). Canada Funds-in-Trust. Ottawa, Department of Fisheries and Oceans Canada, by arrangement with the Food and Agriculture Organization of the United Nations, vol. 1-7: pag. var.

KNER, R. 1865-67. Fische. _In_: Reise der österreichischen Fregatte "Novara" um die Erde in den Jahren 1857-59, unter den Befehlen des Commodore B. von Wüllerstorf-Urbain. Zoologischer Teil. Wien. 1. Abt., 1865:1-109, pl. 1-5; 2.Abt., 1865:111-272, pl. 6-11; 3.Abt., 1867:

273-433, pl. 12-16.

--. 1868. Ueber neue Fische aus dem Museum der Herren Joh.Caes. Godeffroy & Sohn in Hamburg. IV. Folge. Sber. Akad. Wiss. Wien, 58:293-356, 9 pl.

KNER, R.; STEINDACHNER, F. 1866. Neue Fische aus dem Museum der Herren Joh. C. Godeffroy & Sohn in Hamburg. Sber. Akad. Wiss. Wien, 54:356-395, 5 pl.

KNIPOVICH, N. M. 1926. Guide for determination of the fishes of Barents-Sea, White-Sea and Kara-Sea. Trudy Inst. Izuch. Sev., 27:1-183, fig. 1-126.

KNUDSEN, H. 1971. Determination of mortality of Sardinella stocks in Ghanaian waters from tagging experiments and length distributions. Ghanaian Fishery Project, 19 p. (mimeo).

KOEFOED, E. 1927. Fishes from the sea-bottom. Rep. scient. Results Michael Sars N. Atlant. deep Sea Exped., 4(1):1-148, 55 fig., 6 pl.

--. 1944. Pediculati. Rep. scient. Results Michael Sars N. Atlant. deep Sea Exped., 4(2)(1):18 p., 2 fig., 3 pl.

--. 1952. Zeomorphi, Percomorphi, Plectognathi. Rep. scient. Results Michael Sars N. Atlant. deep Sea Exped., 4(2)(2):1-26, 3 fig., 3 pl.

--. 1953. Synentognathi, Solenichthyes, Anacanthini, Berycomorphi, Xenoberyces. Rep. scient. Results Michael Sars N. Atlant. deep Sea Exped., 4(2)(3):1-38, fig. 1-6, 4 pl.

--. 1955. Iniomi (Myctophidae exclusive), Lyomeri, Apodes. Rep. scient. Results Michael Sars N. Atlant. deep Sea Exped., 4(2)(4):1-15, 3 fig., 2 pl.

--. 1956. Isospondyli 1. Gymnophotodermi and Lepidophotodermi. Rep. scient. Results Michael Sars N. Atlant. deep Sea Exped., 4 (2)(5): 1-21, 2 fig., 3 pl.

--. 1958. Isospondyli. 2. Heterophotodermi. 1. Rep.scient. Results Michael Sars N. Atlant. deep Sea Exped., 4(2)(6):1-17, fig. 1-2, 1 pl. (2 fig.).

--. 1960. Isospondyli. 2. Heterophotodermi. 2. Rep. scient. Results Michael Sars N. Atlant. deep Sea Exped., 4(2)(8):1-15, fig. 1-9, 1 pl. (2 fig.).

--. 1962. Isospondyli, 2. Heterophotodermi. 3. Rep. scient. Results Michael Sars N. Atlant. deep Sea Exped., 4(2)(10):1-14, 2 fig., 2 pl.

KOEHLER, R. 1896. Résultats scientifiques de la campagne du "Caudan" dans le Golfe de Gascogne, Août-Septembre, 1895. Poissons. Annls Univ. Lyon, (3)26:475-526, pl. 26-27.

KÖLLIKER, R. A. 1853. Bau von Leptocephalus und Helmichthys. Z. wiss. Zool. 4:360-366.

--. 1853. Weitere Bemerkungen über die Helmichthyiden. Verh. phys.-med. Ges. Würzb., 4(1):100-102.

KOELREUTER, J. G. 1770. Piscium variorum e Museo Petropolitano excerptorum descriptiones. Novi Comment. Acad. Petropol., 8:337-430.

KOKEN, E. 1884. Ueber Fisch-Otolithen, insbesondere über diejenigen der norddeutschen Oligocän-Ablagerungen. Z. dt. geol. Ges., 36(3):500-565, pl. 9-12.

--. 1888. Neue Untersuchungen an tertiären Fisch-Otolithen, Z. dt. geol. Ges., 40:274-305, 3 pl.

KOLOMBATOVIC, G. 1891. Glamoci (Gobii) Spljetskog Pomorskog Okruzja u Dalmaciji. Split, 29 p.

--. 1902. Contribuzione alla fauna dei Vertebrati della Dalmatia. Glasn. hrv. narodosl. Drust., 13:22-37, pl.

KOMAROV, Y. 1967. (The West African sardines - in Russian). Trudy atlant. nauchno-issled. Inst. ryb. Khoz. Oceanogr., 1967:1-39.

KONARE, A. 1975. Les poissons du delta central du Niger. 79 p., fig.

KONOVALENKO, I.I.; PARIN, N.V. 1985. The first record of Nesiarchus

<u>nasutus</u> in the Southeast Pacific. <u>Vop.</u> <u>Ikhtiol.</u>, 25(3):513-515.

KOTLYAR, A. N. 1980. Systematics and distribution of trachichthyid fishes (Trachichthyidae, Beryciformes) of the Indian Ocean. <u>Trudy</u> <u>Inst.</u> <u>Okeanol.</u>, 110:177-224 (in Russian).

--. 1981. Age and growth of <u>Hoplostethus</u> <u>atlanticus</u> Collett and <u>H.</u> <u>mediterraneus</u> Cuvier (Trachichthyidae, Beryciformes), p. 68-88, 9 fig., 8 tab. <u>In</u>: Fishes of the open ocean (N.V. Parin, ed.). Moscow, 119 p.

--. 1983. <u>Acanthochaenus</u> <u>luetkenii</u> (Stephanoberycidae) from the eastern part of the Indian Ocean. <u>J.</u> <u>Ichthyol.</u>, 23(5):146-149.

--. 1984. Descriptions of the fries of the four species in the genus <u>Hoplostethus</u> (Trachichthyidae, Beryciformes). <u>Byull.</u> <u>mosk.</u> <u>Obshch.</u> <u>Ispyt.</u> <u>Prir.</u> <u>(Biol.)</u>, 89(3):33-39 (in Russian).

--. 1986. Systematics and distribution of species of the genus <u>Hoplostethus</u> Cuvier (Beryciformes, Trachichthyidae). <u>Trudy</u> <u>Inst.</u> <u>Okeanol.</u>, 121:97-140, 17 fig. (in Russian).

KOTLYAR, A. N.; LIPSKAYA, N. Ya. 1981. Feeding of <u>Hoplostethus</u> <u>atlanticus</u> Collett (Trachichthyidae, Beryciformes). p. 89-93, 2 tab. <u>In</u>: Fishes of the open ocean (N.V. Parin, ed.). Moscow, 119 p.

KOTTELAT, M. 1984. Catalogue des types du Musée d'Histoire Naturelle de Neuchâtel. <u>Bull.</u> <u>Soc.</u> <u>neuchâtel.</u> <u>Sci.</u> <u>nat.</u>, 107:143-153.

KOTTHAUS, A. 1952. <u>Hoplostethus</u> <u>islandicus</u>, nov. spec. (Acanthopterygia, Abt. Beryciformes, Familie Trachichthyidae) aus den südislandischen Gewässern. <u>Helgoländer</u> <u>wiss.</u> <u>Meeresunters.</u>, 4:62-87.

--. 1953. Der Holotypus von <u>Hoplostethus</u> <u>islandicus</u> Kotthaus. <u>Zool.</u> <u>Anz.</u>, 150:81-83.

--. 1967. Fische des Indischen Ozeans. Ergebnisse der ichthyologischen Untersuchungen während der Expedition des Forschungsschiffes "Meteor" in den Indischen Ocean, Oktober 1964 bis Mai 1965. <u>"Meteor"</u> Forsch.-Ergebn., (D)1:1-84, fig. 1-96. (A. Systematischer Teil I, Isospondyli und Giganturoidei; II, Ordnung Iniomi).

--. 1968. Fische des Indischen Ozeans. Ergebnisse der ichthyologischen Untersuchungen während der Expedition des Forschungsschiffes "Meteor" in den Indischen Ozean, Oktober 1964 bis Mai 1965. <u>"Meteor"</u> <u>Forsch.</u> Ergebn., (D)(3): 14-56, fig. 97-152. (A. Systematischer Teil III, Ostariophysi und Apodes).

--. 1970. Fische des Indischen Ozeans. Ergebnisse der ichthyologischen Untersuchungen während der Expedition des Forschungsschiffes "Meteor" in den Indischen Ozean, Oktober 1964 bis Mai 1965. <u>"Meteor"</u> Forsch.-Ergebn. (D) (5):53-70, fig. 195-215. (A. Systematischer Teil VI, Anacanthini (2), Berycomorphi, Zeomorphi).

--. 1972. Fische des Indischen Ozeans. Ergebnisse der ichthyologischen Untersuchungen während der Expedition des Forschungsschiffes "Meteor" in den Indischen Ozean, Oktober 1964 bis Mai 1965. <u>"Meteor"</u> Forsch.-Ergebn., (D)(12):12-35. (A. Systematischer Teil IX, Iniomi (Nachtrag: Fam. Myctophidae)).

--. 1972. Die meso-und bathypelagischen Fische der "Meteor"-Rossbreiten-Expedition 1970 (2 und 3. Fahrtabschnitt). <u>"Meteor"</u> Forsch.-Ergebn., (D)(11):1-28, 139 fig.

KOTTHAUS, A.; KREFFT, G. 1957. Fischfaunenliste der Fahrten mit F.F.S. "Anton Dohrn" nach Island-Grönland. <u>Ber.</u> <u>dt.</u> <u>wiss.</u> <u>Kommn</u> <u>Meeresforsch.</u>, 14(3):169-191.

KOUMANS, F. 1953. Biological results of the "Snellius" expedition. XVI. The Pisces and Leptocardii. <u>Temminckia</u>, 9:177-275, fig. 1-2.

--. 1953. Gobioidea. <u>In</u>: M. Weber & L. F. de Beaufort (ed.). The Fishes of the Indo-Australian Archipelago. Brill, Leiden, 10:423 p., 95 fig.

KOVALEVSKAYA, N. V. 1964. Study of development of the "two-winged" flying fishes of the genus <u>Exocoetus</u> (Exocoetidae, Pisces). <u>Trudy</u> <u>Inst.</u> <u>Okeanol.</u>, 73:204-223, 7 fig. (in Russian).

--. 1972. New materials on reproduction, development and distribution of larvae and juveniles of flying fishes of the genus *Hirundichthys* (Pisces, Exocoetidae) in the Pacific and Indian oceans. Trudy Inst. Okeanol., 93:42-69, 14 fig. (in Russian).

--. 1977. The larvae and fry of flyingfish of the family Exocoetidae: *Cheilopogon cyanopterus* (Val.), *Ch. spilonotopterus* (Bleeker) and *Ch. longibarbus* Parin. Vop. Ikhtiol., 17(2):291-300, fig. 1-5 (in Russian).

--. 1980. Reproduction, development and distributional patterns of larvae and juveniles of the oceanic flying fishes in the Pacific and Indian oceans. Trudy Inst. Okeanol., 97:212-275, fig. 1-26 (in Russian).

KRAMER, D. 1960. Development of eggs and larvae of Pacific mackerel and distribution and abundance of larvae 1952-56. Fishery Bull. Fish. Wildl. Serv. U.S., 60(174):393-438.

KREFFT, G. 1953. Ichthyologische Mitteilungen aus dem Institut für Seefischerei der Bundesforschungsanstalt für Fischerei. I. Zool. Anz., 150:275-282, 5 fig.

--. 1953. Ichthyologische Mitteilungen aus dem Institut für Seefischerei der Bundesforschungsanstalt für Fischerei. II. 3. Eine neue *Searsia*-Art (Isospondyli, Searsidae) aus isländischen Gewässern. Zool. Anz., 151 (9/10):259-266, fig. 1.

--. 1955. Ichthyologische Mitteilungen aus dem Institut für Seefischerei der Bundesforschungsanstalt für Fischerei IV. Zool. Anz., 154 (7/8): 157-164, fig. 1-2.

--. 1958. Antarktische Fische und Fischlarven aus den Planktonfängen. Wiss. Ergebn. dt. antarkt. Exped., 2:249-256.

--. 1959. Rare fish from distant northern seas area. Germany. A. Records of the Institut für Seefischerei, Hamburg. Annls biol., Copenh. (1957), 14:41-42.

--. 1960. Rare fish. A. Records of the Institut für Seefischeri, Hamburg. Annls biol., Copenh. (1958), 15:70-72.

--. 1961. *Beryx splendens*, ein Erstfund in nordeuropäischen Gewässern nebst Bemerkungen zum Auftreten von *B. decadactylus*. Arch. FischWiss., 12:24-38, fig. 1-5, tab. 1-2.

--. 1961. Rare fish. Germany. Annls biol., Copenh. (1959), 16:90-91.

--. 1963. Rare fish. Germany. Annls biol., Copenh. (1961), 18:82-83.

--. 1964. Rare fish. Germany. Annls biol., Copenh. (1962), 19:79.

--. 1965. Die ichthyologische Ausbeute der ersten Westafrika-Fahrt des fischereitechnischen Forschungsschiffes "Walther Herwig". I. *Raja herwigi* spec. nov., eine neue Rochenart aus dem Seegebiet der Kapverden. Arch. FischWiss., 15(3):209-216, fig. 1-2.

--. 1966. German observations on Rare Fish in 1964. Annls biol., Copenh. (1964), 21:175-178.

--. 1967. German observations on Rare Fish in 1965. Annls biol., Copenh. (1965), 22:183-186.

--. 1967. *Paraholtbyrnia cyanocephala* gen. nov., spec. nov. (Pisces, Salmoniformes, Alepocephaloidei), ein neuer Searside aus dem tropischen Atlantik. Arch. FischWiss., 18(1):1-11, fig. 1-11.

--. 1968. Ergebnisse der Forschungsreisen des FFS "Walther Herwig" nach Südamerika. IV. *Luciosudis* Fraser-Brunner, 1931, ein valides Genus der Familie Scopelosauridae (Osteichthyes, Alepisauroidei). Arch. FischWiss., 19 (2/3):95-102, fig. 1-2.

--. 1968. Knorpelfische (Chondrichthyes) aus dem tropischen Ostatlantik. Atlantide Rep., (10):33-76, pl. 3-4.

--. 1968. Neue und erstmalig nachgewiesene Knorpelfische aus dem Archibenthal des Südwestatlantiks, einschlisslich einer Diskussion einiger *Etmopterus*-Arten südlicher Meere. Arch. FischWiss., 19:1-42, 16 fig.,

8 tab.

--. 1969. Ergebnisse der Forschungsreisen des FFS "Walther Herwig" nach Süd-Amerika. VI. Fische der Familie Centrolophidae (Perciformes, Stromateoidei). Arch. FischWiss., 20(1):1-9, 3 fig.

--. 1970. Ergebnisse der Forschungsreisen des FFS "Walther Herwig" nach Südamerika. XII. Barbantus elongatus spec. nov. (Pisces, Alepocephaloidei), ein weiterer neuer Searside aus dem tropischen Atlantik. Arch. FischWiss., 21(1):22-27, fig. 1-5.

--. 1970. Zur Systematik und Verbreitung der Gattung Lampadena Goode & Bean, 1896 (Osteichthys, Myctophoidei, Myctophidae) im Atlantischen Ozean, mit Beschreibung einer neuen Art. Ber. dt. wiss. Kommn Meeresforsch., 21(1-4):271-289, 5 fig.

--. 1973. Squatinidae, p. 49-50, Chimaeridae, Rhinochimeridae, p. 78-80, Alepocephalidae, p. 86-93, Searsiidae, p. 95-98, Scopelosauridae, p. 168-169. In: Hureau, J.C. & Th. Monod (ed.). Check-list of the fishes of the north-eastern Atlantic and of the Mediterranean/Catalogue des poissons du nord-est Atlantique et de la Méditerranée (Clofnam). Paris, Unesco, 2 vol.

--. 1974. Investigations on midwater fish in the Atlantic Ocean. Ber. dt. wiss. Kommn Meeresforsch., 23(3):226-254, 15 fig.

--. 1978. Distribution patterns of oceanic fishes in the Atlantic Ocean. Revue Trav. Inst. Pêch. marit., 40(3-4):439-460, fig. 1-13.

--. 1978. Fischtypen in der Sammlung des Institutes für Seefischerei, Hamburg. Mitt. Inst. Seefisch., (25):1-20.

--. 1983. Taxonomy and distribution of the fish-genus Ichthyococcus (Bonaparte, 1841) (Photichthyidae Weitzman, 1974) in the Atlantic Ocean. Investigación pesq., 47(2):295-309.

--. 1984. Notosudidae, in Whitehead, P.J.P., Bauchot, M.L., Hureau, J.C., Nielsen, J. & E. Tortonese, Fishes of the North-eastern Atlantic and the Mediterranean/Poissons de l'Atlantique du Nord-Est et de la Méditerranée, Paris, Unesco, 1:421-425, 4 fig., 4 maps.

--. 1985. Alepocephalidae (Osteichthyes, Argentinoidei) dreier Reisen der Fischereiforschungsschiffe "Anton Dohrn" und "Walther Herwig" in den Nordatlantik. Arch. FischWiss., 36(1/2):213-233.

KREFFT, G.; BEKKER, V. E. 1973. Myctophidae, p. 171-198, in Hureau, J. C. & Th. Monod (ed.). Check-list of the fishes of the north-eastern Atlantic and of the Mediterranean/Catalogue des poissons du nord-est Atlantique et de la Méditerranée (Clofnam). Paris, Unesco, 2 vol.

KREFFT, G.; HAEDRICH, R. L. 1978. Bottom fishes from the Denmark Strait and Irminger Sea: species list and individual data. Tech. Rep. Woods Hole oceanogr. Instn, (78-42):1-15.

KREFFT, G.; LÜBBEN, C. 1966. Raja mollis Bigelow & Schroeder, 1950 (Batoidea, Elasmobranchii, Chondrichthyes), ein Erstfund im Nordost-Atlantik. Zool. Anz., 176(6):389-395, fig. 1-2.

KREFFT, G.; MAUL, G. E. 1955. Notosudis lepida n. sp. (Iniomi, Notosudidae), eine neue Fischart aus dem östlichen Nord-Atlantik. Arch. FischWiss., 6(5/6):305-316, fig. 1-4.

KREFFT, G.; STEHMANN, M. 1973. Pristidae, p. 51-52, Rhinobatidae, p. 53-54, Dasyatidae, p. 70-73, in Hureau, J.C. & Th. Monod (ed.). Check-list of the fishes of the north-eastern Atlantic and of the Mediterranean/Catalogue des poissons du nord-est Atlantique et de la Méditerranée (Clofnam). Paris, Unesco, 2 vol.

KREFFT, G.; TORTONESE. 1973. Squalidae, p. 35-48, in Hureau, J.C. & Th. Monod (ed.). Check-list of the fishes of the north-eastern Atlantic and of the Mediterranean/Catalogue des poissons du nord-est Atlantique et de la Méditerranée (Clofnam). Paris, Unesco, 2 vol.

KREFFT, S. (unpublished). Ein Beitrag zur Systematik und Verbreitung der Leuchtsardinengattung Hygophum (Tåning) Bolin, 1939 im Atlantik. MS.

KRØYER, H. N. 1838-53; Danmarks Fiske. Kjøbenhavn, 1, 1838-40: 616 p.;
2, 1843-45:644 p.; 3 (part I), 1846-49:704 p.; 3 (part II), 1852-
53:705-1279 p.

--. 1842-56. Poissons de la Mer du Nord, reçus pendant le voyage scientifique en Scandinavie, en Laponie, au Spitzberg et aux Féroé sur la
corvette "Recherche", exécuté sous la direction de P. Gaimard. Section VII (Zool.), Paris, only plates.

--. 1845. Ichthyologiske Bidrag (Fortsaettelse). 10. Ceratias holboelli
Kr. Naturh. Tidsskr. (1844-45), (2)1:639-649.

--. 1847. Ichthyologiske Bidrag (Fortsaettelse). Naturh. Tidsskr., (2)2
(3):225-290.

--. 1868. To nye fiske for den danske Faune. Tidsskr. Fisk., 2:70-71.

KRUEGER, W. H. et al., 1977. Baseline report of environmental conditions
in deepwater dumpsite 106. NOAA Dumpsite Evaluation Rep. 77-1.

KRUEGER, W. H.; GIBBS, R. H. Jr. 1966. Growth changes and sexual dimorphism in the stomiatoid fish Echiostoma barbatum. Copeia, 1966(1):43-
49.

KRUEGER, W. H.; GIBBS, R. H. Jr.; KLECKNER, R. C.; KELLER, A. A.; KEENE,
M. J. 1977. Distribution and abundance of mesopelagic fishes on cruises 2 and 3 at deepwater dumpsite 106. NOAA Dumpsite Evaluation Rep.,
77-1, 2:377-422, 3:737-794.

KRZELJ, S. 1971. Note sur les stades larvaires d'Harengula rouxi (Poll,
1953) (Pisces, Clupeidae). UNDP (SF)/FAO, Results Regional Fisheries
Survey West Africa, Abidjan, (Sci. Rep. 2):12 p. (mimeo).

--. 1974. Note sur les stades larvaires d'Harengula rouxi (Poll, 1953)
(Pisces, Clupeidae). Revue Zool. afr., 88(1):180-188.

KRZEPTOWSKI, M. 1970. Z badan'ekspedycji pizyrodniczej na wodach szelfu
po'lnocno-zachodniej Afryky. Mater. Zachodniopomorkie, Muzeum Narodo-
we, Szczecin, 16:521-540 (in Polish, abstract in French).

--. 1975. Sardine in Polish catches in the Spanish Sahara region during
the years 1971-1974. ICES, C.M. 1975/J18, Pelagic Fish, Southern
Comm., 18 p. (mimeo).

KUBOTA, T.; UYENO, T. 1972. On the occurence of the lanternfish Electrona rissoi in Japan. Jap. J. Ichthyol., 19(2):125-128, fig. 1-2.

KUKUEV, E. I. 1982. Fish fauna of Corner Mountains and New England submarine ridge in the western North-Atlantic, p. 92-109. In: Insufficiently studied fishes of the open ocean (ed. N.V. Parin). Moscow,
140 p.

KULIKOVA, E. B. 1961. Data on the lanternfishes of the genus Diaphus
(family Scopelidae) from the western part of the Pacific Ocean. Trudy
Inst. Okeanol., 43:5-39 (in Russian).

KUREPINA, M. N.; PAVLOVSKY, E. N. 1946. The brain structure of fishes as
connected with the conditions of habitations. Izv. Akad. Nauk SSSR
(Biol.),(1):5-56, 47 fig. (Russian with English summary).

KURODA, N. 1951. A nominal list with distribution of the fishes of Suruga Bay, inclusive of the freshwater species found near the coast.
Jap. J. Ichthyol., 1(5):314-338; 1(6):376-394.

--. 1974. New additions to the list of fishes of Suruga Bay, Japan (N°
23). Zool. Mag.. Tokyo, 83(1):107-109, 2 fig.

KURONUMA, K. 1941. Notes on rare fishes taken off the Pacific coast of
Japan. Bull. biogeogr. Soc. Japan, 11(8):37-67, 6 fig., 2 pl., 1 tab.

KWEI, E. A. 1964. Migration of Sardinella aurita (Val. et Cuv.). Ghana
J. Sci., 4(1):34-43.

--. 1969. The fat cycle in the sardine, (Sardinella aurita, Cuv. and
Val.) in Ghanaian waters. p. 269-275. In: Proc. Symp. Oceanogr.
Fish. Resources trop. Atlant. (20-28 October 1968, Abidjan). Paris,
430 p.

--. 1978. Food and spawning activity of Caranx hippos (L.) off the coast of Ghana. <u>J. nat. Hist.</u>, 1978, 12:195-215.

KYLE, H. M. 1913. Flat-fishes (Heterosomata). <u>Rep. Dan. oceanogr. Exped.</u> <u>Mediterr.</u>, 2:1-150, 30 fig., 4 pl.

LACEPEDE, B. 1798-1803. Histoire naturelle des Poissons. 5 vol., 4°, Paris. 1, 1798:8 + cxlvii + 532 p., 25 pl., 1 tabl. (inset); 2, 1800: lxiv + 632 p., 20 pl.; 3, 1801:558 p., 34 pl.; 4, 1802: xliv + 728 p., 16 pl.; 5, 1803: xlviii + 803 p., 21 pl.

--. 1803. Histoire naturelle des Poissons. (14 vol. 8°). Paris, 10:401 p.

--. 1804. Mémoire sur plusieurs animaux de la Nouvelle-Hollande dont la description n'a pas encore été publiée. <u>Annls Mus. Hist. nat. Paris</u>, 4:184-211.

--. 1832. Histoire naturelle des quadrupèdes, ovipares, serpents, poissons et cétacés. (F.D. Pillot, ed.). IV. Poissons. Paris, 8, p. 1-398.

LACHNER, E. A. 1955. Populations of the berycoid fish family Polymixiidae. <u>Proc. U.S. natn. Mus.</u>, 105(3356):189-206, 1 pl.

--. 1981. Echeneidae, <u>in</u> Fischer, W., Bianchi, G. & W.B. Scott (ed.), FAO species identification sheets for fishery purposes. Eastern Central Atlantic; fishing areas 34, 47 (in part). Canada funds-in-Trust. Ottawa, Department of Fisheries and Oceans Canada, by arrangement with the Food and Agriculture Organization of the United Nations, vol. 1-7, pag. var.

--. 1986. Echeneidae, <u>in</u> P.J.P. Whitehead, M.L. Bauchot, J.C. Hureau, J. Nielsen & E. Tortonese. Fishes of the North-eastern Atlantic and the Mediterranean/Poissons de l'Atlantique du Nord-Est et de la Méditerranée, Paris, Unesco, 3:1329-1334, 7 + 7 fig.

LADIGES, W. 1958. Die Typen und Typoide der Fischsammlung des Hamburgischen Zoologischen Staatsinstituts und Zoologischen Museums. <u>Mitt. hamb. zool. Mus. Inst.</u>, 56:155-167.

LAFONT, A. 1873. Description d'une nouvelle espèce de Raie. <u>Act. Soc. linn. Bordeaux</u>, (3)28(8):503-504, pl. 15.

LAGOIN, Y.; SALMON, G. 1967. Etude technique et économique comparée de la distribution du poisson de mer dans les pays de l'Ouest Africain [13 parts, separately paged, including Introduction, Regional Study, Mauritania, Senegal, Gambia, Guinea, Liberia, Sierra Leone, Ivory Coast, Ghana, Togo, Benin, Nigeria). Soc. Centr. Equip. Territ. Coop., Paris, 889 p.

--. 1969. Etude technique et économique comparée de la distribution du poisson de mer dans les pays de l'Ouest Africain (Cameroon). Soc. Centr. Equip. Territ. Coop., Paris, 153 p.

--. 1970. Etude technique et économique comparée de la distribution du poisson de mer dans les pays de l'Ouest Africain (Introduction, Sâo Thomé. Equitorial Guinea, Gabon, Congo, Zaire, Angola). Soc. Centr. Equip. Territ. Coop., Paris, 549 p.

LALOY, L. 1910. Biologie de l'anguille. <u>Naturaliste</u>, 32:306-307.

LAMBOEUF, M. 1977. Estimation acoustique quantitative de l'abondance et de la distribution de la sardine marocaine en janvier 1977. <u>Trav. Docums Dév. Pêch. marit.</u>, Casablanca, (22):1-11 (mimeo).

LA MONTE, F. 1944. Notes on breeding grounds of blue marlin and swordfish off Cuba. <u>Copeia</u>, 1944(4):258.

--. 1945. North American game fishes. A guide to fresh and salt water game fishes. New York, 206 p., 73 pl.

--. 1952. Marine game fish of the world. Description, distribution, seasons, tackle. New York, 190 p., 69 pl.

--. 1955. A review and revision of the marlins, genus <u>Makaria</u>. <u>Bull. Am. Mus. nat. Hist.</u>, 107(3):323-358, 12 pl.

LA MONTE, F.; MARCY, D. E. 1941. Swordfish, sailfish, marlin, and spearfish. Ichthyol. Contr. int. Game Fish Ass., New York, 1(2):1-24, 5 pl.

LAMPE, M. 1914. Die Fische der Deutschen-Südpolar-Expedition 1901-03. III. Die Hochsee-und Küstenfische. Dt Südpol-Exped., 15(2):201-256, 7 fig., pl. 11.

LANE, E. D.; STEWART, K. W. 1968. A revision of the genus Hoplunnis Kaup (Apodes, Muraenesocidae), with a description of a new species. Contrib. mar. Sci., 13:51-64, 3 fig.

LARSEN, V. 1973. Nemichthyidae, p. 231-232, in Hureau, J.C. & Th. Monod (ed.). Check-list of the fishes of the northeastern Atlantic and of the Mediterranean/Catalogue des poissons du nord-est Atlantique et de la Méditerranée (Clofnam). Paris, Unesco, 2 vol.

LAST, P. R.; HARRIS, J. G. K. 1981. New locality records and preliminary information on demersal fish faunal assemblages in Tasmanian waters. Pap. Proc. R. Soc. Tasm., 115:189-209.

LAST, P. R.; SCOTT, E. O. G.; TALBOT, F. H. 1983. Fishes of Tasmania. Hobart, viii + 1-563 p., many fig.

LATHAM, J. 1794. An essay of the various species of sawfish. Trans. Linn. Soc. Lond., 2(25):273-282, pl. 26-27.

LATREILLE, P. A. 1804. Tableau méthodique des poissons. p. 71-105. In: Nouv. Dict. Hist. nat., Paris, 24.

LAVENBERG, R. J.; FITCH, J. E. 1966. Annoted list of fishes collected by midwater trawl in the Gulf of California, March-April 1964. Calif. Fish Game, 52(2):92-110, 6 fig.

LAY, G. T.; BENNET, E. T. 1839. Fishes. p. 41-75, pl. 15-23. In: J. Richardson et al.. The zoology of Captain Beechey's voyage... in His Majesty's Ship Blossom... London.

LEA, E. 1913. Muraenoid larvae from the "Michael Sars" North Atlantic Deep-Sea Expedition, 1910. Rep. scient. Results Michael Sars N. Atlant. deep Sea Exped., 3(1):1-48, 6 pl., 38 fig.

LEACH, W. E. 1814. Zoological miscellany; being descriptions of new or interesting animals. 3 vol. London, 1814-1817. 1, 1814:144 p., 60 col. pl.

--. 1818. Description of a swordfish (Xiphias rondeleti) found in the Firth of Forth in June 1811. Mem. Wernerian nat. Hist. Soc., Edinb., 2:58-60.

LEAPLEY, W. T. 1953. First record of the ribbonfish, Trachipterus trachyurus, from the mainland of North America. Copeia, 1953:236, 1 pl.

LEATHERLAND, T. M.; BURTON, J. D.; CULKIN, F.; McCARTNEY, J. M.; MORRIS, R. J. 1973. Concentrations of some trace metals in pelagic organisms and of mercury in Northwest Atlantic ocean water. Deep Sea Res., 20: 679-685.

LEBOUR, M. V. 1919. The young of the Gobiidae from the neighbourhood of Plymouth. J. mar. biol. Ass. U.K., 12:48-80.

LE CLUS, F. 1977. A comparison of four methods used in fecundity determinations of the pilchard Sardinops ocellata. Fish. Bull. S. Afr. 9: 11-15.

LE DANOIS, E. 1913. Contribution à l'étude systématique et biologique des poissons de la Manche occidentale. Annls Inst. océanogr., Monaco, 5(5):1-124, 319 fig.

--. 1954. Résumé de nos connaissances actuelles sur l'albacore ou yellow fin tuna (Thunnus albacore Bonnaterre). Bull. Inst. fr. Afr. noire, (A)16(1):283-294, 1 fig.

LE DANOIS, E.; LE DANOIS, Y. 1963. L'ordre des Scombres. Mém. Inst. fr. Afr. noire, (68):153-192, 19 fig.

LE DANOIS, Y. 1954. Sur le dimorphisme sexuel des poissons de la famille des Diodontides. C. r. hebd. Séances Acad. Sci. Paris, 238:2354-2356.

--. 1959. Etude ostéologique, myologique et systématique des poissons du sous-ordre des Orbiculates. _Annls_ _Inst._ _océanogr._, _Monaco_, (n.s.)36 (1):1-273, 221 fig.

--. 1961. Catalogue des types de poissons du Muséum National d'Histoire Naturelle. Famille des Triacanthidae, Balistidae, Monacanthidae et Aluteridae. _Bull._ _Mus._ _natn._ _Hist._ _nat._, _Paris_, (2)32(6):413-527.

--. 1961. Remarques sur l'ostéologie et la myologie d'un poisson de l'ordre des Jugulaires _Batrachus_ _didactylus_ Bl. & Schn. (Poisson-Crapaud ou Toad-fish). _Bull._ _Inst._ _fr._ _Afr._ _noire_, (A)23:806-854, fig. 1-31.

--. 1961. Catalogue des types de poissons Orbiculates du Muséum National d'Histoire Naturelle II. Familles des Tetraodontidae, Lagocephalidae, Colomesidae, Diodontidae et Triodontidae. _Bull._ _Mus._ _natn._ _Hist._ _nat._, _Paris_, (2)33(5):462-478.

--. 1962. Etude de la myologie et de l'ostéologie de l'uranoscope (_Uranoscopus_ _scaber_ L.) de l'ordre des Jugulaires. _Bull._ _Inst._ _ocean._ _Monaco_, (1229):1-50, 29 fig.

--. 1964. Etude anatomique et systématique des Antennaires, de l'ordre des Pédiculates. _Mém._ _Mus._ _natn._ _Hist._ _nat._, _Paris_, (A)31(1):1-162, 76 fig.

LEE, S.; CHANG, K.; WU, W.; YANG, H. 1977. Formosan ribbonfishes (Perciformes: Trichiuridae). _Bull._ _Inst._ _Zool._ _Acad._ _Sin._, 16(2):77-84.

LE GALL, J. 1931. _Epigonus_ _telescopus_ Risso, 1810. _Faune_ _ichthyol._ _Atlant._ _N._, Cahier 6:217.

--. 1932. _Scomber_ _scombrus_ L., _Scomber_ _colias_ Gmelin. _Faune_ _ichthyol._ _Atlant._ _N._, Cahier 10:279-280.

--. 1933. _Orthopristis_ _(Pristipoma)_ _benetti_ Lowe. _Faune_ _ichthyol._ _Atlant._ _N._, Cahier 13:254.

--. 1934. _Sarda_ _sarda_, _Orcynopis_ _unicolor_. _Faune_ _ichthyol._ _Atlant._ _N._, Cahier 15:287-288.

--. 1949. Germon. Résumé des connaissances acquises sur la biologie du germon. _Revue_ _Trav._ _Off._ _(scient._ _tech.)_ _Pêch._ _marit._, 15(1-4):1-42.

LE GALL, J. Y. 1974. Exposé synoptique des données biologiques sur le germon _Thunnus_ _alalunga_ (Bonnaterre) de l'Océan Atlantique. _FAO_ _Fish._ _Biol._ _Synopses_, (109):70 p.

--. 1974. Annotated bibliography on albacore _Thunnus_ _alalunga_ (Bonnaterre, 1788) of the Atlantic Ocean, 1962-72. _FAO_ _Fish._ _Rep._ 6(4). (Suppl. 1):1-67.

LE GALL, P. 1970. Quelques poissons signalés dans les eaux normandes. _Penn_ _ar_ _Bed_, 7(4), (63):440.

LEGAND, M. 1967. Cycles biologiques des poissons mésopélagiques dans l'est de l'Océan Indien. Première note. _Scopelopsis_ _multipunctatus_ Brauer, _Gonostoma_ sp., _Notolychnus_ _valdiviae_ Brauer. _Cah._ _ORSTOM._ _Océanogr._, 5(4):47-71.

LEGAND, M.; RIVATON, J. 1967. Cycles biologiques des poissons mésopélagiques dans l'est de l'Océan Indien. Deuxième note. Distribution moyenne des principales espèces de l'ichthyofaune. _Cah._ _ORSTOM._ _Océanogr._, 5(4):73-98.

LEGAND, M.; BOURRET, P.; FOURMANOIR, P.; GRANDPERRIN, R.; GUEREDRAT, A.; MICHEL, A.; RANCUREL, P.; REPELIN, R.; ROGER, C. 1972. Relations trophiques et distributions verticales en milieu pélagique dans l'Océan Pacifique intertropical. _Cah._ _ORSTOM._ _Océanogr._, 10(4):303-393.

LEGASPI, V. A. 1956. A contribution to the life history of the nomeid fish _Psenes_ _cyanophrys_. _Bull._ _mar._ _Sci._ _Gulf._ _Caribb._, 6:179-199.

LEGENDRE, R. 1928. Présence du _Tetrapturus_ _belone_ au large de la Bretagne. _Bull._ _Soc._ _zool._ _Fr._, 53:391-392, 1 fig.

--. 1932. La nourriture du germon, _Germo_ _alalunga_ (Bonnaterre). _Archs_ _Zool._ _exp._ _gén._, 74:531-540.

--. 1934. La faune pélagique de l'Atlantique au large du Golfe de Gascogne recueillie dans les estomacs de germons. 1re partie: poissons. <u>Annls Inst. océanogr.</u>, <u>Monaco</u>, 14:249-418, 53 fig.

LE GUEN, J. C.; POINSARD, F.; TROADEC, J. P. 1965. The yellowfin tuna fishery in the eastern tropical Atlantic (preliminary study). <u>Comml Fish. Rev.</u>, 27(8):7-18.

LEIBY, M. M. 1981. Larval morphology of the eels <u>Bascanichthys</u> <u>bascanium</u>, <u>B. scuticaris</u>, <u>Ophichthus</u> <u>melanoporus</u> and <u>O. ophis</u> (Ophichthidae), with a discussion of larval identification methods. <u>Bull. mar. Sci.</u>, 31:46-71.

--. 1982. Leptocephalus larvae of the tribe Sphagebranchini (Pisces, Ophichthidae) in the western North Atlantic. <u>Bull. mar. Sci.</u>, 32(1): 220-236.

--. 1984. Leptocephalus larvae of the tribe Callechelyini (Anguilliformes, Ophichthidae, Ophichthinae) in the western North Atlantic. <u>Bull. mar. Sci.</u>, 34:398-428.

--. (in press). Ophichthidae. <u>In</u>: Fishes of the western North Atlantic. <u>Mem. Sears Fdn mar. Res.</u>, 1(9).

LEIDENFROST, G. 1917. Halak és tüskésbörüek az Adriábol. <u>Allatt. Közl.</u>, 16:13, fig. 1.

LEIGH-SHARPE, W. H. 1920-1926. The comparative morphology of the secondary sexual characters of elasmobranch fishes. The claspers, clasper siphons, and clasper glands. Mem. I, IV, VII, XI. <u>J. Morph.</u> Mem. I, 34, 1920:245-265, fig. 1-12; Mem. IV,36, 1922: 199-220; Mem. VII, 39, 1924:567-577, fig. 1-15; Mem. XI, 42, 1926:349-358, fi. 1-12.

LEIM, A. H.; DAY, L. R. 1959. Records of uncommon und unusual fishes from eastern Canadian waters, 1950-58. <u>J. Fish. Res. Bd Can.</u>, 16(4): 503-514.

LEIM, A. M.; SCOTT, W. B. 1966. Fishes of the Atlantic coast of Canada. <u>Bull. Fish. Res. Bd Can.</u>, (155):485 p., many fig. unnumbered, 4 col. pl.

LEIS, J. M. 1977. Development of the eggs and larvae of the slender mola, <u>Ranzania</u> <u>laevis</u> (Pisces, Molidae). <u>Bull. mar. Sci.</u>, 27(3):448-466, 11 fig.

--. 1978. Systematics and zoogeography of the porcupinefishes (<u>Diodon</u>, Diodontidae, Tetraodontiformes), with comments on egg and larval development. <u>Fishery Bull.</u>, 76(3):535-567, 23 fig.

--. 1981. Diodontidae, <u>in</u> Fischer, W., Bianchi, G. & Scott (ed.), FAO species identification sheets for fishery purposes. Eastern Central Atlantic; fishing areas 34, 47 (in part). Canada Funds-in-Trust. Ottawa, Department of Fisheries and Oceans Canada, by arrangement with the Food and Agriculture Organization of the United Nations, vol. 1-7: pag. var.

LE LOEUFF, P.; INTES, A. 1973. Note sur le régime alimentaire de quelques poissons démersaux de la Côte d'Ivoire. <u>Docums. scient. provis. Centre Rech. Océanogr. Abidjan</u>, 4(2):17-44.

LEMA, T. de. 1976. Ocorrência de várias espécies de peixes tropicais marinhos na costa do Estado de Santa Catarina, Brazil (Osteichthyes, Actinopterygii, Teleostei). <u>Iheringia (Zool.)</u>, 49:39-65, 10 fig.

LENDENFELD, R. von. 1887. Report on the structure of the phosphorescent organs of fishes, p. 277-329, pl. 69-73. <u>In</u>: A. Günther. Report on the deep-sea fishes collected by H.M.S. "Challenger" during the years 1873-76. App. B. <u>Rep. scient. Results Voyage Challenger Zool.</u>, 22.

LEO, G. de (see DE LEO, G.)

LESSON, R. P. 1830. Poissons, 2(1):66-238, Atlas (1826), pl. 1-38. <u>In</u>: L.I. Duperrey, Voyage autour du monde, exécuté par Ordre du Roi, sur la corvette de sa Majesté, "La Coquille", pendant les années 1822, 1823, 1824 et 1825..., Zoologie. Paris.

LESUEUR, C. A. 1821. Description of two new species of _Exocoetus_. _J. Acad. nat. Sci. Philad._, 2(1):8-11.

--. 1821. Observations on several genera and species of fish, belonging to the natural family of the Esoces. _J. Acad. nat. Sci. Philad._, 2(1):124-138, pl. 10.

--. 1825. Descriptions of four new species of _Muraenophis (M. macularia, M. zebra, M. ocellata, M. bengalensis)_. _J. Acad. nat. Sci. Philad._, 5:107-109.

LETACONNOUX, R. 1953. Observations sur la sardine des Açores et nouvelle contribution à l'étude du genre _Sardina_. _Vie Millieu_, 4(1):37-56.

--. 1955. Rapport sur le merlu. _Rapp. P.-v. Réun. Cons. perm. int. Explor. Mer_, 137:40-43.

LEWIS, J. B.; BRYNDRITT, J. K.; FISH, A. G. 1962. The biology of the flyingfish _Hirundichthys affinis_ (Günther). _Bull. mar. Sci. Gulf. Caribb._, 12(1):73-94.

LEYDIG, F. 1879. Ueber die Nebenaugen des _Chauliodus sloani_. _Arch. Anat. Physiol._, 1879:365-382, fig.

--. 1881. Die augenähnlichen Organe der Fische. Bonn, 100 p., 10 pl.

LHOMME, F.; DOMAIN, F.; BOUR, W. 1973. La pêche chalutière à Dakar de 1965 à 1972. _Docum. scient. provis. Centre Rech. oceanogr. Dakar-Thiaroye_, (52):1-21, 8 tab., 15 fig.

LICHTENSTEIN, H. 1823. Verzeichnis der Doubletten des Zoologischen Museums der Königl. Universität zu Berlin, nebst Beschreibung vieler bisher unbekannter Arten von Säugethieren, Vögeln, Amphibien und Fischen. Berlin, i-ix, 1-118.

LIENARD, E. 1840. Description d'une nouvelle espèce du genre mole (_Orthagoriscus_ Schn.) découverte à l'île Maurice. _Revue Zool._, 3:291-292; _Mag. Zool._ (1841), (2)3: pl. 4.

LILLJEBORG, W. 1891. Sveriges och Norges Fiskar. Upsala, 1: xxi + 782 p.; 2:788 p.; 3:830 p.

LIN, S. 1933. A new genus and three new species of marine fish from Hainan Island. _Lingnan Sci. J._, 21(1-4):107-122.

LINCK, H. F. 1790. Versuch einer Eintheilung der Fische nach den Zähnen. _Mag. f. d. Neueste Phys. Naturg._, 6(3):28-38.

LINDBERG, G. U.; HEARD, A.S.; RASS, T. S. 1980. Multilingual dictionary of names of marine food-fishes of world fauna. Leningrad, 562 p., fig. (n. num.).

LINDROTH, P. G. 1798. _Gymnetrus Grillii_. Uptåck och beskrifven af... _K. svenska Vetensk. Akad. Handl._, 19:288-298, pl. 8.

LINNAEUS, C. 1758. Systema naturae... reformata. Cl. III Nantes + Cl. IV Pisces. p. 230-338. _In_: Tomo I (ed. X). Holmiae, 824 p.

--. 1761. Fauna Svecica, sistens animalia Sveciae Regni. (ed.II). Stockholm, 578 p., 2 pl.

--. 1764. Museum S.R.M. Adolphi Friderici Regis Svecorum, Gothorum, Vandalorumque in quo Animalia rariora imprimis et exotica: Aves, Amphibia, Pisces. Tomi secundi prodromus. Holmiae, 49-111, pl. 1-23.

--. 1766. Systema Naturae. Nantes & Pisces p. 394-532. (ed. XII, reformata). Holmiae, 1:532 p.

--. 1789. Cf. J.F. Gmelin, 1789.

LINTON, E. 1907. Notes on parasites of Bermuda fishes. _Proc. U.S. natn. Mus._, 33(1560):85-126.

LIPSKAYA, N. Y.; GORBUNOVA, N. N. 1980. Feeding of larvae of snake mackerel, _Gempylus serpens_. _Okeanologija_, 20(2):314-320 (in Russian).

LLEONART, J. 1981. Análisis de la comunidades epibentónicas del Atlántico Sudoriental a partir de los datos de la campaña "Benguela I". _Result. Exped. cient. B/O Cornide de Saavedra_, (9):37-51, 7 fig., 3 tab.

LLEONART, J.; RUCABADO, J. 1984. Datos pesqueros de la campaña "Benguela

I". Datos informativos, Inst. Investnes Pesc., Barcelona, 9:11-83.
LLORIS, D. 1981. Peces capturados en el SW africano durante la campana "Benguala I" (Noviembre, 1979). Result Exped. cient.. B/O cornide de Saavedra, (9):17-28, 1 fig., 2 tab.
--. 1982. Peces capturados en el SW africano durante la campana "Benguela II" (Julio-Septiembre 1980). Result. Exped. cient.. B/O Cornide de Saavedra, (10):3-15.
--. 1986. Ictiofauna demersal y aspectos biogeogràficos de la costa Sudoccidental de Africa (SWA/Namibia). Monografias Zool. mar., 1:9-432, 228 fig., 201 tab.
LLORIS, D.; RUCABADO, J. A. 1979. Mapas de presencias de especies de peces en las costas del Sahara Occidental. Barcelona, 205 p., 198 fig. (mimeo).
--. 1979. Especies ictiològicas de las expediciones pesqueras realizadas en la plataforma del NW de Africa (1971-1975). Result. Exped. Cient. B/O Cornide de Saavedra, (8):3-151, 13 fig.
--. 1979. Some considerations on the ichthyofauna of NW Africa. Third European ichthyol. Congr., Warszawa, 1979, 7 p. (mimeo).
LLORIS, D.; RUCABADO, J.; CARRILLO, J. 1978. Influencia de la Corriente de Canarias en el ciclo vital de Pagellus acarne (Fam. Sparidae) en el Banco Canario-Sahariano. Miscnea Zool. Barcelona, 4(2):183-190.
LLORIS, D.; RUCABADO, J. A.; GOMES, J. A. 1984. Lesueurigobius sanzoi (De Buen, 1918), (Osteichthyes, Gobiidae), in the south-east Atlantic (Namibia). Thalassas, 2:93-97.
LLORIS, D.; RUCABADO, J.; FUSTE, X.; ALLUE, C.; BAS, C. 1978. Area de afloramiento del N. W. de Africa. Campañas "ATLOR III" (1973) y "ATLOR V" (1974) - Cabo Bojador (26°10'N) a Cabo Blanco (21°45'N) - Pescas de arrastre de fondo. Datos informativos. Inst. Investnes pesq., Barcelona, 4:1-247.
LLOYD, R. E. 1909. A description of the deep-sea fish caught by the R.I.M.S. Ship "Investigator" since the year 1900 with supposed evidence of mutation in Malthopsis. Mem. Indian Mus., 2(3):139-180, 8 fig., 7 pl.
LO BIANCO, S. 1888. Notizie biologiche riguardanti specialmente il periodo di maturità sessuale degli animali del Golfo di Napoli. Mitt. zool. Stn Neapel, 8:385-440.
--. 1903. Le pesche abissali eseguite da F. A. Krupp col yacht "Puritan", nelle adiacenze di Capri ed in altre località del Mediterraneo. Mitt. zool. Stn Neapel, 16:109-279, 3 pl.
--. 1909. Notizie biologiche riguardanti specialmente il periodo di maturità sessuale degli animali del Golfo di Napoli. Mitt. zool. Stn Neapel, 19:513-761.
LOCKET, N. A. 1971. Retinal anatomy in some scopelarchid deep-sea fishes. Proc. R. Soc., (B) 178:1561-184.
LOCKINGTON, W. N. 1879. On a new genus and species of Scombridae. Proc. Acad. nat. Sci. Philad., 31:133-136.
--. 1880. Descriptions of new genera and species of fishes from the coast of California. Proc. U.S. natn. Mus., 2:326-332.
LOHBERGER, K. 1929. Ein neuer Fundort von Mugil grandisquamis C.V. Zool. Anz., 84:84.
LOIR, M. 1978. Une espèce nouvelle de poisson. Etud. Sports sous-marins, 40:18.
LONGHURST, A. R. 1957. The food of the demersal fish of a west African estuary. J. Anim. Ecol., 26:369-387.
--. 1960. A summary survey of the food of West African demersal fish. Bull. Inst. fr. Afr. noire, (A)22(1):276-282, tab. 1-3.
--. 1960. Fish landings in Sierra Leone during 1959. Occ. Pap. Fish. Dev. Res. Unit. Freetown, (1): 2 p., (mimeo).

--. 1960. Local movements of _Ethmalosa_ _fimbriata_ off Sierra Leone from tagging data. _Bull. Inst. fr. Afr. noire_, (A)22:1337-1340.

--. 1961. Report on the fisheries of Nigeria, 1961. Fedl Fish. Serv., Lagos, 42 p. (mimeo).

--. 1963. Notes on the identification of fish commonly caught in trawls. _Occ. Pap. fedl Fish. Serv. Nigeria_, 6:1-8 p.

--. 1964. Bionomics of the Sciaenidae of tropical West Africa. _J. Cons. perm. int. Explor. Mer_, 29:93-114.

--. 1964. A study of the Nigerian trawl fishery. _Bull. Inst. fr. Afr. noire_, (A)26:686-700.

--. 1965. A survey of the fish resources of the eastern Gulf of Guinea. _J. Cons. perm. int. Explor. Mer._, 29(3):302-334.

--. 1965. The biology of west African polynemid fishes. _J. Cons. perm. int. Explor. Mer._, 30(1):58-74, 2 fig.

--. 1969. Synopsis of biological data on West African croakers. _FAO Fish. Synopsis_, (35)(Rev. 1): 50 p., 9 fig., 21 tab.

--. 1969. Species assemblages in tropical demersal fisheries p. 147-168, 2 fig., 9 tab. _In_: Proc. Symp. Oceanogr. Fish. Resources trop. Atlant., Abidjan, 20-28 october 1966. Paris, 430 p.

--. 1971. The clupeoid resources of tropical seas. _Oceanogr. mar. Biol._, 9:349-385.

LONGLEY, W. H. 1934. Studies on West Indian fishes: description of six new species _Yb. Carnegie Instn Wash._, 33:257-260.

--. 1935. Osteological notes and descriptions of new species of fishes. _Yb. Carnegie Instn Wash._, 34:86-89.

LONGLEY, W. H.; HILDEBRAND, S. F. 1940. New genera and species of fishes from Tortugas, Florida. _Pap. Tortugas Lab._, 32:223-285, 1 pl., 28 fig.

--. 1941. Systematic catalogue of the fishes of Tortugas, Florida, with observations on color, habits, and local distribution. _Pap. Tortugas Lab._, 34: xiii + 331 p., 34 pl.

LÖNNBERG, E. 1894. List of fishes observed and collected in South-Florida. _Öfvers. K. Vetensk. Akad. Förh._, 51:93-131.

--. 1895. Notes on fishes collected in the Cameroons by Y. Sjöstedt. _Öfvers. K. Vetensk. Akad. Förh._, 52:179-195.

--. 1896. Linnean type-specimens of birds, reptiles, batrachians and fishes in the zoological museum of the Royal University in Upsala. _Bih. K. svenska Vetensk. Akad. Handl._, 22(4)(1):1-45.

LOPEZ, R. B. 1955. La caballa del mar Argentino. Datos sobre su biologia y pesca. _Revta Asoc. argent. Diet._, 13:25-33.

--. 1957. Pez aguja, "_Scomberesox saurus_" (Walbaum) pescado en Necochea. _Notas Mus. La Plata_, 19 (Zool. 176):145-151, 8 fig.

--. 1959. La caballa del mar Argentino. I. Sistematica, distribucion y pesca. _Comun. Mus. argent. Cienc. nat. Bernardino Rivadavia (Zool.)_, 3(3):95-130.

LOPEZ-JAMAR, E.; IGLESIAS, J.; OTERO, J. J. 1984. Contribution of infauna and mussel-raft epifauna to demersal fish diets. _Mar. Ecol. Progr. Ser._, 15:13-18.

LOSSE, G. F. 1969. _Harengula rouxi_ Poll. UNDP (SF)/FAO, Results Regional Fisheries Survey West Africa, Abidjan (Mem. 69/2): 4 p. (mimeo).

--. 1969. Seasonal distribution of _Sardinella aurita_ and _Sardinella eba_ in West African coastal waters. A preliminary note. UNDP (SF)/FAO, Results Regional Fisheries Survey West Africa, Abidjan, (Rep. 69/5): 6 p. (mimeo) (revised version 14 p., mimeo).

--. 1971. Notes on foreign industrial purse seine fisheries in the Guinea-Sahara region. UNDP (SF)/FAO, Results Regional Fisheries Survey West Africa, Abidjan, (Rep. 71/12): 11p. (mimeo).

--. 1971. Length distributions of some pelagic fishes caught from M/TR

Thue Jr during 1969/70. UNDP (SF)/FAO, Results Regional Fisheries Survey West Africa, Abidjan, (Rep. 71/17): 11 p. (mimeo).

LOSSE, G. F.; SCHMIDT, W.; JOHANESSON, K. A. 1971. Acoustic and biological observation on sardine resources off Sierra Leone with special reference to Sardinella aurita. UNDP (SF)/FAO, Results Regional Fisheries Survey West Africa, Abidjan, (Rep. 71/3): 38 p. (mimeo).

LOUBENS, G. 1964. Travaux en vue du développement de la pêche dans le bassin inférieur de l'Ogooué. Publs Cent. tech. for. trop., (27):1-151.

LOURIE, A. 1969. Occurence of two lantern fishes (Myctophidae) in the open sea off Mount Carmel (Israel). Israel J. Zool., 18:379-380.

LOWE, J. 1874. Fauna and flora of Norfolk. Trans. Norfolk Norwich Nat. Soc., 1:21-56.

LOWE, R. T. 1833. Description of Alepisaurus a new genus of fishes. Proc. zool. Soc. Lond., 1:104.

--. 1834. (Characters of a new genus, and of several new species of fishes from Madera.) Proc. zool. Soc. Lond. (1833), 1:142-144.

--. 1838. Piscium Maderensium species quaedam novae vel minus rite cognitae breviter descriptae. Trans. Camb. phil. Soc. (1836), 6:195-201, pl. 1-6.

--. 1838. (5 Dec. 1838). A synopsis of the fishes of Madeira; with the principal synonyms, Portuguese names, and characters of the new genera and species. Trans. zool. Soc. Lond., 2(3):173-200.

--. 1839. (Oct. 1839). A supplement to a synopsis of the fishes of Madeira. Proc. zool. Soc. Lond., 7:76-92. (also publ. in Trans. zool. Soc. Lond., 1842, 3(1):1-20).

--. 1841. (Jan.) (Description of certain new species of Madeiran fishes, with additional information relating to those already described). Proc. zool. Soc. Lond., 8(89):36-39 (also Ann. Mag. nat. Hist., (Apr.)7:92-94).

--. 1843. Notices of fishes newly observed or discovered in Madeira during the years 1840, 1841 and 1842. Proc. zool. Soc. Lond., 11:81-95.

--. 1843-60. A history of the fishes of Madeira, with original figures from nature of all the species, by C.E.C. Norton and M. Young. London: xvi + 196 p., 28 pl. Part 1, July 1843: i-xvi + 1-20, 1 + 4 pl.; part 2, Sept. 1843:21-52, pl. 5-8; part 3, Nov. 1843:53-84, pl. 9-12; part 4, Jan. 1844:85-116, pl. 13-17; part 5, Oct. 1860:117-196, pl. 18-27.

--. 1846. On a new genus of the family Lophiidae (Les Pectorales Pédiculées Cuv.) discovered in Madeira. Trans. zool. Soc. Lond., 3(4):339-344, pl. 51 (col.)

--. 1846. On a new genus of the family Lophiidae. Proc. zool. Soc. Lond. 14:81-83.

--. 1850. An account of fishes discovered or observed in Madeira since the year 1842. Proc. zool. Soc. Lond., 18:247-253.

--. 1852. An account of fishes discovered or observed in Madeira since the year 1842. Ann. Mag. nat. Hist., (2)10:49-55.

LOZANO, I. J.; LOZANO, G. 1985. Lista preliminar de las especies de los órdenes Beryciformes y Stephanoberyciformes (Pisces) de las Islas Canarias y aguas adyacentes. Vieraea, 14(1-2):191-202.

LOZANO CABO, F. 1948. Relación de una campaña de pesca de arrastre en pareja en la costa del Sáhara Español, y noticia sobre los otros tipos de pescas allí practicados. Boln Inst. esp. Oceanogr., 9:1-33, 10 fig., 2 map.

--. 1950. Datos sobre repartición geográfica de especies de peces de la costa de N.W. Africa. Boln R. Soc. esp. Hist. nat. (Biol.), 48, (1): 5-14.

--. 1950. Estudio de la fauna ictiológica de los bancos de Cabo Blanco y de Arguín, como transitoria entre la paleártica y la tropical. Boln R. Soc. esp. Hist. nat. (Biol.), 48(1):137-150.

--. 1953. Monografía de los centracántidos mediterráneos con un estudio especial de la biología, biometría y anatomía de Spicara smaris (L.). Boln Inst. esp. Oceanogr., (59):1-128, 42 fig.

--. 1958. Los escombridos de la aguas españolas y marroquíes y su pesca. Primera parte. Trab. Inst. esp. Oceanogr., (25):1-254, 187 fig.

--. 1961. Resultados de la expedición Peris-Alvarez a Annobón, 1959. 3. Nota sobre algunos peces marinos procedentes de la Isla de Annobon. Boln R. Soc. esp. Hist. nat. (Biol.), 59:137-152.

--. 1963. Nota preliminar sobre la campaña realizada por el buque oceanográfico francés "Thalassa" en las costas del Sáhara español y Mauritania. Boln R. Soc. esp. Hist. nat. (Biol.), 61:323-352, 12 fig.

--. 1970. Les problèmes ichthyologiques du plateau continental nord-ouest africain. Rapp. P.-v. Réun. Cons. perm. int. Explor. Mer, 159:149-151.

--. 1970. Caractéristiques zoogéographiques de la faune ichtyologique des côtes des Iles Canaries, du Maroc, du Sahara espagnol et de la Mauritanie avec une étude spéciale des poissons côtiers. Rapp. P.-v. Réun. Cons. perm. int. Explor. Mer, 159:152-164.

LOZANO y REY, L. 1916. Sobre el hallazgo de un Orcynopsis unicolor (Geoffr.) en Melilla. Boln R. Soc. esp. Hist. nat., 16(6):298-302.

--. 1934. Algunos peces pelágicos o de profundidad procedentes del Mediterraneo occidental. Boln R. Soc. esp. Hist. nat., 34:85-92.

--. 1934. Las pesquerías del Sáhara español. Publnes Dir. gen. Marruecos y Colonias Edit., Madrid, 91 p.

--. 1935. Los peces fluviales de España. Mems R. Acad. Cienc. exact. fis. nat. Madr. (Cienc. nat.), 5:xi + 390 p., 28 pl.

--. 1947. Peces Ganoideos y Fisóstomos. Mems R. Acad. Cienc. exact. fis. nat. Madr. (Cienc. nat.), 11: xv + 839 p., 190 fig., 20 pl.

--. 1950. Etude systématique des clupéidés et des engraulidés de l'Espagne, du Maroc et du Sahara espagnol. Rapp. P.-v. Réun. Cons. perm. int. Explor. Mer., 126:7-20, 4 pl.

--. 1952. Peces fisoclistos. Subserie Torácicos. Mems R. Acad. Cienc. exact. fis. nat. Madr. (Cien. nat.), 14: Primera parte: Ordenes Bericiformes, Zeiformes, Perciformes, Escorpeniformes y Balistiformes, vii-xv + 3-378, fig. 1-20, pl. 1-30. Segunda parte: Ordenes Labriformes y Escombriformes. (6) + 385-703 + (1), fig. 21-31, pl. 31-51.

--. 1960. Peces fisoclistos. Tercera parte: Subseries Torácicos (Ordenes Equeneiformes y Gobiformes), Pediculados y Asimétricos. Mems R. Acad. Cienc. exact. fis. nat. Madr. (Cien. nat.), 14:xiv + 1-613 + (1), 173 fig., 7 pl.

LUBBOCK, R. 1978. A new hawkfish of the genus Amblycirrhitus Gill, 1862, from Ascension Island/South Atlantic (Pisces: Perciformes: Percoidei: Cirrhitidae). Senckenberg biol., 58, (5/6):261-265, 2 fig.

--. 1980. The shore fishes of Ascension Island. J. Fish Biol., 17:283-303.

LÜTKEN, C. F. 1851. Nogle bemaerkinger om naeseborenes stilling hos dei i gruppe med Ophisurus staaende slaegter af aalefamilien. Vidensk. Meddr dansk naturh. Foren., 1851:1-21, Taf. 1.

--. 1871. Oneirodes eschrichtii Ltk, en ny grønlandsk Tudsefisk. Overs. K. danske Vidensk. Selsk. Forh., 1871:56-74, fig. 1-2, pl. 2.

--. 1878. Til Kundskab om to arktiske Slaegter af DybhavsTudsefiske: Himantolophus og Ceratias. K. danske Vidensk. Selsk. Skr., (5) Nat. math., 11(5):309-348, fig., pl.

--. 1880. Spolia Atlantica. Bidrag til Kundskab om Formforandringer hos Fiske under deres Vaext og Udvikling, searligt hos nogle af Atlanterhavets Højsøfiske. <u>K. danske Vidensk. Selsk. Skr.</u>, (5) Nat. math., 12(6):409-613, fig. n. num., 5 pl. (See also translation, 1881, <u>Ann. Mag. nat. Hist.</u>, (5)7:1-14, 107-123).

--. 1888. Korte Bidrag til nordisk Ichthyographi. VI. En for Grønlands-havet ny Rokke-Art (<u>Raja fyllae</u> n. sp. ad int.) m.m. <u>Vidensk. Meddr dansk naturh. Foren.</u>(1887), 49:1-4, 1 pl.

--. 1892. Korte Bidrag til nordisk Ichthyographi. VII. Nogle sjaeldnere Dybhavsfiske fra Davis-og Danmarksstraedet. <u>Vidensk. Meddr dansk naturh. Foren.</u> (1891), 53:28-35.

--. 1892. Korte Bidrag til nordisk Ichthyographi. VIII. Nogle nordiske Laxesild (Scopeliner). <u>Vidensk. Meddr dansk naturh. Foren.</u> (1891), 53: 203-233.

--. 1892. Spolia Atlantica. Scopelini Musei Zoologici Universitatis Hauniensis. Bidrag til Kundskab om det aabne Havs Laxesild eller Scopeliner. Med et Tillaeg om en anden pelagisk Fiskeslaegt. <u>K. dansk Vidensk. Selsk. Skr.</u>, (6)7:221-297, 3 pl.

--. 1898. Det ichthyologiske udbytte. <u>Dan. Ingolf-Exped.</u>, 2(1):1-38, 2 fig. n. num., 4 pl., 1 map (also Engl. transl.).

LUNA, E. 1950. Il cervelletto di "<u>Corvina nigra</u>". <u>Monitore zool. ital.</u>, 58:111-115, 4 fig.

LUND, W. A. 1970. Movements and migrations of tagged bluefish of New York and southern New England. <u>Trans. Am. Fish. Soc.</u>, 99:719-725.

LUNEL, G. 1865. Révision du genre castagnole (<u>Brama</u>) et description d'une espèce nouvelle, <u>Brama saussurii</u>. <u>Mém. Soc. Phys. Hist. nat. Genève</u>, 18:165-196, 2 pl.

LUTHER, W.; FIEDLER, K. 1968. Peces y demás fauna marina de las costas del Mediterráneo. Barcelona, 374 p., 32 fig., 46 pl. (transl. from German).

LUX, F. G. 1972. Predation by bluefish on flatfishes. <u>Mar. Fish. Rev.</u>, 34(7-8):30-35.

LYTHGOE, J. H.; LYTHGOE, G. 1971. Fishes of the Sea. London, Blandford, 320 p.

McALLISTER, D. E. 1960. List of marine fishes of Canada. <u>Bull. natn. Mus. Can.</u>, (168):1-76.

--. 1968. Evolution of branchiostegals and classification of teleostome fishes. <u>Bull. natn. Mus. Can.</u>, (221):1-239, 21 pl.

--. 1984. Osmeridae, <u>in</u> P. J. P. Whitehead, M. L. Bauchot, J. C. Hureau, J. Nielsen & E. Tortonese, Fishes of the North-eastern Atlantic and the Mediterranean/Poissons de l'Atlantique du Nord-Est et de la Médi-terranée, Paris, Unesco, 1:399-402, 3 + 7 fig., 3 maps.

McALLISTER, D.E.; REES, E. J. S. 1964. A revision of the eelpout genus <u>Melanostigma</u> with a new genus and with comments on <u>Maynea</u>. <u>Bull. natn. Mus. Can.</u>, (199):85-109.

McARDLE, A. F. 1901. Natural history notes from the Royal Indian Marine Survey Ship "Investigator", Commander T.H. Heming, R.N., commanding. Series III, No 5. An account of the trawling operations during the surveying season of 1900-1901. <u>Ann. Mag. nat. Hist.</u>, (7)8:517-526.

McCANN, C.; McKNIGHT, D. G. 1980. The marine fauna of New Zealand: ma-crourid fishes (Pisces: Gadida). <u>Mem. N.Z. oceanogr. Inst.</u>, (61):91 p., 69 fig.

McCLELLAND, J. 1845. Apodal fishes of Bengal. <u>J. nat. Hist. Calcutta</u> (1844), 5(18):150-226.

McCOSKER, J. E. 1977. The osteology, classification and relationships of the eel family Ophichthidae. <u>Proc. Calif. Acad. Sci.</u>, (4)41(1):1-123.

M'COY, F. 1841. On some new or rare fish occurring on the coast of Ire-land. <u>Ann. Mag. nat. Hist.</u>, (1)6:402-408.

McCULLOCH, A. R. 1909. Studies in Australian fishes. No 2. Rec. Aust. Mus., 7(4):315-321.

--. 1914. Biological results of the fishing experiments carried out by the F.I.S. "Endeavour", 1909-1914. III. Report on some fishes obtained by F.I.S. "Endeavour" on the coasts fo Queensland, New South Wales, Victoria, Tasmania, South and South-Western Australia. Pt. II; Biol. Results Fish. Exp. "Endeavour", 2(3):77-165, fig. 1-15, pl. 13-34.

--. 1915. Report on some fishes obtained by the F.I.S. "Endeavour" on the coasts of Queensland, New South Wales, Victoria, Tasmania, South and South-Western Australia. Part III. Biol. Results Fish. Exp. "Endeavour", 3(3):97-170.

--. 1916. Report on some fishes obtained by the F.I.S. "Endeavour" on the coasts of Queensland, New South Wales, Victoria, Tasmania, South and South-Western Australia. Part IV. Biol. Results Fish. Exp. "Endeavour", 4(4):169-199.

--. 1922. Check list of the fish and fish-like animals of New South Wales. Aust. Zool., 2(3):86-130, pl. 25-43.

--. 1923. Fishes from Australia and Lord Howe Island, No 2. Rec. Aust. Mus., 14:113-125, pl. 14-16.

--. 1926. Report on some fishes obtained by the F.I.S. "Endeavour". Biol. Results Fish. Exp. "Endeavour", 5(4):1-155, fig. 1-4, pl. 43-56.

--. 1929. A check-list of the fishes recorded from Australia. Mem. Aust. Mus., 5(3):329-436.

--. 1934. The fishes and fish-like animals of New South Wales. (3rd ed.). With supplement by G.P. Whitley. Sydney, xxvi + 104 p., 43 pl.

McCULLOCH, A. R.; WAITE, E. R. 1918. Some new and little known fishes from South Australia. Rec. S. Aust. Mus., 1:1-78.

McDOWELL, S. B. 1973. Family Halosauridae. In: Fishes of the western North Atlantic. Mem. Sears Fdn mar. Res., 1(6):32-123, fig. 1-13.

--. 1973. Family Notacanthidae. In: Fishes of the western North Atlantic. Mem. Sears Fdn mar. Res., 1(6):124-207, fig. 14-16, tab. 1.

McEACHRAN, J. D. 1984. Anatomical investigations of the New Zealand skates Bathyraja asperula and B. spinifera, with an evaluation of their classification within the Rajoidei (Chondrichthyes). Copeia, 1984(1):45-58, fig. 1-7.

McEACHRAN, J. D.; BRANSTETTER, S. 1984. Squalidae, in P. J. P. Whitehead, M. L. Bauchot, J. C. Hureau, J. Nielsen & E. Tortonese,Fishes of the North-eastern Atlantic and the Mediterranean/Poissons de l'Atlantique du Nord-Est et de la Méditerranée, Paris, Unesco, 1:128-147, 22 fig., 22 maps.

McEACHRAN, J. D.; CAPAPE, C. 1984. Rhinobathidae, in P. J. P. Whitehead, M. L. Bauchot, J. C. Hureau, J. Nielsen & E. Tortonese, Fishes of the North-eastern Atlantic and the Mediterranean/Poissons de l'Atlantique du Nord-Est et de la Méditerranée, Paris, Unesco, 1:156-158, 3 fig., 2 maps.

--. 1984. Dasyatidae, in P. J. P. Whitehead, M. L. Bauchot, J. C. Hureau, J. Nielsen & E. Tortonese,Fishes of the North-eastern Atlantic and the Mediterranean/Poissons de l'Atlantique du Nord-Est et de la Méditerranée, Paris, Unesco, 1:197-202, 6 fig., 6 maps.

--. 1984. Gymnuridae, in P. J. P. Whitehead, M. L. Bauchot, J. C. Hureau, J. Nielsen & E. Tortonese,Fishes of the North-eastern Atlantic and the Mediterranean/Poissons de l'Atlantique du Nord-Est et de la Méditerranée, Paris, Unesco, 1:203-204, 1 fig., 1 map.

--. 1984. Myliobatidae, in P. J. P. Whitehead, M. L. Bauchot, J. C. Hureau, J. Nielsen & E. Tortonese,Fishes of the North-eastern Atlantic and the Mediterranean/Poissons de l'Atlantique du Nord-Est et de

la Méditerranée, Paris, Unesco, 1:205-207, 2 fig., 2 maps.
--. 1984. Rhinopteridae, in P. J. P. Whitehead, M. L. Bauchot, J. C.
Hureau, J. Nielsen & E. Tortonese,Fishes of the North-eastern Atlantic and the Mediterranean/Poissons de l'Atlantique du Nord-Est et de
la Méditerranée, Paris, Unesco, 1:208-209, 1 fig., 1 map.
--. 1984. Mobulidae, in P. J. P. Whitehead, M. L. Bauchot, J. C. Hureau,
J. Nielsen & E. Tortonese, Fishes of the North-eastern Atlantic and
the Mediterranean/Poissons de l'Atlantique du Nord-Est et de la Méditerranée, Paris, Unesco, 1:210-211, 1 fig., 1 map.
McEACHRAN, J. D.; COMPAGNO, J. L. V. 1982. Interrelationships of and
within Breviraja based on anatomical structures. Bull. mar. Sci., 32
(2):399-425, fig. 1-18.
McEACHRAN, J. D.; STEHMANN, M. 1977. Subgeneric placement of the clasper, cranium and pelvic girdle. Copeia, 1977(1):21-25, fig. 1-5.
McGILCHRIST, A. C. 1905. Natural history notes from the R.I.M.S. Inves-
tigator, Capt. T. H. Heming, R.N (retired), commanding. Series III,
No 8. On a new genus of teleostean fish closely allied to Chiasmodus.
Ann. Mag. nat. Hist., (7)15:268-270.
McGINNIS, R. F. 1974. Biogeography of lanternfishes (family Myctophidae)
south of 30° S. Ph.D. thesis, University of Southern California.
McINTOSH, W. C. 1887. Notes from the St. Andrews Marine Laboratory.
VIII. Ann. Mag. nat. Hist., (5)20:300-304.
--. 1888. Report from the St. Andrews Marine Laboratory. V. Rep. Fishery
Bd Scotl., 6:265-276.
McINTOSH, W. C.; MASTERMAN, A. T. 1897. The life-histories of the Bri-
tish marine foodfishes. London, xv + 516 p., illust.
McLAREN, P. I. R. 1952. Fishing of bonga. Niger. Fld, 17(2):79-81.
MacLEAY, W. S. 1854. Illustrated Sydney News. 1(23):179, 1 fig.
--. 1881. On two hitherto undescribed fishes well-known in the Sydney
market (Otolithus teraglin; Synaptura nigra). Proc. Linn. Soc.
N.S.W., 5:48-49.
--. 1881. Descriptive catalogue of the fishes of Australia. Part II.
Proc. Linn. Soc. N.S.W., 5:510-629.
--. 1882. Descriptive catalogue of the fishes of Australia. Part IV.
Proc. Linn. Soc. N. S.W., 6:202-387.
--. 1883. Contribution to a knowledge of the fishes of New Guinea. Proc.
Linn. Soc. N.S.W., 7(3):585-598.
--. 1884. Some results of trawl fishing outside Port Jackson. Proc.
Linn. Soc. N.S.W., 8(4):457-462.
--. 1886. A remarkable fish from Lord Howe Island. Proc. Linn. Soc.
N.S.W., 10:718-720.
McLEISH, K. 1954. A vertical voyage to the bottom of the sea. Life, 37
(13):88-100.
McLELLAN, T. 1977. Feeding strategies of the macrourids. Deep Sea Res.,
24:1019-1036.
MACHADO Y NUNEZ, A. 1857. Catalogo de los peces que habitan o frequentan
las costas de Càdiz y Huelva, con inclusion del Rio Guadalquivir.
Sevilla, 25 p.
MACHIDA, Y. 1984. Order Zeiformes. p. 117-118, pl. 102, 103, 347. In:
The fishes of the Japanese Archipelago. (ed. Masuda, H., K. Amaoka,
C. Araga, T. Uyeno & T. Yoshino). Tokyo, Text: xxii + 437 p.; Plates:
1-370.
MACPHERSON, E. 1979. Ecological overlap between macrourids in the wes-
tern Mediterranean Sea. Mar. Biol., Berl., 53:149-159.
MAGO LECCIA, F. 1965. Nuevas adiciones a la ictiofauna de Venezuela. I.
Acta biol. Venez., 4(13):366-420, 8 fig.
MAIGRET, J. 1972. Campagne experimentale de pêche des sardinelles et
autres espèces pélagiques. Juillet 1970-Octobre 1971. Vol. 1. Obser-

vations concernant l'océanographie et la biologie des espèces. Paris, Soc. Centrale Equip. Territ.-Coop., 148 p.

--. 1973. La pêche des senneurs dans la Baie du Lévrier en 1971. Bull. Lab. Pêch. Nouadhibou, 2:35-56.

--. 1974. Inventaire ichtyologique des côtes mauritaniennes. Bull. Lab. Pêch. Nouadhibou, 3:131-152, 1 fig.

--. 1974. La sardine sur les côtes de Mauritanie (Sardina pilchardus, Walb.). Bull. Inst. fond. Afr. noire, (A)36:714-721.

--. 1975. Notes ichtyologiques. II. Complément à l'inventaire ichtyologique des côtes mauritaniennes. Bull. Lab. Pêch. Nouadhibou, 4:121-123.

MAIGRET, J.; BRULHET, J. 1973. Informations sur la pêche au large des côtes mauritaniennes. Bull. Lab. Pêch. Nouadhibou, 2:113-122.

MAIGRET, J.; LY, B. 1986. Les poissons de mer de Mauritanie. Centre National de Recherches Océanogr. et des Pêches, Nouadhibou, Sciences Nat (Publ.), Compiègne, 213 p., 93 fig., 66 phot.

MAIGRET, J.; SOUKHOVERSIN, V.; CHLIBANOV, V. 1980. Notes ichthyologiques III. Complément à l'inventaire ichthyologique des côtes de Mauritanie n° 2. Bull. Cent. natn. Rech. océanogr. Pêch. Nouadhibou, 9(1):122-129.

MAINGUY, P.; DOUTRE, M. 1959. Variations annuelles de la teneur en matières grasses de trois clupéides du Sénégal (Ethmalosa fimbriata Bowdich, Sardinella eba C.V., Sardinella aurita C.V.). Revue Trav. Inst. Pêch. marit. (1958), 22(3):303-321, 12 fig.

MAKSIMOV, V.P. 1971. The biology of the sailfish (Istiophorus platypterus (Shaw et Nodder)) in the Atlantic Ocean. J. Ichthyol., 11(6):850-855, 1 fig.

MAKUSHOK, V. M. 1986. Stichaeidae, in P.J.P. Whitehead, M.L. Bauchot, J.C. Hureau, J. Nielsen & E. Tortonese. Fishes of the North-eastern Atlantic and the Mediterranean/Poissons de l'Atlantique du Nord-Est et de la Méditerranée, Paris, Unesco, 3:1122-1123, 1 fig.

--. 1986. Pholididae, in P.J.P. Whitehead, M.L. Bauchot, J.C. Hureau, J. Nielsen & E. Tortonese. Fishes of the North-eastern Atlantic and the Mediterranean/Poissons de l'Atlantique du Nord-Est et de la Méditerranée, Paris, Unesco, 3:1124-1125, 1 fig.

--. 1986. Lumpenidae, in P.J.P. Whitehead, M.L. Bauchot, J.C. Hureau, J. Nielsen & E. Tortonese. Fishes of the North-eastern Atlantic and the Mediterranean/Poissons de l'Atlantique du Nord-Est et de la Méditerranée, Paris, Unesco, 3:1126-1129, 4 fig.

MALM, A. W. 1861. Raekke af fiske, krebsjur og bløddjur, som ere nye for den Skandinaviske fauna. Forh. skand. Naturf. Møte, 8:616-624.

--. 1877. Göteborgs och Bohusläns Fauna. Ryggradsdjuren. Göteborg, 674 p., 9 pl.

--. 1874. Om de Svenske gobiider. Forh. skand. Naturf.. Møte, 11:380-386.

MANLOV, J.; KOLAROV, P. 1973. Liste d'espèces de poissons peuplant le shelf nord-ouest africain. Izv. nanchno-izsled. Inst. rib. stop. Okeanogr. (Varna), 12:145-154. (in Bulgarian with French abstract).

MANRIQUEZ, M.; RUCABADO, J. 1976. Area de afloramiento del N.W. de Africa: 23°31'N-26°10'N; octobre de 1975 (Campaña ATLOR VI). Datos informativos Inst. Investnes pesq. Barcelona, 1:1-184.

MAPSTONE, G. M.; WOOD, E. M. 1972. The ethology of Abudefduf luridus and Chromis chromis (Pisces: Pomacentridae) from the Azores. J. Zool., 175:179-199.

MARCHAL, E. 1959. Observations sur quelques Neothunnus albacora (Lowe) capturés au large des côtes de Guinée. Bull. Inst. fr. Afr. noire, (A)21(4):1253-1274.

--. 1961. Description d'une nouvelle espèce de Paracubiceps (Belloc),

Paracubiceps multisquamis n. sp. (Poisson Téléostéen). Bull. Mus. natn. Hist. nat., Paris, (2)33(5):487-491.

--. 1961. Quelques observations complémentaires sur les Scombridae pêchés par le "Pyrrhus". Bull. Inst. fr. Afr. noire, (A)23(1):102-107.

--. 1963. Description des stades post-larvaires et juvéniles de quatre espèces de Scombridae de l'Atlantique tropico-oriental. Mém. Inst. fr. Afr. noire, (68):201-240.

--.1963. Exposé synoptique des données biologiques sur la thonine Euthynnus alleteratus (Rafinesque, 1810) (Atlantique Oriental et Méditerranée); FAO Fish. Rep., (6)2:647-662.

--. 1963. Description des stades post-larvaires et juvéniles de Neothunnus albacora (Lowe) de l'Atlantique tropico-oriental. FAO Fish. Rep., (6)3:1797-1811.

--. 1964. Sur la capture en Côte d'Ivoire de deux spécimens d'Ijimaia loppei Roule, (Ateleopidae, Poissons Téléostéens). Bull. Inst. fr. Afr. noire, (A)26(4):1340-1345, fig. 1-3.

--. 1965. Etude de quelques caractères de Sardinella eba C. et V. de Côte d'Ivoire. Cah. ORSTOM. Océanogr., 3:87-94.

--. 1965. Note sur deux caractères de Sardinella aurita C. et V. de Côte d'Ivoire. Cah. ORSTOM. Océanogr., 3(1):95-99.

--. 1966. Fluctuations de la pêche des sardinelles (Sardinella aurita C. V., Sardinella eba C.V.) en Côte d'Ivoire. Docums scient. provis. Centre Rech. océanogr. Abidjan, (001 S.R.):15 p. (mimeo).

--. 1966. Teneur en matières grasses et teneur en eau chez deux clupéidés de Côte d'Ivoire (Sardinella aurita C.V., S. eba C.V.). Docums scient. provis. Centre Rech. océanogr. Abidjan, (004 S.R.):14 p. (mimeo).

--. 1967. Clé provisoire de détermination des oeufs et larves des clupéidés et engraulidés ouest-africains. Docums scient. provis. Centre Rech. océanogr. Abidjan, (014 S.R.):8 p. (mimeo).

--. 1967. La pêche des sardiniers ivoiriens en 1966. Docums scient. provis. Centre Rech. océanogr. Abidjan, (019 S.R.):26 p. (mimeo).

--. 1969. Oeufs, larves et postlarves de l'anchois du Golfe de Guinée Anchoviella guineenis, Blache et Rossignol, p. 281-287. In: Proc. Symp. Oceanogr. Fish. Resources trop. Atlant. (20-28 October 1968, Abidjan) Paris, 430 p.

--. 1971. La pêche des sardiniers ivoiriens en 1967-1968-1969. UNDP (SF)/FAO, Project IVC/6/288, Devel. Coastal Pelagic Fishery, Abidjan, (Rep. 3/71): 30 p. (mimeo).

--. 1971. La pêche des sardiniers ivoiriens en 1970. UNDP (SF)/FAO, Project IVC/6/288, Devel. Coastal Pelagic Fishery, Abidjan, (Rep. 3/71): 32 p. (mimeo)(also 4/71, 18 p., mimeo).

MARETZKI, A.; CASTILLO, A. del. 1967. A toxin secreted by the soap fish Rypticus saponaceus. A preliminary report. Toxicon, 4:245-250, 1 fig., tab.

MARINARO, J. Y. 1968. Contribution à l'étude des oeufs et larves pélagiques de poissons méditerranéens. IV. Les premiers stades de développement de Helicolenus dactylopterus Delaroche. Pelagos, Alger, (10): 133-138, 5 pl.

--. 1971. Contribution à l'étude des oeufs et larves pélagiques de poissons méditerranéens. V. Oeufs pélagiques de la baie d'Alger. Pelagos, Alger, 3(1):1-115, 18 fig., 27 pl.

MARION, A. F. 1894. Sur la pêche et la reproduction du Siouclet (Atherina hepsetus). Annls Mus. Hist. nat. Marseille, 4(1):93-99.

MARKLE, D. F. 1976. Preliminary studies on the systematics of deep-sea Alepocephaloidea (Pisces: Salmoniformes). Ph. D. thesis, Coll. William and Mary. Williamsburg, Va., 1-225 p., 44 fig.

--. 1978. Taxonomy and distribution of Rouleina attrita and Rouleina _maderensis_ (Pisces: Alepocephalidae). _Fishery Bull._, **76**(1):79-87, 4 fig.

--. 1980. A new species and a review of the deep-sea fish genus _Asquami-ceps_ (Salmoniformes: Alepocephalidae). _Bull. mar. Sci._, **30**(1):45-53, 4 fig.

--. 1981. Alepocephalidae, _in_ Fischer, W., Bianchi, G. & W.B. Scott (ed.), FAO species identification sheets for fishery purposes. Eastern Central Atlantic; fishing areas 34, 47 (in part). Canada Funds-in-Trust. Ottawa, Department of Fisheries and oceans Canada, by arrangement with the Food and Agriculture Organization of the United Nations, vol. 1-7: pag. var.

MARKLE, D. F.; MERRETT, N. R. 1980. The abyssal alepocephalid, _Rinoctes nasutus_ (Pisces: Salmoniformes), a redescription and an evaluation of its systematic position. _J. Zool._, **190**:225-239, 9 fig.

MARKLE, D. F.; MUSICK, J. A. 1974. Benthic fish associations on the continental slope of the Middle Atlantic Bight. _Mar. Biol., Berl._, **26**:225-233.

MARKLE, D. F.; QUERO, J. C. 1984. Alepocephalidae, _in_ P. J. P. Whitehead, M. L. Bauchot, J. C. Hureau, J. Nielsen & E. Tortonese, Fishes of the North-eastern Atlantic and the Mediterranean/Poissons de l'Atlantique du Nord-Est et de la Méditerranée, Paris, Unesco, 1:228-253, 33 + 1 fig., 33 maps.

--. 1984. Leptochilichthyidae, _in_ P. J. P. Whitehead, M. L. Bauchot, J. C. Hureau, J. Nielsen & E. Tortonese, Fishes of the North-eastern Atlantic and the Mediterranean/Poissons de l'Atlantique du Nord-Est et de la Méditerranée, Paris, Unesco, 1:254-255, 2 fig., 2 maps.

MARKLE, D. F.; SEDBERRY, G. R. 1978. A second specimen of the deep-sea fish, _Pachycara obesa_, with a discussion of its classification and a checklist of other Zoarcidae off Virginia. _Copeia_, 1978(1):22-25, 1 fig.

MARKLE, D. F.; WENNER, C. A. 1979. Evidence of demersal spawning in the mesopelagic zoarcid fish, _Melanostigma atlanticum_, with comments on demersal spawning in the alepocephalid fish _Xenodermichthys copei._ _Copeia_, 1979(2):363-366, 1 fig., 1 tab.

MARKLE, G. E. 1974. Distribution of larval swordfish in the Northwest Atlantic Ocean. _NOAA tech. Rep. NMFS SSRF_, (675)(Part 2): 252-260, 4 + 1 fig.

MARSHALL, N. B. 1954. Aspects of deep-sea biology. London, 380 p., 103 fig., 4 pl.

--. 1955. Studies of alepisauroid fishes. "_Discovery_" _Rep._, **27**:303-336, 1 pl.

--. 1960. Swimbladder structure of deep-sea fishes in relation to their systematics and biology. "_Discovery_" _Rep._, **31**:3-121, 3 pl., 47 fig.

--. 1961. A young _Macristium_ and the ctenothrissid fishes. _Bull. Br. Mus. nat. Hist._ (Zool.), **7**(8):353-370.

--. 1962. Observations on the Heteromi, and order of teleost fishes. _Bull. Br. Mus. nat. Hist._, (Zool.), **9**(6):251-270.

--. 1962. The biology of sound-producing fishes. _Symp. zool. Soc. Lond._, (7):45-60.

--. 1963. Diversity, distribution and speciation of deep-sea fishes. p. 181-195, 2 fig. In: J.P. Harding & N. Tebble. Speciation in the sea. _Publs Syst. Ass._, (5):1-199,fig.

--. 1964. Bathypelagic macrourid fishes. _Copeia_, 1964(1):86-93, 3 fig.

--. 1965. Systematic and biological studies of the macrourid fishes (Anacanthini-Teleostei). _Deep Sea Res._, **12**:299-322, 9 fig.

--. 1966. Family Scopelosauridae. _Mem. Sears Fdn mar. Res._, **1**(5):194-203, fig. 50-54.

--. 1967. The olfactory organs of bathypelagic fishes. _Symp. zool. Soc._

Lond., (19):57-70.

--. 1971. La vie des Poissons. Lausanne, I (Encycl. Nat., 8):1-384, fig. 1-302; II (Encycl. Nat., 9):385-768, fig. 303-645 (trad. L. Dilé, révis. scient. M.L. Bauchot).

--. 1972. Swimbladder organisation and depth ranges of deep-sea teleosts. Symp. Soc. expl. Biol., 26:261-272, 1 fig.

--. 1973. Macrouridae, p. 287-299, in Hureau, J.C. & Th. Monod (ed.), Check-list of the Fishes of the north-eastern Atlantic and of the Mediterranean/Catalogue des Poissons du nord-est Atlantique et de la Méditerranée (Clofnam). Paris, Unesco, 2 vol.

MARSHALL, N. B.; BOURNE. D. W. 1907. Deep-sea photography in the study of fishes. p. 251-257. In: J.B. Hersey (ed.): Deep-sea photography. Johns Hopkins oceanogr. Stud., (3).

--. 1964. A photographic survey of benthic fishes in the Red Sea and Gulf of Aden, with observations on their population density, diversity, and habits. Bull. Mus. comp. Zool. Harv., 132(2):225-244.

MARSHALL, N. B.; IWAMOTO, T. 1973. Family Macrouridae. Mem. Sears Fdn mar. Res., 1(6):496-665, 53 fig., 7 tab.

MARSHALL, N. B.; MERRETT, N. R. 1977. The existence of a benthopelagic fauna in the deep-sea. p. 483-497, 2 fig. In: A voyage of Discovery. George Deacon 70th Anniversary Volume (M. W. Angel, ed.). Oxford (suppl. to Deep Sea Res., 24).

MARSHALL, N. B.; STAIGER, J. C. 1975. Aspects of the structure, relationships, and biology of the deep-sea fish Ipnops murrayi Günther (Family Bathypteroidae). Bull. mar. Sci., 25(1):101-111.

MARSHALL, N. B.; TANING, A. V. 1966. The bathypelagic macrourid fish, Macrouroides inflaticeps Smith and Radcliffe. Dana Rep., (69):1-6 , 1 pl.

MARSHALL, T. C. 1964. Fishes of the Great Barrier Reef and coastal waters of Queensland. Sydney, xiii (n. num.) + 1-566 p., 64 pl., 72 col. pl., 12 fig.

--. 1966. Tropical fishes of the Great Barrier Reef. Sydney, 239 p.

MARTEAU, J. 1967. Inventaire des poissons marins en collection au Centre ORSTOM de Pointe Noire. Docums scient. Centre ORSTOM, Pointe Noire, (Océanogr.), (367):1-71.

MARTENS, E. von. 1876. Die preussische Expedition nach Ost-Asien. Zoologische Abteilung. 2 vol., 15 pl.

MARTIN, F. 1956. Ictiología del Archipiélago de Los Roques. p. 87-144. In: A. Mendez et al.: El Archipiélago de Los Roques y la Orchila. Caracas.

MARUYAMA, K.; ONO, K. 1973. On the second record of Barbourisia rufa from Japan. Jap. J. Ichthyol., 20(4):242-244, 1 fig.

MASSUTI, M. 1965. La campaña del "Walther Herwig" en Africa occidental. Publns téc. Jta Estud. Pesca, 4:77-204.

--. 1967. Los fondos de pesca en la plataforma de Marruecos, Sáhara español, Mauritania, Senegal, Republica de Guinea, Ensenada de Biafra y Archipiélago de Cabo Verde. Trab. Inst. esp. Oceanogr., (34):1-99, 105 fig.

MATALLANAS, J. 1975. Contribución al conocimiento de la ictiofauna ibérica. Nota sobre un Zoarcidae (Melanostigma atlanticum Koefoed, 1952) del mar Catalán. Investigación pesq., 39(2):373-377, 3 fig., 1 tab.

MATALLANAS, J.; RUBIO, M. 1979. Catálogo de peces del mar Catalán del Museo del Aquarium de Blanes (Gerona). Investigación pesq., 43(2): 559-564.

MATHER, III, F. J. 1952. Sport fishes of the vicinity of the Gulf of Honduras, certain Caribbean Islands, and Carmen, Mexico. Proc. Gulf Caribb. Fish. Inst., (4th a. Session):118-129.

--. 1954. Northerly occurrences of warmwater fishes in the western At-

lantic. <u>Copeia</u>, 1954(4):292-293.
--. 1960. Recaptures of tuna, marlin and sailfish tagged in the western North Atlantic. <u>Copeia</u>, 1960(2):149-151.
MATHER, III, F. J.; DAY, C. G. 1954. Observations of pelagic fishes of the tropical Atlantic. <u>Copeia</u>, 1954(3):179-188.
MATHIASSON, S. 1961. <u>Paralepis</u> <u>brevis</u>-en for Sverige ny fiskart. <u>Fauna Flora. Upps.</u>, 56:129-132, fig. 1-2.
--. 1963. Några mindre vanliga fiskfynd från Nordsjön (<u>Brama</u> <u>raschi, Pagellus</u> <u>acarne, Lepidopus</u> <u>caudatus</u>). <u>Fauna</u> <u>Flora. Upps.</u>, 58:19-25.
MATSUBARA, K. 1936. A new bramid fish found in Japan. <u>Bull. Jap. Soc. scient. Fish.</u>, 4(5):297-300.
--. 1938. Studies on the deep-sea fishes of Japan, VI-VIII. <u>J. imp. Fish. Inst.. Tokyo</u>, 33(1):37-66.
--. 1938. Studies on the deep-sea fishes of Japan. X. On a new fish <u>Notacanthus</u> <u>fascidens</u>, belonging to Heteromi with special reference to its variation. <u>Bull. Jap. Soc. scient. Fish.</u>, 7(3):131-136.
--. 1940. Studies on the deep-sea fishes of Japan. XIII. On Prof. Nakazawa's collection of fishes referable to Isospondyli, Iniomi and Allotriognathi (1). <u>Suisan</u> <u>KenKiu-shi</u>, Japan, 35(12):314-319.
--. 1941. Studies on the deep-sea fishes of Japan. XIII. On Professor Nakazawa's collection of fishes referable to Isospondyli, Iniomi and Allotriognathi. <u>Suisan</u> <u>KenKiu-shi</u>, Japan, 36(2):34-37.
--. 1943. Studies on the scorpaenoid fishes of Japan. Anatomy, phylogeny and taxonomy (I). <u>Trans. Sigenkag. Kenk.</u>, (1):1-170, fig. 1-66.
--. 1943. Studies on the scorpaenoid fishes of Japan. Anatomy, phylogeny and taxonomy (II). <u>Trans. Sigenkag. Kenk.</u>, (2):171-486, fig. 67-156, 4 pl.
--. 1943. Ichthyological annotations from the depth of the Sea of Japan. III. A review of the scopelid fish, referable to the genus <u>Neoscopelus. J. Sigenkag. Kenk.</u>, 1(1):55-63, fig. 11-15.
--. 1943. Ichthyological annotations from the depth of the Sea of Japan. VI. A new stomiatoid fish, <u>Yarrella</u> <u>surugaensis</u>, belonging to Gonostomatidae. <u>J. Sigenkag. Kenk.</u>, 1(1):73-76, fig. 22.
--. 1952. The capture of the Atlantic lantern-fish, <u>Lampadena</u> <u>nitida</u> (Taaning) in Japan. <u>Jap. J. Ichthyol.</u>, 2:111-112.
--. 1954. The first appearance of deep-sea iniomous fish, <u>Bathypterois</u> <u>atricolor</u> <u>antennatus</u> (Gilbert) in Japan. <u>Jap. J. Ichthyol.</u>, 3(2):62-63.
--. 1955. Fish morphology and hierarchy, Tokyo, 1-3: 1605 p., 536 fig., 135 pl. (in Japanese).
MATSUBARA, K.; IWAI, T. 1952. Studies on some Japanese fishes of the family Gempylidae. <u>Pacif. Sci.</u>, 6(3):193-212, 12 fig.
--. 1958. Anatomy and relationships of the Japanese fishes of the family Gempylidae. <u>Mem. Coll. Agric. Kyoto. Univ.</u> (Fish.), (Spec. No):25-54, 14 fig.
MATSUI, T. 1967. Review of the mackerel genera <u>Scomber</u> and <u>Rastrelliger</u> with description of a new species of <u>Rastrelliger</u>. <u>Copeia</u>, 1967(1): 71-83.
MATSUI, T.; ROSENBLATT, R. H. 1979. Two new searsid fishes of the genera <u>Maulisia</u> and <u>Searsia</u> (Pisces: Salmoniformes). <u>Bull. mar. Sci.</u>, 29(1): 62-78, fig. 1-10.
--. 1984. Review of the deep-sea fish family Platytroctidae (Pisces: Salmoniformes). <u>Bull. Scripps Instn Oceanogr.</u>, 26: i-vii + 1-159, 28 fig.
MATSUMOTO, M. 1959. Descriptions of <u>Euthynnus</u> and <u>Auxis</u> larvae from the Pacific and Atlantic Oceans and adjacent seas. <u>Dana Rep.</u>, (50):34 p., 31 fig.
--. 1960. Notes on the Hawaiian frigate mackerel of the genus <u>Auxis</u>.

Pacif. Sci., 14(2):173-177.

--. 1967. Morphology and distribution of larval wahoo Acanthocybium solandri (Cuvier) in the Central Pacific Ocean. Fishery Bull., 66(2): 299-322.

MATTA, F. 1955. Nota su di uno Stomias boa Risso catturato nel Tirreno mediante rete a strascico. Boll. Pesca Piscic. Idrobiol., 30:116-119.

MATTHEWS, F. D.; DAMKAER, D. M.; KNAPP, L. W.; COLLETTE, B. B. 1977. Food of western North Atlantic tunas (Thunnus) and lancetfishes (Alepisaurus). NOAA tech. Rep. NMFS SSRF, (706):1-19.

MATTHEWS, J. D. 1887. Note on ova, fry and nest of the ballan wrasse. Rep. Fishery Bd Scotl., 5:245-247, pl. 11.

MATTHEWS, J. P. 1960. Synopsis on the biology of the South West African pilchard (Sardinops ocellata Pappe). FAO Proc. Wld scient. Meet. Biol. Sardines, 2:115-135.

--. 1964. The pilchard of South Africa (Sardinops ocellata). Sexual development, condition factor and reproduction 1957-1960. Investl Rep. mar. Res. Lab. S.W. Afr., 10:1-96.

MATTHEWS, J. P.; BERRUTI, A. 1983. Diet of Cape Gannet and Cape Cormoran off Walvis Bay, 1958-1959. S. Afr. J. mar. Sci., 1:61-63.

MATTHEWS, L. H. 1950. Reproduction in the basking shark, Cetorhinus maximus (Gunner). Phil. Trans. R. Soc., (B)234:247-316.

MAUGE, A. 1981. Atherinidae, Chaetodontidae, Pomacanthidae, in Fischer, W., Bianchi, G. & W. B. Scott (ed.), FAO species identification sheets for fishery purposes. Eastern Central Atlantic; fishing areas 34, 47 (in part). Canada Funds-in-Trust. Ottawa, Department of Fisheries and Oceans Canada, by arrangement with the Food and Agriculture Organization of the United Nations, vol. 1-7: pag. var.

MAUL, G. E. 1945. Monografia dos peixes do Museu Municipal do Funchal. Familia Sudidae. Bolm Mus. munic. Funchal, (1, 1):1-38, fig. 1-10.

--. 1946. Monografia dos peixes do Museu Municipal do Funchal. Ordem Iniomi. Bolm Mus. munic. Funchal, (2, 2):5-61, fig. 1-22.

--. 1948. Monografia dos peixes do Museu Municipal do Funchal. Ordem Isospondyli. Bolm Mus. munic. Funchal, (3, 5):5-41, fig. 1-16.

--. 1948. Lista sistemàtica dos peixes assinalados nos mares de Madeira e indice alfabetico. p. 135-181. In: A.C. de Noronha & A.A. Sarmento, Vertebrados da Madeira. 2. Peixes, 2a ed. Funchal, (also sep., 1949).

--. 1948. Quatro peixes novos dos mares da Madeira. Bolm Mus. munic. Funchal, (3, 6):41-55, fig. 17-20, tab.

--. 1949. Nota sobre dois peixes descritos no boletim anterior. Bolm Mus. munic. Funchal, (4, 10):20-22.

--. 1949. Alguns peixes notàveis. Bolm Mus. munic. Funchal, (4, 11):22-42, fig. 8-17.

--. 1950. A espada preta. Publçoes Liga Prot. natur., Lisboa, 4:10 p.

--. 1951. Monografia dos peixes do Museu Municipal do Funchal. Familias Macrouridae e Merlucciidae. Bolm Mus. munic. Funchal, (5, 12):5-55, fig. 1-12.

--. 1951. Notas sobre as duas espécies do género Neoscopelus. Bolm Mus. munic. Funchal, (5, 3):56-63, figs 13-15.

--. 1952. Monografia dos peixes do Museu Municipal do Funchal. Additions to previously revised families. Bolm Mus. munic. Funchal, (6, 16):51-62, fig. 13-17.

--. 1954. Monografia dos peixes do Museu Municipal do Funchal. Ordem Berycomorphi. Bolm Mus. munic. Funchal, (7, 17):5-41, fig. 1-13.

--. 1954. Monografia dos peixes do Museu Municipal do Funchal. Additions to previously revised families. Bolm Mus. munic. Funchal, (7, 18): 41-63, fig. 14-21.

--. 1956. Additions to previously revised orders or families of fishes of the Museu Municipal do Funchal (Stomiatidae, Astronestidae, Paralepididae). **Bolm Mus. munic. Funchal**, (**9**, 24):75-96, fig. 6-18.

--. 1959. **Aulostomus**, a recent spontaneous settler in Madeiran waters. **Bocagiana**, (1):1-18, 1 fig.

--. 1961. The ceratioid fishes in the collection of the Museu Municipal do Funchal (Melanocetidae, Himantolophidae, Oneirodidae, Linophrynidae). **Bolm Mus. munic. Funchal**, (14, 50):87-159, fig. 1-32.

--. 1962. On a small collection of ceratioid fishes from off Dakar and two recently acquired specimens from stomachs of **Aphanopus carbo** taken in Madeira (Melanocetidae, Himantolophidae, Diceratiidae, Oneirodidae, Ceratiidae). **Bolm Mus. munic. Funchal**, (**16**, 54):5-27, fig. 1-10.

--. 1962. Report on the fishes taken in Madeiran and Canarian waters during the summer-autumn cruises of the "Discovery II" 1959 and 1961. I. The ceratioid fishes (Melanocetidae, Himantolophidae, Oneirodidae, Gigantactinidae, Linophrynidae). **Bolm Mus. munic. Funchal**, (16, 56):33-46, fig. 1-6.

--. 1962. On four rare paralepidids from off Dakar, with a discussion on two type specimens of **Omosudis elongatus** Brauer from the Atlantic Ocean. **Bull. Inst. fr. Afr. noire**, (A)24(2):523-550, fig. 1-9.

--. 1964. Observations on young live **Mupus maculatus** (Günther) and **Mupus ovalis** (Valenciennes). **Copeia**, 1964:93-97.

--. 1965. On a new genus and species of paralepidid from Madeira. **Bolm Mus. munic. Funchal**, (**19**, 81):55-61, fig. 1.

--. 1969. On the genus **Cetomimus** (Cetomimidae) with the description of a new species. **Bocagiana**, (18):1-12, 1 fig.

--. 1969. A new subspecies of **Lampadena urophaos** Paxton, 1963 from the Atlantic Ocean. **Bocagiana**, (22):1-8.

--. 1971. On a new goby of the genus **Lesueurigobius** from off the Atlantic coast of Morocco and Madeira. **Bocagiana**, (29):7 p.

--. 1973. Macristiidae, p. 113, **in** Hureau, J.C. & Th. Monod (ed.). Check-list of the fishes of the north-eastern Atlantic and of the Mediterranean/Catalogue des poissons du nord-est Atlantique et de la Méditerranée (Clofnam). Paris, Unesco, 2 vol.

--. 1976. The fishes taken in bottom trawls by R.V. "Meteor" during the 1967 seamounts cruises in the Northeast Atlantic. **"Meteor" Forsch.-Ergebn.**, (D)(22):1-69, 38 fig., 1 pl.

--. 1981. Berycidae, Trachichthyidae, **in** Fischer, W., Bianchi, G. & W.B. Scott (ed.), FAO species identification sheets for fishery purposes. Eastern Central Atlantic; fishing areas 34, 47 (in part). Canada Funds-in-Trust. Ottawa, Department of Fisheries and Oceans Canada, by arrangement with the Food and Agriculture Organization of the United Nations, vol. 1-7: pag. var.

--. 1986. Berycidae, **in** P.J.P. Whitehead, M.L. Bauchot, J.C. Hureau, J. Nielsen & E. Tortonese. Fishes of the North-eastern Atlantic and the Mediterranean/Poissons de l'Atlantique du Nord-Est et de la Méditerranée, Paris, Unesco, 2:740-742, 2 fig.

--. 1986. Trachichthyidae, **in** P.J.P. Whitehead, M.L. Bauchot, J.C. Hureau, J. Nielsen & E. Tortonese. Fishes of the North-eastern Atlantic and the Mediterranean/Poissons de l'Atlantique du Nord-Est et de la Méditerranée, Paris, Unesco, 2:749-752, 3 + 1 fig.

--. 1986. Melamphaidae, **in** P.J.P. Whitehead, M.L. Bauchot, J.C. Hureau, J. Nielsen & E. Tortonese. Fishes of the North-eastern Atlantic and the Mediterranean/Poissons de l'Atlantique du Nord-Est et de la Méditerranée, Paris, Unesco, 2:756-765, 12 + 7 fig.

--. 1986. Stephanoberycidae, **in** P.J.P. Whitehead, M.L. Bauchot, J.C. Hureau, J. Nielsen & E. Tortonese. Fishes of the North-eastern Atlantic and the Mediterranean/Poissons de l'Atlantique du Nord-Est et de la Méditerranée, Paris, Unesco, 2:766, 1 fig.

MAUL, G. E.; KOEFFOED, E. 1950. On a new genus and species of macrourid fish, *Phalacromacrurus pantherinus*. *Ann. Mag. nat. Hist.*, (12)3: 970-976, 1 fig.

MAURIN, C. 1955. Note sur le merlu des côtes du Maroc. *Rapp. P.-v. Réun. Cons. perm. int. Explor. Mer*, 137:45-46, 1 fig.

--. 1962. Etude des fonds chalutables de la Méditerranée occidentale (Ecologie et Pêche). Résultats des campagnes des navires océanographiques "Président Théodore Tissier", 1957 à 1960 et "Thalassa", 1960 et 1961. *Revue Trav. Inst. Pêch. marit.*, 26(2):163-218, 16 fig.

--. 1968. Ecologie ichthyologique des fonds chalutables atlantiques (de la Baie ibéro-marocaine à la Mauritanie) et de la Méditerranée occidentale. *Revue Trav. Inst. Pêch. marit.*, 32(1):1-147, fig. 1-60.

MAURIN, C.; BONNET, M. 1970. Poissons des côtes nord-ouest africaines (campagnes de la "Thalassa", 1962 et 1968). *Revue Trav. Inst. Pêch. marit.*, 34(2):125-170, fig. 1-26.

MAURIN, C.; BONNET, M.; QUERO, J. C. 1978. Poissons des côtes nord-ouest africaines (campagnes de la "Thalassa" 1962, 1968, 1971 et 1973). Clupeiformes, Scopeliformes et Cetomimiformes. *Revue Trav. Inst. Pêch. marit.*, 41(1):5-92, fig. 1-51.

MAURIN, C.; LOZANO CABO, F.; BONNET, M. 1970. Inventaire faunistique des principales espèces ichthyologiques fréquentant les côtes nord-ouest africaines. *Rapp. P.-v. Réun. Cons. perm. int. Explor. Mer.*, 159:15-21.

MAYER, G. F. 1974. A revision of the cardinalfish genus *Epigonus* (Perciformes, Apogonidae), with description of two new species. *Bull. Mus. comp. Zool. Harv.*, 146(3):147-203, 23 fig., 27 tab.

MAYER, G. F.; TORTONESE, E. 1977. *Epigonus trewavasae* Poll, a junior synonym of *Epigonus constanciae* (Giglioli), (Perciformes, Apogonidae). *Breviora*, (443):1-13, 3 tab.

MAYER, R.; NALBANT, T. 1972. Additional species of fishes in the fauna of Peru Trench. Results of the 11th cruise of R/V "Anton Bruun", 1965. *Revue roum. Biol. Zool.*, 17(3):159-165, fig. 1-5.

MAZZARELLI, G. 1909. Gli animali abissali e le correnti sottomarine dello Stretto di Messina. *Riv. mens. Pesca Idrobiol.*, 11(9-12):177-218, 3 fig.

--. 1910. Larve e forme giovanili di Teleostei dello Stretto di Messina. *Riv. mens. Pesca Idrobiol.*, 5(10-12):317-328, pl.

--. 1912. Studi sui pesci bathypelagici dello Stretto di Messina. I. Larve stiloftalmoidi ("periscopiche" di Holt e Byrne) di Scopelidae e loro metamorfosi iniziale. *Riv. mens. Pesca Idrobiol.*, 7:1-26, 129-134, 4 pl.

MEAD, G. W. 1957. On the bramid fishes of the Gulf of Mexico. *Zoologica, N. Y.*, 42:51-61, pl. 1-3.

--. 1958. A catalog of the type specimens of fishes formely in the collections of the Department of Tropical Research, New York Zoological Society. *Zoologica, N.Y.*, 43(4):131-134.

--. 1960. Hermaphroditism in archibenthic and pelagic fishes of the order Iniomi. *Deep Sea Res.*, 6:234-235.

--. 1963. Observations on fishes caught over anoxic waters of the Cariaco Trench, Venezuela. *Deep Sea Res.*, 10:251-257, fig.

--. 1964. Report on the fishes taken in Madeiran and Canarian waters during the summer-autumn cruises of the "Discovery II", 1959 and 1961. II. The identity of the North Atlantic fanfish, *Pteraclis*. *Bolm Mus. munic. Funchal*, (18, 70):114-120, 1 fig., 2 tab.

--. 1966. Families Aulopidae, Bathysauridae, Bathypteroidae, Ipnopidae, Chlorophthalmidae. *In*: Fishes of the western North Atlantic. *Mem. Sears Fdn mar. Res.*, 1(5):19-29, 103-113, 114-146, 147-161, 162-189.

--. 1972. Bramidae. *Dana Rep.*, (81):1-166, 58 fig., 9 pl.

--. 1973. Bramidae, p. 386-388, in Hureau, J.C & Th. Monod (ed.), Check-list of the fishes of the north-eastern Atlantic and of the Mediterranean/Catalogue des poissons du nord-est Atlantique et de la Méditerranée (Clofnam). Paris, Unesco, 2 vol.

MEAD, G. W.; BERTELSEN, E.; COHEN, D. M. 1964. Reproduction among deep-sea fishes. Deep Sea Res., 11:569-596.

MEAD, G. W.; BÖHLKE, J. E. 1953. Scopelarchus linguidens, a new bathypelagic fish from off northern Japan. Jap. J. Ichthyol., 2(6):241-245, 1 fig.

--. 1953. Leptoderma springeri, a new alepocephalid fish from the Gulf of Mexico. Tex. J. Sci., 5(2):265-267.

MEAD, G. W.; HAEDRICH, R. L. 1965. The distribution of the oceanic fish Brama brama. Bull. Mus. comp. Zool. Harv., 134(2):29-44, 8 fig.

MEAD, G. W.; MAUL, G. E. 1958. Taractes asper and the systematic relationships of the Steinegeriidae and Trachyberycidae. Bull. Mus. comp. Zool. Harv., 119:393-417, 1 pl.

MEAD, G. W.; TAYLOR, F. H. C. 1953. A collection of oceanic fishes from off northeastern Japan. J. Fish. Res. Bd Can., 10(8):560-582.

MEEK, A. 1885. A review of the American species of the genus Synodus. Proc. Acad. nat. Sci., Philad., (1884):130-136.

--. 1903. On the egg of a species of Goby. Rep. scient. Invest. Northumb. Sea Fish. Comm., 1903:49, pl.

MEEK, S. E.; HILDEBRAND, S. F. 1928. The marine fishes of Panama. Publs Field Mus. nat. Hist.: Part I (1923)-(215) (Zool. 15): i-xii, 1-330, pl. 1-24. Part II (1925)-(226) (Zool. 15): xiii-xx, 331-707, pl. 25-69. Part III (1928)-(249)(Zool. 15): xxi-xxxi, 709-1045, pl. 72-102.

MEES, G. F. 1959. Additions to the fish fauna of western Australia. I. Fish. Bull. West. Aust., 9(1):1-10.

--. 1962. The subterranean freshwater fauna of Yardie Creek Station, North West Cape, Western Australia. J. Proc. R. Soc. West. Aust., 45: 24-32, 2 fig.

--. 1962. A preliminary revision of the Belonidae. Zool. Verh., Leiden, (54):1-96 p.

--. 1964. Further revisional notes on the Belonidae. Zoöl. Meded., Leiden, 39:311-326.

MEINEL, W. 1962. Über den Suspensionsmodus des Mandibularbogens und den splanchnischen Apparat bei Stylephorus chordatus Shaw, 1791 (Lampridiformes, Stylephoridae). Naturwissenschaften, 49:453.

MEINKEN, H. 1966. Batanga lebretonis microphthalmus, eine neue Eleotris-Unterart aus Westafrika (Pisces - Osteichthyes - Eleotridae). Aquar. Terrar.-Kunde, 13:220-221.

MELLISS, J. C. 1875. Sancta Helena. A physical, historical and topographical description of the island. Pisces, p. 100-133, 3 pl. London, xiv + 426 p., text-fig., 62 pl., 1 map.

MEL'NIKOV, Yu. S. 1980. The feeding of Allocyttus verrucosus (Gilchrist) (Oreosomatidae). Vop. Ikhtiol., 20:490-497, 3 fig., 4 tab. (Engl. transl. in: J. Ichthyol., 20(3):99-105).

--. 1981. Size-age composition and growth pattern of Allocyttus verrucosus (Gilchrist)(Oreosomatidae). Vop. Ikhtiol., 21:380-385, 4 fig., 2 tab. (Engl. transl. in: J. Ichthyol., 21(2):178-184).

MENON, A. G. K. 1977. A systematic monograph of the Tonge Soles of the genus Cynoglossus Hamilton-Buchanan (Pisces: Cynoglossidae). Smiths. Contr. Zool., (238): vi + 129 p., 48 fig., 21 pl.

--. 1981. Cynoglossidae, in Fischer, W., Bianchi, G. & W.B. Scott (ed.), FAO species identification sheets for fishery purposes. Eastern Central Atlantic; fishing areas 34, 47 (in part). Canada Funds-in-Trust. Ottawa, Department of Fisheries and Oceans Canada, by arrangement with the Food and Agriculture Organization of the United Nations, vol. 1-7: pag. var.

MENON, A. G. K.; RAMA RAO, K. V. 1969. Occurrence of flying fish, Hirun-dichthys speculiger (Val.), first in the Indian seas. Bull. natn. Inst. Sci. India, (38):767-770, 2 fig.

MENON, A. G. K.; YAZDANI, G. M. 1968. Catalogue of type-specimens in the Zoological Survey of India, Part 2. Fishes. Rec. zool. Surv. India, 61 (1/2):91-190.

MERRETT, N. R. 1970. Gonad development in billfish (Istiophoridae) from the Indian Ocean. J. Zool., 160:355-370, 4 fig., 8 pl.

--. 1971. Aspects of the biology of billfish (Istiophoridae) from the equatorial western Indian Ocean. J. Zool., 163:351-395, 19 fig.

--. 1978. On the identity and pelagic occurrence of larval and juvenile stages of rattail fishes (family Macrouridae) from 60°N, 20°W and 53°N, 20°W. Deep Sea Res., 25:147-160, 7 fig.

--. 1980. Bathytyphlops sewelli (Pisces: Chlorophthalmidae) a senior synonym of B. azorensis, from the eastern North Atlantic with notes on its biology. Zool. J. Linn. Soc., 68(2):99-109.

MERRETT, N. R.; DOMANSKI, P. A. 1985. Synopsis of catch and analysis data of deep-sea bottom-living fishes collected off the Moroccan slope, eastern North Atlantic (27°-34°N) on "Discovery" Cruise 77. Rep. Inst. oceanogr. Sci., (208): 29 p.

MERRETT, N. R.; MARSHALL, N. B. 1981. Observations on the ecology of deep-sea bottom-living fishes collected off north-west Africa (08°-27°N). Prog. Oceanogr., 9:185-244, fig. 1-25.

MERRETT, N. R.; WHEELER, A. 1983. The correct identification of two trachichthyid fishes from the slope fauna west of Britain, with notes on the abundance and commercial importance of Hoplostethus atlanti-cus. J. nat. Hist., 17(4):569-573.

MERRETT, N. R.; BADCOCK, J.; HERRING, P. J. 1973. The status of Benthal-bella infans (Pisces: Myctophoidei), its development, bioluminescen-ce, general biology and distribution in the eastern North Atlantic. J. Zool. Lond., 170:1-48.

METAXA, T. 1833. Descrizione di una specie de Gymnetrus:53-57, pl. (fig. 7). Memorie di zoologia medicale. Roma, 90 p., 1 col. pl.

METZELAAR, J. 1919. Report on the fishes, collected by Dr. J. Boeke, in the Dutch West Indies, 1904-105. With comparative notes on marine fishes of tropical West Africa. p. 185-315. In: J. Boeke: Rapp... Kolonie Curaçao. 2(1). 's-Gravenhage, 315 p., 64 fig.

--. 1919. Over tropisch atlantische Visschen, p. 1-314. Dissertation, Amsterdam.

MEUSCHEN, C. F. 1778. Museum Gronovianum... Lugduni Batavorum, 252 p.

MEYER, A. 1951. Seltene Fische. Fischereiwelt, 3(7):116-117, fig. 1-2.

MEYER-ROCHOW, V. B. 1968. Mikroskopie auf hoher See. Mikrokosmos, 57: 265-268.

--. 1974. Leptocephali and other transparent fish larvae from the south-eastern Atlantic Ocean. Zool. Anz., 192:240-251, 7 fig.

MEYERS, T. R.; SAWYER, T. K. 1977. Henneguya sp. (Cnidospora: Myxospori-da) parasitic in the heart of the bluefish Pomatomus saltatrix. J. Parasit., 63(5):890-896.

MIKHAILIN, S.V. 1982. Trichiuroid fish fauna in the southern East Atlan-tic Ocean. p. 64-77. In: Fisheries investigations of VNIRO in the eastern tropical Atlantic. Moscow,

--. 1983. On intraspecific variations on the principal meristic charac-ters in Diplospinus multistriatus Maul (Gempylidae, Perciformes). Vop. Ikhtiol., 23(3):366-372.

MILLER, P. J. 1961. Age, growth and reproduction of the rock goby, Go-bius paganellus L., in the Isle of Man. J. mar. biol. Ass. U.K., 41:737-769.

--. 1969. Systematics and biology of the leopard-spotted goby, *Gobius ephippiatus* (Teleostei: Gobiidae), with description of a new genus and notes on the identity of *G. macrolepis* Kolombatovic. *J. mar. biol. Ass. U.K.*, **49**:831-855.

--. 1971. Gobies. p. 259-278. *In*: Lythgoe, J. & G. Lythgoe. Fishes of the Sea. London.

--. 1973. Gobiidae, p. 483-515, *in* Hureau, J.C. & Th. Monod (ed.). Check-list of the fishes of the north-eastern Atlantic and of the Mediterranean/Catalogue des poissons du nord-est Atlantique et de la Méditerranée (Clofnam). Paris, Unesco, 2 vol.

--. 1974. Gobies from Rhodes and the systematic features of *Zebrus zebrus* (Teleostei: Gobiidae). *Zool. J. Linn. Soc.*, **60**:339-362, 2 pl., 8 fig.

--. 1978. The systematic position and origin of *Gobius ocheticus* Norman, 1927 from the Suez Canal. *Zool. J. Linn. Soc.*, **62**:39-58, 1 pl., 5 fig.

--. 1978. The status of the West African fish *Gobius nigricinctus* with reference to New World autochthones and an Old World colour-analogue. *Zool. J. Linn. Soc.*, **64**:17-39.

--. 1981. Eleotridae, Gobiidae, Periophthalmidae, *in* Fischer, W., G. Bianchi, W. B. Scott (ed.), FAO species identification sheets for fishery purposes. Eastern Central Atlantic; fishing areas 34, 47 (in part). Canada Funds-in-Trust. Ottawa, Department of Fisheries and Oceans Canada, by arrangement with the Food and Agriculture Organization of the United Nations, vol. 1-7: pag. var.

--. 1981. The systematic position of a West African gobioid fish, *Eleotris maltzani* Steindachner. *Zool. J. Linn. Soc.*, **73**:273-286.

--. 1984. The gobiid fishes of temperate Macaronesia (eastern Atlantic). *J. Zool. Lond.*, **204**:363-412.

--. 1986. Gobiidae, *in* P.J.P. Whitehead, M.L. Bauchot, J.C. Hureau, J. Nielsen & E. Tortonese. Fishes of the North-eastern Atlantic and the Mediterranean/Poissons de l'Atlantique du Nord-Est et de la Méditerranée, Paris, Unesco, 3:1019-1085, 74 + 86 fig.

--. 1987. Studies on *Silhouettea* Smith, 1959 and an account of *Ebomesogobius* Herre, 1946 (Pisces: Gobiidae). *Senck. biol.*, **68**:241-273, 16 fig., 4 tab.

--. 1988. New species of *Corcyrogobius, Thorogobius*, and *Wheelerigobius* from West Africa (Teleostei: Gobiidae). *J. nat. Hist.*, **22**:1245-1262.

MILLER, P. J.; RICE, A. L.; JOHNSTONE, A. D. F. 1973. A western Scottish population of the leopard-spotted goby, *Thorogobius ephippiatus* (Lowe)(Telostei: Gobioidea). *J. Fish Biol.*, **5**:233-239.

MILLER, P. J.; WRIGHT, J.; WRONGRAT, P. 1989. An Indo-Pacific goby (Teleostei, Gobiidae) from West Africa, with systematic notes on *Butis* and related eleotrinid genera. *J. nat. Hist.*, **23**:311-324.

MILLER, R. R. 1966. Geographical distribution of Central American freshwater fishes. *Copeia*, 1966(4):773-802.

MIRANDA RIBEIRO, A. de. 1907. Fauna brasiliense. Peixes. *Archos Mus. nac. Rio de J.*, 14:131-217, pl. 2-20.

--. 1915. Fauna Brasiliense. Peixes. Tomo V. Eleutherobranchios asipirophoros (Physoclisti). *Archos Mus. nac. Rio de J.*, 17: ca 600 p., many fig. and pl. (This is only the second part of vol. V.; part 1: Resenha historica and part 3: Bibliographia e indice, were published in 1918, *Archos Mus. nac. Rio de J.*, 21:11-35 and 37-227).

--. 1919. A fauna vertebrada da Ilha de Trindade. *Archos Mus. nac. Rio de J.*, 22:171-193, 6 pl.

--. 1923. Peixes (exclus. Characinidae). *Publcoes Comm. Linhas Telegr. estrat. Matto Grosso Amazonas*, (Comissào Rondon, 58):1-15, 17 fig.

MISRA, K. S. 1953. An aid to the identification of the fishes of India,

Burma, and Ceylon. II. Rec. Indian Mus., 50:367-422, fig. 1-30.

MITIANI, F. 1965. No: 422. Lepidocybium flavobrunneum (Smith). 423. Ruvettus pretiosus Cocco. 427. Gempylus serpens Cuvier. 429. Nealotus tripes Johnson. 430. Promethichthys prometheus (Cuvier et Valenciennes). In: New illustrated Encyclopedia of the fauna of Japan. 3. Tokyo:1-763.

MITCHILL, S. L. 1815. The fishes of New York, described and arranged. Trans. lit. phil. Soc., N.Y., 1(5):355-492, 6 pl.

--. 1818. The fishes of New York described and arranged. In a supplement to the memoir on the same subject, printed in the New-York Literary and Philosophical Transactions. Am. mon. Mag. crit. Rev., 2(4):241-248, 1 fig.; (5):321-328.

--. 1824. Description of an extraordinary fish ressembling the Stylephorus of Shaw. Ann. Lyceum nat. Hist., N.Y., 1(1):82-86.

MITO, S. 1961. Pelagic fish eggs from Japanese waters. I. Clupeina, Chanina, Stomiatina, Myctophida, Anguillida, Belonida, and Syngnathida. Sci. Bull. Fac. Agric., Kyushu Univ., 18(3):285-310.

MIYAKE, M.; HAYASHI, S. 1972. Field manual for statistic and sampling of Atlantic tunas and tuna-like fishes. Int. Comm. Conserv. Atlantic Tunas, Madrid, 1:1-11; 2:1-8; 3:1-46; 4:1-15; 5:1-6; many fig.

MOAL, R. A. 1953. Sardina pilchardus sur les côtes de Mauritanie durant l'année 1953. Publ. Lab. Pêch. Nouadhibou, (mimeo).

--. 1957. Etude comparée de quelques charactères de Sardina pilchardus sur la côte occidentale d'Afrique. ICES, C.M.H(87). Comité Sardine, 10 p. (mimeo).

--. 1957. Biology of economically important species: "Courbine" or "Meagre". Oceanography and sea fisheries on the West African coast, CCTA/CSA Symposium, Luanda 21st-27th November 1957. Sea Fisheries I. Africa, (57) O.C.W. 3: 30p. (roneo).

--. 1957. La pêche autochtone en Guinée. Oceanography and sea fisheries on the West African coast, CCTA/CSA Symposium (20-27 November 1957, Luanda), Paper No 5: 17 p. (mimeo).

MOHR, E. W. 1927. Teleostei Physoclisti. 12. Labriformes. In: G. Grimpe & E. Wagler, Tierwelt N.-u. Ostsee, 10(12):h85-h92, 7 fig.

--. 1937. Revision der Centriscidae (Acanthopterygii, Centrisciformes). Dana Rep., (13):1-69, 33 fig., 2 pl.

MOLTENO, C. J. 1948. The South African tunas, a preliminary study of the economic potentialities. S. Afr. Fish. Ind. Res. Inst., Cape Town, 34 p.

MONOD, Th. 1924. Sur le genre Panturichthys Pellegrin. Bull. Mus. natn. Hist. nat., Paris, 30(2):133-134, 1 fig.

--. 1927. Pisces I, pisces marini, p. 643-742. In: Monod, Th. (ed.), Contributions à l'étude de la faune du Cameroun (2ème part). Paris, Soc. Edit. Geogr. Marit. Colon., 742 p.

--. 1928. L'industrie des pêches au Cameroun. Paris, Soc. Edit. Géogr. Marit. Colon., 504 p.

--. 1949. Sur l'appareil branchiospinal de quelques téléostéens tropicaux. Bull. Inst. fr. Afr. noire, 11(1-2):36-76.

--. 1960. A propos du pseudobrachium des Antennarius (Pisces, Lophiiformes). Bull. Inst. fr. Afr. noire, (A)22(2):620-698, 83 fig.

--. 1961. Brevoortia Gill, 1861 et Ethmalosa Regan, 1917. Bull. Inst. fr. Afr. noire, (A)23:506-547.

--. 1962. Sur le rayon marginal de la pectorale du tarpon Atlantique (Megalops atlanticus Val. 1846). Bull. Inst. fr. Afr. noire, (A)24(4):1246-1247.

--. 1968. Le complexe urophore des poissons téléostéens. Mém. Inst. fond. Afr. noire, (81), 1-705, 989 fig., 3 tab.

--. 1973. Pomacentridae, p. 424-425, Scaridae, p. 444-445, Polynemidae,

p. 575, Cephalacanthidae, p. 613-614, in Hureau, J.C. & Th. Monod (ed.). Check-list of the fishes of the north-eastern Atlantic and of the Mediterranean/Catalogue des poissons du nord-est Atlantique et de la Méditerranée (Clofnam). Paris, Unesco, 2 vol.

MONOD, Th.; BUDKER, P. 1942. Sur un nouvel exemplaire de Parkuhlia boulengeri Pellegrin, 1913. Bull. Mus. natn. Hist. nat., Paris, (2)14: 112-117, 4 fig.

MONOD, Th.; HUREAU, J.-C. 1975. Les travaux scientifiques de César Honoré Sarato. Ann. Mus. Hist. nat. Nice (1974), 2:19-47.

--. 1978. Antoine Risso, 1777-1845, volume publié à l'occasion du bicentenaire de sa naissance. Annls Mus. Hist. nat. Nice (1977), 5:1-214, 10 fig., 12 pl.

--. 1978. Essai de bibliographie de Risso. p.159-163 In: Monod, Th. & J.C. Hureau, Antoine Risso 1777-1845, volume publié à l'occasion du bicentenaire de sa naissance. Annls Mus. Hist. nat. Nice (1977), 5.

MONTAGU, G. 1804. Observations on some species of British quadrupeds, birds, and fishes. Trans. Linn. Soc. Lond., 7:274-294.

--. 1811. An account of five rare species of British fishes. Mem. Wernerian nat. Hist. Soc., Edinb. (1808-10), 1:79-101.

--. 1818. An account of several new and rare species of fishes, taken on the south coast of Devonshire, with some remarks on some others of more common occurrence. Mem. Wernerian nat. Hist. Soc., Edinb. (1811-16), 2:413-463, pl. 21-23.

MONTALBAN, M. R. 1927. Pomacentridae of the Philippine Islands. Monogr. Philipp. Bur. Sci., 24:3-117, 19 pl., 38 fig.

MONTALENTI, G. 1937. Maenidae, Mullidae, Scienidae, Cepolidae. In: Uova, larve e stadi giovanili di Teleostei (ed. U. d'Ascona). Fauna Flora Golfo Napoli, 38:383-412, fig. 263-277, pl. 31-33.

MONTEIRO, R. 1956. Contribuçoes para o estudo da biologia dos Clupeidae de Angola. 1. Sardinella aurita C. V. Anais Jta Invest. Ultramar, 9 (2):151-177.

--. 1957. Metric and meristic observations on Sardinella eba, C. and V. Oceanography and sea fisheries on the West African coast, CCTA/CSA Symposium (20-27 November 1957, Luanda), paper No 58 (mimeo).

--. 1958. Heterosomata de Angola. I. Contribuiçao para o estudo das familias Psettodidae, Citharidae, Paralichthyidae e Bothidae. Trabhs Miss. Biol. marit., 19:89-130, 26 fig., 20 tab.

--. 1960. On the occurence of Sardinops ocellata (Pappe) in the waters of central Angola. FAO Proc. Wld scient. Meet. Biol. Sardines, 3: 1105-1118.

--. 1960. Contribuçâo para o estudo da biologia dos Clupeidae de Angola. 2. Observaçôes metricas e meristicas em Sardinella eba C. et V. Notas mimeogr. Cent. Biol. pisc., (15):16 p. (mimeo).

--. 1960. Alguns elementos para o estudo da captura diferencial des artes de pesca em Angola. Notas mimeogr. Cent. Biol. pisc., (10):10 p. (mimeo).

--. 1962. Sobre a ocorrência de Sardinops ocellata (Pappe) em aguas centro-angolanas. Notas mimeogr. Cent. Biol. pisc., (25):15 p. (mimeo).

MONTES, M. L. A. H. de. 1953. Nota sobre alimentaçâo de alevinos da Sardinella legitima au "Verdadeira". Boln Inst. oceanogr. S. Paulo, 4(1-2):161-180.

MOORE, D. 1962. Development, distribution, and comparison of rudder fishes Kyphosus sectatrix (Linnaeus) and K. incisor (Cuvier) in the western North Atlantic. Fishery Bull. Fish Wildl. Serv. U.S., 61:451-480, 22 fig.

--. 1967. Triggerfishes (Balistidae) of the western Atlantic. Bull. mar. Sci., 17:689-722.

--. 1967. Nomenclature of the spotted triggerfish, _Balistes punctatus_ of the eastern Atlantic. _Copeia_, 1967(4):858-861.

MOQUARD, F. 1889. Revision des _Clinus_ de la collection du Museum. _Bull. Soc. philomath. Paris_, (8)1:40-46.

MOREAU, E. 1874. Poissons de France. Notes sur quelques espèces nouvelles des côtes de l'Océan. _Revue Mag. Zool._, 2:115-119, 2 pl.

--. 1881-91. Histoire naturelle des poissons de la France. Paris, 1, 1881:i-vii + 1-480, fig. 1-82; 2, 1881:1-572, fig. 83-145; 3, 1881:1-697, fig. 146-220; Suppl., 1891: 1-144, fig. 221-227.

--. 1888. Le Scopèle de Vérany, _Scopelus Veranyi. Bull. Soc. philomath. Paris_, (7)12:108-111.

--. 1892. Manuel d'ichthyologie française. Paris, viii + 650 p., 3 pl.

MORELAND, J. M. 1956. Notes on four fishes new to the New Zealand fauna. _Rec. Dom. Mus., Wellington_, 3:9-11.

--. 1967. Marine fishes of New Zealand. Wellington, Auckland, Sydney, 1-56 p.

MORI, T. 1952. Check list of the fishes of Korea. _Mem. Hyogo Univ. Agric._, 1(3):1-228, 1 map.

MORICE, J. 1953. Essai systématique sur les familles des Cybiidae, Thunnidae et Katsuwonidae, poissons Scombrides. _Revue Trav. Inst. (scient. tech.) Pêch. marit._, 18(1):35-63.

--. 1953. Un caractère systématique pouvant servir à séparer les espèces de Thunnidae atlantiques. _Revue Trav. Inst. (scient. tech.) Pêch. marit._, 18(1):65-74.

MORROW, J. E. 1954. Fishes from East Africa with new records and descriptions of two new species. _Ann. Mag. nat. Hist._, (12)7:797-820, 2 fig.

--. 1957. On the morphology of the pectoral girdle in the genus _Makaira. Bull. Bingham oceanogr. Coll._, 16(2):88-105, 4 fig.

--. 1957. A redefinition of the subspecies of _Fodiator acutus. Postilla_, (29):1-11.

--. 1958. Names of the blue marlin and black marlin. _Bull. mar. Sci. Gulf Caribb._, 8(4):356-359, 1 fig.

--. 1959. On _Makaira nigricans_ of Lacepède. _Postilla_, (39):1-12, 2 fig.

--. 1959. Distribution of the blue marlin and the black marlin in the Indo-Pacific. _Bull. mar. Sci. Gulf Caribb._, 9(3)321-323.

--. 1959. _Istiompax indicus_ (Cuvier) 1831, a prior name for the black marlin. _Copeia_, 1959(4):347-349, 1 fig.

--. 1961. Taxonomy of the deep-sea fishes of the genus _Chauliodus. Bull. Mus. comp. Zool. Harv._, 125(9):249-294.

--. 1964. Marlins, sailfish and spearfish of the Indian Ocean. _Proc. Symp. Scombroid Fishes, mar. biol. Ass. India_, 1(1):429-440, 1 fig.

--. 1964. Family Chauliodontidae. _Mem. Sears Fdn mar. Res._, 1(4):274-289.

--. 1964. Family Stomiatidae. _Mem. Sears Fdn mar. Res._, 1(4):290-310.

--. 1964. Family Malacosteidae. _Mem. Sears Fdn mar. Res._, 1(4):523-549, 4 fig.

--. 1973. Chauliodontidae, p. 130-131, _in_ Hureau, J.C. & Th. Monod (ed.). Check-list of the fishes of the north-eastern Atlantic and of the Mediterranean/Catalogue des poissons du nord-est Atlantique et de la Méditerranée (Clofnam). Paris, Unesco, 2 vol.

MORROW, J. E.; GIBBS, R. H. Jr. 1964. Family Melanostomiatidae. _Mem. Sears Fdn mar. Res._, 1(4):351-511.

MORROW, J. E.; HARBO, S. J. 1969. A revision of the sailfish genus _Istiophorus. Copeia_, 1969(1):34-44, 15 fig.

MOSER, H. G.; AHLSTROM, E. H. 1970. Development of lanternfishes (family Myctophidae) in the California Current. Part I. Species with narrow-eyed larvae. _Sci. Bull. Los Angeles Cty Mus._, 7:1-145.

--.1972. Development of the lanternfish, Scopelopsis multipunctatus Brauer 1906, with a discussion of its phylogenetic position in the family Myctophidae and its role in a proposed mechanism for the evolution of photophore patterns in lanternfishes. Fishery Bull., 70(3): 541-564.

--. 1974. Role of larval stages in systematic investigations of marine teleosts: the Myctophidae, a case study. Fishery Bull., 72(2):391-413.

MOSHER, C. 1954. Observations on the spawning behavior and the early larval development of the Sargassum fish, Histrio histrio (Linnaeus). Zoologica, N.Y., 39(12):141-152.

MOTAIS, R. 1960. Quelques observations sur la biologie d'un poisson abyssal Trachyrinchus trachyrinchus Risso et sur les conditions de vie en mer profonde. Bull. Inst. océanogr. Monaco, (1165):1-79, 37 fig., 1 map.

MOTOS, L.; IBANEZ, M. 1977. Notas ictiologicas V: Blennius pilicornis Cuvier, 1829 ssp. nov. euskalherriensis, especie nueva para el littoral de la Costa vasca y descripción de una subespecie. Munibe, San Sebastian, 29(3/4):231-236.

MOWBRAY, L. L. 1920. Description of a Thunnus believed to be new. Copeia, 1920(78):9-10.

--. 1925. A new Bahaman razor fish. Mar. Life., 1(1):1-2, fig.

--. 1931. A Bermudian spearfish, family Istiophoridae, Makaira bermudae n. sp. Mowbray, Fauna Bermudensis. (1). Privately published unpaged, fig.

MUJIB, K. A. 1970. The cranial osteology of Brosmiculus imberbis with a note on its position in the family Moridae. Pakistan J. Zool., 2(1): 51-63, fig. 1-3.

MUKHACHEVA, V. A. 1964. On the genus Cyclothone (Gonostomatidae, Pisces) of the Pacific Ocean. Trudy Inst. Okeanol., 73:93-138, fig. 1-17. (in Russian).

--. 1972. On systematics, distribution and biology of the Gonostoma species (Pisces, Gonostomatidae). Trudy Inst. Okeanol., 93:205-249, fig. 1-12 (in Russian).

--. 1974. Cyclothones (Gen. Cyclothone, Fam. Gonostomatidae) of the world ocean and their distribution. Trudy Inst. Okeanol., 96:189-254, fig. 1-10 (in Russian).

--. 1976. Systematics and distribution of Bonapartia Goode et Bean and Margrethia Jespersen et Tåning (Gonostomatidae, Osteichthyes). Trudy Inst. Okeanol., 104:73-91, fig. 1-45. (in Russian).

--. 1978. A review of the species of Diplophos Günther (Gonostomatidae, Osteichthyes) and their vertical and geographical distribution. Trudy Inst. Okeanol., 111:10-27, fig. 1-3. (in Russian).

--. 1980. Geographical variability of Cyclothone acclinidens (Garman) and the taxonomic status of Cyclothone pseudoacclinidens Quéro (Gonostomatidae). Vop. Ikhtiol., 20(3):546-548.

--. 1980. A review of the genus Ichthyococcus Bonaparte (Photichthyidae) Vop. Ikhtiol., 20(6):771-786.

--. 1981. Geographical distribution and variability of Maurolicus muelleri (Gmelin)(Sternoptychidae, Osteichthyes) p. 41-46, 1 fig. In: Fishes of the open ocean. (N.V. Parin, ed.). Moscow, 119 p.

MÜLLER, J. 1843. Beiträge zur Kenntnis der natürlichen Familien der Fische. Arch. Naturgesch., 9(1):292-330, 381-384.

--. 1844. Über den Bau und die Grenzen der Ganoiden and über das natürliche System der Fische. Abh. preuss. Akad. Wiss., 1844:117-216.

MÜLLER, J.; HENLE, F. G. J. 1837. Über die Gattungen der Plagiostomen. Arch. Naturg., 3:394-401.

--. 1837. Gattungen der Haifische und Rochen, nach ihrer Arbeit "Über die Naturgeschichte der Knorpelfische". Ber. preuss. Akad. Wiss., 1837:111-118.

--. 1838. On the generic characters of cartilaginous fishes with descriptions of new genera. Mag. nat. Hist., 2:33-37, 88-91.

--. 1841. Systematische Beschreibung der Plagiostomen. Berlin, xxii + 204 p., 60 pl.

MÜLLER, J.; TROSCHEL, F. H. 1848. Fishes. p. 665-678. In: R. H. Schomburgk "The History of Barbados", London, xx + 772 p., 7 pl. (reprinted in Ann. Mag. nat. Hist., (2)2:11-20).

MÜLLER, J. W. 1864. (Fische, p. 89-109), in: Reisen in den Vereinigten Staaten, Canada und Mexico. Leipzig, 2.

MUNK, O. 1964. The eyes of the three benthic deep-sea fishes caught at great depths. Galathea Rep., 7:137-149.

--. 1965. Ocular degeneration in deep-sea fishes. Galathea Rep., 8:21-32, 7 pl.

--. 1966. On the retina of Diretmus argenteus Johnson, 1963 (Diretmidae, Pisces). Vidensk. Meddr dansk naturh. Foren, 129:73-80. fig. 1-3, pl. 4-5.

--. 1966. Ocular anatomy of some deep-sea teleosts. Dana Rep., (70):1-63, 16 pl.

MUNK, O.; BERTELSEN, E. 1980. On the esca light organ and its associated light-guiding structures in the deep-sea anglerfish Chaenophryne draco (Pisces, Ceratioidei). Vidensk. Meddr dansk naturh. Foren., 142:103-129.

MUNOZ-CHAPULI, R. 1982. Catches of sharks in eastern Atlantic (10°N-45°N). Abstracts, Fourth European Congress of Ichthyologists. Hamburg, (210).

MUNRO, I. S. R. 1943. Revision of Australian species of Scomberomorus. Mem. Qd Mus., 12(2):65-95.

--. 1948. The rare gempylid fish, Lepidocybium flavobrunneum (Smith). Proc. R. Soc. Qd, 60(3):31-41, 3 fig., 1 pl.

--. 1950. Revision of Bregmaceros with descriptions of larval stages from Australasia. Proc. R. Soc. Qd, 61(5):37-53, fig. 1-10.

--. 1955. The marine and freshwater fishes of Ceylon. Canberra, xvi + 351 p., 19 fig., 56 pl.

--. 1956. Handbook of Australian fishes. No 1:1-8, fig. 1-56. In: Fish. Newsl., 15(7):13-20.

--. 1957. Handbook of Australian fishes. No 18:73-76, fig. 505-531. In: Fish. Newsl., 16(12):17-20.

--. 1958. Handbook of Australian fishes. No 28:113-116, 23 fig. In: Fish. Newsl., 17(10):17-20.

--. 1967. The fishes of New Guinea. Port Moresby, xxxvii + 651 p., 23 fig., 6 col. pl., 78 pl.

MURRAY, J.; HJORT, J. 1912. Atlanterhavet fra over flaten til havdipets mørke efter undersøkelser med dampskibet "Michael Sars". Kristiana, xii + 595 p., 493 fig.

--. 1912. The depths of the ocean. A general account of the modern science of oceanography based largely on the scientific researches of the Norwegian steamer "Michael Sars" in the North Atlantic. London, xx + 821 p., 575 fig. + n. num. fig., 4 map, 9 pl.

MUTA, K. 1964. Report on the biological survey of the sardine (Sardinella aurita C.V.). II. Age determination and growth of the sardine S. aurita, by scales in Ghana waters. Fisheries Res. Unit, Tema, 21 p. (mimeo).

--. 1966. General bionomics of the sardine in Ghana waters. Proc. Symp. Oceanogr. Fish. Resources trop. Atlant. (Abidjan). Paper No 35 (mimeo).

--. 1966. Age determination and age of Sardinella aurita C. V. by scales in Ghana waters. Proc. Symp. Oceanogr. Fish. Resources trop. Atlant. (Abidjan). Paper No 36 (mimeo).

MUUS, B. J.; DAHLSTROM, P. 1966. Guide des Poissons de Mer et Pêche. Neuchâtel, 244 p., num. fig.

MYERS, G. S. 1932. A rare deep-sea scombroid fish, Xenogramma carinatum Waite, on the coast of Southern California. Trans. S. Diego Soc. nat. Hist., 7(11):111-118, pl. 7.

--. 1934. Reports on the collections obtained by the first Johnson-Smithsonian deep-sea expedition to the Puerto-Rican deep. Three new deep-water fishes from the West Indies. Smithson. misc. Collns, 91 (9):1-12, 2 fig., 1 pl.

--. 1936. A note on the stephanoberycid fishes. Copeia, 1936(2):118.

--. 1937. The deep-sea zeomorph fishes of the family Grammicolepidae. Proc. U.S. natn. Mus., 84(3008):145-156, pl. 5-7.

--. 1940. A note on Monognathus. Copeia, 1940 (2):141.

--. 1946. On a recently proposed new family of deep-sea fishes (Barbou-risiidae, Parr, 1945). Copeia, 1946(1):41-42.

--. 1960. A new zeomorph fish of the family Oreosomatidae from the coast of California, with notes on the family. Stanford ichthyol. Bull., 7:89-98, 1 fig., 1 tab.

MYERS, G. S.; STOREY, M. H. 1939. Hesperomyrus fryi, a new genus and species of echelid eels from California. Stanford ichthyol. Bull., 1(4):156-159.

NAFPAKTITIS, B. G. 1966. Two new species from the myctophid genus Diaphus from the Atlantic Ocean. Bull. Mus. comp. Zool. Harv., 133(9): 401-424.

--. 1968. Taxonomy and distribution of the lanternfishes, genera Lobianchia and Diaphus in the North Atlantic. Dana Rep., (73):1-131, 69 fig., 2 pl.

--. 1973. A review of the lanternfishes (Family Myctophidae) described by A. Vedel Tåning. Dana Rep., (83):1-46.

--. 1974. A new record and a new species of lanternfish, genus Diaphus (Family Myctophidae), from the North Atlantic Ocean. Contr. Sci. (254):1-6.

--. 1975. Review of the lanternfish genus Notoscopelus (Family Myctophidae) in the North Atlantic and Mediterranean. Bull. mar. Sci., 25 (1):75-87, 6 fig., 1 tab.

--. 1977. Family Neoscopelidae. In: Fishes of the western North Atlantic. Mem. Sears Fdn mar. Res., 1(7):1-12, fig. 1-5.

--. 1978. Systematics and distribution of lanternfishes of the genera Lobianchia and Diaphus (Myctophidae) in the Indian Ocean. Sci. Bull. nat. Hist. Mus. Los Angeles Cty, 30:1-92, 82 fig.

--. 1981. Myctophidae, in Fischer, W., Bianchi, G. & W.B. Scott (ed.), FAO species identification sheets for fishery purposes. Eastern Central Atlantic; fishing areas 34, 47 (in part). Canada Funds-in-Trust. Ottawa, Department of Fisheries and Oceans Canada, by arrangement with the Food and Agriculture Organization of the United Nations, vol. 1-7: pag. var.

NAFPAKTITIS, B. G.; NAFPAKTITIS, M. 1969. Lanternfishes (Family Myctophidae) collected during Cruises 3 and 6 of the R/V/ "Anton Bruun" in the Indian Ocean. Sci. Bull. nat. Hist. Mus. Los Angeles Cty, 5:1-79.

NAFPAKTITIS, B. G.; PAXTON, J. K. 1968. Review of the lanternfish genus Lampadena with a description of a new species. Contr. Sci., (138):1-29, 10 fig.

NAFPAKTITIS, B. G.; BACKUS, R. H.; CRADDOCK, J. E.; HAEDRICH, R. L.; ROBINSON, B. H.; KARNELLA, C. 1977. Family Myctophidae. In: Fishes of the western North Atlantic. <u>Mem.</u> <u>Sears</u> <u>Fdn</u> <u>mar.</u> <u>Res.</u>, 1(7):13-265.

NAIR, R. V. 1952. Studies on some post-larval fishes of the Madras plankton. <u>Proc.</u> <u>Indian</u> <u>Acad.</u> <u>Sci.</u>, (B)35(5):225-244, fig.

NAKABO, T. 1982. Revision of genera of the dragonets (Pisces: Calliony-midae). <u>Publs Seto</u> <u>mar.</u> <u>biol.</u> <u>Lab.</u>, 27(1/3):77-131.

NAKAMURA, E. L. 1970. Observations on the biology of the myctophid <u>Dia-phus garmani</u>. <u>Copeia</u>, 1970 (2):374-377.

NAKAMURA, H. 1938. Report of an investigation of the spearfishes of Formosan waters. <u>Rep.</u> <u>Taiwan</u> <u>Govt</u> <u>gen.</u> <u>Fish.</u> <u>Exp.</u> <u>Sta.</u>, 10:1-34, 2 fig., 15 pl. (in Japanese, Engl. transl. in: <u>Spec.</u> <u>scient.</u> <u>Rep.</u> <u>U.S.</u> <u>Fish.</u> <u>Fish Wildl.</u> <u>Serv.</u>, 153:1-46, 1955).

NAKAMURA, H.; KAMIMURA, T.; YABUTA, Y.; SUDA, A.; UEYANAGI, S.; KIKAWA, S.; HONMA, M.; YUKINAWA, M.; MORIKAWA, S. 1951. Notes on the life-history of the swordfish, <u>Xiphias gladius</u> Linnaeus. <u>Jap.</u> <u>J.</u> <u>Ichthyol.</u> 1(4):264-271, 4 fig.

NAKAMURA, I. 1965. Relationships of fishes referable to the subfamily Thunninae on the basis of the axial skeleton. <u>Bull.</u> <u>Misaki</u> <u>mar.</u> <u>biol.</u> <u>Inst.</u>, (8):7-38.

--. 1974. Some aspects of the systematics and distribution of bill-fishes. <u>NOAA</u> <u>Tech.</u> <u>Rep.</u> <u>NMFS</u> <u>SSRF</u>, 675(2):45-53, 8 fig.

--. 1975. Synopsis of the biology of the black marlin, <u>Makaira indica</u> (Cuvier), 1831. <u>NOAA</u> <u>Tech.</u> <u>Rep.</u> <u>NMFS</u> <u>SSRF</u>, 675(3):17-27, 4 fig.

--. 1978. Gempylidae, <u>in</u> Fischer, W. (ed.), FAO species identification sheets for fishery purposes. Western Central Atlantic (Fishing Area 31). Rome, FAO, vol. 1-7, pag. var.

--. 1981. Gempylidae, Istiophoridae, Trichiuridae, <u>in</u> Fischer, W., Bian-chi, G. & W.B. Scott (ed.), FAO species identifications sheets for fishery purposes. Eastern Central Atlantic; fishing areas 34, 47 (in part). Canada Funds-in-Trust. Ottawa, Department of Fisheries and Oceans Canada, by arrangement with the Food and Agriculture Organiza-tion of the United Nations, vol. 1-7: pag. var.

--. 1985. FAO species catalogue. Vol. 5. Billfishes of the world. An annotated and illustrated catalogue of marlins, sailfishes, spear-fishes and swordfishes known to date. <u>FAO</u> <u>Fish.</u> <u>Synop.</u>, (125) 5:65 p.

--. 1986. Istiophoridae, <u>in</u> P.J.P. Whitehead, M.L. Bauchot, J.C. Hureau, J. Nielsen & E. Tortonese. Fishes of the North-eastern Atlantic and the Mediterranean/Poissons de l'Atlantique du Nord-Est et de la Médi-terranée, Paris, Unesco, 2:1000-1005, 6 fig.

--. 1986. Xiphiidae, <u>in</u> P.J.P. Whitehead, M.L. Bauchot, J.C. Hureau, J. Nielsen & E. Tortonese. Fishes of the North-eastern Atlantic and the Mediterranean/Poissons de l'Atlantique du Nord-Est et de la Méditer-ranée, Paris, Unesco, 2:1006-1007, 1 fig.

--. 1986. Trichiuridae. <u>In</u>: Smiths' Sea Fishes. Ed.: Smith, M.M. & P.C. Heemstra. Johannesburg, Macmillan South Africa (Publ.) Ltd., 1047 p., 144 pl.

NAKAMURA, I.; NAKANO, S. 1978. Dispersal of the shortbill spearfish, <u>Tetrapturus angustirostris</u>, to the Atlantic Ocean. <u>Copeia</u>, 1978(2): 330-333, 1 fig.

NAKAMURA, I.; PAXTON, J. R. 1977. A juvenile gempylid fish, <u>Nealotus tripes</u>, from eastern Australia. <u>Aust.</u> <u>Zool.</u>, 19(2):179-184.

NAKAMURA, I.; FUJII, E.; ARAI, T. 1983. The gempylid <u>Nesiarchus nasutus</u> from Japan and Sulu Sea. <u>Jap.</u> <u>J.</u> <u>Ichthyol.</u>, 29(4):408-415.

NAKAMURA, I.; IWAI, T.; MATSUBARA, K. 1968. A review of the sailfish, spearfish, marlin and swordfish of the world. <u>Spec.</u> <u>Rep.</u> <u>Misaki</u> <u>mar.</u> <u>biol.</u> <u>Inst.</u>, (4):1-95, 26 fig. (in Japanese).

NAKAMURA, I.; WEBB, D.F.; TUNNICLIFFE, G.A. 1981. First record of rare gempylid fish, <u>Nesiarchus nasutus</u> (Teleostei, Gempylidae), from New Zealand. <u>Rec.</u> <u>Canterbury</u> <u>Mus.</u>, 9(7):337-344.

NALBANT, T. 1965. Sur les chaetodons de l'Atlantique avec la description d'un nouveau genre Bauchotia (Pisces, Chaetodontidae). Bull. Mus. natn. Hist. nat., Paris, 36(5):584-589.

--. 1973. Studies on chaetodont fishes with some remarks on their taxonomy (Pisces, Perciformes, Chaetodontidae). Trav. Mus. Hist. nat. "Gr. Antipa", 13:303-331, 22 fig.

--. 1986. Chaetodontidae, in P.J.P. Whitehead, M.L. Bauchot, J.C. Hureau, J. Nielsen & E. Tortonese. Fishes of the North-eastern Atlantic and the Mediterranean/Poissons de l'Atlantique du Nord-Est et de la Méditerranée, Paris, Unesco, 2:914-915, 1 fig.

NARDO, G. D. 1824. Descrizione di un pesce raro dell'Adriatico, con osservazioni ed aggiunte all'Adriatica Ittiologia (1), presentata dal Sig. G. Domenico Nardo al Sig. Giuseppe Cernazai di Udine. G. Fis. Chim. Stor. nat. Med. Arti Pavia, (Dec. II)7:116-123.

--. 1824. Osservazioni ed aggiunte all'Adriatica Ittiologia pubblicata dal Sig. Cav. Fortunato Luigi Naccari presentate dal Sig. Domenico Nardo al Sig. Giuseppe Cernazai di Udine. G. Fis. Chim. Stor. nat. Med. Arti Pavia, (Dec. II)7:222-234.

--. 1824. Fine delle osservazioni ed aggiunte all'Adriatica Ittiologia pubblicata dal Sig. Cav. Fortunato Luigi Naccari presentate dal Sig. Domenico Nardo al Sig. Giuseppe Cernazai di Udine. G. Fis. Chim. Stor. nat. Med. Arti Pavia, (Dec. II)7:249-263.

--. 1824. Lettera del Signor Fortunato Luigi Naccari, viceconsole di S.M. il re delle due Sicilie, ec. ec. Al Signor Gio. Domenico Nardo intorno alcuni pesci dell'Adriatico. G. Sci. Lett. Prov. Venete, 6(35):253-256.

--. 1827. Prodromus observationum et disquisitionum Adriaticae ichthyologiae. G. Fis. Chim. Storia nat. Med. Arti Pavia, 10:22-40 (also publ. in Isis (Oken), 1826:473-488).

--. 1827. De Proctostego, novo piscum genere, specimen ichthyologicum anatomicum. Patavii, 17 p., 1 pl.

--. 1833. De Skeponopodo, novo piscium genere, et de Guebucu marcgravii, specie illi cognata. Isis (Oken), 26(4):416-420.

--. 1840. Considerazioni sulla famiglia dei pesci Mola e sui caratteri che li distinguono. Annali Sci. Lombardo-Veneto, 10:105-112.

--. 1847. Sinonimia moderna delle specie registrate nell'opera inedita dell'Ab. Stefano Chiereghin intitolata: "Descrizione de'crostacei, de'testacei et de'pesci che abitano le lagune e golfo veneto rappresentati in figure à chiaro-scuro e colori". Venezia.

NASH, R. D. M. 1982. The biology of Fries' goby, Lesueurigobius friesii (Malm), in the Firth of Clyde, Scotland, and a comparison with other stocks. J. Fish Biol., 21:69-85.

--. 1982. The diel behaviour of small demersal fish on soft sediments on the west coast of Scotland using a variety of techniques: with special reference to Lesueurigobius friesii (Pisces; Gobiidae). Pubbl. Staz. zool. Napoli (mar. Ecol.), 3:161-178.

NAVARRO, F. de P. 1932. Nuevos estudios sobre la alacha (Sardinella aurita C. V.) de Baleares y de Canarias. Notas Resúm. Inst. esp. Ocenogr., (2)(60):1-35.

--. 1942. Nota preliminar sobre los peces de la costa de Africa, desde el Cabo Bojador a la Bahia de Tanit (Resultados de una campaña industrial de pesca de arrastre). Boln R. Soc. esp. Hist. nat., 40:189-214, 1 fig., 1 pl. (also: Notas Resúm. Inst. esp. Oceanogr., (2)(107):1-28).

NAVARRO, F. de P. et al. 1943. La pesca de arrastre en los fondos del Cabo Blanco y de Banco Arguin (Africa Sahariana). Trab. Inst. esp. Oceanogr., 18:1-225, 33 fig., 38 pl.

NAVARRO, F. de P.; LOZANO CABO, F. 1950. Carta de pesca de la costa del

Sahara desde el Cabo Juby al Cabo Barbas (En tres hojas, en escala 1:500.000). Trab. Inst. esp. Oceanogr., 21:1-24, 3 map.

NDIAYE, B.; DIOUF, O.; SECK, M.; SAGNA, A.; SENE, T. 1982. Répertoire synonymique des principales espèces de poissons débarquées par la pêche artisanale sénégalaise. Archive 108, Centre Rech. Océanogr. Dakar-Thiaroye, 14 p.

NEFEDOV, G. N. 1962. On the age composition and growth rate of Sardinella aurita in the region of Dakar and Takoradi. Trudy azov.-chernomorsk. nauchno-issled. morsk. ryb. Khoz. Okeanogr., 20:44-57 (in Russian).

NEHRING, D.; HOLZLÖHNER, S. 1982. Investigations on the relationship between environmental conditions and distribution of Sardina pilchardus in the shelf area off northwest Africa. Rapp. P.-v. Réun. Cons. perm. int. Explor. Mer., 180:342-344.

NEILL, P. 1808. A list of fishes found in the Firth of Forth, and rivers and lakes near Edinburgh, with remarks. Mem. Wernerian nat. hist. Soc., Edinb., 1:527-555.

NEKRASSOV, V. V. 1978. Systematic position of horse mackerel of the genus Trachurus from the western part of the Indian Ocean. Vop. Ikhtiol., 18(1):15-19.

NELLEN, W. 1973. Kinds and abundance of fish larvae in the neuston of patches of upwelled water off West Africa. ICES, C.M. 1973/J 25: 6 p. (mimeo).

--. 1973. Fischlarven des Indischen Ozeans. Ergebnisse der Fischbrutuntersuchungen während der ersten Expedition des Forschungsschiffes "Meteor" in den Indischen Ozean und den Persischen Golf, Oktober, 1964 bis April, 1965. "Meteor" Forsch.-Ergebn., (D)14:1-66, fig. 1-47.

NELSON, G. J. 1967. Gill arches of teleostean fishes of the family Clupeidae. Copeia, 1967 (2):389-399.

--. 1967. Notes on the systematic status of the eels Neenchelys and Myroconger. Pacif. Sci., 21(4):562-563.

NELSON, J. S. 1978. Bembrops morelandi, a new percophidid fish from New Zealand, with notes on the other members of the genus. Rec. natn. Mus. N.Z., 1(14):237-241, 2 fig.

--.1982. Two new South Pacific fishes of the genus Ebinania and contributions to the systematics of Psychrolutidae (Scorpaeniformes). Can. J. Zool., 60:1470-1504, 21 fig.

NELSON, J. S.; CHIRICHIGNO, N.; BALBONTIN, F. 1985. New material of Psychrolutes sio (Scorpaeniformes, Psychrolutidae) from the eastern Pacific of South America and comments on the taxonomy of Psychrolutes inermis and Psychrolutes macrocephalus from the eastern Atlantic of Africa. Can. J. Zool., 63:444-451.

NEPGEN (de V.), C. S. 1977. The biology of the hottentot Pachymetopon blochii (Val.) and the silverfish Argyrozona argyrozona (Val.) along the Cape south-west coast. Investl Rep. Sea Fish. Brch S. Afr. (105): 1-35.

--. 1979. Trends in the line fishery for snoek Thyrsites atun off the South-western Cape, and in size composition, length-weight relationship and condition. Fish. Bull. S. Afr., (12):35-43.

NESTEROV, A. A. 1974. Age and growth of the North Atlantic saury (Scomberesox saurus). Vop. Ikhtiol., 14(3):460-466, 2 fig.

--. 1981. The feeding of Scomberesox saurus (Scomberesocidae) and its trophic relationships in the epipelagic zone of the ocean. J. Ichthyol., 21(2):55-69.

NESTEROV, A. A.; SHIGANOVA, T. A. 1976. The eggs and larvae of the Atlantic saury, Scomberesox saurus, of the North Atlantic. Vop. Ikhtiol., 16, 2(97):315-322 (= J. Ichthol., 16:277-283.

NEYELOV, A. V. 1973. Cottunculidae, p. 603-604, in Hureau, J.C. & Th. Monod (ed.). Check-list of the fishes of the north-eastern Atlantic and of the Mediterranean/Catalogue des poissons du nord-est Atlantique et de la Méditerranée (Clofnam). Paris, Unesco, 2 vol.

NICHOLS, J. T. 1920. A contribution to the ichthyology of the Bermuda. Proc. biol. Soc. Wash., 33:59-64.

--. 1920. On the range and geographic variation of Caranx hippos. Copeia, 1920 (83):44-45.

--. 1921. A list of Turk Islands fishes with a description of a new flatfish. Bull. Am. Mus. nat. Hist., 44(2):21-24, pl. 3.

--. 1923. Two new fishes from the Pacific Ocean. Am. Mus. Novit., (94): 1-3, 2 fig.

--. 1935. Variation in Pacific Trachurops crumenophthalmus. Am. Mus. Novit., (815):1-6.

NICHOLS, J. T.; BREDER, C. M., Jr. 1928. An annotated list of the Synentognathi. Zoologica, N.Y., 8(7):423-448.

NICHOLS, J. T.; FIRTH, F. E. 1939. Rare fishes off the Atlantic coast including a new grammicolepid. Proc. biol. Soc. Wash., 52:85-88, 1 fig.

NICHOLS, J. T.; GRISCOM, L. 1917. Fresh-water fishes of the Congo Basin obtained by the American Museum Congo Expedition, 1909-1915. Bull. Am. Mus. nat. Hist., 37(25):653-756, 31 fig., pl. 64-83.

NICHOLS, J. T.; LA MONTE, F. R. 1935. The Tahitian black marlin, or silver marlin swordfish. Am. Mus. Novit., (807):1-2, 1 fig.

--. 1935. How many marlins are there ? Nat. Hist., N.Y., 36(4):327-330, 5 fig.

--. 1941. Differences in marlins based on weights and measurements. Ichthyol. Contr. int. Game Fish Ass., 1(1):1-8, 3 fig.

--. 1943. On a large Xenogramma from Peru. Copeia, 1943 (1):50.

NICHOLS, J. T.; MURPHY, R. C. 1914. Fishes from South Trinidad Islet. Bull. Am. Mus. nat. Hit., 33(20):261-266, fig. 1-3.

--. 1922. On a collection of a marine fishes from Peru. Bull. Am. Mus. nat. Hist., 46(9):501-516.

--. 1944. A collection of fishes from the Panama Bight, Pacific Ocean. Bull. Am. Mus. nat. Hist., 83:221-260.

NICOL, J. A. C. 1960. Studies on luminescence. On the subocular light organs of stomiatoid fishes. J. mar. biol. Ass. U. K., 39:529-548, fig. 1-10.

--. 1969. Bioluminescence, p. 355-400 , fig. 1-18, 1 tab. In: W.S. Hoar & D.J. Randall. Fish Physiology 3. New York, xvi + 1-485.

NIELAND, H. 1979. Qualitative and quantitative aspects of the food of Ethmalosa fimbriata (Bowdich) in the Ebrié lagoon (Ivory Coast). Docums scient. provis. Centre Rech. océanogr. Abidjan, 11(1):85-95.

--. 1980. Die Nahrung von Sardinen, Sardinellen und Maifischen vor der Westküste Afrikas. Ber. Inst. Meereskd. Christian-Albrechts-Univ. Kiel, (75):1-137.

--. 1982. The food of Sardinella aurita (Val.) and Sardinella eba (Val.) off the Coast of Senegal. Rapp. P.-v. Réun. Cons. perm. int. Explor. Mer, 180:369-373.

NIELSEN, J. G. 1961. Psettodoidea and Pleuronectoidea (Pisces, Heterosomata). Atlantide Rep., (6):101-127, 8 fig., 1 pl.

--. 1963. Soleoidea (Pisces, Heterosomata). Atlantide Rep., (7):7-35, 9 fig., 1 pl.

--. 1965. On the genera Acanthonus and Typhlonus (Pisces, Brotulidae). Galathea Rep., 8:33-47, fig. 1-10, tab. 1-2, pl. 11.

--. 1968. Redescription and reassignment of Parabrotula and Leucobrotula (Pisces, Zoarcidae). Vidensk. Meddr dansk naturh. Foren., 131:225-249, 12 fig.

--. 1969. Systematics and biology of the Aphyonidae (Pisces, Ophidioidea). Galathea Rep:, 10:1-88, 57 fig., 4 pl.

--. 1971. Redescription of the genus Selachophidium (Pisces, Brotulidae) with two new species. Arch. FischWiss., 22:17-33.

--. 1972. Rare northeast Atlantic aphyonid fishes (Ophidioidei). "Meteor" Forsch.-Ergebn., (D)(12):52-55, 3 fig.

--. 1972. Additional notes on Atlantic Bathylaconidae (Pisces, Isospondyli) with a new genus. Arch. FischWisss., 23(1):29-36, 5 fig.

--. 1973. Bathylaconidae, p. 85, Aulopidae, Synodontidae, Bathysauridae, Bathypteroidae, Ipnopidae, Chlorophthalmidae, p. 160-167, Neoscopelidae, p. 170, Stylephoridae, Polymixiidae, Berycidae, p. 335-337, Caristiidae, Trachichthyidae, Holocentridae, p. 339-342, Stephanoberycidae, p. 347, Parabrotulidae, Brotulidae, Ophidiidae, Aphyonidae, p. 548-556, Citharidae, Scophthalmidae, Bothidae, Pleuronectidae, p. 615-627, in Hureau, J.C. & Th. Monod. (ed.). Check-list of the fishes of the north-eastern Atlantic and of the Mediterranean/ Catalogue des poissons du nord-est Atlantique et de la Méditerranée (Clofnam). Paris, Unesco, 2 vol.

--. 1974. Fish types in the Zoological Museum of Copenhagen. Zool. Mus., Copenhagen, 115 p.

--. 1975. List of ophidioid fishes from the 14th cruise of the "Akademik Kurchatov" with a new species of Aphyonus. Trudy Inst. Okeanol., 100:348-353, 1 fig.

--. 1977. The deepest living fish Abyssobrotula galatheae, a new genus and species of oviparous ophidioids (Pisces, Brotulidae). Galathea Rep., 14:41-48.

--. 1981. Citharidae, Ophiidiidae, Psettodidae, Scophthalmidae, in Fischer, W., Bianchi, G. & W.B. Scott (ed.), FAO species identification sheets for fishery purposes. Eastern C..ntral Atlantic; fishing areas 34, 47 (in part). Canada Funds-in-Trust. Ottawa, Department of Fisheries and Oceans Canada, by arrangement with the Food and Agriculture Organization of the United Nations, vol. 1-7: pag. var.

--. 1986. Nemichthyidae, in P.J.P. Whitehead, M.L. Bauchot, J.C Hureau, J. Nielsen & E. Tortonese. Fishes of the North-eastern Atlantic and the Mediterranean/Poissons de l'Atlantique du Nord-Est et de la Méditerranée, Paris, Unesco, 2:551-554, 5 + 2 fig.

--. 1986. Parabrotulidae, in P.J.P. Whitehead, M.L. Bauchot, J.C. Hureau, J. Nielsen & E. Tortonese. Fishes of the North-eastern Atlantic and the Mediterranean/Poissons de l'Atlantique du Nord-Est et de la Méditerranée, Paris, Unesco, 3:1151-1152, 2 fig.

--. 1986. Bythitidae, in P.J.P. Whitehead, M.L. Bauchot, J.C. Hureau, J. Nielsen & E. Tortonese. Fishes of the North-eastern Atlantic and the Mediterranean/Poissons de l'Atlantique du Nord-Est et de la Méditerranée, Paris, Unesco, 3:1153-1157, 6 fig.

--. 1986. Ophidiidae, in P.J.P. Whitehead, M.L. Bauchot, J.C. Hureau, J. Nielsen & E. Tortonese. Fishes of the North-eastern Atlantic and the Mediterranean/Poissons de l'Atlantique du Nord-Est et de la Méditerranée, Paris, Unesco, 3:1158-1166, 12 fig.

--. 1986. Aphyonidae, in P.J.P. Whitehead, M.L. Bauchot, J.C. Hureau, J. Nielsen & E. Tortonese. Fishes of the North-eastern Atlantic and the Mediterranean/Poissons de l'Atlantique du Nord-Est et de la Méditerranée, Paris, Unesco, 3:1167-1171, 7 fig.

--. 1986. Citharidae, in P.J.P. Whitehead, M.L. Bauchot, J.C. Hureau, J. Nielsen & E. Tortonese. Fishes of the North-eastern Atlantic and the Mediterranean/Poissons de l'Atlantique du Nord-Est et de la Méditerranée, Paris, Unesco, 3:1286, 1 fig.

--. 1986. Scophthalmidae, in P.J.P. Whitehead, M.L. Bauchot, J.C. Hureau, J. Nielsen & E. Tortonese. Fishes of the North-eastern Atlantic

and the Mediterranean/Poissons de l'Atlantique du Nord-Est et de la Méditerranée, Paris, Unesco, 3:1287-1293, 7 + 1 fig.
--. 1986. Bothidae, in P.J.P. Whitehead, M.L. Bauchot, J.C. Hureau, J. Nielsen & E. Tortonese. Fishes of the North-eastern Atlantic and the Mediterranean/Poissons de l'Atlantique du Nord-Est et de la Méditerranée, Paris, Unesco, 3:1294-1298, 6 + 2 fig.
--. 1986. Pleuronectidae, in P.J.P. Whitehead, M.L. Bauchot, J.C. Hureau, J. Nielsen & E. Tortonese. Fishes of the North-eastern Atlantic and the Mediterranean/Poissons de l'Atlantique du Nord-Est et de la Méditerranée, Paris, Unesco, 3:1299-1307, 9 + 4 fig.
NIELSEN, J. G.; BERTELSEN, E. 1985. The gulper-eel family Saccopharyngidae (Pisces, Anguilliformes). Steenstrupia, 11(6):157-206, 28 fig.
--. 1986. Eurypharyngidae, in P.J.P. Whitehead, M.L. Bauchot, J.C. Hureau, J. Nielsen & E. Tortonese. Fishes of the North-eastern Atlantic and the Mediterranean/Poissons de l'Atlantique du Nord-Est et de la Méditerranée, Paris, Unesco, 2:534, 1 fig.
NIELSEN, J. G.; COHEN, D. M. 1968. Redescription of Bellotia apoda Giglioli, 1883. (Pisces, Ophidioidea). Proc. Linn. Soc. Lond., 179:99-106, 3 fig.
NIELSEN, J. G.; EAGLE, R. J. 1974. Descriptions of a new species of Barathronus (Pisces, Aphyonidae) and four specimens of Sciadonus sp. from the East Pacific. J. Fish. Res. Bd Can., 31:1067-1072, 5 fig.
NIELSEN, J. G.; HUREAU, J.-C. 1980. Revision of the ophidiid genus Spectrunculus Jordan & Thompson, 1914, a senior synonym of Parabassogigas Nybelin,1957 (Pisces, Ophidiiformes). Steenstrupia, 6(11):149-169.
NIELSEN, J. G.; LARSEN, V. 1968. Synopsis of the Bathylaconidae (Pisces, Isospondyli) with a new eastern Pacific species. Galathea Rep., 9:221-238, 10 fig., pl. 13-15.
--. 1970. Bathylaconidae from the Atlantic. Arch. FischWiss., 21:28-39, 7 fig.
NIELSEN, J. G.; NYBELIN, O. 1963. Brotulidae (Pisces, Percomorphi) from tropical West Africa. Atlantide Rep., 7:195-213, 13 fig., 1 pl.
NIELSEN, J. G.; SMITH, D. G. 1978. The eel family Nemichthyidae. Dana Rep., (88):1-71, 36 fig., 2 pl.
NIEUHOF, J. 1682. Gedenkwaerdige zee en lantreize door de voornaemste lanschappen van West en Oostindien. 2 pts, Amsterdam, (1): xii + 240; (2): iv + 308 p., 42 pl. (Fishes, 268-281, fig.).
NIJSSEN, H. 1966. The occurence of Cynoglossus browni Chabanaud, 1949, in the North Sea (Pisces, Pleuronectiformes). Beaufortia, 13(155): 87-90, 2 fig.
--. 1972. A lantern fish, Hygophum benoiti (Cocco, 1838), washed ashore in the Netherlands (Pisces, Myctophoidei, Myctophidae). Bull. zool. Mus. Amsterdam, 2(14):147-148, fig., pl.
NIJSSEN, H.; TUIJL, L. van; ISBRÜCKER, I. J. H. 1982. A catalogue of the type-specimens of recent fishes in the Institute of Taxonomic Zoology (Zoölogisch Museum), University of Amsterdan, The Netherlands. Versl. tech. Gegevens, (33):1-173.
NIKOLSKY, A. M. 1923. Chto takoe serdinka y tyul'ka cchakovskikh ribakov. Byull. vseukr. gosud. chernomorsko-azov. nauchno-promysl. opyt Sta. Bugchanpos, (6-7):1-5.
--. 1923. Novyi vid sel'di iz Chernogo morya. Byull. vseukr. gosud. chernomorsko-azov. nauchno-promysl. opyt. Sta. Bugchanpos, (8-9):4-6.
--. 1954. Chastnaya ikhtiologiya. Moscow, 450 p., 312 fig.
NILSSON, S. 1832. Prodromus ichthyologiae Scandinavicae. Lund, iv + 124 p.
--. 1855. Skandinavisk fauna. Fjerde Delen: Fiskarna. 4(1). Lund, xxxiv + 768 p.
NINNI, E. 1931. Atherina boyeri. Atherina hepsetus. Faune ichthyol.

Atlant. N., Cahier 6:344, 345.
--. 1932. Il Temnodon saltator (C. V.) nell'Adriatica e sua distribuzione geografica. Atti Soc. ital. Sci. nat., 71:201-209.
--. 1937. Atherina presbyter. Faune ichthyol. Atlant. N., Cahier 17:343.
--. 1938. I. Gobius dei mari e delle acque interne d'Italia. Memorie R.
Com. talassogr. ital., (242):169 p.
NISHIKAWA, Y.; UEYANAGI, S. 1974. The distribution of the larvae of
swordfish, Xiphias gladius, in the Indian and Pacific Oceans. NOAA
tech. Rep. NMFS SSRF, (675)(2):261-264, 4 fig.
NISHIKAWA, Y; KIKAWA, S.; HONMA, M.; UEYANAGI, S. 1978. Distribution
atlas of larval tunas, billfishes and related species. Results of
larval survey by R/V Shunyo Maru, and Shoyo Maru, 1956-1975. S. ser.
Far Seas Fish. Res. Lab., 9: i-ii, 1-99, 1 fig., many maps.
NISHIMURA, S. 1957. What is the spawner of the so-called "Macrurus egg"
from the adjacent waters of Japan (Maurolicus japonicus Ishikawa).
Rep. Japan Sea reg. Fish. Res. Lab., (3):1-11, 1 pl. (5 fig.).
NOBRE, A. 1935. Fauna marinha de Portugal. I. Vertebrados. Porto, lxxxiv
+ Mamiferos: p. 1-21, fig.; Reptis: p. 1-5, fig., Peixes: p. 1-574,
77 pl.
NOLF, D. 1977. Les otolithes des téléostéens de l'oligo-miocène belge.
Annls Soc. r. zool. Belg., 106(1):3-119, 3 fig., 18 pl.
--. 1980. Etude monographique des otolithes des Ophidiiformes actuels et
révision des espèces fossiles (Pisces, Teleostei). Meded. Werkgr.
Tert. Kwart. Geol., 17:71-195, 8 fig., 20 pl.
NOLF, D.; CAPPETTA, H. 1980. Les otolithes de téléostéens du Miocène de
Montpeyroux (Hérault, France). Paleovertebrata, Montpellier, 10(1):1-
78, 1 fig., 4 pl.
NOLF, D.; TAVERNE, L. 1977. Contribution à l'étude des otolithes des
poissons. V. L'origine des Sciaenidae (Teleostei, Perciformes). Biol.
Jaarb., 45:150-161.
NORCROSS, J. J.; RICHARDSON, S. L.; MASSMAN, W. H.; JOSEPH, E. B. 1979.
Development of young bluefish, Pomatomus saltatrix, and distribution
of eggs and young in Virginian coastal waters. Trans. Am. Fish. Soc.,
103(3):477-497.
NORDGARD, O. 1924. Paralepis borealis Reinhardt. A new fish for the
Norwegian fauna. K. norske Vidensk. Selsk. Skr., 4:1-7, fig. 1.
NORDMANN, A. von. 1840. Observations sur la faune pontique. p. 355-635.
In: A. de Démidoff. Voyage dans la Russie méridionale et la Crimée
par la Hongrie, la Valachie et la Moldavie, exécuté en 1837 sous la
direction de M. Anatole de Demidoff. Paris, 3. Atlas (1842):32 col.
pl.
NORMAN, J. R. 1922. Fishes from Tobago. Ann. Mag. nat. Hist., (9)9:533-
535.
--. 1922. Two new fishes from New Britain and Japan. Ann. Mag. nat.
Hist., (9)10:217-218, fig.
--. 1922. A new eel from Tobago. Ann. Mag. nat. Hist., (9)10:296-297.
--. 1923. A revision of the clupeid fishes of the genus Ilisha and allied genera. Ann. Mag. nat. Hist., (9)11(61):1-22.
--. 1926. A synopsis of the rays of the family Rhinobatidae, with a
revision of the genus Rhinobatus. Proc. zool. Soc. Lond., 1926(62):
941-982, fig. 1-30.
--. 1929. The teleostean fishes of the family Chiasmodontidae. Ann. Mag.
nat. Hist., (10)3:529-542.
--. 1929. A preliminary revision of the berycoid fishes of the genus
Melamphaes. Ann. Mag. nat. Hist., (10)4:153-168.
--. 1929. Notes on the myctophid fishes of the genera Diaphus and Myctophum in the British Museum. Ann. Mag. nat. Hist., (10)4(23):510-515,
3 fig.

--. 1930. A new ray of the genus Rhinobatus from the Gold Coast. Ann. Mag. nat. Hist., (10)6:226-228, 1 fig.

--. 1930. Oceanic fishes and flatfishes collected in 1925-27. "Discovery" Rep., 2:261-370, 47 fig., 2 pl.

--. 1931. Four new fishes from the Gold Coast. Ann. Mag. nat. Hist., (10)7:352-359, 4 fig.

--. 1931. Notes on flatfishes (Heterosomata). I. Notes on flatfishes of the family of Bothinae in the British Museum, with descriptions of three new species. Ann. Mag. nat. Hist., (10)8:507-510.

--. 1934. A systematic monograph of the flatfishes (Heterosomata). 1 (Psettodidae, Bothidae, Pleuronectidae). London, viii + 459 p., 317 fig.

--. 1935. Coast fishes. Part I: The South Atlantic. "Discovery" Rep., 12:1-58, 15 fig.

--. 1935. A revision of the lizard-fishes of the genera Synodus, Trachinocephalus and Saurida. Proc. zool. Soc. Lond., 1935:99-135, 18 fig.

--. 1935. The European and South African sea breams of the genus Spondyliosoma and related genera, with notes on Dichistius and Tripterodon. Ann. S. Afr. Mus., 32(1):5-22, fig. 1-6, pl. 2.

--. 1935. Notes on the fishes of the family Scorpaenidae recorded from the British coasts, with a description of a new species. Proc. zool. Soc. Lond., 1935:611-614, 1 fig.

--. 1935. The carangid fishes of the genus Decapterus Bleeker. Ann. Mag. nat. Hist., (10)16:252-264.

--. 1937. Fishes. Rep. Br. Aust. N.Z. Antarct. Res. Exped., (B)1(2):50-88, 11 fig.

--. 1937. Coast Fishes. Part 2. The Patagonian Region. "Discovery" Rep., 16:1-150, 76 fig., 5 pl.

--. 1939. Fishes. Scient. Rep. John Murray Exped., 7(1):1-116, 41 fig.

--. 1943. Notes on the blennioid fishes, I. A provisional synopsis of the genera of the family Blenniidae. Ann. Mag. nat. Hist. (11)10(72): 793-812.

--. 1966. A draft synopsis of the orders, families and genera of recent fishes and fish-like vertebrates. London, 649 p.

NORMAN, J. R.; FRASER, F. C. 1948. Giant fishes, whales and dolphins. London, xxii + 360 p., 97 fig., 8 pl.

NORMAN, J. R.; IRVINE, F. R. 1947. Marine fishes p. 81-220. In: Irvine, F. R. The fishes and fisheries of the Gold Coast. London, 352 p., 217 fig.

NORONHA, A. C. 1926. Description of a new genus and species of deep water gempyloid fish, Diplogonurus maderensis. Ann. Carneg. Mus., 16 (3-4):381-383, 1 fig.

NORONHA, A. C., de; SARMENTO, A. A. 1934. Os peixes dos mares da Madeira. Funchal, 146 p., fig.

--. 1948. Vertebrados da Madeira. 2. Peixes (2a ed.). Funchal, 181 p.

NOVIKOV, N. P.; KODOLOV, L. S.; GAVRILOV, G. M. 1981. Preliminary list of fishes of the Emperor Underwater Ridge p. 32-35. In: Fishes of the open ocean (ed. N.V. Parin). Moscow, 119 p.

NOVIKOVA, N. F. 1967. Idiacanthids of the Indian and Pacific Oceans. Trudy Inst. Okeanol., 84:159-206, fig. 1-6.

NUNES, A. de A. 1953. Peixes da Madeira. Funchal, 274 p., 25 pl.

NUNES-RUIVO, L. 1956. Copépodes parasitas de peixes dos mares de Angola. Anais Jta Invest. Ultramar, 9(2):9-45.

--. 1962. Copépodes parasitas de peixes das costas de Angola (lista faunistica). Notas mimeogr. Cent. Biol. pisc. (33):1-28.

--. 1962. Copépodes parasites de poissons des côtes d'Angola. Mems Jta Invest. Ultramar, (2)(33):67-86.

NYBELIN, O. 1942. In: K. A. Andersson, Fiskar och Fiske i Norden. Stockholm, 1: xvi + 540 p., num. fig. and pl.

--. 1946. Sur quelques espèces du genre Anotopterus Zugmayer et sur la position systématique de la famille des Anotopteridae. Göteborgs K. Vetensk.-o. VitterhSamh. Handl. (6), (B)5(1):3-16, fig. 1-4.

--. 1947. Notice préliminaire sur quelques espèces nouvelles de poissons. Ark. Zool., 38(B)(2):1-6, 4 fig.

--. 1948. Fishes collected by the "Skagerak" Expedition in the eastern Atlantic, 1946. Göteborgs K. Vetensk.-o. VitterhSamh. Handl. (6), (B)5 (16):1-95, 9 fig., 6 pl. (also publ. in Meddn Göteborgs Mus., zool. Avd., 121:3-95, 6 pl.).

--. 1957. Deep-sea bottom-fishes. Rep. Swed. deep Sea Exped., 2, Zool. (20):247-345, 50 fig., 7 pl.

--. 1963. Zur Morphologie und Terminologie des Schwanzskelettes der Actinopterygier. Ark. Zool., (2)15(35):485-516, 22 fig.

--. 1971. On the caudal skeleton in Elops with remarks on other teleostean fishes. Acta R. Soc. Sci. Litt. Gothoburg. (Zool.), 7:1-52, 6 fig., 12 pl.

--. 1979. Contributions to taxonomy and morphology of the genus Elops (Pisces, Teleostei). Acta R. Soc. Sci. Litt. Gothoburg. (Zool.) (12):1-37.

NYBELIN, O.; POLL, M. 1958. Brotulidae des côtes africaines de l'Atlantique Sud avec description d'une espèce nouvelle de Lamprogrammus. Bull. Inst. r. Sci. nat. Belg., 34(19):1-7, 1 fig., 2 tab.

O'DAY, W. T.; NAFPAKTITIS, B. 1967. A study of the effects of expatriation on the gonads of two myctophid fishes in the North Atlantic Ocean. Bull. Mus. comp. Zool. Harv., 136(5):70-90, 9 fig., 2 pl.

OELSCHLÄGER, H. A. 1981. Lamprididae, in Fischer, W., Bianchi, G. & W.B. Scott (ed.), (FAO species identification sheets for fishery purposes. Eastern Central Atlantic; fishing areas 34, 47 (in part). Canada Funds-in-Trust. Ottawa, Department of Fisheries and Oceans Canada, by arrangement with the Food and Agriculture Organization of the United Nations, vol. 1-7: pag. var.

OGILBY, J. D. 1888. Description of a new genus and species of deep-sea fish from Lord Howe Island. Proc. Linn. Soc. N.S.W., (2)3:1313.

--. 1893. Description of a new pelagic fish from New Zealand. Rec. Aust. Mus., 2(5):64-65.

--. 1897. Some new genera and species of fishes. Proc. Linn. Soc. N.S. W., 22(2):245-257.

--. 1897. On a Trachypterus from New South Wales. Proc. Linn. Soc. N.S.W., 22(3):646-659.--. 1898. New genera and species of fishes. Proc. Linn. Soc. N.S.W., 23 (1):32-41.

--. 1898. New genera and species of fishes. Proc. Linn. Soc. N.S.W., 23: 280-299.

--. 1898. A contribution to the zoology of New Caledonia. Proc. Linn. Soc. N.S.W., 23:762-770.

--. 1899. Additions to the fauna of Lord Howe Island. Proc. Linn. Soc. N.S.W., 24:730-745.

--. 1908. Descriptions of new Queensland fishes. Proc. R. Soc. Qd, 21: 87-98.

--. 1911. Description of new or insufficiently described fishes from Queensland waters. Ann. Qd Mus., 10:36-58.

--. 1918. Edible fishes of Queensland. Parts X-XIV. Mem. Qd Mus., 6:45-90.

--. 1920. Alteration of generic name. Proc. R. Soc. Qd, 31:45.

--. 1954. The commercial fishes and fisheries of Queensland. Revised and illustrated by T.C. Marshall. Brisbane, viii + 121 p., 122 fig., 1 pl.

OKADA, Y. 1955. Fishes of Japan. Illustrations and descriptions of fishes of Japan. Tokyo, 434 + 28 p., 391 fig.

OKADA, Y.; MATSUBARA, K. 1938. Keys to the fishes and fish-like animals of Japan. Including Kuril Islands, southern Sakhalin, Bonin Islands, Ryukyu Islands, Korea and Formosa. Tokyo, xl + 584 p., 113 pl. (in Japanese).

OKAMURA, O. 1970. Studies on the macrourid fishes of Japan. Rep. Usa mar. biol. Stn Kochi Univ., 17(1-2):1-179, 85 fig.

--. 1970. Macrourina (Pisces). Fauna Japonica. Tokyo, 216 p., 87 fig., 44 pl.

OKEN, L. 1816. Lehrbuch der Naturgeschichte. 3 pt in 6 vol., Leipzig, Jena, 3(2): xvi + 1270 p., pl.

--. 1817. V Kl. Fische. Isis, 8(148):1181-1183 (misprinted 1781-1783).

--. 1817. Isis, 1:1575.

OKERA, W. 1978. Fishes taken by the beach seines fishery at Lumley, Freetown (Sierra Leone). J. Fish Biol., 12:81-88.

OKIYAMA, M. 1972. Morphology and identification of the young ipnopid, "Macristiella" from the tropical western Pacific. Jap. J. Ichthyol., 19(3):145-153.

--. 1973. The larval taxonomy of the primitive myctophiform fishes, p. 609-621, 6 fig., 4 tab. In: The early life history of fish (J.H.S. Blaxter, ed.). Berlin, Heidelberg, New-York, 1-765 p.

--. 1979. Morphology of juveniles and phyletic relationships among the Myctophiformes. Oceanogr. Sci., 11(2):131-145.

OLFERS, I. F. J. M. 1831. Die Gattung Torpedo in ihren naturhistorischen und antiquarischen Beziehungen erläutert. Berlin, 35 p., 3 pl.

OLIVER, M.; MASSUTI, M. 1952. El raó, Xyrichthys novacula (fam. Labridae). Notas biológicas y biométricas. Boln Inst. esp. Oceanogr., 48:1-13, 5 fig.

OLIVIER, G. A. 1791. Encyclopédie Méthodique. Histoire Naturelle. 6: Insectes 1-704, Paris.

OLSEN, A. M. 1958. New fish records and notes on some uncommon Tasmanian species. Pap. Proc. R. Soc. Tasmania, 92:155-159.

OLSON, S.; LEFEVERE, S. 1969. Some observations on length and sex distribution and catch effort on bonga Ethmalosa fimbriata (Bowdich). p. 277-279. In: Proc. Symp. Oceanogr. Fish. Resources trop. Atlant. (20-28 October 1968, Abidjan). Paris, 430 p.

OLST, J. C. van (see VAN OLST, J. C.).

OPPENHEIMER, J. M. 1954. Transplantation experiments on the eggs of Gobius. Pubbl. Staz. zool. Napoli, 25:18-25.

ORTEA, J. A.; HOZ, M. M. de la. 1979. Peces marinos de Asturias. Salinas, Ayalga ediciones, 230 p., fig.

ORTON, G. L. 1963. Notes on larval anatomy of fishes of the order Lyomeri. Copeia, 1963:6-15, 4 fig.

--. 1964. The eggs of scomberesocid fishes. Copeia, 1964(1):144-150.

--. 1964. Identification of Leptocephalus acuticeps as the larva of the eel genus Avocettina. Pacif. Sci., 18(2):186-201, 4 fig.

OSBECK, P. 1757. Dagbok en Ostindisk resa etc. L. Frefing, Stockholm (Engl. transl. by J. R. Forster, 1771).

--. 1765. Reise nach Ostindien und China. (German transl. by J.G. Georgi). Rostock: xxvi + 552 p + 13, 13 pl.

O'SHAUGHNESSY, A. W. E. 1875. Descriptions of new species of Gobiidae in the collection of the British Museum. Ann. Mag. nat. Hist., (4)15:144-148.

OSORIO, B. 1890. Estudos ichthylogicos àcerca da fauna dos dominios portuguezes na Africa. 1a nota. Ilhas de Cabo Verde. Jorn. Sci. math.

phys. nat., (2)1(4):277-282.

--. 1890. Estudos ichthyologicos àcerca da fauna dos dominios portugue-zes na Africa. 2a nota. Peixes maritimos d'Angola. Jorn. Sci. math. phys. nat., (2)2(5):50-60.

--. 1891. Estudos ichthyologicos àcerca de fauna dos dominios portugue-zes na Africa. 3a nota. Peixes maritimos das ilhas de S. Thomé, do Principe e ilheo das Rolas. Jorn. Sci. math. phys. nat., (2)2(6):97-139.

--. 1892. Estudos ichthyologicos àcerca de fauna dos dominios portugue-zes na Africa. 3a nota. Peixes maritimos das ilhas de S. Thomé, do Principe e ilhe das Rolas. (continuaçâo). Jorn. Sci. math. phys. nat., (2)2(7):205-209.

--. 1893. Estudos ichthyologicos àcerca de fauna dos dominios portugue-zes na Africa. 2a nota. Peixes maritimos d'Angola. (continuaçô). Jorn. Sci. math. phys. nat., (2)3(10):128-135.

--. 1893. Estudos ichthyologicos àcerca de fauna dos dominios portugue-zes na Africa. 3a nota. Peixes maritimos das ilhas de S. Thomé, do Principe e ilhe das Rolas. (continuaçâo). Jorn. Sci. math. phys. nat., (2)3(10):136-140.

--. 1893. Estudos ichthyologicos àcerca de fauna dos dominios portugue-zes na Africa. 3a nota. Peixes maritimos das ilhas de S. Thomé, do Principe e ilhe das Rolas. (continuaçâo). Jorn. Sci. math. phys. nat., (2)3(11):173-182.

--. 1893. Estudos ichthyologicos àcerca da fauna dos dominios portugue-zes na Africa. 4a nota. Peixes de Bissau. Jorn. Sci. math. phys. nat., (2)3(11):183-184.

--. 1894. D'algumas especies a juntar ao catàlogo dos peixes de Portugal de Capello. Jorn. Sci. math. phys. nat., (2)3(11):186-188.

--. 1895. Peixes da ilha d'Anno Bom. Jorn. Sci. math. phys. nat., (2)3(12):243-247.

--. 1895. Peixes de Dahomy. Jorn. Sci. math. phys. nat., (2)3(12):252-253.

--. 1895. Peixes e crustàceos da ilha de Fernào do Pô e de Elobey. Jorn. Sci. math. phys. nat., (2)4(13):55-58.

--. 1895. Les poissons d'eau douce des Iles du Golfe de Guinée. Jorn. Sci. math. Phys. nat. Acad. Lisboa, (2)4(13):59-64.

--. 1898. Da distribuiçâo goegraphica dos peixes e crustàceos colhidos nas possessôes portuguezas d'Africa occidental e existentes no Museu Nacional de Lisboa. Jorn. Sci. math. phys. nat. (2)5(19):185-202.

--. 1905. Breve noticia àcerca de alguns peixes e crustàceos colhidos nas possessôes portuguezas da Africa Occidental. Jorn. Sci. math. phys. nat., (2)7(26):97-102.

--. 1906. Indicaçôes de algumas espécies que devem ser accrescentadas à fauna ichtyologica da ilha de S. Thomé. Jorn. Sci. math. phys. nat., (2)7(27):156-158.

--. 1906. Description d'un poisson des profondeurs appartenant à un genre nouveau et trouvé sur les côtes du Portugal. Jorn. Sci. math. phys. nat., (2)7(27):172-174, fig. 1-2, pl. 1.

--. 1909. Contribuiçâo para o conhecimento da fauna bathypelàgica visin-ha das costas de Portugal. Mems Mus. Bocage, 1:1-35, 3 pl.

--. 1909. Peixes colhidos nas visinhanças do archipélago de Cabo Verde. Mems Mus. Bocage, 1:51-77, 2 pl.

--. 1909. Nova contribuição para conhecimento da fauna bathypelagica visinha das costas de Portugal. Mems Mus. Bocage, 1:89-93, 1 pl.

--. 1909. Peixes d'àgua doce da Guiné Portugueza. Mems Mus. Bocage, 1:95-107, 1 pl.

--. 1912. Nova contribuição para o conhecimento da fauna bathypelagica visinha das costas de Portugal. Mems Mus. Bocage, 4:89-93, fig. 1-5.

--. 1917. Nota sobre algumas espécies de peixes que vivem no Atlântico Ocidental. Archos Univ. Lisb., 4:103-131, pl. 29-36.

OSORIO, J. 1954. Glossàrio de nomes dos peixes. Publçoes Gabin. Estud. Pesca, (20):1-249.

OSTROUMOFF, A. A. 1896. Zwei neue Relicten-Gattungen im Azowschen Meere. Zool. Anz., 19:30.

--. 1897. Résultats scientifiques de l'expédition "Atmanai". III. Poissons de la mer d'Azov. Izv. imp. Akad. Nauk., (7)5:251-267, 2 fig.

ØSTVEDT, O. J. 1969. On the catch statistics of Sardinella in Ghana. p. 265-268. In: Proc. Symp. Oceanogr. Fish. Resources trop. Atlant. (20-28 october 1968, Abidjan). Paris, 430 p.

O'TOOLE, M. J. 1976. Incidental collections of small and juvenile fishes from egg and larval surveys off South West Africa (1972-1974). Fish. Bull. S. Afr., 8:23-33.

--. 1978. Development, distribution and relative abundance of the larvae and early juveniles of the pelagic goby, Sufflogobius bibarbatus (von Bonde) off South West Africa, 1972-1974. Investl Rep. (S. Afr. Ass. mar. biol. Res.),(116):28 p.

OTTERSTRØM, C. V. 1914. Fisk II. Bløfinnefisk. Danmarks Fauna. Copenhagen, 351 p., 150 fig.

OTTO, A. W. 1821. Conspectus animalium quorundam maritimorum nondum editorum. Inaug. Dissert. Vratislaviae, 20 p.

--. 1821. Animalium quorundum maritimorum nondum editorum. Pars prior. Isis(Oken), 2(9):5 p.

OVTCHINNIKOV, V. V.; VOZMITIEL, E. A. 1971. The behaviour of certain species of fish in the zone of light source activity. Trudy atlant. nauchno-issled. Inst. ryb. Khoz. Okeanogr., (36):24-28 (in Russian).

--. 1971. Comportement de quelques espèces de poissons sous l'effet de sources lumineuses. UNDP(SF)/FAO Projet SEN/66/508, Survey Devel. pelag. Fish Resources, Vop. Povedenia Ryb. Trudy Bypusk, 36 (also: Scient. Transl. Centre Rech. océanogr. Dakar-Thiaroye, (1):8 p., mimeo).

OWEN, R. 1853. Descriptive catalogue of the osteological series contained in the Museum of the Royal College of Surgeons of England, I, Pisces, Reptilia, Aves, Marsupialia. London, 350 p.

PADOA, E. 1956. Scombriformes, Carangiformes. In: Uova, larve e stadi giovanili di Teleostei. (ed. U. d'Ancona). Fauna Flora Golfo Napoli, 38:471-576, fig. 296-433, pl. 36-37.

--. 1956. Triglidae, Peristediidae, Dactylopteridae, Gobiidae, Echeneidae, Jugulares, Gobiesocidae, Heterosomata, Pediculati. In: Uova, larve e stadi giovanili di Teleostei (ed. U. d'Ancona). Faunı Flora Golfo Napoli, 38(3):627-888, fig. 504-785, pl. 39-50.

PAES DA FRANCA, M. L.; VASCONCELOS, M. de S. (see FRANCA M.L.P. da; VASCONCELOS, M. de S.)

PAKHORUKOV, N. P. 1976. Preliminary list of the bathyal bottom fishes of the Rio Grande Rise. Trudy Inst. Okeanol., 104:318-331.

--. 1981. Deep-sea bottom fishes of the Whale Ridge and adjacent area. p. 19-31, 2 tab. In: Fishes of the open ocean. (N.V. Parin, ed.). Moscow, 119 p.

PALLAS, P. S. 1767-78. Spicilegia Zoologica. Berlin, 1 (13 fasc.). (Pisces: (7), 1769:1-42, 6 pl.; (8), 1770:1-54, 5 pl.).

--. 1814. Zoographia Rosso-Asiatica, sistens omnium animalium in extenso Imperio Rossico et adjacentibus maribus observatorum recensionem, domicilia, mores et descriptiones, anatomen atque icones plurimorum. Petropoli, 3:7 + 428 + 125 p. (see Svetovidov, A. N., 1981).

PALMER, G. 1961. The dealfishes (Trachipteridae) of the Mediterranean and north-east Atlantic. <u>Bull. Br. Mus. nat. Hist.</u> (Zool.), 7:335-351, 1 fig., pl. 62.

--. 1966. Additional notes on some umbrinine fishes (Family Sciaenidae). <u>Ann. Mag. nat. Hist.</u>, (13)9:423-427, pl. 6-7.

--. 1973. Regalecidae, Trachipteridae, p. 329-332, Radiicephalidae, p. 333, Lophotidae, p. 334, <u>in</u> Hureau, J.C. & Th. Monod (ed.). Checklist of the fishes of the north-eastern Atlantic and of the Mediterranean/Catalogue des poissons du nord-est Atlantique et de la Méditerranée (Clofnam). Paris, Unesco, 2 vol.

--. 1986. Lamprididae, <u>in</u> P.J.P. Whitehead, M.L. Bauchot, J.C. Hureau, J. Nielsen & E. Tortonese. Fishes of the North-eastern Atlantic and the Mediterranean/Poissons de l'Atlantique du Nord-Est et de la Méditerranée, Paris, Unesco, 2:725-726, 1 fig.

--. 1986. Regalecidae, <u>in</u> P.J.P. Whitehead, M.L. Bauchot, J.C. Hureau, J. Nielsen & E. Tortonese. Fishes of the North-eastern Atlantic and the Mediterranean/Poissons de l'Atlantique du Nord-Est et de la Méditerranée, Paris, Unesco, 2:727-728, 1 fig.

--. 1986. Trachipteridae, <u>in</u> P.J.P. Whitehead, M.L. Bauchot, J.C. Hureau, J. Nielsen & E. Tortonese. Fishes of the North-eastern Atlantic and the Mediterranean/Poissons de l'Atlantique du Nord-Est et de la Méditerranée, Paris, Unesco, 2:729-732, 3 fig.

--. 1986. Radiicephalidae, <u>in</u> P.J.P. Whitehead, M.L. Bauchot, J.C. Hureau, J. Nielsen & E. Tortonese. Fishes of the north-eastern Atlantic and the Mediterranean/Poissons de l'Atlantique du Nord-Est et de la Méditerranée, Paris, Unesco, 2:733, 1 fig.

--. 1986. Lophotidae, <u>in</u> P.J.P. Whitehead, M.L. Bauchot, J.C. Hureau, J. Nielsen & E. Tortonese. Fishes of the North-eastern Atlantic and the Mediterranean/Poissons de l'Atlantique du Nord-Est et de la Méditerranée, Paris, Unesco, 2:734-735, 1 fig.

PALMER, G.; OELSCHLÄGER, H. A. 1976. Use of the name <u>Lampris guttatus</u> (Brünnich, 1788) in preference to <u>Lampris regius</u> (Bonnaterre, 1788) for the opah. <u>Copeia</u>, 1976(2):366-367.

PALOMBI, A.; SANTARELLI, M. 1961. Gli animali comestibili dei mari d'Italia (2nd ed.). Milano, 437 p., illustr.

PALOMERA, I.; FORTUNO, J. M. 1981. Larvas leptocefalas de peces anguilliformes en la costa noroccidental de Africa. <u>Result. Exped. cient., B/O Cornide de Saavedra</u>, (9):3-15.

PALOMERA, I.; RUBIES, P. 1982. Kinds and distribution of fish eggs and larvae off northwest Africa in April-May 1973. <u>Rapp. P.-v. Réun. Cons. perm. int. Explor. Mer.</u>, 180:356-358.

PAPPE, C. W. L. 1853. Synopsis of the edible fishes of the Cape of Good Hope. Cape Town, 34 p.

PAPPENHEIM, P. 1914. Die Fische der deutschen Südpolar Expedition 1901-1903. II. Die Tiefsee-Fische. <u>Dt. Südpol.-Exped.</u>, 15(7):163-200, 10 fig., 2 pl.

PARAISO, F. 1960. Remarques sur deux <u>Promicrops ditobo</u> Roux et Collignon, 1954, pêchés sur les Côtes du Sénégal. <u>Bull. Inst. fr. Afr. noire</u>, (A)22(4):1347-1357.

PARIN, N.V. 1958. Rare pelagic fishes of the northwestern part of the Pacific Ocean (<u>Taractes steindachneri, Palinurichthys japonicus</u> and <u>Centrolophus lockingtoni</u>). <u>Vop. Ikhtiol.</u>, (11):162-170.

--. 1959. On the resemblance in the geographic distribution of sardines, <u>Sardina pilchardus</u> and <u>Sardinops sagax</u>, and flyingfish, <u>Cypselurus pinnatibarbatus</u>. <u>Dokl. Akad. Nauk SSSR</u>, 124(5):1130-1132, 2 fig.

--. 1960. The flying fishes (Exocoetidae) of the northwest Pacific. <u>Trudy Inst. Okeanol.</u>, 31:205-285, 25 fig. (Engl. transl. in Israel Program Scient. Transl., Jerusalem, Cat. No. 618).

--. 1961. Contribution to the knowledge of the flyingfish fauna (Exocoetidae) of the Pacific and Indian Oceans. Trudy Inst. Okeanol., 42:40-91, fig. 1-19 (in Russian).

--. 1961. The basis of classification of the flying fishes (families Oxyporhamphidae and Exocoetidae). Trudy Inst. Okeanol., 43:92-183 (in Russian).

--. 1962. Some features of the distribution of mass pelagic fishes in the zone of the Pacific equatorial currents. (Data of the 34th voyage of the "Vityaz"). Okeanologija, 2(6):1075-1082 (in Russian).

--. 1964. Taxonomic status, geographic variation and distribution of the oceanic halfbeak, Euleptorhamphus viridis (Van Hasselt) (Hemirhamphidae). Trudy Inst. Okeanol., 73:185-203 (in Russian).

--. 1967. Materials on distribution and biology of the snake mackerel, Gempylus serpens (Pisces, Gempylidae) in the Pacific and Indian Oceans. Vop. Ikhtiol., 7(6):990-1000, 3 fig. (in Russian).

--. 1967. Synentognathous fishes of the open ocean. p.43-66, fig. In: Tikhii Okean, 7: Biologia Tikhovo Okeana (ed. V.G. Kort). (3)(Fishes of the open waters). Moscow. (in Russian).

--. 1967. Diurnal variations in the larval occurrence of some oceanic fishes near the ocean surface. Okeanologija, 7(1):148-156.

--. 1967. Review of the marine Belonidae of the western Pacific and Indian Oceans. Trudy Inst. Okeanol., 84:3-83 (in Russian).

--. 1968. Scomberesocidae (Pisces, Synentognathi) of the eastern Atlantic Ocean. Atlantide Rep., (10):275-290, 5 fig.

--. 1968. Ikhtiofauna okeanskoi epipelagiali. Moskva, 185 p., 56 fig., 14 tab. (Engl. transl. 1970: Ichthyofauna of the epipelagic zone. Jerusalem, 205 p., 56 fig.

--. 1971. Distributional pattern of midwater fishes of the Peru Current zone. Trudy Inst. Okeanol., 89:81-95 (in Russian).

--. 1973. Belonidae, Scomberesocidae, Exocoetidae, Hemiramphidae, p. 258-269, in Hureau, J.C. & Th. Monod (ed.). Check-list of the fishes of the north-eastern Atlantic and of the Mediterranean/Catalogue des poissons du nord-est Atlantique et de la Méditerranée (Clofnam). Paris, Unesco, 2 vol.

--. 1975. Change of pelagic ichthyocoenoses along the section at the equator in the Pacific Ocean between 97 and 155° W. Trudy Inst. Okeanol., 102:313-334 (in Russian).

--. 1976. Comparative analysis of the mesopelagic ichthyocoens on four polygones in the western tropical Pacific Ocean. Trudy Inst. Okeanol., 104:195-205 (in Russian).

--. 1978. New records of midwater fishes from off New Guinea and Tonga Islands with description of two new species from the genera Eustomias (Fam. Melanostomiatidae) and Benthodesmus (Fam. Trichiuridae). Trudy Inst. Okeanol., 111:156-168, 4 fig.

--. 1983. Aphanopus mikhailini, sp.n. and A. intermedius, sp. n. (Trichiuridae, Perciformes), two new scabbardfishes from the temperate waters of southern hemisphere and tropical Atlantic Ocean. Vop. Ikhtiol., 23(3):355-365.

--. 1986. Scomberesocidae, in P.J.P. Whitehead, M.L. Bauchot, J.C. Hureau, J. Nielsen & E. Tortonese. Fishes of the North-eastern Atlantic and the Mediterranean/Poissons de l'Atlantique du Nord-Est et de la Méditerranée, Paris, Unesco, 2:610-611, 2 fig.

--. 1986. Exocoetidae, in P.J.P. Whitehead, M.L. Bauchot, J.C. Hureau, J. Nielsen & E. Tortonese. Fishes of the North-eastern Atlantic and the Mediterranean/Poissons de l'Atlantique du Nord-Est et de la Méditerranée, Paris, Unesco, 2:612-619, 8 fig.

--. 1986. Gempylidae, in P.J.P. Whitehead, M.L. Bauchot, J.C. Hureau, J. Nielsen & E. Tortonese. Fishes of the North-eastern Atlantic and the Mediterranean/Poissons de l'Atlantique du Nord-Est et de la Méditerranée, Paris, Unesco, 2:967-973, 7 fig.

--. 1986. Scombrolabracidae, in P.J.P. Whitehead, M.L. Bauchot, J.C. Hureau, J. Nielsen & E. Tortonese. Fishes of the North-eastern Atlantic and the Mediterranean/Poissons de l'Atlantique du Nord-Est et de la Méditerranée, Paris, Unesco, 2:974-975, 1 fig.

--. 1986. Trichiuridae, in P.J.P. Whitehead, M.L. Bauchot, J.C. Hureau, J. Nielsen & E. Tortonese. Fishes of the North-eastern Atlantic and the Mediterranean/Poissons de l'Atlantique du Nord-Est et de la Méditerranée, Paris, Unesco, 2:976-980, 4 fig.

PARIN, N. V.; ABRAMOV, A. A. 1986. Materials for a revision of the genus Epigonus Rafinesque (Perciformes, Epigonidae): species from the submarine ridges of the southern East Pacific and preliminary review of the "Epigonus robustus species-group". Trudy Inst. Okeanol., 121: 173-194, 7 fig. (in Russian).

PARIN, N. V.; ANDRIASHEV, A. P. 1972. Ichthyological studies during the 11th cruise of the Research Vessel "Akademik Kurchatov" in the South Atlantic. Vop. Ikhtiol., 12(5):960-964.

PARIN, N. V.; ASTAKOV, D. A. 1982. Studies on the acustico-lateralis system of beloniform fishes in connection with their systematics. Copeia, 1982(2):276-291.

PARIN, N. V.; BEKKER, V. E. 1972. Materials on taxonomy and distribution of some trichiuroid fishes (Pisces, Trichiuroidae: Scombrolabracidae, Gempylidae and Trichiuridae). Trudy Inst. Okeanol., 93:110-204, 28 fig. (in Russian).

--. 1973. Gempylidae, Scombrolabracidae, Trichiuridae, p. 457-464, in Hureau, J.C. & Th. Monod (ed.). Check-list of the fishes of the north-eastern Atlantic and of the Mediterranean/Catalogue des Poissons du nord-est Atlantique et de la Méditerranée Clofnam). Paris, Unesco, 2 vol.

PARIN, N. V.; BELYANINA, T. N. 1972. New data on the distribution and morphology of fish of the genus Macristiella. Vop. Ikhtiol., 12(1): 1018-1021.

PARIN, N. V.; BORODULINA, O. D. 1986. Preliminary review of the bentho-pelagic fish genus Antigonia Lowe (Zeiformes, Caproidae). Trudy Inst. Okeanol., 121:141-172, 13 fig. (in Russian).

PARIN, N. V.; GOLOVAN, G. A. 1976. Pelagic deep-sea fishes of the families characteristic of the open ocean collected over the continental slope off West Africa. Trudy Inst. Okeanol., 104:250-276, 3 fig., 2 tab. (in Russian).

PARIN, N. V.; GORBUNOVA, N. N. 1963. Development of eggs and larvae of the flying fish, Cheilopogon unicolor (Cuv. & Val.). Trudy Inst. Okeanol., 62:62-67, 2 fig.

--. 1964. On the reproduction and development of some Indian Ocean synentognathous fishes (Beloniformes, Pisces) (based on materials of the R/V "Vityaz" expedition). Trudy Inst. Okeanol., 73:224-234, 4 fig.

PARIN, N. V.; MIKHAILIN, S. V. 1981. A new cutlassfish (Lepidopus dubius Parin and Mikhailin)(Trichiuridae), from the eastern tropical Atlantic. Vop. Ikhtiol., 21(3):403-410, 3 fig., 3 tab. (in Russian).

PARIN, N. V.; NOVIKOVA, N. S. 1974. Taxonomy of viperfishes (Chauliodontidae, Osteichthyes) and their distribution in the world ocean. Trudy Inst. Okeanol., 96:255-315, 19 fig. (in Russian).

PARIN, N. V.; POKHILSKAYA, G. N. 1974. A review of the Indo-Pacific species of the genus Eustomias (Melanostomiatidae, Osteichthyes). Trudy Inst. Okeanol., 96:316-368 (in Russian).

--. 1978. On the taxonomy and distribution of the mesopelagic fish genus Melanostomias (Melanostomiatidae, Osteichthyes). Trudy Inst. Okeanol., 111:61-86.

PARIN, N. V.; SHCHERBACHEV, Y. N. 1973. On the record of a rare deep-sea fish, _Bathyprion_ _danae_ Marshall (suborder Alepocephaloidei) off Japanese coast. _Trudy_ _Inst._ _Okeanol._, **91**:255-256.

PARIN, N. V.; SOKOLOVSKY, A. S. 1976. Species composition of the family Melanostomiatidae in the Kuroshio Current zone. _Trudy_ _Inst._ _Okeanol._, **104**:237-249 (in Russian).

PARIN, N. V.; COLLETTE, B. B.; SHCHERBACHEV. 1980. Preliminary review of the marine halfbeaks (Hemiranphidae, Beloniformes) of the tropical Indo-West-Pacific. _Trudy_ _Inst._ _Okeanol._, **97**:7-172, 47 fig. (in Russian).

PARIN, N. V.; NESIS, K. N.; VINOGRADOV, M. Y. 1969. Data on the feeding of _Alepisaurus_ in the Indian Ocean. _Vop._ _Ikhtiol._, **9**(3):526-538.

PARIN, N. V.; SAZONOV, Y. I.; MIKHAILIN, S. V. 1978. Deep-sea pelagic fishes in the collection of R/V "Fiolent" in the Gulf of Guinea and adjacent areas. _Trudy_ _Inst._ _Okeanol._, **111**:169-184, 3 fig., 5 tab. (in Russian).

PARIN, N. V.; ANDRIASHEV, A. P.; BORODULINA, O. D.; TCHUVASOV, V. M. 1974. Midwater fishes of the southwestern Atlantic Ocean. _Trudy_ _Inst._ _Okeanol._, **98**:76-140, 15 fig., 1 pl. (in Russian).

PARIN, N. V.; BEKKER, V. E.; BORODULINA, O. D.; TCHUVASOV, V. M. 1973. Deep-sea pelagic fishes of the south-eastern Pacific ocean. _Trudy_ _Inst._ _Okeanol._, **94**:71-172, 45 fig. (in Russian).

PARIN, N. V.; POKHILSKAYA, G. N.; SAZONOV, Y. J.; FEDORYAKO, B. J. 1976. Rare and poorly known midwater fishes from the central and eastern equatorial Pacific Ocean. _Trudy_ _Inst._ _Okeanol._, **104**:206-236, 8 fig. (in Russian).

PARIN, N. V.; BEKKER, V. E.; BORODULINA, O. D.; KARMOVSKAYA, E. S.; FEDORYAKO, B. J.; SHCHERBACHEV, Y. N.; POKHILSKAYA, G. N.; TCHUVASOV, V. M. 1977. Midwater fishes in the western tropical Pacific Ocean and the seas of Indo-Australian Archipelago. _Trudy_ _Inst._ _Okeanol._, **107**: 68-188, 42 fig. (in Russian).

PARR, A. E. 1927. Scientific results of the third oceanographic expedition of the "Pawnee" 1927. Ceratioidea. _Bull._ _Bingham_ _oceanogr._ _Coll._, 3(1):1-34, fig. 1-13.

—. 1927. The stomiatoid fishes of the suborder Gymnophotodermi (Astronesthidae, Melanostomiatidae, Idiacanthidae) with a complete review of the species. _Bull._ _Bingham_ _oceanogr._ _Coll._, 3(2):1-123, 62 fig.

—. 1928. Deepsea fishes of the order Iniomi from the waters around the Bahama and Bermuda islands, with annotated keys to the Sudidae, Myctophidae, Scopelarchidae, Evermannellidae, Omosudidae, Cetomimidae and Rondeletidae of the world. _Bull._ _Bingham_ _oceanogr._ _Coll._, 3(3): 1-193, fig. 1-43.

—. 1929. Notes on the species of myctophine fishes represented by type specimens in the United States National Museum. _Proc._ _U.S._ _natn._ _Mus._, 76(10):1-47, fig. 1-21.

—. 1929. Notes on a collection of _Myctophum_ _glaciale_ and _Lampanyctus_ _pusillus._ _Proc._ _New_ _Engl._ _zool._ _Club._, 11:57-62, fig. 1-2.

—. 1929. A contribution to the osteology and classification of the orders Iniomi and Xenoberyces, with description of a new genus and species of the family Scopelarchidae, from the western coast of Mexico, and some notes on the visceral anatomy of _Rondeletia._ _Occ._ _Pap._ _Bingham_ _oceanogr._ _Colln_, 2:1-45, 19 fig.

—. 1930. Teleostean shore and shallow-water fishes from the Bahamas and Turks Island. _Bull._ _Bingham_ _oceanogr._ _Coll._, 3(4):1-148, 38 fig.

—. 1930. A note on _Evermannella_ _atrata_ _atlantica_ Parr, the genus _Coccorella_ Roule, and the classification of the Iniomi. _Ann._ _Mag._ _nat._ _Hist._, (10)6:154-156.

—. 1930. On the osteology and classification of the pediculate fishes of the genera _Aceratias,_ _Rhynchoceratias,_ _Haplophryne,_ _Laevoceratias,_

Allector and _Lipactis_. _Occ._ _Pap._ _Bingham_ _oceanogr._ _Colln_, 2:1-23.

--. 1930. On the probable identity, life history and anatomy of the free-living and attached males of the ceratioid fishes. _Copeia_, 1930 (4):129-135.

--. 1931. On the genera _Paralepis_ and _Lestidium_ and the taxonomic status of some of their species. _Copeia_, 1931(4):152-158, 3 fig.

--. 1931. Scientific results of the second oceanographic expedition of the "Pawnee" 1926. Deepsea fishes from off the western coast of North and Central America with keys to the genera _Stomias_, _Diplophos_, _Melamphaes_, and _Bregmaceros_, and a revision of the _macropterus_ group of the genus _Lampanyctus_. _Bull._ _Bingham_ _oceanogr._ _Coll._, 2(4):1-53, 18 fig.

--. 1931. A substitute name for _Dolichodon_ Parr, a genus of deep-sea fishes. _Copeia_, 1931(4):162.

--. 1932. Scientific results of the third oceanographic expedition of the "Pawnee" 1927. Deep-sea eels, exclusive of larval forms. _Bull._ _Bingham_ _oceanogr._ _Coll._, 3(5):1-41, fig. 1-5.

--. 1933. Deep-sea Berycomorphi and Percomorphi from the waters around the Bahama and Bermuda Islands. _Bull._ _Bingham_ _oceanogr._ _Coll._, 3(6): 1-51, fig. 1-22.

--. 1933. Two new records of deep-sea fishes from New England with description of a new genus and species. _Copeia_, 1933(4):176-179.

--. 1934. Studies of Myctophinae in the Museum of Comparative Zoology. I. Revision of type specimens. II. Myctophinae collected by C. O'D. Iselin in the northern Atlantic in 1928. _Bull._ _Mus._ _comp._ _Zool._ _Harv._, 77(2):41-65, fig. 1-7.

--. 1934. Report on experimental use of a triangular, trawl for bathypelagic collecting. With an account of the fishes obtained and a revision of the family Cetomimidae. _Bull._ _Bingham_ _oceanogr._ _Coll._, 4(6): 1-59, fig. 1-21.

--. 1937. Concluding report on fishes with species index for articles 1-7. (Fishes of the third oceanographic expedition of the "Pawnee"). _Bull._ _Bingham_ _oceanogr._ _Coll._, 3(7):1-79, fig. 1-22.

--. 1945. Barbourisidae, a new family of deep sea fishes. _Copeia_, 1945 (3):127-129, 1 pl.

--. 1946. On taxonomic questions related to the classification of _Barbourisia_, the Cetomimidae and the Iniomi. _Copeia_, 1946(4):260-262.

--. 1946. The Macrouridae of the western North Atlantic and central American seas. _Bull._ _Bingham_ _oceanogr._ _Coll._, 10(1):1-99, 28 fig.

--. 1948. The classification of the fishes of the genera _Bathylaco_ and _Macromastax_, possible intermediates between the Isospondyli and Iniomi. _Copeia_, 1948(1):48-54, 2 fig.

--. 1951. Preliminary revision of the Alepocephalidae, with the introduction of a new family, Searsidae. _Am._ _Mus._ _Novit._, (1531):1-21.

--. 1952. Revision of the genus _Talismania_, with description of a new species from the Gulf of Mexico. _J._ _Wash._ _Acad._ _Sci._, 42(8):268-271, 1 fig.

--. 1952. Revision of the species currently referred to _Alepocephalus_, _Halisauriceps_, _Bathytroctes_ and _Bajacalifornia_ with introduction of two new genera. _Bull._ _Mus._ _comp._ _Zool._ _Harv._, 107(4):254-269.

--. 1954. Review of the deep-sea fishes of the genus _Asquamiceps_ Zugmayer, with descriptions of two new species. _Am._ _Mus._ _Novit._, (1655):1-8.

--. 1960. The fishes of the family _Searsidae_. _Dana_ _Rep._, (51):1-109, 73 fig.

PARRA, D. A. 1787. Descripción de diferentes piezas de historia natural, las más del ramo marítimo, representadas en setenta y cinco laminas. Havana, 195 p., 75 pl.

PARROTT, A. W. 1960. The queer and the rare fishes of New Zealand. London, 192 p., 71 fig.

PAUCA, M. 1930. Fische aus der Walfischbay, Südwestafrika. Annln naturh. Mus. Wien, 44:33-37.

PAULIN, C. D. 1978. New records of anglerfishes (Antennariidae) from New Zealand. N.Z. Jl Zool., 5:485-491.

PAULIN, C. D.; HABIB, G. 1980. First record of Lepidocybium flavobrunneum (Pisces, Gempylidae) from New Zealand. N.Z. Jl mar. Freshwat. Res., 14(4):405-407.

PAULY, D. 1975. On the ecology of a small West-African lagoon. Ber. dt. wiss. Kommn. Meeresforsch., 24:46-62.

PAXTON, J. R. 1963. A new lanternfish (family Myctophidae) of the genus Lampadena from the eastern Pacific Ocean. Copeia, 1963(1):29-33.

--. 1967. A distributional analysis for the lanternfishes (family Myctophidae) of the San Pedro Basin, California. Copeia, 1967(2):422-440, fig. 1-16.

--. 1972. Osteology and relationships of the lanternfishes (family Myctophidae). Sci. Bull. nat. Hist. Mus. Los Angeles Cty, 13:1-81.

--. 1973. Cetomimidae, Ateleopodidae, p. 214-215, in Hureau, J.C. & Th. Monod (ed.). Check-list of the fishes of the north-eastern Atlantic and of the Mediterranean/Catalogue des poissons du nord-est Atlantique et de la Méditerranée (Clofnam). Paris, Unesco, 2 vol.

--. 1974. Myctophid fish recorded from New Zealand as Lampanyctus guentheri reidentified as L. australis (note). N.Z. Jl mar. Freshwat. Res., 8(4):711-712.

--. 1974. Morphology and distribution patterns of the whalefishes of the family Rondeletiidae. J. mar. biol. Ass. India, 15(1):175-188, 7 fig.

--. 1986. Cetomimidae, in P.J.P. Whitehead, M.L. Bauchot, J.C. Hureau, J. Nielsen & E. Tortonese. Fishes of the North-eastern Atlantic and the Mediterranean/Poissons de l'Atlantique du Nord-Est et de la Méditerranée, Paris, Unesco, 2:524-525, 2 fig.

--. 1986. Rondeletiidae, in P.J.P. Whitehead, M.L. Bauchot, J.C. Hureau, J. Nielsen & E. Tortonese. Fishes of the North-eastern Atlantic and the Mediterranean/Poissons de l'Atlantique du Nord-Est et de la Méditerranée, Paris, Unesco, 2:526-527, 1 fig.

--. 1986. Ateleopodidae, in P.J.P. Whitehead, M.L. Bauchot, J.C. Hureau, J. Nielsen & E. Tortonese. Fishes of the North-eastern Atlantic and the Mediterranean/Poissons de l'Atlantique du Nord-Est et de la Méditerranée, Paris, Unesco, 2:528-529, 2 fig.

PAXTON, J. R.; LAVENBERG, R. J. 1973. Feeding mortality in a deep-sea angler fish (Diceratias bispinosus) due to a macrourid fish (Ventrifossa sp.) Aust. Zool., 18:47-51.

PAYNE, A. I. 1975. The relative abundance and feeding habits of the grey mullet species occurring in an estuary in Sierra Leone, West Africa. Aquaculture, 5:108 (abstract).

PAZ, R. de la. 1975. Systématique et phylogenèse des Sparidae du genre Diplodus Raf. (Pisces, Teleostei). Trav. Docums ORSTOM, (45):1-96, 48 fig.

--. 1981. Note critique sur la nomenclature des Diplodus du groupe Sargus (Pisces, Sparidae). Bull. Mus. natn. Hist. nat., Paris, (4)3(A)(3):931-934.

PAZ, R. de la; BAUCHOT, M. L.; DAGET, J. 1974. Les Diplodus (Perciformes, Sparidae) du groupe Sargus: sytématique et phylogénie. Ichthyologia, 5(1):113-128.

PEARCY, W. G. 1964. Some distributional features of mesopelagic fishes off Oregon. J. mar. Res., 22(1):83-102, 7 fig.

PEARCY, W. G.; AMBLER, J. W. 1974. Food habits of deep-sea macrourid fishes off the Oregon coast. Deep Sea Res., 21:745-759.

PEAVOT, H. 1937. List of dates of publication of the early parts of the Society's "Transactions". Proc. zool. Soc. Lond., 107:83-84.

PECHKURENKOV, V. L. 1963. Peculiarities of scale cover in West African Sardinella aurita Val. Vop. Ikhtiol., 3:131-143. (in Russian).

PELLEGRIN, J. 1904. Characinidés nouveaux de la Casamance. Bull. Mus. natn. Hist. nat., Paris, 10:218-221.

--. 1905. Mission des pêcheries de la côte occidentale d'Afrique dirigée par M. Gruvel: Poissons. Bull. Soc. zool. Fr., 30:135-141.

--. 1906. Mission des pêcheries de la côte occidentale d'Afrique. Poissons. Act. Soc. linn. Bordeaux, 60:17-57, 15 fig., pl.

--. 1907. Mission des pêcheries de la côte occidentale d'Afrique. 2e Memoire. Act. Soc. Linn. Bordeaux, 62:71-102, 17 fig.

--. 1907. Mission des pêcheries de la Côte occidentale d'Afrique, dirigée par M. Gruvel, Poissons (2e note). Bull. Soc. zool. Fr., 32:83-89.

--. 1907. Sur une collection de poissons recueillis par M.E. Hang à Ngomo (Ogôué). Bull. Soc. philomath. Paris, (9)9:17-42, 1 pl.

--. 1908. Etude scientifique des matériaux ichtyologiques recueillis par la mission des pêcheries de la côte occidentale d'Afrique. Congr. natn. Pêches marit., Bordeaux, p.413-420.

--. 1909. Poissons des pêcheries de la côte occidentale d'Afrique. C. r. Ass. fr. Avanc. Sci., 38:662-668.

--. 1912. Reptiles, batraciens et poissons du Maroc (Mission de Mme Camille du Cast). Bull. Soc. zool. Fr., 37(7):255-262.

--. 1912. Poissons des côtes de l'Angola. Mission de M. Gruvel (4e note). Bull. Soc. zool. Fr., 37:290-296.

--. 1912. Sur la dentition des Diables de mer. Bull. Soc. Philomath., Paris, (10)4(1-2):1-8.

--. 1913. Poissons des côtes de Mauritanie. Mission de M. Gruvel. Bull. Soc. zool. Fr., 38:116-118.

--. 1913. Poissons marins de Guinée, de la Côte d'Ivoire, du Dahomey et du Congo. Mission de M. Gruvel. Bull. Soc. zool. Fr., 38(5):151-158.

--. 1913. Sur un nouveau genre de centrarchidés du Gabon. C. r. hebd. Séanc. Acad. Sci., Paris, 156:1488-1489.

--. 1914. Sur un poisson apode nouveau de la côte de Mauritanie. 9e Int. Congr. Zool., Monaco, (1):7-8.

--. 1914. Mission Gruvel sur la côte occidentale d'Afrique (1905-12). Poissons. Annls Inst. océanogr., Monaco, 6:1-99, 2 pl., 15 fig.

--. 1914. Sur une collection de poissons de Madagascar. Bull. Soc. zool. Fr., 39:221-234.

--. 1915. Les poissons du Bassin de l'Ogôué. C. r. Ass. fr. Avanc. Sci. (1914), 43:500-505.

--. 1920. Poissons des lagunes de la Côte d'Ivoire. Description de deux espèces nouvelles. Bull. Soc. zool. Fr., 45:115-121.

--. 1921. Les poissons des eaux douces de l'Afrique du Nord française (Maroc, Algérie, Tunisie, Sahara). Mém. Soc. Sci. nat. Maroc, 1(2):1-216.

--. 1921. Les poissons des eaux douces de l'Afrique du Nord française et leur distribution géographique. C. r. Ass. fr. Avanc. Sci., (Strasbourg), p.269-273.

--. 1922. Les poissons des eaux douces de l'Afrique occidentale. C. r. Ass. fr. Avanc. Sci., (Rouen), 1921:633-638.

--. 1922. Poissons recueillis par M. Ch. Alluaud dans la région du Sous (Maroc). Bull. Soc. Sci. nat. Maroc, 2(3-4):103-106.

--. 1923. Les poissons des eaux douces de l'Afrique occidentale (du Sénégal au Niger). Paris, 373 p., 76 fig.

--. 1923. Nouvelle contribution à la faune ichtyologique des eaux douces du Maroc. C. r. hebd. Séanc. Acad. Sci., Paris, 176:787-789.

--. 1924. Les poissons du Sahara. <u>C. r. Ass. fr. Avanc. Sci.. (Bordeaux)</u>, p.585-588.

--. 1926. Contribution à l'étude de la faune ichtyologique du Niger et de la Guinée Française d'après les envois de M. Jean Thomas. <u>Bull. Com. Etud. hist. scient. Afr. occid. fr.</u>, 9:52-77, 4 fig.

--. 1927. Les silurides du bassin du Congo. <u>C. r. Ass. fr. Avanc. Sci.</u> (1926), **50**:426-429.

--. 1927. Les poissons d'eau douce de l'Afrique du Nord et du Sahara. <u>C. r. Séanc. Acad. Sci. colon.. Paris</u>, **6**:235-240.

--. 1928. Description de cichlidés et d'un mugilidé nouveaux du Congo Belge. <u>Revue Zool. Bot. afr.</u>, **15**:52-57.

--. 1929. Les poissons euryhalins de l'Afrique du Nord française. <u>Bull. Soc. Océanogr. Fr.</u>, **9**:909-912.

--. 1934. Description d'un poisson nouveau de la côte occidentale d'Afrique de la famille Muraenesocidae. <u>Bull. Soc. zool. Fr.</u>, **59**:45-48.

--. 1935. Poissons de Guinée française recueillis par M. Waterlot. Description d'une espèce et de deux variétés nouvelles. <u>Bull. Soc. Zool. Fr.</u>, **60**:462-466.

--. 1936. La présence de l'Arnis de Heudelot au Maroc. <u>Bull. Soc. Sci. nat. Maroc</u>, **16**:146.

PENNANT, T. 1776. British Zoology. (4th ed.) 3 Chester, 425 p., 73 pl.

--. 1787. Arctic Zoology, supplement. London, viii + 163 p.

--. 1812. British Zoology. 3. Reptiles, Fishes. (5th ed.). Chester, viii + 546 p., 85 pl.

PENRITH, M. J. 1964. A marked extension of the known range of <u>Tetrapturus angustirostris</u> in the Indian Ocean. <u>Copeia</u>, 1964(1):231-232.

--. 1967. The fishes of Tristan da Cunha, Gough Island and the Vema Seamount. <u>Ann. S. Afr. Mus.</u>, **48**(22):523-548, 2 fig.

--. 1967. Ceratioid angler-fishes from South Africa. <u>Jnl nat. Hist.</u>, 1: 185-188.

--. 1969. New records of deep-water fishes from South West Africa. <u>Cimbebasia</u>, (A)1(3):59-75, 8 tab.

--. 1972. Earliest description and bame for the whale shark. <u>Copeia</u>, 1972:362.

--. 1976. Distribution of shallow water marine fishes around southern Africa. <u>Cimbebasia</u>, (A)4(7):137-154, 1 fig.

--. 1978. An annotated check-list of the inshore fishes of southern Angola. <u>Cimbebasia</u>, (A)4(11):179-190, 1 fig.

--. 1982. Notes on marine fishes collected in the vicinity of Bosluisbaai. <u>Madoqua</u>, 13(2):159-168.

PENRITH, M. J.; PENRITH, M. L. 1971. The status of <u>Batrichthys apatius</u> (Cuvier and Valenciennes) (Pisces: Batrachoididae) with notes on four western southern African species of batrachoid fishes. <u>Cimbebasia</u>, (A)2(3):45-52, 2 fig.

--. 1972. The Blenniidae of western southern Africa. <u>Cimbebasia</u>, (A)2 (5): 65-90.

PENRITH, M. L. 1970. Report on a small collection of fishes from the Kunene River mouth. <u>Cimbebasia</u>, (A)1(7):165-176, 1 pl., 2 fig.

PENRITH, M. L.; KENSLEY, B. F. 1970. The constitution of the intertidal fauna of rocky shores of South West Africa. Part 1. Lüderitzbucht. <u>Cimbebasia</u>, (A)1(9):191-239, 4 fig., 3 pl.

PERON, F. 1807. Voyage de Découvertes aux Terres Australes. Paris, 1:1-496.

PERONICI, A. 1966. Lo sport della pesca. II. La fauna delle acque interne e del mare Firenze, xi + 537 p., 99 fig.

PERUGIA, A. 1891. Intorno ad alcuni pesci raccolti al Congo dal Capitano Giacomo Bove. <u>Annali Mus. civ. Stor. nat. Genova</u>, (2)10:967-977.

PERTSEVA-OSTROUMOVA, T. A. 1964. On some morphological peculiarities of myctophid larvae (Myctophidae, Pisces). Trudy Inst. Okeanol., 73:76-92, 10 fig. (in Russian, English summary).

--. 1967. Larvae of primitive myctophids, Protomyctophum and Electrona (Pisces, Myctophidae). Trudy Inst. Okeanol., 84:222-237 (in Russian).

--. 1974. New data on lantern-fish larvae (Myctophidae, Pisces) with oval eyes from the Indian and Pacific oceans. Trudy Inst. Okeanol., 96:77-142 (in Russian).

PETERS, K. M. 1983. Larval and early juvenile development of the frillifin goby, Bathygobius soporator (Perciformes: Gobiidae). Northeast Gulf Sci., 6:137-153.

PETERS, W. C. H. 1844. Über einige neue Fische und Amphibien aus Angola und Mozambique. Mber. k. preuss. Akad. Wiss. Berl., (1843-1844):32-37.

--. 1859. Eine neue vom Herrn Jagor im atlantischen Meere gefangene Art der Gattung Leptocephalus, und über andere neue Fische des Zoologischen Museums. Mber. k. preuss. Akad. Wiss. Berl., (1859):411-413.

--. 1865. Über einige neue Säugethiere, Amphibien und Fische. Mber. k. preuss. Akad. Wiss. Berl., (1864):381-399.

--. 1869. Mittheilung über eine neue Nagergattung, Chiropodomys penicillatus, sowie über einige neue oder weniger bekannte Amphibien und Fische. Mber. k. preuss. Akad. Wiss. Berl., (1868):448-460.

--. 1876. Über eine neue, mit Halieutaea verwandte Fischgattung, Dibranchus, aus dem atlantischen Ocean. Mber. k. preuss. Akad. Wiss. Berl., (1875):736-742, plate with 5 fig.

--. 1877. Über die von Professor Dr. Reinhold Buchholz in Westafrika gesammelten Fische. Mber. k. preuss. Akad. Wiss. Berl., (1876):244-252, 1 pl.

--. 1877. Übersicht der während der von 1874 bis 1876 unter dem Commando des Herrn Kapitän z. S. Freiherrn von Schleinitz ausgeführten Reise S.M.S. "Gazelle" gesammelten und der kais. Admiralität der kgl. Akademie der Wissenschaften übersandten Fische. Mber. k. preuss. Akad. Wiss. Berl., (1876):831-854.

PETERSEN, C. G. J. 1891. On the eggs and breeding of our Gobiidae. Rep. Dan. biol. Stn, 2:1-9.

--. 1917. On the development of our common gobies (Gobius) from the egg to the adult stages, etc. Rep. Dan. biol. Stn, 24:3-16.

--. 1919. Our gobies (Gobiidae). From the egg to the adult stages. Rep. Dan. biol. Stn, 26:47-66.

PEZOLD, F. L.; CASHNER, R. C. 1983. A character analysis of Gobionellus boleosoma and G. shufeldti (Pisces: Gobiidae) from the north-central Gulf of Mexico. Northeast Gulf Sci., 6:71-77.

PFAFF, J. R. 1933. Report on the fishes collected by M. Harry Madsen during Professor O. Olufsen's expedition to French Sudan in the years 1927-28. Vidensk. Meddr dansk. naturh. Foren., 94:273-315, 13 fig., pl. 6, 10 tab.

PHAM-THUOC; SZYPULA, J. 1973. Biological characteristics of gilt sardine Sardinella aurita Cuv. et Val. 1847 from north-west African coast. Acta ichthyol. piscat., 3(1):19-36.

PHILIBOSIAN, R.; IMSAND, S. 1977. New depth range, locality and a literature correction for the gempylid fish, Promethichthys prometheus. Fla Scient., 40(3):261-262.

PHILIPPI, R. A. 1857. Über einige chilenische Vögel und Fische. Arch. Naturg., 23:262-272.

PHILLIPPS, W. J. 1932. Notes on new fishes from New Zealand. N.Z. Jl Sci. Technol., 13(4):226-234.

PHLEGER, C. F. 1972. Cholesterol and hyperbaric oxygen in swimbladders of deep sea fishes. Diss. Abstr., 33B(1):348.

PIETSCH, T. W. 1972. A review of the monotypic deep-sea anglerfish family Centrophrynidae. Copeia, 1972(1):17-47.

--. 1972. Systematics and distribution of ceratioid fishes of the genus Dolopichthys (family Oneirodidae), with the description of a new species. Arch. FischWiss., 23(1):1-28.

--. 1974. Osteology and relationships of deep-sea anglerfishes of the family Oneirodidae with a review of the genus Oneirodes Lütken. Sci. Bull. nat. Hist. Mus. Los Angeles Cty, 18:1-113, 116 fig., 24 tab.

--. 1974. Systematics and distribution of ceratioid anglerfishes of the genus Lophodolos (family Oneirodidae). Breviora, (425):1-19, 9 fig.

--. 1975. Systematics and distribution of ceratioid anglerfishes of the genus Chaenophryne (family Oneirodidae). Bull. Mus. comp. Zool. Harv., 147(2):75-100.

--. 1976. Dimorphism, parasitism and sex: reproductive strategies among deepsea ceratioid anglerfishes. Copeia, 1976(4):781-793.

--. 1978. The feeding mechanism of Stylephorus chordatus (Teleostei: Lampridiformes): functional and ecological implications. Copeia, 1978(2):255-262.

--. 1979. Systematics and distribution of ceratioid anglerfishes of the family Caulophrynidae with the description of a new genus and species from the Banda Sea. Contr. Sci., (310):1-25, 22 fig., 1 tab.

--. 1981. Antennariidae, in Fischer, W., Bianchi, G. & W.B. Scott (ed.), FAO species identification sheets for fishery purposes. Eastern Central Atlantic; fishing areas 34, 47 (in part). Canada Funds-in-Trust Ottawa, Department of Fisheries and Oceans Canada, by arrangement with the Food and Agriculture Organization of the United Nations, vol. 1-7: pag. var.

--. 1986. Antennariidae, in P.J.P. Whitehead, M.L. Bauchot, J.C. Hureau, J. Nielsen & E. Tortonese. Fishes of the Northeastern Atlantic and the Mediterranean/Poissons de l'Atlantique du Nord-Est et de la Méditerranée, Unesco, 3:1364-1368, 4 fig.

PIETSCH, T. W.; DUZER, J. P. van. 1980. Systematics and distribution of ceratioid anglerfishes of the family Melanocetidae with the description of a new species from the eastern North Pacific Ocean. Fishery Bull., 78(1):59-87.

PIETSCH, T. W.; NAFPAKTITIS, B. G. 1971. A male Melanocetus johnsoni attached to a female Centrophryne spinulosa. Copeia, 1971:322-324.

PIETSCH, T. W.; SEIGEL, J. A. 1980. Ceratioid anglerfishes of the Philippine Archipelago, with descriptions of five new species. Fishery Bull., 78(2):379-399.

PIETSCHMANN, V. 1906. Ichthyologische Ergebnisse einer Reise nach Island, an die atlantische Küste von Marokko und in die westliche Hälfte des Mittelmeeres. Annln naturh. Mus. Wien, 21:72-148, fig. 1-7, pl. 5-6.

--. 1913. Fische des Wiesbadener Museums. Jb. nassau. Ver. Naturk., 66:170-201, 2 pl.

--. 1925. Bandfische und "grosse Seeschlange". Veröff. naturh. Mus. Wien, 5:1-22, fig.

--. 1926. Ein neuer Tiefseefisch aus der Ordnung der Pediculati. Anz. Akad. Wiss. Wien, 63(11):88-89.

--. 1930. Phrynichthys wedli Pietschmann nov. gen. et spec., ein Tiefsee Pediculate. Annln naturh. Mus. Wien, 44:419-422, 1 fig.

PINCHUK, V. I.; PERMITIN, Yu. Ye. 1970. New data on dogfish sharks of the family Squalidae in the southeastern Atlantic. J. Ichthyol., 3:273-276.

PINTO, S.Y. 1970. Observaçoes ictiológicas. VI. Antobrantia, novo gênero de ofictídeo do Brasil (Actinopterygii, Anguilliformes, Ophichthyidae). Atas Soc. Biol. Rio de J., 14(1-2):13-15.

PIOTROVSKY, A. S. 1979. On the range on the black scabbardfish Aphanopus _carbo_ Lowe (Trichiuridae) in the Indian Ocean. _Vop. Ikhtiol._, **19**(5): 931-932 (in Russian).

PISO, G. 1648. Historia naturalis Braziliae. Lugduni Batavorum, 122 p., fig.

PLANAS, A.; VIVES, F. 1955. Contribución a la sistematica de los Centracantidos con un estudio especial de la biometrica y biologia de la xucla (_Spicara chryselis_ L.). _Investigación pesq._, 1:87-135, 21 fig..

--. 1956. Sobre la presencia de _Trachypterus arcticus_ (Brünn) en el Mediterraneo. _Investigación pesq._, 5:135-138, fig.

PLAYFAIR, R. L.; GUENTHER, A. 1866. The fishes of Zanzibar. London, xiv + 153 p., some fig., 21 pl.

PLESSIS, C. G. du. 1960. Trends in the pilchard fishery of the Union of South Africa, 1943-58. _FAO Proc. Wld scient. Meet. Biol. Sardines_, 3:631-666.

PLIYA, J. 1980. La pêche dans le sud-ouest du Bénin. Etude de géographie appliquée sur la pêche continentale et maritime. Paris, xv + 296 p.

PODOSINNIKOV, A. Yu. 1976. Early ontogenesis of _Chaetodon hoefleri_ (S.). _Biol. Morya. Kiev_, **38**:25-27, fig.

POEY, F. 1851-54. Memorias sobre la historia natural de la isla de Cuba, acompañadas de sumarios latinos y extractos en frances. Habana, 1:463 p., 34 pl. (publ. dates: (1) p.1-40, pl.1-8, Nov. 1851; (2) p. 41-120, pl. 9-14, Apr. 1852; (3) p. 121-200, pl. 15-22, Oct. 1852; (4) p. 201-280, pl. 23-30, May 1853; (5) p. 281-463, pl. 31-34, June 1854).

--. 1853. XVI. Quironectos cubanos. p. 214-221, pl. 17. _In_: Memorias sobre la historia natural de la isla de Cuba, acompañadas de sumarios latinos y extractos en frances. Habana, 1, 1851-1854.

--. 1854. XXXII. Los guajacones, pecesillos de agua dulce. p. 374-392, pl. 31, 32. _In_: Memorias sobre la historia natural de la isla de Cuba, acompañadas de sumarios latinos y extractos en frances. Habana, 1, 1851-54.

--. 1858-61. Memorias sobre la historia natural de la isla de Cuba, acompañadas de sumario latinos y extractos en frances. Habana, 2: 442 p., 19 pl. (publ. dates: (1) p. 1-96, pl. 1-9, Oct. 1858; (2) p. 97-336, pl. 10-12+14, July 1860; (3) p. 337-442, pl. 13+15-19, June 1861).

--. 1860-61. XLIX. Poissons de Cuba, espèces nouvelles. p. 115-356, pl. 12-19. _In_: Memorias sobre la historia natural de la isla de Cuba, acompañadas de sumarios latinos y extractos en frances. Habana 2, 1858-1861.

--. 1861. L. Conspectus piscium cubensium. p. 357-404, LIII. Apéndice. p. 415-427. _In_: Memorias sobre la historia natural de la isla de Cuba, acompañadas de sumarios latinos y extractos en frances. Habana, 2, 1858-1861.

--. 1864. Descriptions de poissons nouveaux ou peu connus. _Proc. Acad. nat. Sci. Philad._, (1863)15:180-188.

--. 1865-68. Repertorio fisico-natural de la isla de Cuba. Habana, 1: 420 p., 5 pl. (publ. dates: (1) p. 1-264, 1865; (2) p. 265-420, 1866); 2: 484 p., 4 pl. (publ. dates: (1) p. 1-48, 1866; (2) p. 49-264, 1867; (3) p. 265-484, 1868).

--. 1865-66. Revista de los tipos Cuvierianos y Valenciennianos correspondientes à los peces de la isla de Cuba. p. 193-203, 265-278, 308-338, 369-383, 410-411. _In_: Repertorio fisico-natural de la Isla de Cuba. Habana, 1, 1865-66.

--. 1867-1868. Pesces cubanos, especies nuevos. p. 229-268, 2 pl. _In_: Repertorio fisico-natural de la isla de Cuba. Habana, 2, 1866-68.

--. 1868. Synopsis piscium cubensium. p. 279-468, pl. _In_: Repertorio fisico-natural de la isla de Cuba. Habana, 2, 1866-68.

--. 1870. Review of the fish of Cuba belonging to the genus _Trisotropus_, with an introductory note by J. Carson Brevoort. _Ann. Lyceum nat. Hist. N.Y._, 9:301-309.

--. 1870. New species of Cuban fish. _Ann. Lyceum nat. Hist. N.Y._, 9: 317-322.

--. 1871. Genres des poissons de la faune de Cuba, appartenant à la famille Percidae, avec une note d'introduction par J. Carson Brevoort. _Ann. Lyceum nat. Hist. N.Y._, 10:27-29, 1 pl.

--. 1873. _Grammicolepis brachiusculus_, tipo de una nueva familia en la clase de los peces. _An. Soc. esp. Hist. nat._, 2:403-406, pl. 12.

--. 1875-1877. Enumeratio piscium cubensium. _An. Soc. esp. Hist. nat._, 4: 75-112, pl. 5-8, Apr. 1875 (separate p. 1-38); 113-161, Oct. 1875 (sep. p. 39-88); 5: 131-176, May 1876 (sep. p. 89-134); 177-218, pl. 7-10, Oct. 1876 (sep. p. 135-176); 373-404, pl. 13-14, Dec. 1876 (sep. p. 177-208); 6: 139-154, Oct. 1877 (sep. 209-224).

--. 1876. Poissons de l'ile de Cuba: espèces nouvelles décrites. _Ann. Lyceum nat. Hist. N.Y._, 11(6):58-70, pl. 7-10.

--. 1880. Revisio piscium cubensum. _An. Soc. esp. Hist. nat._, 9:243-261, 5 pl.

--. 1881. Peces. p. 317-350, pl. _In_: J. Gundlach: Apuntes para la fauna puerto-riqueña. 3a parte. _An. Soc. esp. Hist. nat._, 10.

--. 1883. Ictiologia cubana e historia natural de los peces de la isla de Cuba. _Grammicolepis brachiusculus_. (MS not published, deposited at Madrid Nat. Hist. Mus.):641-644. (reference according to Clofnam 2:120).

POGGI, 1881. Guidebook of Canary Isle. Guia de Santa Cruz de Teneriffe.

POINSARD, F. 1969. La pêche au chalut à Pointe-Noire. p. 381-390, 7 fig. _In_: Proc. Symp. Oceanogr. Fish. Resources trop. Atlant., Abidjan, 20-28 october 1966. Paris, 430 p.

POLJAKOV, G. D.; FILIPI, N. D.; BASHO, K. 1958. Peshgit e Shgiperise. (Fish of Albania). Tirane, 286 p.

POLL, M. 1941. Poissons marins (Faune de Belgique). Bruxelles, 452 p., 267 fig., 2 maps n. num. (inset), 2 pl.

--. 1949. Poissons (XIe, XIVe et XVIIe croisières). _In_: Résultats scientifiques des croisières du Navire-Ecole Belge "Mercator". IV. _Mém. Inst. r. Sci. nat. Belg._, (2)33: 173-269, 27 fig., 1 map.

--. 1950. Description de deux poissons percomorphes nouveaux des eaux côtières africaines de l'Atlantique Sud. (1948-1949). _Bull. Inst. r. Sci. nat. Belg._, 26(49):1-14.

--. 1951-1959. Poissons. Résult. scient. Exped. océanogr. belg. Eaux côt. afr. Atlant. Sud (1948-49). 4(1-3B). (1): I. Généralités, II. Sélaciens et Chimères, 1951:1-154, 66 fig., 13 pl. - (2): III. Téléostéens, Malacopterygiens, 1953:1-258, 104 fig., 8 pl. - (3A): IV. Téléostéens Acanthoptérygiens (1ère partie), 1954:1-390, 107 fig., 9 pl. - (3 B): V. Téléostéens Acanthoptérygiens (2e partie), 1959:1-417, 127 fig., 7 pl.

--. 1969. Le prolongement caudal de la vessie hydrostatique des poissons actinopterygiens. _Bull. Acad. r. Belg. Cl. Sci._, (5)55:486-503.

--. 1971. Révision systématique des daurades du genre _Dentex_ de la côte africaine tropicale occidentale et de la Méditerranée. _Mém. Acad. r. Belg. Cl. Sci._, (2)40(1):1-51, 10 fig., 2 map.

POLL, M.; CADENAT, J. 1954. Description de _Erythrocles monodi_ sp. n. de la côte occidentale d'Afrique tropicale. _Bull. Inst. r. Sci. nat. Belg._, 30(22):1-4, fig.

POLL, M.; GOSSE, J. P. 1963. Contribution à l'étude systématique de la faune ichthyologique du Congo Central. _Annls Mus. r. Afr. cent. Ser. 8vo_ (Sci. zool.), (116): 43-101, pl. 1-4.

POLL, M.; ROUX, C. 1955. Description de _Trigla gabonensis_ sp. n. _Bull._

Inst. r. Sci. nat. Belg., 31(43):1-6.

POLLARD, D. A.; PICHOT, P. 1973. The systematic status of the Mediterranean centracanthid fishes of the genus Spicara, and in particular S. chryselis (Valenciennes), as indicated by electrophoretic studies of their eye-lens proteins. J. Fish Biol., 3:59-72.

POLLEN, F. P. L. 1874. Les Pêches à Madagascar et ses dépendances. In: F.P.L. Pollen & D.C. van Dam, 1868-1877. Recherches sur la faune de Madagascar, d'après les découvertes de F.P.L. Pollen et D.C. van Dam. Leyden 4:89 p. (separate).

POPOV, A. M. 1931. K poznaniyu fauny Okhotskogo Morya. Issled. Morei SSSR, 14:121-154.

POPOVA, V. P. 1961. Migrations of sardines in the region of Takoradi. Ryb. Khoz., 37(3):24-30 (in Russian).

POPTA, C. M. L. 1916. Eene vormavariatie van Trygon pastinaca (L.). Zoöl. Meded., Leiden, 2(1):62, pl. 2.

POREBSKI, J.; BIELASZEWSKA, E. 1975. The occurrence of eggs and larvae of the frostfish Lepidopus caudatus (Euph.) in the waters of the S.W. African shelf, based on investigations conducted from the R/V "Prof. Siedlecki". Colln scient. Pap., Madrid JCSEAF, 2:107-112.

PORUMB, J. J. 1968. Le rôle des jeunes Pomatomus saltator L. dans la chaîne trophique de la Mer Noire. Rapp. P.-v. Réun. Commn int. Explor. scient. Mer Méditerr., 19(2):303-305.

POST, A. 1968. Die Paralepididae (Pisces) der 15. Forschungsfahrt des FFS "Walther Herwig". Zool. Anz., 180(1/2):139-146, fig. 1-4.

--. 1968. Ergebnisse der Forschungsreisen des FFS "Walter Herwig" nach Südamerika. V. Notolepis rissoi (Bonaparte, 1841)(Osteichthyes, Iniomi, Paralepididae). Arch. FischWiss., 19(2/3):103-113, fig. 1-2.

--. 1969. Ein Paralepidide (Osteichthyes, Iniomi) aus dem Mittelmeer ohne Zähne und Kiemenreusenfortsätze. Zool. Anz., 183(5/6):355-357, fig. 1.

--. 1969. Ergebnisse der Forschungsreisen des FFS "Walther Herwig" nach Südamerika. VII. Pontosudis quadrimaculata spec. nov. (Osteichthyes, Iniomi, Paralepididae). Arch. FischWiss., 20(1):10-14, fig. 1-4.

--. 1969. Ergebnisse der Forschungsreisen des FFS "Walther Herwig" nach Südameika. VIII. Dolichosudis fuliginosa gen. nov. spec. nov. (Osteichthyes, Iniomi, Paralepididae). Arch. FischWiss., 20(1):15-21, fig. 1-4.

--. 1970. Ergebnisse der Forschungsreisen des FFS "Walther Herwig" nach Südamerika. XIV. Macroparalepis (Osteichthyes, Iniomi, Paralepididae) Arch. FischWiss., 21(3):165-204, fig. 1-14.

--. 1970. Ergebnisse der Forschungsreisen des FFS "Walther Herwig" nach Südamerika. XV. Stemonosudis siliquiventer spec. nov. (Osteichthyes, Iniomi, Paralepididae). Arch. FischWiss., 21(3):205-212, fig. 1-5.

--. 1972. Catalogue of type-specimens and description of lectotypes of the fish-family Paralepididae (Osteichthyes, Myctophoidei). Arch. FischWiss., 23(2):136-165.

--. 1972. Ergebnisse der Forschungsreisen des FFS "Walther Herwig" nach Südamerika. XXII. Pontosudis, ein Synonym von Uncisudis. Arch. Fisch Wiss., 23(1):43-46.

--. 1973. Chromosomes of two fish-species of the genus Diretmus (Osteichthyes, Beryciformes, Diretmidae). p. 103-111, fig. 1-17. In: J.H. Schröder (ed.): Genetics and Mutagenesis of Fish. Berlin, Heidelberg, New York, 356 p.

--. 1973. Ergebnisse der Forschungsreisen des FFS "Walther Herwig" nach Südamerika. XXVII. Macroparalepis Ege, 1933 (Osteichthyes, Myctophoidei, Paralepididae). Revision der Gattung und Beschreibung zweier neuer Arten. Arch. FischWiss., 23(3):202-242, fig. 1-7.

--. 1974. Analyses taxonomiques des Paralepididae, partie des principa-

les proies du germon (<u>Thunnus</u> <u>alalunga</u> Bonnaterre, 1788). <u>ICES</u>, C.
M., 1974/J4, 4 p. (mimeo).

--. 1976. Ergebnisse der Forschungsreisen des FFS "Walther Herwig" nach
Südamerika. XLII. <u>Diretmus</u> Johnson, 1963 (Beryciformes, Berycoidei,
Diretmidae). 2. Morphologie, Entwicklung, Verbreitung. <u>Arch. Fisch
Wiss.</u>, **26**(2/3):87-114, fig. 1-8.

--. 1976. Ergebnisse der Forschungsreisen des FFS "Walther Herwig" nach
Südamerika. XLVI. <u>Diretmus</u> Johnson, 1863 (Beryciformes, Berycoidei,
Diretmidae). 3. Morphologie und Histologie eines subopercularen Drü-
senkomplexes. <u>Arch. FischWiss.</u>, **27**(2):89-96, fig. 1-10.

--. 1981. Diretmidae, Paralepididae, <u>in</u> Fischer, W., Bianchi, G. & W.B.
Scott (ed.), FAO species identification sheets for fishery purposes.
Eastern Central Atlantic; fishing areas 34, 47 (in part). Canada
Funds-in-Trust. Ottawa, Department of Fisheries and Oceans Canada, by
arrangement with the Food and Agriculture Organization of the United
Nations, vol. 1-7: pag. var.

--. 1984. Alepisauridae, <u>in</u> Whitehead, P.J.P., Bauchot, M.L., Hureau,
J.C., Nielsen, J. & E. Tortonese, Fishes of the North-eastern Atlantic and the Mediterranean/Poissons de l'Atlantique du Nord-Est et de
la Méditerranée, Paris, Unesco, 1:494-495, 1 fig., 1 map.

--. 1984. Omosudidae, <u>in</u> Whitehead, P.J.P., Bauchot, M.L., Hureau, J.C.,
Nielsen J. & E. Tortonese, Fishes of the North-eastern Atlantic and
the Mediterranean/Poissons de l'Atlantique du Nord-Est et de la Méditerranée, Paris, Unesco, 1:496-497, 1 fig., 1 map.

--. 1984. Paralepididae, <u>in</u> Whitehead, P.J.P., Bauchot, M.L., Hureau,
J.C., Nielsen, J. & E. Tortonese, Fishes of the North-eastern Atlantic and the Mediterranean/Poissons de l'Atlantique du Nord-Est et de
la Méditerranée, Paris, Unesco, 1:498-508, 14 fig., 15 maps.

--. 1984. Anotopteridae, <u>in</u> Whitehead, P.J.P., Bauchot, M.L., Hureau,
J.C., Nielsen, J. & E. Tortonese, Fishes of the North-eastern Atlantic and the Mediterranean/Poissons de l'Atlantique du Nord-Est et de
la Méditerranée, Paris, Unesco,1:509-510, 1 + 1 fig., 1 map.

--. 1985. Paralepididae, p. 391-400. <u>In</u>: Fischer, W.& Hureau, J.C. (ed.)
FAO Species identification sheets for fishery purposes. Southern
Ocean (Fishing Areas 48, 58 and 88). Prepared and printed with the
support of the Commission for the Conservation of Antarctic Marine
Living Resources (CCAMLR). Rome, vol. 2

--. 1986. Diretmidae, <u>in</u> P.J.P. Whitehead, M.L. Bauchot, J.C. Hureau, J.
Nielsen & E. Tortonese. Fishes of the North-eastern Atlantic and the
Mediterranean/Poissons de l'Atlantique du Nord-Est et de la Méditer-
ranée, Paris, Unesco, 2:743-746, 3 + 3 fig.

--. 1986. Caristiidae, <u>in</u> P.J.P. Whitehead, M.L. Bauchot, J.C. Hureau,
J. Nielsen & E. Tortonese. Fishes of the North-eastern Atlantic and
the Mediterranean/Poissons de l'Atlantique du Nord-Est et de la Méditerranée, Paris, Unesco, 2:747-748, 2 fig.

--. 1986. Anoplogasteridae, <u>in</u> P.J.P. Whitehead, M.L. Bauchot, J.C. Hureau, J. Nielsen & E. Tortonese. Fishes of the North-eastern Atlantic
and the Mediterranean/Poissons de l'Atlantique du Nord-Est et de la
Méditerranée, Paris, Unesco, 2:767-768, 1 fig.

POST, A.; HECHT, T. 1977. Results of the research cruises of the F.R.V.
"Walther Herwig" to South America. XLIX. The otoliths of <u>Diretmus</u>
Johnson, 1863 (Osteichthyes, Beryciformes, Diretmidae). <u>Mitt. hamb.
zool. Mus. Inst.</u>, 74:165-170, fig. 1-3.

POST, A.; QUERO, J.C. 1977. Paralepidides. p. 78-79. <u>In</u>: Maurin, C.,
Bonnet, M. & Quéro, J.C. Poissons des côtes nord-ouest africaines
(Campagnes de la "Thalassa" 1962, 1968, 1971 et 1973). <u>Revue Trav.
Inst. Pêch. marit.</u>, 41(1):5-92, fig. 1-51.

POSTEL, E. 1949. Pêche sur les côtes d'Afrique occidentale. I. Campagne du chalutier "Gérard-Tréca". Avril-Septembre 1949. Gouvern. Gén. Afr. Occidentale fr., Inspect. Gén. Elevage, Dakar, 40 p., 4 + 25 + 9 fig.

--. 1950. Note sur Ethmalosa frimbriata (Bowdich). Bull. Serv. Elev. Industr. Anim. Afr. occid. fr., 3(1):49-59.

--. 1954. Contribution à l'étude des thonidés de l'Atlantique tropical. J. Cons. perm. int. Explor. Mer, 19(3):356-363.

--. 1955. Les faciès bionomiques des côtes de Guinée françaises. Rapp. P.-v. Réun. Cons. perm. int. Explor. Mer, 137:10-13.

--. 1955. Résumé des connaissances acquises sur les clupéidés de l'ouest africain. Rapp. P.-v. Réun. Cons. perm. int. Explor. Mer., 137:14-16.

--. 1955. Contribution à l'étude des thonidés de l'Atlantique tropical (deuxième note). Rapp. P.-v. Réun. Cons. perm. int. Explor. Mer, 137: 31-32.

--. 1955. Le merlu noir (Merluccius senegalensis). Rapp. P.-v. Réun. Cons. perm. int. Explor. Mer., 137:49-52, 3 fig.

--. 1955. Contribution à l'étude de la biologie de quelques Scombridae de l'Atlantique tropico-oriental. Annls Stn océanogr. Salammbô, (10): 1-167.

--. 1955. Considérations biométriques sur la denture des Cybiidés (dents maxillaires). Bull. Stn océanogr. Salammbô, (51):57-67.

--. 1956. Les affinités tropicales de la faune ichthyologique du Golfe de Gabès. Bull. Stn océanogr. Salammbô, 53:50-63, 3 fig.

--. 1956. Essai sur la palomette, Orcynopsis unicolor (Geoffroy Saint-Hilaire, 1809). Bull. Inst. fr. Afr. noire, (A)18(4):1220-1248.

--. 1959-60. Liste commentée des poissons signalés dans l'Atlantique tropico-oriental nord, du Cap Spartel au Cap Roxo, suivie d'un bref aperçu sur leur répartition bathymétique et géographique. Bull. Soc. scient. Bretagne, 34(1/2):129-170(1959); (3/4):241-281(1960).

--. 1960. Rapport sur la sardinelle (Sardinella aurita Valenciennes) (Atlantique africain). FAO Proc. Wld scient. Meet. Biol. Sardines, 2: 55-95.

--. 1963. Exposé synoptique des données biologiques sur la bonite à ventre rayé Katsuwonus pelamis (Linné) 1758 (Atlantique et Méditerranée). FAO Fish Rep., (6)2:515-537.

--. 1963. Exposé synoptique sur la biologie du germon Germo alalunga (Cetti) 1777 (Atlantique oriental). FAO Fish. Rep., (6)2:931-975.

--. 1964. Les thonides d'Afrique du Nord. Proc. Symp. Scombroid fishes. mar. biol. Ass. India, 1(1):211-220.

--. 1969. Synopsis of the biology of the sardine (Sardinella aurita Valenciennes). African Atlantic. UNDP (SF)/FAO, Results Regional Fisheries Survey W. Africa, Abidjan (Rep. 69/8)(unpaged, mimeo).

--. 1973. Scomberomoridae, p. 473-475, in Hureau, J.C. & Th. Monod (ed.). Check-list of the fishes of the north-eastern Atlantic and of the Mediterranean/Catalogue des poissons du nord-est Atlantique et de la Méditerranée (Clofnam). Paris, Unesco, 2 vol.

POSTEL, E.; ROUX, Ch. 1964. Scorpaena folgori, poisson téléostéen nouveau des Iles du Cap Vert. Bull. Mus. natn. Hist. nat., Paris, (2)36 (2):165-171, 2 fig.

POSTOLAKII, A. I. 1965. A new finding of Notusudis lepida (Pisces, Notosudidae) in the waters of West Greenland. Zool. Zh., 44:622-624, fig. 1-2. (in Russian, English summary).

POTTHOFF, T.; RICHARDS, W. J.; UEYANAGI, S. 1980. Developments of Scombrolabrax heterolepis (Pisces, Scombrolabracidae) and comments on familial relationships. Bull. mar. Sci., 30(2):329-357.

PRAET, M. van (see VAN PRAET, M.)

PROBATOV, A. N. 1959. The second scientific-commercial expedition to the coast of West Africa. Ryb. Khoz., 35(1):7-13. (in Russian).

--. 1960. On the morphology of Sardinella aurita terrassae. Trudy balt.

nauchno-issled. Inst. morsk. ryb. Khoz. Okeanogr., 5:137-139. (in Russian).

--. 1960. Second scientific fishing expedition in the Central Atlantic (West Africa) from 31-1-1958 to 18-6-1968 on the fishing vessels "Kazan" and "Alaseja". Trudy atlant. nauchno-issled. Inst. ryb. Khoz. Okeanogr., 5:3-59. (in Russian).

PROBATOV, A. N.; PUPYSHEV, V. A. 1969. Behaviour of Sardinella aurita (Val.) on the western coast of Africa. In: Fishing regions off the west coast of Africa. Trudy atlant. nauchno-issled. Inst. ryb. Khoz. Okeanogr., 22:221-239. (in Russian).

PROCE, M. de. 1822. Sur plusieurs espèces nouvelles de poissons et de crustacés observées. Bull. Soc. philomath., Paris, (3)9:129-134.

PROSVIROV, E. S. 1960. Results of the 1959 commercial expedition in central Atlantic waters. Fat content of sardinelles biological analyses. Ryb. Khoz., 36(8):14-19. (in Russian).

PROSVIROV, E. S.; OSETINSKAYA, I. I. 1962. On the state of the commercial stock of Sardinella aurita in the Takoradi and Dakar regions. Trudy balt. nauchno-issled. Inst. morsk. ryb. Khoz. Okeanogr., 9:81-87. (in Russian).

PROSVIROV, E. S.; RJABIKOV, O. G. 1961. Some problems of the biology and fishing of Sardinella aurita in the region of Takoradi and Dakar. Trudy balt. nauchno-issled. Inst. morsk. ryb. Khoz. Okeanogr., 7:3-16. (in Russian).

--. 1961. On behavioural characteristics of West African sardines in the region of Takoradi. Ryb. Khoz., 37(1):7-8. (in Russian).

--. 1962. Morphological features of Sardinella aurita in the Takoradi and Dakar regions. Trudy balt. nauchno-issled. Inst. morsk. ryb. Khoz. Okeanogr., 8:24-26. (in Russian).

--. 1962. On the question of the condition of the stock of Sardinella aurita in the Takoradi and Dakar regions. Trudy balt. nauchno-issled. Inst. morsk. ryb. Khoz. Okeanogr., 9:81. (in Russian).

PUGH, W. L. 1973. Notes on the occurrence of Diaphus taaningi in the Caribbean Sea. Copeia, 1973(2):362-363.

PUTNAM, F. W. 1871. Note on the occurrence of Euleptorhamphus longirostris on the coast of Massachusetts. Proc. Boston Soc. nat. Hist., 13:236-240.

PUYO, J. 1957. Deux nouveaux poissons du Sénégal. Bull. Soc. Hist. nat. Toulouse, 92:182-188.

PYLAIE, de la. 1835. Recherches, en France, sur les poissons de l'océan pendant les années 1832 et 1833. Congrès scient. France (Poitiers, 1834):524-534.

QUENSEL, C. 1806. Försök att närmare bestämma och naturligare uppställa svenska artena af flunderslägtet. K. svenska Vetensk.-Akad. Handl., 27: 44-56, 203-233.

QUERO, J. C. 1969. Liste des poissons capturés au cours des pêches pélagiques profondes de la "Thalassa" dans l'Atlantique nord-est. ICES, C. M. 1969/L25: 5 p. (mimeo).

--. 1970. Les poissons de la famille des searsidés capturés dans l'Atlantique nord-est. Campagnes du "Président-Théodore Tissier" et de la "Thalassa". Revue Trav. Inst. Pêch. marit., 34(3):261-276, fig. 1-13.

--. 1970. Observations françaises sur les poissons rares en 1968 et 1969. Annls biol., Copenh. (1969), 26:280-282.

--. 1973. Grammicolepididae. p. 351. In: Hureau, J.C. & Th. Monod (ed.). Check-list of the fishes of the north-eastern Atlantic and of the Mediterranean/Catalogue des poissons du nord-est Atlantique et de la Méditerranée (Clofnam). Paris, Unesco, 2 vol.

--. 1973. Les merlus d'Afrique du Sud et leur pêche. Revue Trav. Inst. Pêch. marit., 37(1):117-136, fig. 1-13.

--. 1973. Sur la capture de trois espèces de Gempylides (Pisces, Percomorphi, Trichiuroidea) par les chalutiers de la Rochelle. Annls Soc. Sci. nat. Charente-Marit., 5(5-9):337-343.

--. 1974. Hoplosthetus cadenati sp. nov., Pisces, Beryciformes, Trachichthyidae, poisson nouveau de l'Atlantique est. Revue Trav. Inst. Pêch. marit., 38(1):103-109, 1 fig.

--. 1975. Capture d'une espèce nouvelle pour la faune de l'Atlantique nord-est Photostylus pycnopterus Beebe, 1933. Poissons, Clupéiformes, Alepocéphalidés. Revue Trav. Inst. Pêch. marit., 38(4):435-436, fig. 1.

--. 1975. Rouleina mollis (Koehler, 1896). Poissons, Clupéiformes, Alepocéphalidés, en remplacement de Rouleina attrita (Vaillant, 1888) nomen dubium. Revue Trav. Inst. Pêch. marit., 38(4):437-438.

--. 1975. Talismania grandisquamis sp. nov. Poissons, Clupéiformes, Alepocéphalidés, espèce nouvelle de l'Atlantique centre-est. Revue Trav. Inst. Pêch. marit., 38(4):439-442, 1 fig.

--. 1975. Etude des stades juvéniles de Paraholtbyrnia cyanocephala Krefft, 1967. Poissons, Clupéiformes, Searsiidés. Revue Trav. Inst. Pêch. marit., 38(4):443-447, fig. 1-3.

--. 1975. Cyclothone pseudoacclinidens sp. nov. Poissons, Clupeiformes, Gonostomatidés, espèce nouvelle de l'Atlantique. Revue Trav. Inst. Pêch. marit., 38(4):449-457, fig. 1-5.

--. 1978. Famille des Bathylaconidés :9; Famille des Alepocephalidés :9-17, fig. 4-8; Famille des Searsiidés :17-20, fig. 9-11; Famille des Gonostomatidés :29-37, fig. 17-23; Famille des Sternoptychidés :37-39, fig. 24; Famille des Astronesthidés :39-41, fig. 25; Famille des Chauliodontidés :41-42, fig. 26; Famille des Stomiatidés :42-43; Famille des Melanostomiatidés :43-45; Famille des Malacosteidés :45; Famille des Idiacanthidés :45-46; Famille des Microstomatidés :49; Famille des Bathylagidés :49-50, fig. 31; Famille des Notosudidés :60; Famille des Neoscopélidés :60-62, fig. 41-42; Famille des Myctophidés :62-76, fig. 43-47; Famille des Scopelarchidés :76-77; Famille des Alepisauridés :77; Famille des Omosudidés :77-78; Famille des Cetomimidés :80; Famille des Barbourisiidés :80; Famille des Rondelitiidés :80, fig. 48. In: Maurin, C., Bonnet, M. & J.C. Quéro. Poissons des côtes nord-ouest africaines (Campagnes de la "Thalassa" 1962, 1968, 1971 et 1973). Clupéiformes, Scopéliformes et Cétomimiformes. Revue Trav. Inst. Pêch. marit., 41(1):5-92, fig. 1-51.

--. 1979. Répartition des Zeidés (Pisces, Zeiformes) capturés dans l'Atlantique oriental entre le Cap Vert (15°N) et le Cap Juby (28°N) au cours des campagnes de la "Thalassa". Bull. Off. natn. Pêch. Tunisie, 1978,2(1-2):49-61, fig. 1-8.

--. 1979. Observations d'un photophore maxillaire (MXO), nouveau pour les Searsiidae (Pisces, Clupeiformes), chez Sagamichthys schnakenbecki (Krefft, 1973). Cybium, 3(3):99-100, 2 fig.

--. 1979. Remarques sur le Grammicolepis brachiusculus (Pisces, Zeiformes, Grammicolepididae) observé au port de La Rochelle. Annls Soc. Sci. nat. Charente-Marit., 6(6):573-576, 2 fig.

--. 1981. Ateleopodidae, Cepolidae, Soleidae, Zeidae, in Fischer, W., Bianchi, G. & W.B. Scott (ed.), FAO species identification sheets for fishery purposes. Eastern Central Atlantic; fishing areas 34, 47 (in part). Canada Funds-in-Trust. Ottawa, Department of Fisheries and Oceans Canada, by arrangement with the Food and Agriculture Organization of the United Nations, vol. 1-7: pag. var.

--. 1982. Zeiformes. p. 49-62, fig. 36-48. In: Maurin, C. & J.C. Quéro (ed.). Poissons des côtes nord-ouest africaines (Campagnes de la

"Thalassa" 1962, 1968, 1971 et 1973). <u>Revue</u> <u>Trav.</u> <u>Inst.</u> <u>Pêch.</u> <u>marit.</u>, 45:5-71, 53 fig., 1 pl.

--. 1982. Melamphaeidae and Trachichthyidae. <u>In</u>: Maurin, C. & J.C. Quéro (ed.). Poissons des côtes nord-ouest africaines (Campagnes de la "Thalassa" 1962, 1968, 1971 et 1973). <u>Rev.</u> <u>Trav.</u> <u>Inst.</u> <u>Pêch.</u> <u>marit.</u>, 1981, 45(1):21-26; 29-43.

--. 1984. Odontaspididae, <u>in</u> Whitehead, P.J.P., Bauchot, M.L., Hureau, J.C., Nielsen, J. & E. Tortonese, Fishes of the North-eastern Atlantic and the Mediterranean/Poissons de l'Atlantique du Nord-Est et de la Méditerarnée, Paris, Unesco, 1:78-81, 3 fig., 3 maps.

--. 1984. Mitsukurinidae, <u>in</u> Whitehead, P.J.P., Bauchot, M.L., Hureau, J.C., Nielsen, J. & E. Tortonese, Fishes of the North-eastern Atlantic and the Mediterranean/Poissons de l'Atlantique du Nord-Est et de la Méditerranée, Paris, Unesco, 1:82, 1 fig., 1 map.

--. 1984. Lamnidae, <u>in</u> Whitehead, P.J.P., Bauchot, M.L., Hureau, J.C., Nielsen, J. & E. Tortonese, Fishes of the North-eastern Atlantic and the Mediterranean/Poissons de l'Atlantique du Nord-Est et de la Méditerranée, Paris, Unesco, 1:83-88, 4 fig., 4 maps.

--. 1984. Cetorhinidae, <u>in</u> Whitehead, P.J.P., Bauchot, M.L., Hureau, J.C., Nielsen, J. & E. Tortonese, Fishes of the North-eastern Atlantic and the Mediterranean/Poissons de l'Atlantique du Nord-Est et de la Méditerranée, Paris, Unesco, 1:89-90, 1 fig., 1 map.

--. 1984. Alopiidae, <u>in</u> Whitehead, P.J.P., Bauchot, M.L., Hureau, J.C., Nielsen, J. & E. Tortonese, Fishes of the North-eastern Atlantic and the Mediterranean/Poissons de l'Atlantique du Nord-Est et de la Méditerranée, Paris, Unesco, 1:91-92, 2 fig., 2 maps.

--. 1984. Ginglymostomatidae, <u>in</u> Whitehead, P.J.P., Bauchot, M.L., Hureau, J.C., Nielsen, J. & E. Tortonese, Fishes of the North-eastern Altantic and the Mediterranean/Poissons de l'Atlantique du Nord-Est et de la Méditerranée, Paris, Unesco, 1:93-94, 1 fig., 1 map.

--. 1984. Scyliorhinidae, <u>in</u> Whitehead, P.J.P. Bauchot, M.L., Hureau, J.C., Nielsen, J. & E. Tortonese, Fishes of the North-eastern Atlantic and the Mediterranean/Poissons de l'Atlantique du Nord-Est et de la Méditerranée, Paris, Unesco, 1:95-100, 6 fig., 6 map.

--. 1984. Pseudotriakidae, <u>in</u> Whitehead, P.J.P., Bauchot, M.L., Hureau, J.C., Nielsen, J. & E. Tortonese, Fishes of the North-eastern Atlantic and the Mediterranean/Poissons de l'Atlantique du Nord-Est et de la Méditerranée, Paris, Unesco, 1:101, 1 fig., 1 map.

--. 1984. Sphyrnidae, <u>in</u> Whitehead, P.J.P., Bauchot, M.L., Hureau, J.C., Nielsen, J. & E. Tortonese, Fishes of the North-eastern Atlantic and the Mediterranean/Poissons de l'Atlantique du Nord-Est et de la Méditerranée, Paris, Unesco, 1:122-125, 4 fig., 4 maps.

--. 1984. Oxynotidae, <u>in</u> Whitehead, P.J.P., Bauchot, M.L., Hureau, J.C., Nielsen, J. & E. Tortonese, Fishes of the North-eastern Atlantic and the Mediterranean/Poissons de l'Atlantique du Nord-Est et de la Méditerranée, Paris, Unesco, 1:126-127, 3 maps.

--. 1986. Capture de trois espèces nouvelles pour la faune ichtyologique irlandaise. <u>Cybium</u>, 10(2):203-204, 1 fig.

--. 1986. Zeidae, <u>in</u> P.J.P. Whitehead, M.L. Bauchot, J.C. Hureau, J. Nielsen & E. Tortonese. Fishes of the North-eastern Atlantic and the Mediterranean/Poissons de l'Atlantique du Nord-Est et de la Méditerranée, Paris, Unesco, 2:769-772, 3 fig.

--. 1986. Grammicolepididae, <u>in</u> P.J.P. Whitehead, M.L. Bauchot, J.C. Hureau, J. Nielsen & E. Tortonese. Fishes of the North-eastern Atlantic and the Mediterranean/Poissons de l'Atlantique du Nord-Est et de la Méditerranée, Paris, Unesco, 2:773-774, 1 fig.

--. 1986. Caproidae, <u>in</u> P.J.P. Whitehead, M.L. Bauchot, J.C. Hureau, J. Nielsen & E. Tortonese. Fishes of the North-eastern Atlantic and the

Mediterranean/Poissons de l'Atlantique du Nord-Est et de la Méditerranée, Paris, Unesco, 2:777-779, 2 fig.

QUERO, J. C.; VAYNE, J. J. 1978. Clé de détermination des poissons marins de l'Atlantique du nord-est. I. Agnathes, Sélaciens et Holocéphales. Inst. (scient. tech.) Pêches marit., La Rochelle:1-102.

--. 1979. Clé de détermination des poissons marins de l'Atlantique du
nord-est (entre le 80° et le 30° parallèle nord). II. Pleuronectiformes. Inst. (scient. tech.) Pêch. marit., La Rochelle, 41 p. fig.

QUERO, J.C.; DESOUTTER, M.; LAGARDERE, F. 1986. Soleidae, in P.J.P. Whitehead, M.L. Bauchot, J.C. Hureau, J. Nielsen & E. Tortonese. Fishes
of the North-eastern Atlantic and the Mediterranean/Poissons de l'Atlantique du Nord-Est et de la Méditerranée, Paris, Unesco, 3:1308-
1324, 17 + 3 fig.

--. 1986. Cynoglossidae, in P.J.P. Whitehead, M.L. Bauchot, J.C. Hureau,
J. Nielsen & E. Tortonese. Fishes of the North-eastern Atlantic and
the Mediterranean/Poissons de l'Atlantique du Nord-Est et de la Méditerranée, Paris, Unesco, 3:1325-1328, 4 + 2 fig.

QUERO, J. C.; HARAMBILLET, G.; PERCIER, A.; POUVREAU, B. 1979. Données
sur la faune ichthyologique du Golfe de Gascogne. 2. Captures de
Tripterygion atlanticus (Tripterygiidae, Perciformes). Cybium, 3(2):
97-100.

QUERO, J.C.; MATSUI, T.; ROSENBLATT; R. H.; SAZONOV, Y. I. 1984. Searsiidae, in Whitehead, P.J.P., Bauchot, M.L., Hureau, J.C., Nielsen, J.
& E. Tortonese, Fishes of the North-eastern Atlantic and the Mediter-
ranean/Poissons de l'Atlantique du Nord-Est et de la Méditerranée,
Paris, Unesco, 1:256-267, 11 + 12 fig., 11 maps.

QUIGNARD, J. P. 1965. Redescription de Lappanella fasciata (Cocco, 1833)
= Ctenolabrus iris Valenciennes, 1839 et diagnose du genre Lappanella
Jordan, 1890. (Téléostéens, Perciformes Labridae). Bull. Mus. natn.
Hist. nat., Paris, (2)36(5):578-583.

--. 1966. Recherches sur les Labridés (Poissons, Téléostéens, Perciformes) des côtes européennes. Systématique et biologie. Naturalia mons-
pel. (Zool.), 5:247 p., 87 fig., 79 tab.

--. 1967. Le nid, l'oeuf et la larve du Labridé, Symphodus (Crenilabrus)
mediterraneus (Linné, 1758). Revue Trav. Inst. Pêch. marit., 31(4):
359-362, 6 fig., 1 tab.

--. 1968. Rapport entre la présence d'une "gibbosité frontale" chez les
Labridae (Poissons, Téléostéens) et le parasite Leposphilus labrei
Hesse, 1866 (Copépode, Philichthyidae. Annls Parasit. hum. comp., 43
(1):51-57, 7 fig., 1 tab.

QUIGNARD, J. P.; CAPAPE, C. 1971. Etude du nombre de vertèbres chez
trente et une espèces de Sélaciens des côtes de Tunisie. Bull. Inst.
natn. scient. tech. Océanogr. Pêche Salammbô, 2(2):157-162, 2 tab.

QUIGNARD, J.P.; PRAS, A. 1986. Pomacentridae, in P.J.P. Whitehead, M.L.
Bauchot, J.C. Hureau, J. Nielsen & E. Tortonese. Fishes of the North-
eastern Atlantic and the Mediterranean/Poissons de l'Atlantique du
Nord-Est et de la Méditerranée, Paris, Unesco, 2:916-918, 2 fig.

--. 1986. Labridae, in P.J.P. Whitehead, M.L. Bauchot, J.C. Hureau, J.
Nielsen & E. Tortonese. Fishes of the North-eastern Atlantic and the
Mediterranean/Poissons de l'Atlantique du Nord-Est et de la Méditer-
ranée, Paris, Unesco, 2:919-942, 24 fig.

--. 1986. Scaridae, in P.J.P. Whitehead, M.L. Bauchot, J.C. Hureau, J.
Nielsen & E. Tortonese. Fishes of the North-eastern Atlantic and the
Mediterranean/Poissons de l'Atlantique du Nord-Est et de la Méditer-
ranée, Paris, Unesco, 2:943-944, 1 fig.

--. 1986. Atherinidae, in P.J.P. Whitehead, M.L. Bauchot, J.C. Hureau,
J. Nielsen & E. Tortonese. Fishes of the North-eastern Atlantic and
the Mediterranean/Poissons de l'Atlantique du Nord-Est et de la Médi-

terranée, Paris, Unesco, 3:1207-1210, 4 + 5 fig.
QUOY, J. R. C.; GAIMARD, P. 1824. Zoologie. Poissons:183-401, pl. 43-65. In: L. de Freycinet. Voyage autour du monde... exécuté sur les corvettes de S. M. "L'Uranie" et "La Physicienne" pendant les années 1817, 1818, 1819 et 1820. Paris, 712 p., 96 pl.
--. 1834. Zoologie. Poissons:647-720, 20 pl. In: Voyage de découvertes de "L'Astrolabe", exécuté par ordre du Roi, pendant les années 1826-1829, sous le commandement de M. J. Dumont d'Urville. Zoologie. Paris, 9(3):1-954, pl.

RACKETT, T. 1804. Description of the Esox saurus. Trans. Linn. Soc. Lond., 7:60, fig.
RADCLIFFE, L. 1916. The sharks and rays of Beaufort, North Carolina. Bull. Bur. Fish., Wash., 34:239-284.
--. 1926. "Opah" and "skilligalle" landed at Boston fish pier. Copeia, (151):112.
RAE, B. B. 1951. Exotic fishes. Annls biol., Copenh., 7:38.
RAE, B. B.; LAMONT, J. M. 1965. Rare fish. Scotland. Annls biol., Copenh. (1963), 20:109-110.
RAFFAELE, F. 1888. Le uova galleggianti e le larve dei Teleostei nel Golfo di Napoli. Mitt. zool. Stn Neapel, 8:1-84, 5 pl.
-- 1889. Note intorno alle specie mediterranee del genere Scopelus. Mitt. zool. Stn Neapel, 9:179-186, pl. 7, fig. 1-11.
RAFINESQUE, C. S. 1810. Caratteri di alcuni nuovi generi e nuove specie di animali e piante della Sicilia, con varie osservazioni sopra i medesimi. Palermo, 105 p., 20 pl.
--. 1810. Indice d'ittiologia siciliana ossia catalogo metodico dei nomi latini, italiani, e siciliani dei pesci, che si rinvengono in Sicilia. Messina, 70 p., 2 pl. (Reprint, 1967, Asher-Amsterdam).
--. 1814. Précis des découvertes et travaux somniologiques de Mr. C.S. Rafinesque-Schmaltz entre 1800 et 1814; ou choix raisonné de ses principales découvertes en zoologie et en botanique pour servir d'introduction à ses ouvrages futurs. Palermo, 55 p.
--. 1814. Specchio delle scienze, o giornale enciclopedico di Sicilia, deposito letterario delle moderne cognizioni, scoperte ed osservazione sopra le scienze ed arte... 2 vol. Palermo, 216 p., 2 pl.
--. 1815. Analyse de la nature, ou tableau de l'univers et des corps organisés. Palermo, 224 p.
--. 1818. Second decade of new North American fishes. Am. mon. Mag. crit. Rev., 2(3):204-206.
--. 1820 (April). Fishes of the River Ohio (continued). West. Rev. Misc. Mag., 2(3):169-177.
--. 1820 (December). Ichthyologia Ohiensis, or natural history of the fishes inhabiting the river Ohio and its tributary streams, preceded by a physical description of the Ohio and its branches. Lexington, Kentucky, 90 p. (reprint of 1899 by R.E. Call, Lexington, 175 p. with bio- and bibliographic data).
RAINBOTH, W. J. 1983. An illustrated key to the fishes of the Gambia river and estuary. Univ. Michigan Great Lakes and Marine Waters Center, International Programs Report, (3): 613 p.
RAITT, D. F. S. 1964. A comparison of Gadiculus from Scottish and Mediterranean waters. J. mar. biol. Ass. U.K., 44:693-809.
--. (ed.). 1970. Local french names of West African marine fishes. UNDP (SF)/FAO Results Regional Fisheries Survey West Africa, Abidjan, (Rep. 70/3)(unpaged, mimeo).
RAITT, D. F. S.; LOSSE, G. F.; SCHMIDT, W.; HOFF, I. 1970. Preliminary results of acoustic/fishing surveys in West African coastal waters.

Tech. conference fish finding, purse seining and aimed trawling (May, 1970, Reykjavik), FAO, FII/ FF/70/31: 16 p. (mimeo).

RAJU, S. N. 1974. Three new species of the genus *Monognathus* and the leptocephali of the order Saccopharyngiformes. *Fishery Bull.*, 72(2): 547-562, 4 fig.

RAMALHO, A. 1929. Contribution à l'étude des races de la sardine au Portugal, au Maroc et aux Açores. *Rapp. P.-v. Réun. Cons. perm. int. Explor. Mer.*, 54:46-55.

--. 1931. *Lagocephalus lagocephalus* (L., 1758), *L. laevigatus* (L., 1766). *Faune ichthyol. Atlant. N.*, Cahier 8:406, 407.

RAMSAY, E. P. 1881. Notes on *Histiophorus gladius. Proc. Linn. Soc. N. S. W.*, 5:295-297.

RAMSAY, E. P.; OGILBY, J. D. 1886. On an undescribed *Sciaena* from the New South Wales coast. *Proc. Linn. Soc. N.S.W.*, (2)1:941.

--. 1888. Descriptions of new Australian fishes. *Proc. Linn. Soc. N.S.W.*, (2)2:241-243, 561-564.

RANDALL, J. E. 1956. A revision of the surgeon fish genus *Acanthurus. Pacif. Sci.*, 10(2):159-253, 23 fig., 3 pl.

--. 1963. Review of the hawkfishes (Family Cirrhitidae). *Proc. U.S. natn. Mus.*, 114(3472): 389-451, pl. 1-16.

--. 1963. Notes on the systematics of parrotfishes (Scaridae), with emphasis on sexual dichromatism. *Copeia*, 1963(2):225-237, 4 fig., 3 col. pl.

--. 1963. An analysis of the fish population of artificial and natural reefs in the Virgin Islands. *Caribb. J. Sci.*, 3(1):31-47.

--. 1964. A revision of the filefish genera *Amanses* and *Cantherhines. Copeia*, 1964(2):331-361.

--. 1965. A review of the razorfish genus *Hemipteronotus* (Labridae) of the Atlantic Ocean. *Copeia*, 1965(4):487-501.

--. 1966. The West Indian blenniid fishes of the genus *Hypleurochilus* with the description of a new species. *Proc. biol. Soc. Wash.*, 79:57-72.

--. 1967. Food habits of reef fishes of the West Indies. *Stud. trop. Oceanogr.*, (5):665-847.

--. 1968. Caribbean Reef Fishes. Jersey City, 318 p., 324 fig.

--. 1971. The nominal triggerfishes (Balistidae) *Pachynathus nycteris* and *Oncobalistes erythropterus*, junior synonyms of *Melichthys vidua. Copeia*, 1971(3):462-469.

--. 1981. Acanthuridae, Scaridae, *in* Fischer, W., Bianchi, G. & W.B. Scott (ed.), FAO species identification sheets for fishery purposes. Eastern Central Atlantic; fishing areas 34, 47 (in part). Canada Funds-in-Trust. Ottawa, Department of Fisheries and Oceans Canada, by arrangement with the Food and Agriculture Organization of the United Nations, vol. 1-7: pag. var.

RANDALL, J. E.; ORMOND, R. F. G. 1978. On the Red Sea parrotfishes of Forsskål, *Scarus psittacus* and *S. ferrugineus. Zool. J. Linn. Soc.*, 63(3):239-248, 1 fig.

RANDALL, J. E.; RANDALL, H. A. 1963. The spawning and early development of the Atlantic parrot fish, *Sparisoma rubripinne*, with notes on other scarid and labrid fishes. *Zoologica. N.Y.*, 48(2):49-60, 2 fig., 2 pl.

RANDALL, J. E.; AIDA, K.; HIBIYA, T.; MITSUURA, N.; KAMIYA, H.; HASHIMO-TO, Y. 1971. Grammistin, the skin toxin of soapfishes, and its significance in the classification of the Grammistidae. *Publs Seto mar. biol. Lab.*, 19(2/3):157-190, 23 fig., 4 tab.

RANNOU, M.; GABORIT-REZZOUK, M. 1976. Contribution à l'étude des Bathypteroidae (Pisces: Iniomi) de l'Atlantique et de la Méditerranée. *Bull. Mus. natn Hist. nat.. Paris*, (3)(375)(Zool. 263):453-466.

RANNOU, M.; NIELSEN, J. G.; HUREAU, J. C. 1974. Note sur quelques Aphyonidae de l'Atlantique Nord (Téléostéens, Ophidioidei). Bull. Mus.
natn. Hist. nat., Paris, (3)(247)(Zool. 171):1249-1257, 4 fig.
RANZANI, C. 1818. Descrizione di un pesce il quale appartiene ad un
nuovo genere della famiglia dei Tenioidi del Signor G. Cuvier. Opus-
coli scient., Bologna, 2:133-137, pl. 6.
--. 1838. De novis speciebus piscium; dissertatio prima. Nuovi Annali
Sci. nat. Bologna, 1:80.
--. 1840. De novis speciebus piscium. Dissertatio prima. Novi Comment.
Acad. Sci.. Inst. Bonon. (1840), 4:65-83, pl. 8-13.
--. 1842. De novis speciebus piscium. Dissertatio secunda. Novi Comment.
Acad. Sci. Inst. Bonon., 5:1-21, pl. 1-6.
--. 1842. De nonnullis novis speciebus piscium. Opusculum tertium. Novi
Comment. Acad. Sci. Inst. Bonon., 5:307-338, pl. 23-28.
--. 1843. De novis speciebus piscium. Dissertatio IV. Novi Comment.
Acad. Sci. Inst. Bonon. (1842), 5:339-365, pl. 29-38.
RANZI, S. 1933. Sparidae, Lobotidae. In: Uova, larve e stadi giovanili
di Teleostei. (ed. U. d'Ancona). Fauna Flora Golfo Napoli, 38:332-
383, fig. 257-262, pl. 22-30.
RAO, V. V. 1975. Pogonymus goslinei sp. nov. (Pisces: Callionymidae)
from Ennore Estuary, Madras. Matsya, 1:27-29.
RASQUIN, P. 1958. Ovarian morphology and early embryology of the pediculate fishes Antennarius and Histrio. Bull. Am. Mus. nat. Hist., 114
(4):327-372.
RASS, T. S. 1955. Deep-water fishes of the Kurile-Kamchatka Trench. Trudy Inst. Okeanol., 12: 328-339, 5 fig. (English transl., Vanderbilt
Fdn, Stanford Univ.).
--. 1960. On the geographical distribution of bathypelagic fishes of the
family Myctophidae in the Pacific Ocean. Trudy Inst. Okeanol., 41:
146-152 (in Russian).
--. 1961. Prososcopa stilbia Rass gen. n., sp. n. - a new deep sea fish
from the Indian Ocean. Zool. Zh., 40:1858-1860, fig. 1-2.
--. 1962. Prososcopa stilbia Rass, 1961 equals Xenophthalmichthys danae
Regan, 1925 (Xenophthalmichthyidae, Pisces). Zool. Zh., 41:1578-1579.
--. 1971. Deep-sea fish in the Caribbean Sea and the Gulf of Mexico (the
American Mediterranean region). p. 509-526. Symp. Investig. Resources
Caribb. Sea adj. Reg., Paris.
RASS, T. S.; GRIGORASH, V. A.; SPANOVSKAYA, V. D.; SHCHERBACHEV, Y. N.
1975. Deep-sea bottom fishes caught during 14th cruise of the R/V
"Akademik Kurchatov". Trudy Inst. Okeanol., 100:337-347.
RATHKE, M. H. 1837. Beitrag zur Fauna der Krym. Mém. prés. div. Sav.
Acad. imp. Sci. St. Pétersb., 3:309-354, 10 taf.
RAUTHER, M. 1945. Über die Schwimmblase und die zu ihr in Beziehung
tretenden somatischen Muskeln bei den Triglidae and anderen Scleropa-
rei. Zool. Jb. (Anat.), 69(2):159-250, 50 fig.
RAY, J. 1686. De lingua piscium. Appendix ad historiam naturalem pis-
cium... p. 8-13. In: F. Willoughby, Historia piscium... libri qua-
tuor... Totum opus recognovit, coaptavit, supplevit librum etiam
primum & secundum integras adjecit Johannes Raius. Oxonii, 343 p.,
186 pl.
RAYBURN, R. 1975. Food of deep-sea demersal fishes of the northwestern
Gulf of Mexico. M.S. Thesis, Texas A & M Univ., May 1975, 118 p.
RE, P. 1980. The eggs and newly hatched larvae of Abudefduf luridus
(Cuvier, 1830)(Pisces: Pomacentridae) from the Azores. Archos Mus.
Bocage, (2)7(8):109-116.
REAY, P. J. 1986. Ammodytidae, in P.J.P. Whitehead, M.L. Bauchot, J.C.
Hureau, J. Nielsen & E. Tortonese. Fishes of the North-eastern Atlan-
tic and the Mediterranean/Poissons de l'Atlantique du Nord-Est et de

la Méditerranée, Paris, Unesco, 2:945-950, 8 + 6 fig.

REBERT, J. P. 1979. Un essai d'interprétation de l'influence des conditions du milieu sur les rendements de la pêche industrielle dakaroise de _Sardinella_ _aurita_. Annex II. p. 100-103. _In_: Anon., 1979.

REES, E. I. S. 1963. The batfish, _Dibranchus_ _atlanticus_ Peters, on the Canadian Atlantic slopes. _J._ _Fish._ _Res._ _Bd_ _Can._, 20(6):1513-1517.

REGAN, C. T. 1903. Descriptions de poissons nouveaux faisant partie de la collection du Musée d'Histoire naturelle de Genève. _Revue_ _suisse_ _Zool._, 2:413-418, pl. 13-14.

--. 1903. On the systematic position and classification of the gadoid or anacanthine fishes. _Ann._ _Mag._ _nat._ _Hist._, (7)11(65):459-466.

--. 1908. A collection of fishes from Natal, Zululand and Cape Colony. _Ann._ _Natal_ _Mus._, 1(3):241-255, pl. 37-42.

--. 1908. Report on the marine fishes collected by Mr. J. Stanley Gardiner in the Indian Ocean. _Trans._ _Linn._ _Soc._ _Lond._, (2)12(3):217-255, pl. 23-32.

--. 1908. The systematic position of _Stylophorus_ _caudatus._ _Ann._ _Mag._ _nat._ _Hist._, (8)2:447-449.

--. 1909. A revision of the fishes of the genus _Elops._ _Ann._ _Mag._ _nat._ _Hist._, (8)3:37-40.

--. 1911. The classification of the teleostean fishes of the order Synentognathi. _Ann._ _Mag._ _nat._ _Hist._, (8)7(36):327-335, 4 fig.

--. 1911. The anatomy and classification of the teleostean fishes of the order Iniomi. _Ann._ _Mag._ _nat._ _Hist._, (8)7(13):120-133.

--. 1911. On the systematic position of _Macristium_ _chavesi._ _Ann._ _Mag._ _nat._ _Hist._, (8)7(19):204-205.

--. 1912. XXVIII. The classification of the teleostean fishes of the order Pediculati. _Ann._ _Mag._ _nat._ _Hist._, (8)9:277-289.

--. 1912. Description of two new eels from West Africa, belonging to a new genus and family. _Ann._ _Mag._ _nat._ _Hist._, (8)10:323-324.

--. 1912. The osteology and classification of the teleostean fishes of the order Apodes. _Ann._ _Mag._ _nat._ _Hist._, (8)10:75-86.

--. 1913. A deep-sea angler-fish, _Melanocetus_ _johnsonii._ _Proc._ _zool._ _Soc._ _Lond._, 1913:1096-1097.

--. 1913. A revision of the myxinoids of the genus _Myxine._ _Ann._ _Mag._ _nat._ _Hist._, (8)11:395-398.

--. 1913. The Antarctic fishes of the Scottish National Antarctic Expedition. _Trans._ _R._ _Soc._ _Edinb._, 49(15):229-292, fig. 1-6, 11 pl.

--. 1914. Fishes. _Br._ _Antarct._ _Terra_ _Nova_ _Exped._ _1910_, Zool., 1(1):1-54, 13 pl.

--. 1914. Diagnosis of new marine fishes collected by the British Antarctic "Terra Nova" Expedition. _Ann._ _Mag._ _nat._ _Hist._, (8)13:11-17.

--. 1914. A synopsis of the fishes of the family Macrorhamphosidae. _Ann._ _Mag._ _nat._ _Hist._, (8)13:17-21.

--. 1915. A collection of fishes from Lagos. _Ann._ _Mag._ _nat._ _Hist._, (8) 15:124-130.

--. 1916. Larval and post-larval fishes. _Br._ _Antarct._ _Terra_ _Nova_ _Exped._ _1910_, Zool., 1(4):125-155, 5 fig., 10 pl.

--. 1916. Fishes from Natal, collected by Mr. Romer Robinson. _Ann._ _Durban_ _Mus._, 1(3):167-170.

--. 1916. The British fishes of the subfamily Clupeinae and related species in other seas. _Ann._ _Mag._ _nat._ _Hist._, (8)18:1-19, 3 pl.

--. 1917. A revision of the clupeoid fishes of the genera _Pomolobus,_ _Brevoortia_ and _Dorosoma_, and their allies. _Ann._ _Mag._ _nat._ _Hist._, (8) 19:297-316.

--. 1917. A revision of the clupeoid fishes of the genera _Sardinella,_ _Harengula_, etc. _Ann._ _Mag._ _nat._ _Hist._, (8)19:377-395.

--. 1920. A revision of the flat-fishes (Heterosomata) of Natal. _Ann._

Durban Mus., 2:205-222, 5 fig.

--. 1921. New fishes from deep water off the coast of Natal. Ann. Mag. nat. Hist., (9)7:412-420.

--. 1923. The classification of the stomiatoid fishes. Ann. Mag. nat. Hist., (9)11:612-614.

--. 1924. A young swordfish (Xiphias gladius), with a note on Clupeolabrus. Ann. Mag. nat. Hist., (9)13:224-225, 1 fig.

--. 1924. The morphology of a rare fish, Stylophorus chordatus Shaw; based on specimens collected in the Atlantic by the "Dana" Expeditions 1920-22. Proc. R. Soc., (B)96:193-207, fig. 1-12.

--. 1925. The fishes of the genus Gigantura, A. Brauer; based on specimens collected in the Atlantic by the "Dana" expeditions, 1920-22. Ann. Mag. nat. Hist., (9)15:53-59.

--. 1925. Description of a new salmonoid fish from the Caribbean Sea, obtained by the "Dana" Expeditions, 1920-22. Ann. Mag. nat. Hist., (9)15:59-60.

--. 1925. New ceratioid fishes from the North Atlantic, the Caribbean Sea, and the Gulf of Panama collected by the "Dana". Ann. Mag. nat. Hist., (9)15(89):561-567.

--. 1925. Dwarfed males parasitic on the females in oceanic angler-fishes (Pediculti, Ceratioidea). Proc. R. Soc., (B)97(684):386-400, fig. 1-9, pl. 20.

--. 1926. The Pediculate fishes of the suborder Ceratioidea. Oceanogrl Rep. Danish "Dana" Exped. 1920-22, (2): 45 p., 27 fig., 13 pl.

--. 1930. A ceratioid fish (Caulophryne polynema, sp. n.), female with male, from off Madeira. J. Linn. Soc., Zool., 37:191-195, fig. 1-3.

REGAN, C. T.; TREWAVAS, E. 1929. Description of a new stomiatoid fish of the genus Borostomias from the Atlantic. Ann. Mag. nat. Hist., (10)4: 96.

--. 1929. The fishes of the families Astronesthidae and Chauliodontidae. Oceanogrl Rep. Danish "Dana" Exped. 1920-22, (5):1-39, 25 fig., 7 pl.

--. 1930. The fishes of the families Stomiatidae and Malacosteidae. Oceanogrl Rep. Danish"Dana" Exped. 1920-22, (6):143 p., 138 fig., 14 pl.

--. 1932. Deep-sea anglerfishes (Ceratioidea). Dana Rep., (2):113 p., 172 fig., 10 pl.

REICHENOW, A. 1875. Über die ichthyologischen Sammlungen der deutschen Expedition nach der Loango-Küste. Sber. Ges. naturf. Freunde, Berl., 1875:146-148.

--. 1878. Über von der deutschen Expedition zur Loango-Küste eingegangene Reptilien und Fische. Sber. Ges. naturf. Freunde Berl., 1878: 92-93.

--. 1878. Übersicht der Fische aus Chinchoxo und anderen Gegenden Westafrikas, welche die Afrikanische Gesellschaft dem Berliner Zoologischen Museum übersandt hat. Mber. k. preuss. Akad. Wiss. Berl., (1877):621-624.

REID, E. D. 1935. Two new fishes of the families Dactyloscopidae and Clinidae from Ecuador and the Galapagos. Copeia, 1935:163-166.

--. 1936. Revision of the fishes of the family Microdesmidae, with description of a new species. Proc. U. S. natn. Mus., 84:55-72, pl. 2, fig. 9-12.

--. 1940. A new genus and species of eel from the Puerto Rican Deep. Smithson. misc. Collns, 91(31):1-5, 1 fig.

REID, J. 1849. An account of a specimen of the vaagmaer or Vogmarus islandicus (Trachypterus Bogmarus of Cuvier and Valenciennes) thrown ashore in the Firth of Forth. Ann. Mag. nat. Hist., (2)3(18):456-477, pl. 16.

REINHARDT, J. C. H. 1831. (Bidrag til vor kundskab om Grønlands Fiske), p. 18-24. In: H.C. Örsted (ed.). Overs. K. danske Vidensk. Selsk.

Forh. (1830-31):1-40, (also publ. in K. danske Vidensk. Selsk. natur-vid. math. Afh., 1832, 5:lxxiv-lxxvi).

--. 1837. (Ichtyologiske bidrag til den grønlandske fauna) p. 8-12. In: Örsted, H.C. (ed.). Overs. K. danske Vidensk. Selsk. Forh., (1835-1836):1-32.

--. 1837. Ichthyologiske bidrag til den grønlandske fauna. København, 114 p., 8 col. pl. (also publ. in K. danske Vidensk. Selsk. naturvid. math. Afh., 1838, 7:83-196, 221-228, 8 pl.).

REINSCH, H. H. 1968. Fund von Fluss-Aalen Anguilla anguilla (L.) im Nordatlantik. Arch. FischWiss., 19(1):62-63.

REINTJES, J. W. 1964. Annoted bibliography on biology of menhadens and menhadenlike fishes of the world. Fishery Bull., 63(3):531-549.

REIZER, C. 1968. Division des recherches piscicoles du Sénégal (Richard-Toll). Rapp. a. Cent. tech. for. trop., Nogent-sur-Marne (1967): 207-320.

--. 1971. Contribution à l'étude hydrobiologique du Bas-Sénégal. Pre-mières recommendations d'aménagement halieutique. Rapp. a. Cent. tech. for. trop., Nogent-sur-Marne (1971):1-176.

REPELIN, R. 1972. Etude préliminaire des amphipodes du bol alimentaire de poissons pélagiques provenant de pêches à la longue ligne. Cah. ORSTOM. Oceanogr., 10(1):47-55, 5 fig., 5 tab.

RETZIUS, A. J. 1768. Lampris, en ny fiskslägt beskrifven. K. svenska Vetensk-Akad. Handl., 20:91-100.

--. 1785. Tetrodon mola beskrifven. K. svenska Vetensk-Akad. Handl., 6: 115-121.

RETZIUS, M. G. 1881. Das Gehörorgan der Wirbelthiere. I. Das Gehörorgan der Fische und Amphibien. Stockholm, xi + 222 p., 2 fig., 35 pl.

REYNOLDS, J. D. 1966. The clupeids of Lake Volta and their possible exploitation. Univ. of Ghana, Volta Basin Research Project, tech. Rep. X15: 6 p. (mimeo).

RIBEIRO, A. M. (see MIRANDA RIBEIRO, A. de)

RICHARD, J. 1905. Campagne scientifique du yacht "Princesse Alice" en 1904. Observations sur la faune bathypélagique. Bull. Inst. océa-nogr. Monaco, (41):1-30.

--. 1908. Campagne scientifique de la Princesse-Alice en 1908, liste des stations avec cartes. Bull. Inst. océanogr. Monaco., (126):1-11, 1 map.

RICHARD, P.; SHIRLEY, J. (see PHILIBOSIAN, R.; IMSAND, S.)

RICHARDS, S. W. 1975. Age, growth and food of bluefish (Pomatomus salta-trix) from east-central Long Island Sound from July through November. Trans. Am. Fish. Soc., 105(4):523-525.

RICHARDS, W. J. 1968. Eastern Atlantic Triglidae. Atlantide Rep., (10): 77-114, 13 fig., 1 pl.

--. 1969. Elopoid leptocephali from Angolan waters. Copeia, 1969:515-518, 4 fig., 2 pl.

--. 1981. Triglidae, in Fischer, W., Bianchi, G. & W.B. Scott (ed.), FAO species identification sheets for fishery purposes. Eastern Central Atlantic; fishing areas 34, 47 (in part). Canada Funds-in-Trust. Ottawa, Department of Fisheries and Oceans Canada, by arrangement with the Food and Agriculture Organization of the United Nations, vol. 1-7: pag. var.

RICHARDS, W. J.; KLAWE, W. L. 1972. Indexed bibliography of the eggs and young of tunas and other scombrids (Pisces, Scombridae) 1880-1970. NOAA tech. Rep. NMF SSRF, (652):1-107.

RICHARDS, W. J.; MILLER, P. J. 1981. Peristediidae, in Fischer, W., Bianchi, G. & W.B. Scott (ed.), FAO species identification sheets for fishery purposes. Eastern Central Atlantic; fishing areas 34, 47 (in part). Canada Funds-in-Trust. Ottawa, Department of Fisheries and

Oceans Canada, by arrangement with the Food and Agriculture Organization of the United Nations, vol. 1-7: pag. var.

RICHARDS, W. J.; RANDALL, J. E. 1967. First Atlantic records of the narrow-corseleted frigate mackerel, Auxis thazard. Copeia, 1967(1): 245-247.

RICHARDS, W. J. SAKSENA, V. P. 1978. Notes on the distribution of gurnard fishes (Genus Lepidotrigla, Family Triglidae) off the north-west coast of Africa. Bull. mar. Sci., 28(4):792-794.

--. 1980. Description of larvae and early juveniles of laboratory-reared gray snapper, Lutjanus griseus (Linnaeus) (Pisces, Lutjanidae). Bull. mar. Sci., 30(2):515-521, 6 fig.

--. 1980. Trigla (Trigloporus) africana Smith, a synonym of Chelidonichthys lastoviza (Bonnaterre)(Pisces: Triglidae). Copeia, 1980 (1): 156-157.

RICHARDSON, J. 1839. Description of fishes collected at Port Arthur in Van Diemen's Land. Proc. zool. Soc. Lond., 7:95-100.

--. 1842. Report on the present state of the ichthyology of New Zealand. Rep. Br. Ass. Advmt Sci. (12th meet.), 1842:12-30.

--. 1844. Ichthyology. In: R.B. Hinds (ed.). The zoology of the voyage H.M.S. "Sulphur", under the command of Capt. Sir E. Belcher, during the years 1836-1842. London, 2:71-150, 30 pl.

--. 1844-48. Ichthyology of the voyage of H.M.S. Erebus and Terror. In: Richardson, J. & J.E. Gray (ed.). The zoology of the voyage of H.M.S. "Erebus" and "Terror" under the command of Capt. Sir. J.C. Ross during 1839-43, 2:viii + 139 p., fig., 62 pl.

--. 1846. Description of six fishes. (in appendix). In: J.L. Stokes. Discoveries in Australia; with an account of the coast and rivers explored and surveyed during the voyage of H.M.S. "Beagle" in 1837-43... also a narrative of Capt. O. Stanley's visits to the Arafura sea. 2 vol. London, 1, fig.

--. 1846. Report on the ichthyology of the Seas of China and Japan. Rep. Br. Ass. Advmt Sci. (15th meet.), 1845:187-320 (Reprint, Lochem, 1972).

--. 1848. Fishes. 1-28, 10 pl. In: A. Adams. The zoology of the voyage of H.M.S. "Samarang", under the command of Captain Sir Edward Belcher, during the years 1843-1846. London.

--. 1855. Appendix: account of the Fish. p. 347-376, pl. 23-30. In: E. Belcher. The last of the Arctic voyages; being a narrative of the expedition of the H.M.S. "Assistance", under the command of Captain Sir Edward Belcher, C. B., in search of Sir John Franklin, during the years 1852-53-54, with notes on the natural history by Sir John Richardson... 2 vol. London, 2, 820 p.

--. 1856. Ichthyology. Encycl. Britannica (ed. 8). London, 12:137.

--. 1859. Second supplement to the first volume of "The history of British fishes" by William Yarrell. London, 36 p. (at end of 3rd ed.).

RIEDEL, D. 1960. Sardine production off the Atlantic coast of Europe and Morocco since 1920. FAO Proc. Wld scient. Meet. Biol. Sardines, 3: 877-911.

RIEDL, R. 1963. Fauna und Flora der Adria. Hamburg, 640 p., fig., 8 col. pl., 221 pl.

RIMBAULT, R. 1963. Notes sur certaines espèces ichthyologiques capturées au cours des campagnes de l'Institut des Pêches en Méditerranée (1957-1961). Revue Trav. Inst. Pêch. marit., 27(2):161-176, 23 fig.

RISSO, A. 1810. Ichthyologie de Nice, ou histoire naturelle des poissons du département des Alpes Maritimes. Paris, xxxvi + 388 p., pl. 1-11 (Reprint 1966, Amsterdam.).

--. 1820. Mémoire sur deux nouvelles espèces de poissons du genre Scopelus, observées dans la mer de Nice (Scopelus augustidens. S. balbo).

Memorie R. Accad. Sci. Torino, 25:262-269.

--. 1820. Mémoire sur un nouveau genre de poisson nommé Alepocéphale vivant dans les grandes profondeurs de la mer de Nice. Memorie R. Accad. Sci. Torino., 25:270-272, 4 fig.

--. 1820. Mémoire sur quelques poissons observés dans la mer de Nice (1). J. Phys. Chim. Hist. nat., 91:241-255.

--. 1827. Histoire naturelle des principales productions de l'Europe méridionale et particulièrement de celles des environs de Nice et des Alpes Maritimes. Paris, Strasbourg, 3:xvi + 486 p., 16 pl. [for the publication dates, see Monod, Th. and Hureau, J.C., 1978].

--. 1840. Note sur une nouvelle espèce du genre Gymnetrus. Arch. Naturgesch., 6:13-16.

RITZHAUPT, H. 1961. Ein Beitrag zur Biologie von Sardinella aurita im Seegebiet von Takoradi. Fisch.-Forsch., wiss. Schr. Reihe, 4(1-2): 26-29.

RITZHAUPT, H.; LAMBERT, K. 1968. Weitere Ergbnisse der Such- und Fangreise mit dem Fang- und Verarbeitungsschiff "Rudolf Leonhard" (17. Juli bis 31. Oktober 1966) und der Untersuchungsfahrt mit dem FFS "Ernst Haeckel" (3. Januar bis 5. April 1967) nach Südwestafrika. Fisch.-Forsch., wiss. Schr. Reihe, 6(1):7-28, fig. 1-49.

RIVAS, L. R. 1951. A preliminary review of the western North Atlantic fishes of the family Scombridae. Bull. mar. Sci. Gulf. Caribb., 1(3): 209-230.

--. 1956. The occurrence and taxonomic relationships of the blue marlin (Makaria ampla Poey) in the Pacific Ocean. Bull. mar. Sci. Gulf Caribb., 6(1):59-73, 2 fig.

--. 1961. A review of the tuna fishes of the subgenera Parathunnus and Neothunnus (genus Thunnus). Annali Mus. civ. Stor. nat. Giacomo Doria, 72:126-143.

--. 1974 (MS). Occurrence of the black marlin (Makaira indica) in the Atlantic Ocean.

--. 1975. Synopsis of biological data on blue marlin, Makaira nigricans Lacepède, 1802. NOAA tech. Rep. NMFS SSRF, (675)(3):1-16, 2 fig.

RIVAS, L. R.; WARLEN, S. M. 1967. Systematics and biology of the bonefish, Albula nemoptera (Fowler). Fishery Bull., 66(2):251-258.

RIVERO, L. H. 1936. Some new, rare and little known fishes from Cuba. Proc. Boston Soc. nat. Hist., 41:41-76.

RJABIKOV, O. G.; SARNOV, A. A. 1960. Distribution of the sardinelle (Sardinella aurita) in the Dakar and Takoradi regions, in relation to certain hydrological factors. Trudy balt. nauchno-issled. Inst. morsk. ryb. Khoz. Okeanogr., 5:100-104 (in Russian).

ROBERTSON, D. A. 1975. A key to the planctonic eggs of some New Zealand marine teleosts. Occ. Publs Fish. Res. Div., Wellington, (9):19 p.

--. 1977. Planktonic eggs of the lanternfish Lampanyctodes hectoris (family Myctophidae). Deep Sea Res., 24(9):849-852.

ROBERTSON, D. A.; GRIMES, P. J.; McMILLAN, P. J. 1984. Orange roughy on Chatham rise; results of trawl survey, August-September 1982. Occ. Publs Fish. Res. Div., Wellington, (46):1-27.

ROBINS, C. H. 1971. The comparative morphology of the synaphobranchid eels of the Straits of Florida. Proc. Acad. nat. Sci. Philad., 123 (7):153-204, 12 fig.

ROBINS, C. H.; ROBINS, C. R. 1970. The eel family Dysommidae (including the Dysomminidae and Nettodaridae), its osteology and composition, including a new genus and species. Proc. Acad. nat. Sci. Philad., 122 (6):293-335, 10 fig.

--. 1976. New genera and species of dysommine and synaphobranchine eels (Synaphobranchidae) with an analysis of the Dysomminae. Proc. Acad. nat. Sci. Philad., 127(18):249-280, 9 fig.

ROBINS, C. R. 1957. A study of the bluefin tuna populations from the eastern and western North Atlantic based on an analysis of fin-ray and gill-raker counts. Charles F. Johnson oceanic gamefish Invest., Univ. Miami, 57-13:1-7 (mimeo).

--. 1958. Check list of the game and commercial fishes of Florida and the West Indies with a standardization of common names. Educ. Ser. Fla St. Bd Conserv., (12):1-46.

--. 1966. The R/V Pillsbury deep-sea biological expedition to the Gulf of Guinea, 1964-65. 6. Microdesmus aethiopicus at Fernando Poo. Stud. trop. Oceanogr., (4)(1):125-127.

--. 1970. The R/V Pillsbury deep-sea biological expedition to the Gulf of Guinea, 1964-65. 15. A new goby from Annobon Island, Gulf of Guinea. Stud. trop. Oceanogr., (4)(2):281-284.

--. 1975. Synopsis of biological data on the longbill spearfish, Tetrapturus pfluegeri Robins and de Sylva. NOAA tech. Rep. NMFS SSRF, (675) (3):28-38, 4 fig.

ROBINS, C. R.; SYLVA, D.P. de. 1961. Description and relationships of the longbill spearfish, Tetrapturus belone, based on western North Atlantic specimens. Bull. mar. Sci. Gulf Caribb., 10(4):383-413, 5 fig.

--. 1963. A new western Atlantic spearfish, Tetrapturus pfluegeri, with a redescription of the Mediterranean spearfish Tetrapturus belone. Bull. mar. Sci. Gulf Caribb., 13(1):84-122, 5 fig.

ROBINS, C. R.; LACHNER, E. A. 1966. The status of Ctenogobius Gill (Pisces: Gobiidae). Copeia, 1966:867-869.

ROBINS, C. R.; MANNING, R. B. 1958. The status and distribution of the fishes of the family Microdesmidae in the western Atlantic. J. Wash. Acad. Sci., 48(9):301-304.

ROBINS, C. R.; NIELSEN, J. G. 1970. Snyderidia bothrops, a new tropical, amphi-Atlantic species (Pisces, Carapidae). Stud. trop. Oceanogr., (4)(2):285-293.

ROBINS, C. R.; ROBINS, C. H. 1966. The R/V Pillsbury deep-sea biological expedition to the Gulf of Guinea, 1964-65. 5: Xenoconger olokun, a new xenocongrid eel from the Gulf of Guinea. Stud. trop. Oceanogr., (4)(1):117-124, 3 fig.

ROBINS, C. R.; STARCK II, W. A. 1961. Materials for a revision of Serranus and related fish genera. Proc. Acad. nat. Sci. Philad., 113(11): 259-314.

ROBINS, C. R.; BAILEY, R. M.; BOND, C. E.; BROOKER, J. R.; LACHNER, E. A.; LEA, R. N.; SCOTT, W. B. 1980. A list of common and scientific names of fishes from the United States and Canada (4th ed.). Spec. Publ. Am. Fish. Soc., (12):1-174.

ROCHEBRUNE, A. T. de. 1880. Description de quelques nouvelles espèces de poissons propres à la Sénégambie. Bull. Soc. philomath. Paris, (7)4: 159-169.

--. 1882. Faune de la Sénégambie. Poissons. Act. Soc. linn. Bordeaux, 36(4)6:23-190, col. pl. 1-6 (also separate, Paris, Bordeaux, 1883:166 p., 6 pl.).

RODRIGUEZ, A. V. 1980. Sobre la biología y pesca de la castañeta (Brama brama). Investigación pesq., 44(2):241-252.

RODRIGUEZ-RODA, J. 1957. Avance al estudio de la sardina (S. pilchardus Walb.) de Barbate y Larache y algunas consideraciones generales. Inst. Investigación pesq., 3rd Reun. Prod. Pesq., p. 29-31 (mimeo).

--. 1958. Estudio comparativo del crecimiento de las sardinas, Sardina pilchardus Walbaum de Barbate (costa sudatlántica española) y Larache (costa atlántica de Marruecos). Investigación pesq., 13:3-36.

--. 1959. Sardinas parasitadas por el copépodo Perodema cylindricum Heller, en las zonas de Barbate y Laranche. Investigación pesq., 15:

125-129.

--. 1960. Estudios sobre la Sardina pilchardus Walb. de Barbate (Costa sudatlàntica española) y Larache (Costa marroqui). Boln R. Soc. esp. Hist. nat., 58:363-370.

--. 1960. Nombres vulgares y cientificos de las principales especies comerciales de peces de la regiòn sudatlàntica española. Investigación pesq., 17:109-125.

--. 1961. Captura de un Harriotta raleighana Goode y Bean, 1894 en aguas de Cabo Blanco (Africa Occidental). Investigación pesq., 20:79-82, fig. 1-2.

--. 1966. Estudio de la bacoreta, Euthynnus alleteratus (Raf.), bonito, Sarda sarda (Bloch) y melva Auxis thazard (Lac.), capturados por las almadrabas españolas. Investigación pesq., 30:247-292.

--. 1979. Edad y crecimiento de la bacoreta, Euthynnus alleteratus (Raf.) de la costa sudatlàntica de España. Investigación pesq., 43 (3):591-599.

RÖHL, E. 1942. Fauna descriptiva de Venezuela. Peces, p. 353-413, fig. 172-230. Caracas, 432 p., 230 fig.

ROFEN, R. R. 1959. The whale-fishes: families Cetomimidae, Barbourisiidae and Rondeletiidae (Order Cetunculi). Galathea Rep., 1:255-260, 4 fig.

--. 1960. Reidentification of the bathypelagic fishes of the family Paralepididae collected by the Snellius Expedition in the East Indies. Temminckia, 10:200-208, 6 pl., 2 fig.

--. 1963. Diagnoses of new species of alepisauroid fishes of the family Evermannellidae. Aquatica, (1):1-2.

--. 1963. Diagnoses of new genera and species of alepisauroid fishes of the family Paralepididae. Aquatica, (2):1-7, 1 fig.

--. 1963. Diagnoses of new species and a new genus of alepisauroid fishes of the family Scopelarchidae. Aquatica, (3):1-4.

--. 1966. Family Paralepididae. In: Fishes of the western North Atlantic. Mem. Sears Fdn mar. Res., 1(5):205-461, fig. 55-162.

--. 1966. Family Omosudidae. In: Fishes of the western North Atlantic. Mem. Sears Fdn mar. Res., 1(5):462-481, fig. 163-172.

--. 1966. Family Anotopteridae. In: Fishes of the western North Atlantic. Mem. Sears Fdn mar. Res., 1(5):498-510, fig. 177-184.

--. 1966. Family Evermannellidae. In: Fishes of the western North Atlantic. Mem. Sears Fdn mar. Res., 1(5):511-565, fig. 185-204.

--. 1966. Family Scopelarchidae.In: Fishes of the western North Atlantic. Mem. Sears Fdn mar. Res., 1(5):566-602, fig. 205-218.

ROGNES, K. 1973. Head skeleton and jaw mechanism in Labrinae (Teleostei: Labridae) from Norwegian waters. Årbok Univ. Bergen (mat.-naturv. Ser.), (1971-4):1-149, 3 + 108 fig.

ROHR, B. A. 1968. The searsiid fish, Platytroctes apus, from the western tropical Atlantic and Gulf of Mexico. Copeia, 1968:624-625.

ROLLEFSEN, G. (ed.) et al. 1960. Havet og våre Fisker. Bergen, 1:156 p. + 111 p., with num. fig. and pl. (containing: RASMUSSEN, Th.; DANNEVIG, G. Norges saltvannsfisker i plansjer og tekst (med et tillegg om matnyttige skjell)).

ROMAN, B. 1970. Peces de Rio Muni, Guinea Ecuatorial (aguas dulces y salobres). Barcelona, 295 p., 118 fig.

RONDELET, G. 1554. Libri de piscibus marinis, in quibus verae piscium effigies expressae sunt... Lugduni, 1-583, 23 pl.

ROSA, H., Jr. 1950. Scientific and common names applied to tunas, mackerels and spear fishes of the world with notes on their geographic distribution. Washington, 235 p.

ROSEN, D. E. 1971. The Macristiidae, a ctenothrissiform family based on juvenile and larval scopelomorph fishes. Am. Mus. Novit., (2452):1-

22, 13 fig., 2 tab.

--. 1973. Interrelationships of higher euteleostean fishes, p. 397-513, 129 fig. In: Interrelationships of fishes (P.H. Greenwood, R.S. Rosen & C. Patterson, ed.). Zool. J. Linn. Soc., 53, suppl. (1): xvi + 1-536.

ROSEN, D. E.; GREENWOOD, P. H. 1976. A fourth neotropical species of synbranchid eel and the phylogeny and systematics of synbranchiform fishes. Bull. Am. Mus. nat. Hist., 157:1-70, 67 fig., 4 tab.

ROSENBLATT, R. H.; BUTLER, J. E. 1977; The ribbonfish genus Desmodema with the description of a new species (Pisces, Trachipteridae). Fishery Bull., 75(4):843-856, 8 fig.

ROSENTHAL, H. 1970. Anfütterung und Wachstum der Larven und Jungfische des Hornhechtes Belone belone. Helgoländer wiss. Meeresunters, 21(3): 320-332.

ROSENTHAL, H.; FONDS, M. 1973. Biological observations during rearing experiments with the garfish Belone belone. Mar. Biol., Berl., 21(1):203-218.

ROSS, S. W.; LINK, G. W., Jr.; MacPHERSON, K. A. 1981. New records of marine fishes from the Carolinas, with notes on additional species. Brimleyana, (6):61-72, fig.

ROSSIGNOL, M. 1952. Note sur les muges des côtes marocaines. Annls biol., Copenh., 8:89-90.

--. 1955. Premières observations sur la biologie des sardinelles dans la région de Pointe-Noire (Sardinella aurita Val., Sardinella eba Val.). Rapp. P.-v. Réun. Cons. perm. int. Explor. Mer., 137:17-21.

--. 1955. Les sardinelles de la région de Pointe-Noire, perspectives économiques qu'elles offrent. Docums scient. provis. Centre Rech. océanogr. Pointe-Noire, (n.s)(33):12 p. (mimeo; also Sci. Pêche, 3(3-6).

--. 1959. Contribution à l'étude biologique des sardinelles. Etude de la variabilité d'un caractère méristique: le nombre de branchiospines. Revue Trav. Inst. Pêch. marit., 23(2):211-223.

ROSSIGNOL, M.; BLACHE, J. 1961. Sur le statut spécifique de deux poissons pélagiques du Golfe de Guinée, Anchoviella guineensis nov. sp. (Clupeiformes, Engraulidae), Atherina lopesiana nov. sp. (Mugiliformes, Atherinidae). Bull. Mus. natn. Hist. nat., Paris, (2)33:285-293, 2 fig.

--. 1961. Sur un poisson Stromateidae nouveau du Golfe de Guinée, Psenes benardi nov. sp. Bull. Mus. natn. Hist. nat., Paris, (2)33(4):384-386, 1 fig.

ROSSIGNOL, M.; BLACHE, J.; REPELIN, R. 1962. Fonds de pêches le long des côtes de la République du Gabon. [Cah. ORSTOM] Centre océanogr. Pointe Noire, [1]:1-15, 2 map.

ROST, H. 1954. Nesiarchus nasutus Johnson (Gempylidae) a fish new to the Norwegian fauna. Astarte, 7:1-8.

ROUGHLEY, T. C. 1951. Fish and fisheries of Australia. Sydney, London, 343 p., frontispiece & 80 pl. (1966. Rev'd reprint. Sydney, Melbourne & London, 328 p., 88 pl.).

ROULE, L. 1902. Atherina riqueti, nov. sp., nouvelle espèce d'athérine vivant dans les eaux douces; Zool. Anz., 25:262-267, 4 fig.

--. 1913. Notice préliminaire sur Grimaldichthys profundissimus nov. gen., nov. sp., poisson abyssal recueilli à 6035 mètres de profondeur dans l'Ocean Atlantique par S.A.S. le Prince de Monaco. Bull. Inst. océanogr. Monaco, (261):1-8, fig.

--. 1914. Sur les poissons abyssaux appartenant à la famille des eurypharyngidés. C. r. hebd. Séanc. Acad. Sci., Paris, 158:1821-1823.

--. 1915. Sur un nouveau genre de poisson apodes, et sur quelques particularités de la biologie de ces êtres. C. r. hebd. Séanc. Acad. Sci.,

Paris, **160**:283-284.

--. 1916. Considération sur les deux espèces abyssales du genre Solea dans l'Atlantique paléarctique et sur le sous-genre nouveau Bathysolea. Bull. Mus. natn. Hist. nat., Paris, 22:8-10.

--. 1916. Notice préliminaire sur quelques espèces nouvelles ou rares des poissons provenant des croisières de S.A.S. le Prince de Monaco. Bull. Inst. océanogr. Monaco, (320):1-32.

--. 1919. Poissons provenant des campagnes du yacht "Princesse Alice" (1891-1913) et du yacht "Hirondelle II" (1914). Résult. Camp. scient. Prince Albert I, 52:1-191, 7 pl.

--. 1922. Sur un genre de poisson abyssal japonais très rare, nouvellement retrouvé dans l'océan atlantique nord-africain. C. r. hebd. Séanc. Acad. Sci., Paris, **174**:640-642.

--. 1922. Description de Scombrolabrax heterolepis nov. gen., nov. sp., poisson abyssal nouveau de l'île de Madère. Bull. Inst. océanogr. Monaco, (408):1-8, 4 fig.

--. 1923. Un cas probable de mutation chez les poissons. C. r. Séanc. Soc. Biol., **89**:1027-1028.

--. 1924. Etude sur l'ontogenèse et la croissance avec hypermétamorphose de Luvarus imperialis Raf. (poisson rapporté à l'ordre des Scombriformes). Annls Inst. océanogr., Monaco, (n.s.) 1:119-157, 18 fig.

--. 1925. Les poissons des eaux douces de la France. Paris, xvi + 228 p., fig., 37 pl.

--. 1929. Description de poissons abyssaux provenant de l'île Madère et des parages du Maroc. Bull. Inst. océanogr. Monaco, (546):1-19, 3 fig.

--. 1934. Les poissons et le monde vivant des eaux. L'abime des grands fonds marins. Paris, 7:1-322, fig., pl.

--. 1935. Nouvelles observations sur quelques espèces de poissons abyssaux provenant de Madère. Bull. Inst. océanogr. Monaco, (674):1-6.

--. 1935. Observation sur la présence dans le golfe de Gascogne d'un poisson abyssal de l'Atlantique moyen (Nesiarchus nasutus Johnson), espèce nouvelle pour la faune française. Bull. Soc. zool. Fr., **60**: 439-441.

ROULE, L.; ANGEL, F. 1921. Notice préliminaire (2e) sur les larves et les alevins de poissons recueillis par S.A.S. le Prince de Monaco dans ses croisières de 1896 à 1901. Bull. Inst. océanogr. Monaco, (397):1-8.

--. 1924. Notice préliminaire (4e et dernière) sur les larves et les alevins de Poissons recueillis par S.A.S le Prince Albert I de Monaco dans ses croisières (années 1905 à 1915). Bull. Inst. océanogr. Monaco, (451):1-7.

--. 1930. Larves et alevins de poissons provenant des croisières du Prince Albert I de Monaco. Résult. Camp. scient. Prince Albert I, **79**:148 p., 6 pl. (165 fig.).

--. 1931. Observations et rectifications concernant divers poissons recueillis par S.A.S. le Prince Albert I de Monaco au cours des campagnes de 1911 à 1914. Bull. Inst. océanogr. Monaco, (581):1-8.

--. 1933. Poissons provenant des campagnes du Prince Albert I de Monaco. Résult. Camp. scient. Prince Albert I, **86**:1-115, 4 pl.

ROULE, L.; BERTIN, L. 1924. Notice préliminaire sur la collection des nemichthydés recueillie par l'expédition du Dana (1921-1922), suivie de considérations sur la classification de cette section des poissons Apodes. Bull. Mus. natn. Hist. nat. Paris, **30**:61-67.

--. 1929. Les poissons apodes appartenant au sous-ordre des Nemichthydiformes. Oceanogrl Rep. "Dana" Exped. 1920-1922, (4):1-113, 57 fig., 9 pl.

ROUX, C. 1950. Généralités sur le littoral de l'A.E.F. Observations biologiques sur quelques espèces de poissons. Congr. Pêch. Pêcheries Un. fr. Outre-Mer, Marseille, (2):188-192.

--. 1954. Description de deux espèces nouvelles de poissons des côtes d'Afrique Equatoriale Française, Dentex polli et Scorpaena gaillardae. Bull. Mus. natn. Hist. nat., Paris, (2)26(4):486-472, 1 fig.

--. 1955. Activités du Centre d'Océanographie de Pointe-Noire de 1950 à 1953. Rapp. P.-v. Réun. Cons. perm. int. Explor. Mer., 137:66-68.

--. 1957. Poissons marins. Première partie. p. 137-253. In: J. Collignon, M. Rossignol & C. Roux. Mollusques, crustacés, poissons marins des côtes d'A.E.F. en collection au Centre d'océanographie de l'Institut d'Etudes Centrafricaines de Pointe-Noire. Paris, 369 p.

--. 1960. Note sur le tarpon (Megalops atlanticus C. et V.) des côtes de la République du Congo. Bull. Mus. natn. Hist. nat., Paris, (2)32(4): 314-319.

--. 1966. Une nouvelle espèce de poisson apode de la famille des Ophichthidae: Caecula (Sphagebranchus) monodi sp. nov. Bull. Mus. natn. Hist. nat., Paris, (2)37(4):593-598.

--. 1971. Révision des poissons marins de la famille des Batrachoididae de la côte occidentale africaine. Bull. Mus. natn. Hist. nat., Paris, (2)42(4):626-643, fig. 1-9.

--. 1971. Les poissons de la famille des Batrachoididae (Téléostéens marins) de la campagne de chalutage du Golfe de Guinée. Bull. Mus. natn. Hist. nat., Paris, (2)42(5):849-852.

--. 1971. Comparaison entre Batrachoides liberiensis (Steindachner, 1867) et Batrachoides pacifici (Günther, 1861)(Poissons téléostéens, Batrachoididae). Bull. Mus. natn. Hist. nat., Paris, (3)(6)(Zool. 6):345-348.

--. 1981. Batrachoididae, Gerreidae, Lethrinidae, Pomadasyidae, Trachinidae, Uranoscopidae, in Fischer, W., G. Bianchi & W.B. Scott (ed.), FAO species identification sheets for fishery purposes. Eastern Central Atlantic (Fishing Areas 34 and 47 in part). Dept. Fish. Oceans Canada, Ottawa & FAO, Rome, vol. 1-7, pag. var.

--. 1984. Squatinidae, in Whitehead, P.J.P., Bauchot, M.L., Hureau, J.C., Nielsen, J. & E. Tortonese. Fishes of the north-eastern Atlantic and the Mediterranean/Poissons de l'Atlantique du Nord-Est et de la Méditerranée, Paris, Unesco, 1:148-150, 3 fig., 3 maps.

--. 1986. Lobotidae, in Whitehead P.J.P., M.L. Bauchot, J.C. Hureau, J. Nielsen & E. Tortonese. Fishes of the North-eastern Atlantic and the Mediterranean/Poissons de l'Atlantique du Nord-Est et de la Méditerranée, Paris, Unesco, 2:854-855, 1 fig.

--. 1986. Leiognathidae, in Whitehead P.J.P., M.L. Bauchot, J.C. Hureau, J. Nielsen & E. Tortonese. Fishes of the North-eastern Atlantic and the Mediterranean/Poissons de l'Atlantique du Nord-Est et de la Méditerranée, Paris, Unesco, 2:856-857, 1 fig.

--. 1986. Dactylopteridae, in Whitehead P.J.P., M.L. Bauchot, J.C. Hureau, J. Nielsen & E. Tortonese. Fishes of the North-eastern Atlantic and the Mediterranean/Poissons de l'Atlantique du Nord-Est et de la Méditerranée, Paris, Unesco, 3:1284-1285, 1 fig.

--. 1986. Batrachoididae, in Whitehead P.J.P., M.L. Bauchot, J.C. Hureau, J. Nielsen & E. Tortonese. Fishes of the North-eastern Atlantic and the Mediterranean/Poissons de l'Atlantique du Nord-Est et de la Méditerranée, Paris, Unesco, 3:1360-1361, 1 fig.

ROUX, C.; COLLIGNON, J. 1950. Clef de détermination des principaux poissons marins des côtes de l'A.E.F. Bull. Inst. Etud. centrafr. (num. hors série), 27 p.

--. 1954. Description d'une nouvelle espèce de poisson, de la famille des Serranidae, observée sur les côtes de l'Afrique Equatoriale Française: Promicrops ditobo. Bull. Mus. natn. Hist. nat., Paris, 26(4):

473-475.

--. 1957. Poissons marins. Deuxième partie. Clef pour la détermination des principaux poissons marins fréquentant les côtes de l'A.E.F. p. 255-368, 121 fig., 3 phot. non num. In: J. Collignon, M. Rossignol, C. Roux. Mollusques, crustacés, poissons marins des côtes d'A.E.F. en collection au Centre d'Océanographie de l'Institut d'Etudes Centrafricaines de Pointe-Noire. Paris, 369 p.

ROUX, C.; HUREAU, J.C. 1969. Campagne d'essais du "Jean Charcot" (3-8 décembre 1968). 2. Liste des Poissons. Bull. Mus. natn Hist. nat., Paris, (2)41(4):1021-1026.

ROUX, C.; WHITLEY, G. P. 1971. Perulibatrachus, nouveau nom de genre de poissons téléostéens de la famille des Batrachoididae, en remplacement de Parabatrachus Roux, 1970. Bull. Mus. natn. Hist. nat., Paris, (3)(6)(Zool. 6):349-350.

ROYCE, W. F. 1957. Observations on the spear fishes of the central Pacific. Fishery Bull. Fish. Wildl. Serv. U. S., 57(124):497-554, 27 fig.

--. 1965. A morphometric study of yellowfin tuna Thunnus albacores (Bonnaterre). Fishery Bull., 63(2):395-443.

RUBIES, P.; PALOMERA, I. 1977. Abundance and distribution of sardine eggs and larvae off northwest Africa, April-May 1973. ICES, C.M., 1977/L7:8 p. (mimeo).

RUBINOFF, I. 1966. Gymnothorax galetae, a new moray eel from the Atlantic coast of Panama. Breviora, (240):1-4, fig. 1.

RUCABADO, J. A.; BAS, C. 1984. Resultados de las expediciones oceanografico-pesqueras "Benguela I" (1979) y "Benguela II" (1980) realizadas en el Atlantico Sudoriental (Namibia). Datos informativos, Inst. Investnes pesq., Barcelona, 9:1-248.

RUCABADO, J.; LLORIS, D. 1977. Sobre la presencia de Diplodus senegalensis Cadenat, 1964, en el area de afloramiento del NW de Africa (23°-26° lat. N.). Result. Exped. cient. B/O Cornide Saavedra, (6):291-296, 4 fig.

--. 1986. Denominaciones vernaculas en castellano de la ictiofauna demersal de las costas de Namibia (expediciones Benguela I a Benguela IV). Colln scient. Pap., Madrid, ICSEAF, 13(2):197-204.

RUDOMETKINA, G. P. 1980. Larval Gonostoma atlanticum Norman, 1930 (Gonostomatidae) from the tropical Atlantic Ocean. Vop. Ikhtiol., 20(6): 938-940.

--. 1981. Postembryonic development of Diplophos taenia Günther, 1873 (Gonostomatidae) from the Atlantic Ocean. Vop. Ikhtiol., 21(4):711-718.

RUIVO, L. N. (see NUNES-RUIVO, L.)

RULEV, N. N. 1959. Techno-chemical characteristics of sardines caught in waters off the West African coast. Ryb. Khoz., 36(6):62-68 (in Russian).

RÜPPELL, E. 1828. Atlas zu der Reise im nördlichen Afrika von Eduard Rüppell. Zoologie 4. Fische des Rothen Meeres. Frankfurt am Main, 141 + 3 p., 35 col. pl.

--. 1835-1838. Neue Wirbeltiere zu der Fauna von Abyssinien gehörig. Fische des Rothen Meeres. Frankfurt am Main, 148 p., 33 pl.

RUSSELL, F. S. 1976. The eggs and planktonic stages of British marine fishes. London, i-xvi + 524 p., 137 fig., 6 tab.

RUSSELL, P. 1803. Descriptions and figures of two hundred fishes, collected at Vizagapatam on the coast of Coromandel. London, 1:vii + (1) + 78 + (4) p., fig. 1-100; 2:85 + (4) p., fig. 101-208.

RUTTER, C. M. 1897. A collection of fishes obtained in Swatow, China, by Miss Adele M. Fielde. Proc. Acad. nat. Sci. Philad., 1897:56-90.

SABROSKY, C. W., 1966. Comments on the type-species of _Sciaena_ Linnaeus 1758. Z.N.(S.) 850. _Bull. zool. Nom._, 23:3.

SADOWSKI, S. 1960. Sardinela wod tropikalnych Afryki zachodniej. _Gospod. rybna_, 12(6).

SADOWSKY, V. 1970. First record of broad-snouted seven-gilled shark from Cananeia, coast of Brazil. _Bolm Inst. Oceanogr. S. Paulo_, 18:33-35.

SAEMUNDSSON, B. 1908. Oversigt over Islands fiske. _Skr. Kommn Havunders., Kbh._, 5:1-140, 4 fig., 1 pl.

--. 1922. Zoologiske Meddelelser fra Island. XIV. _Vidensk. Meddr dansk naturh. Foren._, 74:159-201, pl. 3-5.

--. 1926. Fiskarnir (Pisces Islandiae). _Islensk Dýr_, 1: xvi + 528 p., 266 fig.

--. 1927. Synopsis of the fishes of Iceland. Reykjavik, 66 p.

--. 1932. _Centrophorus calceus. C. squamosus. C. jonsonii._ Faune ichthyol. _Atlant. N._, Cahier 10:24-26.

--. 1932. _Regalecus glesne._ Faune ichthyol. _Atlant. N._, Cahier 10:198.

--. 1939. Zoologiske Meddelelser fra Island XVII. 6 Fiske, nye for Island og Tilføjelser om andre, tidligere kendte. _Vidensk. Meddr dansk naturh. Foren._, 102:183-212.

--. 1949. Marine Pisces. _Zoology Iceland_, 4(72):1-150.

SAGER, G. 1983. Jährliche Waschstumsvariationen der Grundel _Lesueurigobius friesii_ vor Schottland nach Daten von Gibson und Essi (1978). _Anat. Anz._, 154:245-254.

SAKAMOTO, K. 1932. Two new genera and species of cottoid fishes from Japan. _J. imp. Fish. Inst., Tokyo_, 27:1-6, 2 fig.

SALDANHA, L. 1965. Sobre três espécies de teleósteos (Nettastomidae e Notacanthidae) novas para a costa de Portugal. _Notas Estud. Inst. Biol. mar., Lisb._, 32(3):1-18.

--. 1966. Sobre a ocorrência na costa de Sesimbra de jovens de _Chromis chromis_ (L.)(Pisces, Pomacentridae). _Archos Mus. Bocage_, (2)1, Notas e supl. (4):11-18, fig. 1-3, pl. 1.

--. 1969. _Ophisichthys dubius_ Osorio, 1917, synonyme de _Panturichthys mauritanicus_ Pellegrin, 1913 (Pisces, Anguilliformi, Heterenchelyidae). _Archos Mus. Bocage_, (2)2(10):131-137, 4 fig.

--. 1977. Poissons capturés et photographiés lors des plongées du bathyscaphe Archimède aux Açores, 1969. _Archos Mus. Bocage_, (2)6(3): 35-50, 5 fig.

--. 1982. Anguilliformes, p. 7-20. _In_ Maurin, C. & Quéro, J.C., Poissons des côtes nord-ouest africaines (campagnes de la "Thalassa" 1962, 1968, 1971 et 1973). _Rev. Trav. Inst. Pêch. marit._ (1981), 45(1):5-71.

--. 1986. Heterenchelyidae, _in_ P.J.P. Whitehead, M.L. Bauchot, J.C. Hureau, J. Nielsen & E. Tortonese. Fishes of the North-eastern Atlantic and the Mediterranean/Poissons de l'Atlantique du Nord-Est et de la Méditerranée, Paris, Unesco, 2:545-547, 2 + 2 fig.

--. 1986. Xenocongridae, _in_ P.J.P. Whitehead, M.L. Bauchot, J.C. Hureau, J. Nielsen & E. Tortonese. Fishes of the North-eastern Atlantic and the Mediterranean/Poissons de l'Atlantique du Nord-Est et de la Méditerranée, Paris, Unesco, 2:555-556, 1 + 1 fig.

--. 1986. Nettastomatidae, _in_ P.J.P. Whitehead, M.L. Bauchot, J.C. Hureau & E. Tortonese. Fishes of the North-eastern Atlantic and the Mediterranean/Poissons de l'Atlantique du Nord-Est et de la Méditerranée, Paris, Unesco, 2:562-566, 4 + 4 fig.

SALDANHA, L.; BAUCHOT, M. L. 1986. Cyemidae, _in_ P.J.P. Whitehead, M.L. Bauchot, J.C. Hureau, J. Nielsen & E. Tortonese. Fishes of the North-eastern Atlantic and the Mediterranean/Poissons de l'Atlantique du Nord-Est et de la Méditerranée, Paris, Unesco, 2:557-558, 1 + 1 fig.

--. 1986. Synaphobranchidae, _in_ P.J.P. Whitehead, M.L. Bauchot, J.C.

Hureau, J. Nielsen & E. Tortonese. Fishes of the North-eastern Atlantic and the Mediterranean/Poissons de l'Atlantique du Nord-Est et de la Méditerranée, Paris, Unesco, 2:586-592, 6 + 6 fig.

SALDANHA, L.; BLACHE, J. 1968. Contribution à la connaissance de *Facciolella physonema* (Facciolà, 1914)(Pisces, Anguilliformi, Nettastomidae). *Revta Fac. Ciênc. Univ. Lisb.*, (2)(C)15(2):181-202, 15 fig.

SALZEN, E. A. 1957. A trawling survey off the Gold Coast. *J. Cons. perm. int. Expl. Mer.*, 23(1):72-82, 2 fig.

--. 1958. Observations on the biology of the West African shad, *Ethmalosa fimbriata* (Bowdich). *Bull. Inst. fr. Afr. noire*, (A)20(4):1388-1426.

SAMYSHEV, E. Z.; SCHETINKIN, S. V. 1973. Feeding patterns of some species of Myctophidae and *Maurolicus muelleri* caught in the sound-dispersing layers in the north-western African area. *Annls biol., Copenh.* (1971), 28:212-215, fig. 153-154.

SANCHES, J. G. 1966. Peixes de Angola (Teleosteos). *Notas mimeogr. Cent. Biol. pisc.*, (46):iii + 227 p., 201 fig.

SANTAELLA, E.; BRAVO de LAGUNA, J.; SANTOS, A. 1975. Resultados de una campañà de prospección pesquera en la isla de La Palma (Islas Canarias). Crustáceos decápodos y peces. *Boln Inst. esp. Oceanogr.*, (193):1-35, 5 fig., 8 tab.

SANZ ECHEVERRIA, J. 1926. Datos sobre el otolito sagitta de los peces de España. *Boln R. Soc. esp. Hist. nat.*, 26(1):145-160, 71 fig.

--. 1932. Sobre otolitos de los Apogonidos. *Boln R. Soc. esp. Hist. nat.*, 32:151-154, 1 pl.

--. 1937. Los otolitos de los lábridos de España (primera parte). *Boln Soc. ibér. Cienc. nat.*, 37:29-45, pl. 2 (24 fig.).

--. 1950. Notas sobre otolitos de peces procedentes de las costas del Sàhara. Segunda parte. *Boln Inst. esp. Oceanogr.*, (27):1-14, 3 pl.

SANZO, L. 1905. Uova e larve di Murenoidi. *Atti Accad. pelorit.*, 19(2):311-315.

--. 1910. Uova e larve di pesce spada (*Xiphias gladius* L.). *Riv. mens. Pesca Pavia*, 5(12):206-209, 2 fig.

--. 1911. Distribuzione delle papille cutanee (organi ciatiformi) e suo valore sistematico nei Gobi. *Mitt. zool. Stn Neapel*, 20:249-328.

--. 1912. Comparsa degli organi luminosi in una serie di larve di *Gonostoma denudatum. Memorie R. Com. talassogr. ital.*, (9):1-23, fig. 1-4, 1 pl.

--. 1912. Larva di *Stomias boa* Risso. *Memorie R. Com. talossogr. ital.*, (10):1-6.

--. 1913. Larva di *Ichthyococcus ovatus. Memorie R. Com. talassogr. ital.*, (27):1-7, 1 pl.

--. 1913. Stadi post-embrionali di *Vinciguerria attenuata* e *V. poweriae. Memorie R. Com. talassogr. ital.*, (35):7 p., 1 pl.

--. 1915. Contributo alla conoscenza degli stadi larvali negli Scopelini Müller. (*Bathophilus nigerrimus* Gigl., *Scopelus caninianus* C. e V., *Sc. humboldtii* Risso). *Atti Accad. naz. Lincei Memorie*, (5)10:714-720.

--. 1915. Stadi larvali di *Bathophilus nigerrimus* Gigl. *Memorie R. Com. talassogr. ital.*, (48):1-10.

--. 1915. Notizie ittiologiche. II. Stadi larvali di *Stomias boa* Risso. III. Stadi larvali di *Bathophilus nigerrimus* Gigl. V. Stadi larvali di *Chauliodus sloani* Bl. *Monitore zool. ital.*, 26:131-144.

--. 1915. Contributo alla conoscenza dello sviluppo negli Scopelini Müller (*Saurus griseus* Lowe, *Chlorophthalmus agassizii* Bp., *Aulopus filamentosus* Cuv.). *Memorie R. Com. talassogr. ital.*, (49):1-21.

--. 1917. Stadi larvali di *Paralepis hyalina*, C.V. *Memorie R. Com. talassogr. ital.*, (59):3-7, fig. 1-5.

--. 1917. Sviluppo larvale di Paralepis rissoi Bp. Memorie R. Com. talassogr. ital., (62):9 p.

--. 1918. Nuovo contributo alla conoscenza dello sviluppo larvale di Stomias boa Risso. Atti Accad. naz. Lincei Rc., (5)27:77-82.

--. 1918. Contributo alla conoscenza dello sviluppo post-embrionale negli Scopelini Müller. Memorie R. Com. talassogr. ital., (66):1-55.

--. 1918. Uova e larve di Trachypterus cristatus Bp. Memorie R. Com. talassogr. ital., (64):1-16, pl.

--. 1919. Contributo alla conoscenza degli stadi larvali di Orthagoriscus Bl. Memorie R. Com. talassogr. ital., (69):7 p., 1 pl.

--. 1922. Uova e larve di Xiphias gladius L. Memorie R. Com. talassogr. ital., (79):1-17, 2 pl.

--. 1925. Uova e larve di Regalecus glesne Asc. Memorie R. Com. talassogr. ital., (118):1-7, pl.

--. 1928. Uova, sviluppo embrionale, stadi larvali, post-larvali e giovanili di Sternoptychidae e Stomiatidae. 1. Argyropelecus hemigymnus. Cocco. Monografie R. Com. talassogr. ital., 2:1-68, pl. 1-7.

--. 1929. Uova e larve di Tonno (Orcynus thynnus Ltkn). Atti Accad. naz. Lincei Rc., (6)9:104-106.

--. 1930. Giovanissimo stadio larvale di Xiphias gladius L. di mm. 6.4. Memorie R. Com. talassogr. ital., (170):1-8, 3 fig.

--. 1930. Uova, sviluppo embrionale, stadi larvali, post-larvali e giovanili di Sternoptychidae e Stomiatidae. 2. Ichthyococcus ovatus. Monografie R. Com. talassogr. ital., 2:71-119, pl. 8-9.

--. 1931. Salmonoidei, Stomiatoidei. In: Uova, larve e stadi giovanili di Teleostei. (ed. U. d'Ancona). Fauna Flora Golfo Napoli, 38:21-92, fig. 31-58, pl. 2-7.

--. 1932. Uova e primi larvali di Pelamys sarda C. V. Memorie R. Com. talassogr. ital. (188):3-9.

--. 1932. Uova, stadi larvali e giovanili di Centrolophus pompilus C. V. Memorie R. Com. talassogr. ital., (196):1-16, fig. 1-12.

--. 1933. Macrouridae. In: Uova, larve e stadi giovanili di Teleostei (ed. U. d'Ancona). Fauna Flora Golfo Napoli, 38:255-265, fig. 227-230, pl. 16.

--. 1933. Uova, stadi larvali e giovanili di Dactylopterus volitans L. Memorie R. Com. talassogr. ital., (207):26 p., 1 pl., 22 fig.

--. 1935. Uova, sviluppo embrionale, stadi larvali, post-larvali e giovanili di Sternoptychidae e Stomiatidae. 3. Maurolicus pennanti. Monografie R. Com. talassogr. ital., 2:123-180, pl. 10-12.

--. 1938. Uova ovariche e sviluppo larvale di Aulopus filamentosus Cuv. Memorie R. Com. talassogr. ital., (254):3-7.

--. 1938. Uova, stadi embrionali, prelarve e larve di Saurenchelys cancrivora Peters. Memorie R. Com. talassogr. ital., (249):1-10, 12 fig.

--. 1939. Rarissimi stadi larvali di Teleostei. Archo zool. ital., 26: 121-151, 2 pl.

--. 1940. Uova e larva appena schiusa di Lophotes cepedianus (Giorna). Memorie R. Com. talassogr. ital., (272):1-6 pl.

--. 1940. Sviluppo embrionale e larva appena schiusa di Scombresox saurus (Walb.). Memorie R. Com. talassogr. ital., (276):1-6, 8 fig.

SARDOU, J. 1966. Oeuf et développement embryonnaire de Trachypterus taenia (Ordre des Lampridiformes Allotriognathes, famille des Trachypteridae). Vie Milieu, (A)17:199-215, 4 pl.

--. 1970. Périodes de ponte de quelques téléostéens dans la région de Villefranche-sur-Mer. Journées Etud. planctonol., Monaco, C.I.E.S.M. :141-145, 1 fig.

--. 1981. Contribution à la connaissance de la faune ichthyologique liguro-provençale: Melanostigma atlanticum Koefoed, 1952 (Zoarcidae), poisson nouveau pour la faune française. Annls Inst. océanogr., Mona-

co, 57(1):25-30.
SARENAS, A. M. 1954. A revision of the Philippine Myctophidae. Philipp. J. Sci., 82:375-427.
SAROSKIN, V. L. 1962. On the age of the West African Sardine (Sardinella aurita Valenciennes. Vop. Ikhtiol., 2(3)(24):487-491 (in Russian).
SASSI, A. 1846. Pesci, p. 111-147. In: Descrizione di Genova e del Genovesato. 3 vol. Genova, 1:xvi + 182 + 180 p., 7 pl.
SAUSKAN, V. I.; SEMENOV, G. N. 1969. Results of research on Scomberesox saurus in the north-western Atlantic in 1968. Annls biol., Copenh., 25:250-252.
SAUTER, H. 1905. Notes from the Owston Collection. I. A new ateleopodid fish from the Sagami Sea. Annotnes zool. jap., 5(4):233-238, 1 fig.
SAUVAGE, H. E. 1873. Note sur le Sebastes minutus. Annls Sci. nat. (Zool.), (5)17(5):1 p.
--. 1878. Description de poissons nouveaux ou imparfaitement connus de la collection du Muséum d'Histoire Naturelle. Nouv. Archs Mus. Hist. nat., Paris, (2)1:109-158, 2 pl.
--. 1879. Mémoire sur la faune ichthyologique de l'île Saint-Paul. Archs Zool. exp. gén., 8:1-46.
--. 1879. Description de quelques poissons d'espèces nouvelles de la collection du Muséum d'Histoire Naturelle. Bull. Soc. philomath. Paris, (7)3:204-212.
--. 1880. Description des gobioides nouveaux ou peu connus de la collection du Muséum d'Histoire Naturelle. Bull. Soc. philomath. Paris, (7)4:40-58.
--. 1880. Description de quelques blennioides de la collection du Muséum d'Histoire Naturelle. Bull. Soc. philomath. Paris, (7)4:215-220.
--. 1880. Etude sur la faune ichthyologique de l'Ogôué. Nouv. Archs Mus. Hist. nat., Paris, (2)3:5-56, pl. 1-3.
--. 1882. Notice sur les poissons du Territoire d'Assinie (Côte d'Or). Bull. Soc. zool. Fr., 7:313-325, pl. 5.
--. 1884. Notice sur une collection de reptiles et de poissons recueillie à Majumba, Congo. Bull. Soc. Zool. Fr., 9:199-208.
SAZONOV, Y. I. 1976. New species of fishes of the family Searsiidae (Salmoniformes, Alepocephaloidei) from the Pacific Ocean. Trudy Inst. Okeanol., 104:13-25, fig. 1-4. (in Russian).
--. 1976. Materials on the systematics and distribution of fishes of the family Searsiidae (Salmoniformes, Alepocephaloidei). Trudy Inst. Okeanol., 104:26-72, fig. 1-22. (in Russian).
--. 1978. Species of Searsiidae (Osteichthyes, Salmoniformes) firstly recorded from the Indian Ocean. Trudy Inst. Okeanol., 111:100-107, fig. 1-2. (in Russian).
--. 1986. Morphology and classification of fishes of the family Platytroctidae (Salmoniformes, Alepocephaloidei). Trudy Inst. Okeanol., 121:51-96, 20 fig. (in Russian).
SAZONOV, Y. I.; GOLOVAN, G. A. 1976. New species of fishes of the family Searsiidae (Salmoniformes, Alepocephaloidei) from the eastern Atlantic Ocean. Trudy Inst. Okeanol., 104:7-12, fig. 1-3. (in Russian).
SAZONOV, Y. I.; IVANOV, A. N. 1979. New species from the family Alepocephalidae, order Salmoniformes, from submarine ridges of the tropical region of the Indian Ocean. Vop. Ikhtiol., 19(6):1006-1013.
--. 1980. Slickheads (Alepocephalidae and Leptochilichthyidae) from thalassobathyal zone of the Indian Ocean. Trudy Inst. Okeanol., 110:7-104.
SAZONOV, Y. I.; TRUNOV, I. A. 1978. New data on the fishes of the family Searsiidae (Salmoniformes, Alepocephaloidei) from the south-eastern Atlantic. Trudy Inst. Okeanol., 111:87-99, fig. 1-6. (in Russian).
SCACCINI, A. 1941. Primo elenco di pesci raccolti in Atlantico nelle

acque della Mauritania, del Sahara Spagnolo e delle Canarie. <u>Thalas-</u><u>sia</u> (Venezia), 4(10):1-49, 22 fig.

SCHEFFERS, W. J. 1966. The fishery resources of Gambia. UNDP (SF)/FAO Project GAM/72/006, Devel. Inshore Fishery, (Tech. Paper 1)(24 p., mimeo).

--. 1971. Note préliminaire sur quelques aspects de la biologie de <u>Eth-</u><u>malosa</u> <u>fimbriata</u> (Bowdich) dans les eaux sénégambiennes. UNDP (SF)/ FAO Project SEN/66/508. Survey Devel. pelag. Fish Resources (Tech. Rep. 9/71)(24 p., mimeo).

SCHEFFERS, W. J.; CONAND, F. 1976. A study on <u>Ethmalosa</u> <u>fimbriata</u> (Bow-dich) in the Senegambian region. 3rd note: the biology of <u>Ethmalosa</u> in the Gambian waters. <u>Docum. scient. provis. Centre Rech. océanogr. Dakar-Thiaroye</u>, (59):19 p. (mimeo).

SCHEFFERS, W. J.; CORREA, J. B. 1971. Investigations on the biology and fisheries of bonga (<u>Ethmalosa</u> <u>fimbriata</u> Bowdich) in Senegambia. Sam-pling of bonga (<u>Ethmalosa</u> <u>fimbriata</u>) in the Gambia from 17-23 January 1971. UNDP (SF)/FAO Project SEN/66/508 Survey Devel. pelag. Fish Resources, (Tech. Rep. 1/71) (mimeo).

--. 1971. Investigations on the biology and fisheries of bonga (<u>Ethmalo-</u><u>sa</u> <u>fimbriata</u> Bowdich) in Senegambia. Sampling of bonga (<u>Ethmalosa</u> <u>fimbriata</u>) in the Gambia for March 1971. UNDP (SF)/FAO Project SEN/ 66/508 Survey Devel. pelag. Fish Resources, (Tech. Rep. 15/71).

SCHEFFERS, W. J.; SAYNA, A. 1970. Investigations on the biology and fisheries of bonga (<u>Ethmalosa</u> <u>fimbriata</u>) in the Gambia from 8 to 14 November 1970. UNDP (SF)/FAO. Project SEN/66/508 Survey Devel. pelag. Fish. Resources, (Tech. Rep. 16/70)(12 p., mimeo).

SCHEFFERS, W. J.; TAYLOR-THOMAS, A. O. 1974. Data on landings of <u>Eth-</u><u>malosa</u> <u>fimbriata</u> by the artisanal fishery in three villages along the southern Atlantic coast of Gambia, during the period November 1973, to April 1974. UNDP (SF)/FAO, Project GAM/72/006, Devel. Inshore Fishery, (Fish. Publ.)(9 p., mimeo).

SCHEFFERS, W. J.; CORREA, J. B.; DIAGNE, O. 1971. Investigations on the biology and fisheries of bonga (<u>Ethmalosa</u> <u>fimbriata</u> Bowdich) in the Senegambia during May 1971. UNDP (SF)/FAO Project SEN/66/508 Survey Devel. pelag. Fish Resources, (Tech. Rep. 10/71)(32 p., mimeo)(see also 5/71).

SCHINZ, H. R. 1821-25. (ed.). Das Thierreich, eingetheilt nach dem Bau der Thiere, als Grundlage ihrer Naturgeschichte und der vergleichen-den Anatomie von dem Herrn Ritter von Cuvier, aus dem Französischen frei übersetzt und mit vielen Zusätzen versehen. 4 vol. Stuttgart & Tübingen. Fische. 2:1822, p. 189-553.

SCHIØDTE, J. C. 1868. Om Øiestillingens udvikling hos flynderfiskene. <u>Naturh. Tidsskr.</u>, (3)5:269-275, pl. 11.

SCHLEGEL, H. 1842. Uebersicht der in Japan vorkommenden Fische. <u>Ber. Versamm. dt. Naturf.</u>, 1842:208.

SCHMELTZ, J. D. E. 1866. Museum Godeffroy. Cat., 3 Hamburg, xii + 52 p.

--. 1879. Museum Godeffroy, Cat., 7. Hamburg, vii + 99 p.

SCHMIDT, (E.) J. 1906. Contribution to the life history of the eel (<u>An-</u><u>guilla</u> <u>vulgaris</u> Flem.). <u>Rapp. P.-v. Réun. Cons. perm. int. Explor. Mer.</u>, 5(4):137-264, 267-273, 6 fig., 4 pl., 3 map.

--. 1909. On the occurrence of leptocephali (larval muraenoids) in the Atlantic W. of Europe. <u>Meddr Kommn Havunders.</u> (Fisk.), 3(6):1-19, 6 fig., 2 pl.

--. 1909. On the distribution of the fresh-water eels (<u>Anguilla</u>) throughout the world. I. Atlantic Ocean and adjacent regions. <u>Meddr Kommn Havunders.</u> (Fisk.), 3(7):1-45, 1 fig., 1 pl.

--. 1912. Danish researches in the Atlantic and Mediterranean on the life-history of the fresh-water eel (<u>Anguilla</u> <u>vulgaris</u>) with notes on other species. <u>Int. Revue ges. Hydrobiol. Hydrogr.</u>, 5:317-342.

--. 1912. Contributions to the biology of some north Atlantic species of eels. <u>Vidensk. Meddr dansk naturh. Foren.</u>, 64:39-51, 1 pl.

--. 1913. On the identification of muraenoid larvae in their early ("preleptocephaline") stages. <u>Meddr Kommn Havunders.</u> (Fisk.), 4(2): 1-13, 1 pl. (8 fig.).

--. 1913. Aalens fortplantning och lekplaster. <u>Svensk Fisk-Tidskr.</u>, (1):17-28, 2 fig.

--. 1916. On the early larval stages of the freshwater eels (<u>Anguilla</u>) and some other North Atlantic muraenoids. <u>Meddr Kommn Havunders</u>. (Fisk.), 5(4):1-20, 14 fig., 4 pl.

--. 1918. Argentinidae, Microstomidae, Opisthoproctidae. Mediterranean Odontostomidae. <u>Rep. Dan. oceanogr. Exped. Mediterr.</u>, 1908-1910, 2, Biol.,(A5):1-40, 23 fig.

--. 1921. New studies on sun-fishes made during the "Dana" Expedition, 1920. <u>Nature. Lond.</u>, 107:76-79, fig.

--. 1926. Further studies on sun-fishes made during the "Dana" Expedition, 1921. <u>Nature. Lond.</u>, 117:80-81, fig.

--. 1930. <u>Nessorhamphus</u>, a new cosmopolitan genus of oceanic eels. <u>Vidensk. Meddr dansk naturh. Foren</u>, 90:371-375, pl. 4-5.

--. 1931. Oceanographical expedition of the Dana, 1928-1930. <u>Nature. Lond.</u>, 127:444-446, 487-490, 6 fig.

--. 1932. Dana's togt omkring jorden 1928-1930. Copenhagen, 368 p., 278 fig.

SCHMIDT, W. 1968. Vergleichend morphologische Studie über die Otolithen mariner Knochenfische. <u>Arch. FischWiss.</u>, 19 (Beiheft 1):1-96, 25 pl.

--. 1969. Seasonal distribution of <u>Sardinella aurita</u> and <u>S. eba</u> in West African coastal waters. UNDP(SF)/FAO, Results Regional Fish Survey W. Africa, Abidjan, (Rep. 69/5)(unpaged, mimeo).

--. 1971. Sex, maturity, feeding and fatness of <u>Sardinella</u> and some other pelagic species caught from Thue Jr during 1969/70. UNDP(SF)/ FAO, Results Regional Fish. Survey W. Africa, Abidjan, (Rep. 71/18) (94 p., mimeo).

--. 1971. Observations on deep-scattering <u>Sardinella aurita</u> from the offshore waters of Mauritania. UNDP(SF)/FAO, Results Regional Fish. Survey W. Africa, Abidjan, (Rep. 71/6)(17 p., mimeo).

--. 1972. Results of the UNDP(SF)/FAO Regional fisheries survey in West Africa. Report n° 1. Deep scattering <u>Sardinella aurita</u> off Mauritania. <u>Mar. Biol., Berl.</u>, 16(2):91-101.

SCHNAKENBECK, W. 1929. Über einige Meeresfische aus Südwestafrika. <u>Mitt. zool. StInst. Hamb.</u>, 44:23-46, fig. 1-9.

--. 1938. <u>Bathygadus melanobranchus. Faune ichthyol. Atlant. N.</u> Cahier 12:165.

SCHNEIDER, J. G. 1801. M.E. Blochii Systema ichtyologiae iconibus cx illustratum. Post obitum auctoris opus inchoatum absolvit, correxit, interpolavit J.G. Schneider, Saxo. Berolini, lx + 584 p., 110 pl.

SCHOLANDER, P. F.; DAM, L. van. 1953. Composition of the swimbladder gas in deep sea fishes. <u>Biol. Bull. mar. biol. Lab. Woods Hole</u>, 104:75-86, 5 fig.

SCHRANK, F. von Paula. 1798. Fauna Boica: durchgedachte Geschichte der in Baiern einheimischen und zahmen Thiere. Fische: p. 295-340, Nürnberg, 1:xii + 720 p.

SCHREINER, C.; MIRANDA RIBEIRO, A. de. 1902. A collecçào de peixes do Museu Nacional do Rio de Janeiro. Part 1. <u>Archos Mus. nac., Rio de J.</u>, 12:1-41.

SCHROEDER, W. C. 1940. Some deep-sea fishes from the North Atlantic. <u>Copeia</u>, 1940:231-238.

SCHULTZ, L. P. 1937. A new species of deep-sea fish, <u>Argyropelecus an-</u>

trorsospinus, of the family Sternoptychidae. Smithson. misc. Collns, 91(27):5-8.
--. 1938. Review of the fishes of the genera Polyipnus and Argyropelecus (family Sternoptychidae), with descriptions of three new species. Proc. U.S. natn. Mus., 86:135-155, fig. 42-45.
--. 1940. Two new genera and three new species of cheilodipterid fishes, with notes on other genera of the family. Proc. U.S. natn. Mus., 88: 403-423.
--. 1943. Fishes of the Phoenix and Samoa Islands, collected in 1939 during the expedition of the U.S.S. "Bushnell". Bull. U.S. natn. Mus., (180):1-316, 9 pl., 27 fig.
--. 1946. A revision of the genera of mullets, fishes of the family Mugilidae, with descriptions of three new genera. Proc. U.S. natn. Mus., 96:377-395, fig. 28-32.
--. 1948. A revision of six subfamilies of atherine fishes, with descriptions of new genera and species. Proc. U.S. natn. Mus., 98:1-48, 9 fig., 2 pl.
--. 1949. A further contribution to the ichthyology of Venezuela. Proc. U.S. natn. Mus., 99:1-211, 3 pl., 20 fig.
--. 1953. Family Serranidae: Groupers, sea basses. p. 328-388. In: L. P. Schultz et al.: Fishes of the Marshall and Marianas Islands. Families from Asymmetrontidae through Siganidae. Bull. U.S. natn. Mus., Bull. 202, 1: xxxii + 685 p., 90 fig., 74 pl.
--. 1957. The frogfishes of the family Antennariidae. Proc. U.S. natn. Mus., 107(3383):47-105, 8 fig., 14 pl.
--. 1958. Review of the parrotfishes, family Scaridae. Bull. U.S. natn. Mus., (214): v + 143 p., 31 fig., 27 pl.
--. 1960. Suborder Scombrina. Family Scombridae: Tunas. p. 410-417. In: L.P. Schultz et al. Fishes of the Marshall and Marianas Islands. Families from Mullidae through Stromateidae. Bull. U.S. natn. Mus. Bull. 202, 2: ix + 428 p., fig. 91-132, pl. 75-123.
--. 1961. Revision of the marine silverhatchetfishes (family Sternoptychidae). Proc. U.S. natn. Mus., 112:587-649.
--. 1964. Family Sternoptychidae. In: Fishes of the western North Atlantic. Mem. Sears Fdn mar. Res., 1(4):241-273, fig. 62-72.
--. 1964. Three new species of frogfishes from the Indian and Pacific Oceans with notes on other species (family Antennariidae). Proc. U.S. natn. Mus., 116:171-182.
--. 1968. A new subspecies of parrot fish Nicholsina ustus collettei from the eastern Atlantic Ocean. Proc. U.S. natn. Mus., 124(3642):1-5, 2 fig.
--. 1969. The taxonomic status of the controversial genera and species of parrotfishes with a descriptive list. (Family Scaridae). Smithson. Contrib. Zool. (17):1-49, 2 fig., 8 pl.
SCHULTZ, L. P.; REID, E. D. 1939. A revision of the soap fishes of the genus Rhypticus. Proc. U.S. natn. Mus., 87(3074):261-270.
SCHULTZ, L. P.; SPRINGER, S. 1956. Lepidocybium flavo-brunneum, a rare gempylid fish new to the fauna of the Gulf of Mexico. Copeia, 1956 (1):65.
SCHWARTZ, F. 1961. Maryland billfishes. Md Conserv., 38(4):20-24, 10 fig.
SCOPOLI, J. A. 1777. Introductio ad historiam naturalem... subinde ad leges naturae. Prague, x + 506 p.
SCORDIA, C. 1936. Intorno alla accertata presenza dello Scomberesox saurus (Walb.) nel Mediterraneo. Boll. Zool., 7:217-220.
SCOTT, T. 1906. Observations on the otoliths of some teleostean fishes. 24. A. Rep. Fishery Bd Scotl. (1905), 3:48-82, 5 pl.
SCOTT, T. D. 1962. The marine and fresh water fishes of South Australia.

Adelaide, 338 p., fig.

SCOTT, W. B. 1981. Molidae, in Fischer, W., Bianchi, G. & W.B. Scott (ed.), FAO species identification sheets for fishery purposes. Eastern Central Atlantic; fishing areas 34, 47 (in part). Canada Funds-in-Trust. Ottawa, Department of Fisheries and Oceans Canada, by arrangement with the Food and Agriculture Organization of the United Nations, vol. 1-7: pag. var.

SCOTT, W. B.; TIBBO, S. N. 1968. Food and feeding habits of swordfish, Xiphias gladius, in the western North Atlantic. J. Fish. Res. Bd Can., 25(5):903-919, 3 fig.

--. 1974. Food and feeding habits of swordfish, Xiphias gladius Linnaeus, in the North-west Atlantic Ocean. NOAA tech. Rep. NMFS SSRF, (675):138-141, 1 fig.

SCOTTON, L. N.; SYLVA, D. P. de. 1972. Fish babies. Sea Front., 18:194-201.

SEALE, A. 1901. Report of a Mission to Guam, Part. 2, Fishes. Occ. Pap. Bernice P. Bishop Mus. (1900), 1(3):61-128.

--. 1901. New Hawaiian fishes. Occ. Pap. Bernice P. Bishop Mus., 1(4):3-15.

--. 1906. Fishes of the South Pacific. Occ. Pap. Bernice P. Bishop Mus., 4(1):1-89, fig. 1-23.

--. 1917. New species of apodal fishes. Bull. Mus. comp. Zool. Harv., 61, 4:79-94.

SEBASTIAN, M. J.; VEDAVYASA RAO, P. 1963. On the feeding habits of the snake mackerel Gempylus serpens (Cuvier), with some remarks on the specimens collected off the Indian coast. J. mar. biol. Ass. India, 5(2): 322.

SECCA, M. 1945. A pesca em Cabo Verde. Divisâo de Propagande dos Serviços de Estadistico da Colonia de Cabo Verde, 154 p.

SEDBERRY, G.; MUSICK, J. A. 1978. Feeding strategies of some demersal fishes of the continental slope and rise off the Mid-Atlantic coast of the USA. Mar. Biol., Berl., 44:357-375.

SEDLETSKAYA, V. A. 1972. Spawning conditions and the structure of spawning concentrations of sardine Sardinella aurita Val. off the northwest coast of Africa. Vop. Ikhtiol., 12(1):65-71 (Engl. transl. J. Ichthyol., 12(1):55-60).

--. 1979. Aires et période de reproduction de la sardinelle ronde (Sardinella aurita) et des principales autres espèces de poissons pélagiques côtiers de l'Atlantique centre-est. Annex 12, p. 104-113. In: Anon., 1979.

SELLA, M. 1911. Contributo alla conoscenza della riproduzione e dello sviluppo del pesce spada (Xiphias gladius L.). Memorie R. Com. talassogr. ital., (2):1-16, 5 fig.

SERBETIS, K. 1951. A new form of Xiphias, Xiphias thermaicus n. sp. Praktika Akad. Athenon, 22:269-273 (in Greek with Italian summary).

SERENKO, V. A.; BEKKER, V. E. 1975. A new find of the rare species Diaphus adenomus (Myctophidae, Osteichthyes). J. Ichthyol., 15(5):821-825.

SERET, B.; McEACHRAN, D. 1987. Catalogue critique des types de poissons du Museum National d'Histoire naturelle. Poissons batoïdes (Chondrichthyes, Elasmobranchii, Batoidea). Bull. Mus. natn. Hist. nat., Paris, (4) 8 A (4)(1986 Suppl.):3-50.

SERET, B.; OPIC, P. 1981. Poissons de mer de l'ouest africain tropical. Init.-Documns tech. ORSTOM, Paris, (49): vi + 416 p., num. col. and black and white fig.

SERRA, J. A.; ALBUQUERQUE, R. M. 1957. On species differences in Crenilabrus (Pisces, Perciformes). Revta port. Zool. Biol. ger., 1(1): 1-27.

SERVENTY, D. L. 1956. The southern bluefin tuna _Thunnus thynnus maccoyii_ (Castelnau), in Australian waters. _Aust. J. mar. Freshwat. Res._, 7 (1):1-43.

SGANO, T. 1981. Kyphosidae, _in_ Fischer, W., Bianchi, G. & W.B. Scott (ed.), FAO species identification sheets for fishery purposes; Eastern Central Atlantic; fishing areas 34, 47 (in part). Canada Funds-in-Trust. Ottawa, Department of Fisheries and Oceans Canada, by arrangement with the Food and Agriculture Organization of the United Nations, vol. 1-7: pag. var.

SHAKLEE, J. B.; TAMARU, C. S. 1981. Biochemical and morphological evolution of Hawaiian bonefishes (_Albula_). _Syst. Zool._, 30(2):125-146.

SHANKLIN, W. M. 1934. The cerebella of three deep-sea fish. _Acta zool., Stockh._, 15(2/3):409-430.

--. 1935. On diencephalic and mesencephalic nuclei and fiber paths in the brains of three deep-sea fish. _Phil. Trans. R. Soc._, (B)224(256): 361-419.

SHAPIRO, S. 1938. A study of proportional changes during the post-larval growth of the blue marlin. (_Makaira nigricans ampla_ Poey). _Am. Mus. Novit._, (995):1-20, 8 fig.

SHAUGHNESSY, W. A. E. (see O'SHAUGHNESSY, W. A. E.)

SHAW, G. 1791. Description of the _Stylephorus chordatus_, a new fish. _Trans. Linn. Soc. Lond._, 1:90-92.

--. 1803-1804. Pisces. _In_: General zoology or systematic natural history. London, 4(1): v + 1- 186, pl. 1-25; 4(2): xiii + 187-632, pl. 26-92; 5(1): v + 1-250, pl. 93-132; 5(2): vi + 251-463, pl. 133-182.

SHAW, G.; NODDER, F. P. 1794. The Naturalist's Miscellany, or coloured figures of natural objects, drawn and described from nature. London, 5: pl. 147-182.

--. 1811. The Naturalist's Miscellany, or coloured figures of natural objects, drawn and described from nature. London, 23: pl. 973-1020.

SHCHERBACHEV, Y. N. 1978. Long-nosed chimaeras (Rhinochimaeridae, Chimaeriformes) from the waters of South Africa. _Trudy Inst. Okeanol._, 111:7-9.

SHCHERBACHEV, Y. N.; NOVIKOVA, N. S. 1976. Material on distribution and taxonomy of the mesopelagic fishes of the family Stomiatidae (Osteichthyes). _Trudy Inst. Okeanol._, 104:92-112.

SHCHERBACHEV, Y. N.; LEVITSKY, V. N.; PORTSEV, P. J. 1978. On records of the rare species of deep-sea fishes from off southern Africa. _Trudy Inst. Okeanol._, 111:185-194, 1 fig. (in Russian).

SHCHERBACHEV, Y. N.; PARIN, N. V.; PAKHORUKOV, N. P.; PIOTROVSKY, A. S. 1986. Mesobenthic and mesobenthopelagic fishes from submarine rises in the western Indian Ocean. _Trudy Inst. Okeanol._, 121:195-214.

SHELTON, P. A.; DAVIES, S. L. 1979. Occurrence of lightfish off the Cape coast. _S. Afr. Shipp. News Fishg Ind. Rev._, 34(6):28-29.

SHEPHERD, C. E. 1910. Comparisons of otoliths found in fishes. _Zoologist_, (4)14:292-298, 2 pl.

--. 1922. Sur quelques erreurs dans la détermination des otolithes fossiles. _Bull. Soc. géol. Fr._, 22(4):138-141, 1 pl.

SHIINO, S. M. 1972. List of English names of Japanese fishes with proposition of new names. _Sci. Rep. Shima Marineld_, (1):1-210, 1 pl.

--. 1976. List of common fishes of the world, those prevailing among English-speaking nations. _Sci. Rep. Shima Marineland_, (4):1-262.

SHIMIZU, T. 1983. (Zeiformes). p. 286-293, fig. _In_: Fishes trawled off Suriname and French Guiana. (ed. Uyeno, T., K. Matsuura & E. Fujii). Tokyo, 519 p., 10 fig., col. photos.

SHIOHAMA, T.; MYOJIN, M.; SAKAMOTO, H. 1965. The catch statistic data for the Japanese tuna long-line fishery in the Atlantic Ocean and some simple considerations on it. _Rep. Nankai reg. Fish. Res. Lab._,

(21):1-131, 6 fig., many maps.
SHIPP, R. L. 1974. The pufferfishes (Tetraodontidae) of the Atlantic Ocean. Publs Gulf Coast Res. Lab. Mus., 4:1-163, 29 fig., 11 tab.
--. 1981. Tetraodontidae, in Fischer, W., Bianchi, G. & W.B. Scott (ed.) FAO species identification sheets for fishery purposes. Eastern Central Atlantic; fishing areas 34, 47 (in part). Canada Funds-in-Trust. Ottawa, Department of Fisheries and Oceans Canada, by arrangement with the Food and Agriculture Organization of the United Nations, vol. 1-7: pag. var.
SHOEMAKER, N. H. 1958. A female ceratid angler, Cryptopsaras couesii Gill, from the Gulf of Mexico, bearing three parasitic males. Copeia, 1958:143-145.
SHORES, D. L. 1969. Postlarval Sudis (Pisces: Paralepididae) in the Atlantic Ocean. Breviora, (334):1-4, fig. 1-4.
SHUFELDT, R. W. 1888. Further studies on Grammicolepis brachiusculus Poey. J. Morph., 2:271-296, 14 fig.
SHUL'MAN, G. L.; DEMIDOV, V. F. 1961. Variations in fatness of Dakar sardines (Sardinella aurita Valenciennes) during the pre-spawning period of the annual cycle. Zool. Zh., 40(10):1532-1535 (in Russian).
--. 1962. Variations in fatness of the Dakar sardine (Sardinella aurita Valenciennes) during the pre-spawning period of the annual cycle. Referat. Zh. (Biol.), :6170 (Engl. transl.).
SICHER, E. 1896. Pesci nuovi e poco noti della Sicilia. Atti Soc. veneto-trent. Sci. nat., (2)2:419-431.
SILAS, E. G.; SELVARAJ, D.; REGUNATHAN, A. 1969. Rare chimaeroid and elasmobranch fishes from the continental slope off the west coast of India. Curr. Sci., 38:105-106, fig. 1.
SILVESTER, Ch. F. 1915. Fishes new to the fauna of Porto-Rico. Yb Carnegie Instn Wash., 14:214-217.
SIVALINGAM, S. 1975. On the grey mullets of the Nigerian coast, prospects of their culture and results of trials. Aquaculture, 5:345-357.
SIVERTSEN, E. 1945. Fishes of Tristan da Cunha with remarks on age and growth based on scale readings. Results Norw. scient. Exped. Tristan da Cunha, (12):1-44, 17 fig., 8 pl.
SLASTENENKO, E. P. 1934. I Blennius del Mar Nero. Pubbl. Staz. zool. Napoli, 14:95-109.
--. 1935. The Scorpaena of the Black Sea (in Russian and English). Dokl. Akad. Nauk SSSR, (2)1(1):74-80.
--. 1936. Notes sur quelques poissons de la Mer Noire. Annls scient. Univ. Jassy, 22:297-305, pl. 1-8.
--. 1939. Les poissons de la Mer Noire et de la Mer d'Azov. Annls scient. Univ. Jassy (II), 25(1):1-196.
SLOANE, H. 1707. A voyage to Jamaica. In: A voyage to the islands Madeira, Barbados, Nieves, S. Christophers and Jamaica, with the natural history of the herbs and trees,... fishes, of the last of those islands. 2 vol. London, 1: cliv + 265 p., 158 pl.
SMALL, G. J. 1981. A review of the bathyal fish genus Antimora (Moridae: Gadiformes). Proc. Calif. Acad. Sci., 42(13):341-348, 5 fig.
SMITH, A. 1828. Descriptions of new, or imperfectly known objects of the animal kingdom, found in the south of Africa. S. Afr. Comml Advertiser, 3(145):2.
--. 1838. On the necessity for a revision of the groups included in the Linnaean genus Squalus. Proc. zool. Soc. Lond., 1837:85-86 (also 1838 in: Ann. nat. Hist., 1:72-74).
--. 1849. Illustrations of the zoology of South Africa, consisting chiefly of figures and descriptions of the objects of natural history collected during an expedition into the interior of South Africa in 1834-1836. 4. Pisces. London, 77 p. n. num., 31 col. pl.

SMITH, C. L. 1971. A revision of the American groupers: <u>Epinephelus</u> and allied genera. <u>Bull. Am. Mus. nat. Hist.</u>, 146(2):67-242, fig. 1-41.

--. 1981. Anthiidae, Lobotidae, Moronidae, Serranidae, <u>in</u> Fischer, W., Bianchi, G. & W.B. Scott (ed.), FAO species identification sheets for fishery purposes. Eastern Central Atlantic; fishing areas 34, 47 (in part). Canada Funds-in-Trust. Ottawa, Department of Fisheries and Oceans Canada, by arrangement with the Food and Agriculture Organization of the United Nations, vol. 1-7: pag. var.

SMITH, C. L.; BAILEY, R. M. 1962. The subocular shelf of fishes. <u>J. Morph.</u>, 110:1-17.

SMITH, C. L.; ATZ, E. H.; TYLER, J. C. 1971. Aspects of oral brooding in the cardinalfish <u>Cheilodipterus affinis</u> Poey. <u>Am. Mus. Nov.</u>, (2456): 1-11, 3 fig.

SMITH, D. G. 1971. Osteology and relationships of the congrid eels of the western North Atlantic. PhD Diss., Univ. Miami, 163 p., 63 fig.

--. 1974. Dysommid eel larvae in the western North Atlantic. <u>Copeia</u>, 1974(4):671-680, 8 fig.

--. 1979. Guide to the leptocephali (Elopiformes, Anguilliformes and Notacanthiformes). <u>NOAA tech. Rep. NMFS Circ.</u>, (424):i-iv, 1-39, 54 fig.

--. 1980. Early larvae of the tarpon, <u>Megalops atlantica</u> Valenciennes (Pisces: Elopidae), with notes on spawning in the Gulf of Mexico and the Yucatan Channel. <u>Bull. mar. Sci.</u>, 30(1):136-141.

--. 1981. Congridae, Muraenesocidae, <u>in</u> Fischer, W., Bianchi, G. & W.B. Scott (ed.), FAO species identification sheets for fishery purposes. Eastern Central Atlantic; fishing areas 34, 47 (in part). Canada Funds-in-Trust. Ottawa, Department of Fisheries and Oceans Canada, by arrangement with the Food and Agriculture Organization of the United Nations, vol. 1-7: pag. var.

--. 1984. A redescription of the rare eel <u>Myroconger compressus</u> (Pisces: Myrocongridae), with notes on its osteology, relationships, and distribution. <u>Copeia</u>, 1984(3):585-594, fig. 1-15.

SMITH, D. G.; KANAZAWA, R. H. 1977. Eight new species and a new genus of congrid eels from the western North Atlantic with redescriptions of <u>Ariosoma analis, Hildebrandia guppyi</u>, and <u>Rhechias vicinalis. Bull. mar. Sci.</u>, 27(3):530-543, fig. 1-12.

SMITH, H. M. 1901. Additions to the fish fauna in 1899. <u>In</u>: Contributions from the Biological Laboratory of the U.S. Fish. Commission, Woods Hole, Massachusetts. Biological Notes. No. 1. <u>Bull. U.S. Fish. Commn</u> (1899), 19:309-310.

--. 1907. Fishes of North Carolina. <u>N. Carol. geol. econ. Surv.</u>, 2:1-453.

SMITH, H. M.; RADCLIFFE, L. 1912. <u>In</u>: L. Radcliffe, 1912. Scientific results of the Philippine cruise of the fisheries steamer "Albatross" 1907-1910. No 16. New pediculate fishes from the Philippine Islands and contiguous waters. <u>Proc. U.S. natn. Mus.</u>, 42:199-214.

--. 1913. <u>In</u>: L. Radcliffe, 1913. Descriptions of seven new genera and thirty-one new species of fishes of the families Brotulidae and Carapidae from the Philippine Islands and the Dutch East Indies. <u>Proc. U.S. natn. Mus.</u>, 44:135-176, 11 pl.

SMITH, J. A. 1859. Notice of the ukpam, a large species (probably new) of stingray (<u>Trygon</u> Cuv.), found in the Old Calabar River, Africa. <u>Proc. R. phys. Soc. Edinb.</u>, 2:64-69.

--. 1867. Notice of a species of pipe-fish of the genus <u>Doryichthys</u> Kaup, probably new, recently brought from Old Calabar. <u>Proc. R. phys. Soc. Edinb.</u>, 3:227-228.

SMITH, J. L. B. 1933. An interesting new myctophid fish from South Africa. <u>Trans. R. Soc. S. Afr.</u>, 21(2):125-127.

--. 1934. Marine fishes of seven genera new to South Africa. Trans. R. Soc. S. Afr., 22:89-100.

--. 1934. The Triglidae of South Africa. Trans. R. Soc. S. Afr., 22:321-336, 8 pl.

--. 1935. New and little known fishes from South Africa Rec. Albany Mus., 4:169-235, 5 fig., pl. 18-23.

--. 1938. The South African fishes of the families Sparidae and Dentici-dae. Trans. R. Soc. S. Afr., 26(3):225-305, 25 fig., pl. 18-29.

--. 1947. New species and new records of South African marine fishes. Ann. Mag. nat. Hist., (11)13:793-821.

--. 1949. The sea fishes of southern Africa. Johannesburg, xvi + 550 p., 103 pl. (102 col.), 519 fig.; (2nd ed.) 1950, 550 p., 103 pl. (102 col.), 519 fig.; (3rd ed.) 1953, 564 p., 107 pl. (104 col.), 550 fig.; (4th ed.) 1961 + (5th ed.) 1965, 580 p., 111 pl. (104 col.), 557 fig.

--. 1949. Forty-two fishes new to South Africa, with notes on others. Ann. Mag. nat. Hist., (12)2:97-111.

--. 1949. The stromateid fishes of South Africa. Ann. Mag. nat. Hist., (12)2:839-851.

--. 1956. Self-inflation in a gobioid fish. Nature. Lond., 177:714.

--. 1956. The parrot fishes of the family Callyodontidae of the western Indian Ocean. Ichthyol. Bull. Rhodes Univ., (1):1-23, 1 pl.

--. 1956. An extraordinary fish from South Africa. Ann. Mag. nat. Hist., (12)9:54-57.

--. 1956. Swordfish, marlin and sailfish in South and East Africa. Ich-thyol. Bull. Rhodes Univ., (2):25-33, 2 fig., 2 pl.

--. 1956. A marlin in Angola, with a note on Makaira herschelii (Gray), 1838. Nature. Lond., 177:1246-1247.

--. 1956. The fishes of the family Sphyraenidae in the western Indian Ocean. Ichthyol. Bull. Rhodes Univ., (2):37-46, pl. 1-2.

--. 1957. Four interesting new fishes from South Africa. S. Afr. J. Sci., 53(8):219-222.

--. 1957. The fishes of the family Scorpaenidae in the western Indian Ocean. Part I. The sub-family Scorpaeninae. Ichthyol. Bull. Rhodes Univ., (4):49-69, 5 fig., 4 pl.

--. 1958. Fishes of the families Tetrarogidae, Caracanthidae and Synan-ciidae, from the western Indian Ocean, with further notes on scorpae-nid fishes. Ichthyol. Bull. Rhodes Univ., (12):167-181, pl. 7-8.

--. 1960. Fishes of the family Gobiidae in South Africa. Ichthyol. Bull. Rhodes Univ., (18):299-314, 13 fig.

--. 1960. Two interesting fishes from South Africa. S. Afr. J. Sci., 56:91-92, 1 fig.

--. 1960. A new grammicolepid fish from South Africa. Ann. Mag. nat. Hist., (13)3:232-235, pl. 3 + 4.

--. 1960. Oreosoma atlanticum Cuvier, 1829, in South Africa, with notes on other zeid fishes. Ann. Mag. nat. Hist., (13)3:565-569, pl. 8.

--. 1961. Fishes of the family Apogonidae of the western Indian Ocean and the Red Sea. Ichthyol. Bull. Rhodes Univ., (22):373-418, 11 fig., pl. 46-52.

--. 1961. A new stromateid fish from South Africa, and another new to that area. S. Afr. J. Sci., 57:158-160.

--. 1962. Fishes from the Cape described by Lichtenstein in 1823. S. Afr. J. Sci., 58:39-40.

--. 1962. The sparid genus Lithognathus Swainson, 1839 with description of an interesting new species. S. Afr. J. Sci., 58:109-114, 2 pl.

--. 1963. A new apogonid fish from deeper water of the Gulf of Guinea. Ann. Mag. nat. Hist., (13)6:621-624, 1 fig.

--. 1963. Fishes of the families Draconettidae and Callionymidae from the Red Sea and the western Indian Ocean. <u>Ichthyol. Bull. Rhodes Univ.</u>, (28):547-564, pl. 83-87, 8 fig.

--. 1964. An interesting new fish of the family Chiasmodontidae from South Africa with redescriptions of <u>Odontonema kerberti</u> Weber, 1913. <u>Ann. Mag. nat. Hist.</u>, (13)7:567-574.

--. 1964. Scombroid fishes of South Africa.<u>Proc. Symp. Scombroid Fishes mar. biol. Ass. India</u>,1(1):165-183, 1 fig., 10 pl.

--. 1965. New records and descriptions of fishes from Southwest Africa. <u>Occ. Pap. Dep. Ichthyol.. Rhodes Univ.</u>, (3):13-23, pl. 2-5.

--. 1966. An interesting new callionymid fish from Madagascar and the first record of a clingfish from there. <u>Ann. Mag. nat. Hist.</u>, (13)8 (89 & 90):321-324, 1 fig.

--. 1966. A new stromateid fish from South Africa with illustration of the unique rare <u>Centrolophus huttoni</u> Waite, 1910. <u>Ann. Mag. nat. Hist.</u>, (13)9:1-3, 1 pl.

--. 1966. Hitherto unknown early developmental larval stadia of the West African albulid fish, <u>Pterothrissus belloci</u> Cadenat, 1937. <u>Occ. Pap. Dep. Ichthyol.. Rhodes Univ.</u>, (6):57-61, 1 fig.

--. 1966. Interesting fishes from South Africa. <u>Occ. Pap. Dep. Ichthyol.. Rhodes Univ.</u>, (8):83-94, 1 fig., pl. 15-17.

--. 1968. New and interesting fishes from deepish water off Durban, Natal and southern Mozambique. <u>Investl Rep. (S. Afr. Ass. mar. biol. Res.)</u>, (19):1-30, 2 fig., 2 tab., 6 pl.

SMITH, J. L. B.; SMITH, M. M. 1966. Fishes of the Tsitsikama Coastal National Park. Johannesburg, 161 p., 2 fig., 136 col. fig.

SMITH, M. M. 1975. Common and scientific names of the fishes of southern Africa. <u>Spec. Publ. J.L.B. Smith Inst. Ichthyol.</u>, (14):1-178.

SMITH, M. M.; HEEMSTRA, P. C. 1986. Smith's Sea Fishes. Johannesburg, Macmillan South Africa, 1047 p., 270 fig., 144 pl.

SMITH-VANIZ, W. F. 1981. Dactylopteridae, <u>in</u> Fischer, W., Bianchi, G. & W.B. Scott (ed.), FAO species identification sheets for fishery purposes. Eastern Central Atlantic; fishing areas 34, 47 (in part). Canada Funds-in-Trust. Ottawa, Department of Fisheries and Oceans Canada, by arrangement with the Food and Agriculture Organization of the United Nations, vol. 1-7: pag. var.

--. 1986. Carangidae, <u>in</u> P.J.P. Whitehead, M.L. Bauchot, J.C. Hureau, J. Nielsen & E. Tortonese. Fishes of the North-eastern Atlantic and the Mediterranean/Poissons de l'Atlantique du Nord-Est et de la Méditerranée, Paris, Unesco, 2:815-844, 11 + 25 + 2 fig.

SMITH-VANIZ, W. F.; BERRY, F. H. 1981. Carangidae, <u>in</u> Fischer, W., Bianchi, G. & W.B. Scott (ed.), FAO species identification sheets for fishery purposes. Eastern Central Atlantic; fishing areas 34, 47 (in part). Canada Funds-in-Trust. Ottawa, Department of Fisheries and Oceans Canada, by arrangement with the Food and Agriculture Organization of the United Nations, vol. 1-7: pag. var.

SMITH-VANIZ, W. F.; SPRINGER, V. G. 1971. Synopsis of the tribe Salariini, with description of five new genera and three new species (Pisces: Blenniidae). <u>Smithson. Contr. Zool.</u>, (73):1-72.

SMITH-VANIZ, W. F.; STAIGER, J. C. 1973. Comparative revision of <u>Scomberoides. Oligoplites. Parona</u> and <u>Hypacanthus</u> with comments on the phylogenetic position of <u>Campogramma</u> (Pisces: Carangidae). <u>Proc. Calif. Acad. Sci.</u>, 39(13):185-256.

SMITT, F. A. 1892. Skandinaviens fiskar. Stockholm, 1-266 p., 134 fig. (2nd ed. of Fries, B.F., Eckström, C.U. and C.J. Sundevall, Skandinaviens fiskar... 1836-57; revd. by F.A. Smitt).

--. 1893-1895. A history of Scandinavian fishes by B. Fries, C. U. Ekström and C. Sundevall with coloured plates by W. von Wright and text illustrations. 2nd ed. rev. and compl., Stockholm, 1, 1893:1-566, fig. 1-134, pl. 1-27. 2, 1895:567-1240, fig. 135-380, pl. 27A-53.

--. 1900. Preliminary notes on the arrangement of the genus Gobius with an enumeration of its European species. Öfvers. K. Vetensk. Akad. Förh., (1899):543-555.

SNYDER, J. O. 1904. A catalogue of the shore fishes collected by the steamer "Albatross" about the Hawaiian Islands in 1902. Bull. U.S. Fish. Commn (1902), 22:513-538, pl. 1-13.

SOLANDER, D. C. (1768). Manuscript lists and descriptions of animals and plants collected during Capt. Cook's first voyage, 1768-1771. ii. Pisces Australiae. (Ms, preserved in BMNH.).

--. (1772). Pisces Islandici. Zoological manuscripts. (Ms, preserved in BMNH).

SOLDATOV, V. K. 1939. Some new or rare species of fishes of our northern seas. p. 151-166, 3 fig. N. M. Knipowitch meml. vol., Moscow, Leningrad (in Russian, English summary).

SOLDATOV, V.K.; PAVLENKO, M. 1915. Description of a new species of the family Rajidae from Peter the Great Bay and from the Okhotsk Sea. Ezheg. zool. Muz., 20(1):162, 1 pl.

SOLJAN, T. 1948. Fauna i flora jadrana. 1. Ribe. Zagreb, 437 p., 1350 fig.

--. 1963. Fishes of the Adriatic (Ribe Jadrana). Fauna flora adriatica, 1 (revised and enlarged for the English edition), 428 p., many unnumbered fig.

SOLOMON-RAJU, N.; ROSENBLATT, R. H. 1971. New record of the parasitic eel Simenchelys parasiticus from the central North Pacific with notes on its metamorphic form. Copeia, 1971(2):312-314, 1 fig.

SOMIYA, H. 1977. Bacterial bioluminescence in chlorophthalmid deep-sea fish: a possible interrelationship between the light organ and the eyes. Experientia, 33:906-908.

--. 1979. "Yellow lens" eyes and luminous organs of Echiostoma barbatum (Stomiatoidei, Melanostomiatidae). Jap. J. Ichthyol., 25(4):269-272.

SONNIER, F.; TEERLING, J.; HOESE, H. D. 1976. Observations on the offshore reef and platform fish fauna of Louisiana. Copeia, 1976(1):105-111.

SOOT-RYEN, T. 1937. Aphanopus Schmidti Saemundsson, a fish new to the Norwegian fauna. Nyt Mag. naturvid., 76:237-243.

SOURIE, R. 1954. Contribution à l'étude écologique des côtes rocheuses du Sénégal. Mém. Inst. fr. Afr. noire, (38):1-342.

SOUTH, J. F. 1845. Thunnus. p. 620-622. In: Encyclopedia Metropolitana (ed. by Smedley, Rose & Rose). London, 25.

SPARTA, A. 1927. Contributo alla conoscenza dello sviluppo post-larvale in Regalecus glesne. Asc. Memorie R. Com. talassogr. ital., (135):1-8, 1 pl.

--. 1929. Stadi post-larvali di Tetragonurus cuvieri Risso. Memorie R. Com. talassogr. ital., (168):1-8.

--. 1933. Uova e larve di Julis giofredi Risso e Julis pavo ottenute da fecondazione artificiale e stadi post-embrionali raccolti nel plancton. Memorie R. Com. talassogr. ital., (203):3-16, 23 fig.

--. 1934. Uove e larve di Gobiidae. 1. Gobius paganellus L. Memorie R. Com. talassogr. ital., (211):10 p.

--. 1934. Uova e larve di Gobiidae. 2. Gobius jozo L. Memorie R. Com. talassogr. ital., (216):8 p.

--. 1936. Uova e larve di Xyrichthys novacula Cuvier. Memorie R. Com. talassogr. ital., (238):3-5, 8 fig.

--. 1939. Contributo alla conoscenza dello sviluppo embrionale e post-embrionale nei Murenoidi. 5. Nuova specie di uovo di Muraenoidi (Conger conger?). Memorie R. Com. talassogr. ital., (264):1-6, 1 pl.

--. 1939. Contributo alla conoscenza dello sviluppo embrionale e post-embrionale nei Murenoidi. 6. Ophisurus serpens (L.). Memorie R. Com. talassogr. ital., (267):1-12, 1 pl.

--. 1940. Contributo alla conoscenza dello sviluppo embrionale e post-embrionale nei Murenoidi. 8. Echelus myrus L. Memorie R. Com. talassogr. ital., (273):1-6, 8 fig.

--. 1941. Contributo alla conoscenza dello sviluppo embrionale e post-embrionale negli Scorpenidi. I. Scorpaena porcus L. Archo Oceanogr. Limnol., 1:109-116, pl. 1.

--. 1941. Contributo alla conoscenza di uova, stadi embrionali e post-embrionali negli Scorpenidi. II. Scorpaena scrofa L. III. S. ustulata Lowe. IV. S. dactyloptera Delaroche. Archo Oceanogr. Limnol., 1:203-210, fig. 1-15, pl. 1.

--. 1946. Contributo alla conoscenza dello sviluppo post-embrionale in Cubiceps gracilis Lowe. Boll. Pesca Piscic. Idrobiol., 22(1):17-21, 5 fig.

--. 1948. Uovo e larva alla schiusa di Stomiatidae riferibilia Bathophilus nigerrimus Gigl. Boll. Pesca Piscic. Idrobiol., 24:194-196.

--. 1950. Uova e larve di Gobiidae. V. Gobius capito C. V. Boll. Pesca Piscic. Idrobiol., (n. s.)5(1):8 p.

--. 1952. Contributo alla conoscenza dello sviluppo larvale di Myctophum metopoclampum Cocco. Boll. Pesca Piscic. Idrobiol., (n.s.)7:5-10, fig. 1-6.

--. 1954. Des stades jeunes rares de téléostéens marins (Uraleptus maraldii Risso, Temnodon saltator C.V., Lophotes cepedianus Giorna). Rapp. P.-v. Réun. Commn int. Explor. scient. Mer Méditerr., 12:79-84, 1 pl.

--. 1956. Pharyngognathi, Scorpaenidae. In: Uova, larve e stadi giovanili di Teleostei. (U. d'Ancona ed.). Fauna Flora Golfo Napoli, 38:576-626, fig. 434-503, pl. 38-39.

SPINOLA, M. 1807. Lettre sur quelques poissons peu connus du Golfe de Gênes, adressée à M. Faujas de Saint-Fond. Annls Mus. Hist. nat., Paris, 10:366-380, pl. 28.

SPIX, J. B.; AGASSIZ, L. 1829-1831. Selecta genera et species piscium quos in itinere per Brasiliam annis 1817-20... peracto collegit et pingendos curavit Dr. J.B. de Spix... digressit, descripsit et observationibus anatomicis illustravit Dr. L. Agassiz, praefatus est et editit itineris socius Dr. C.F. Ph. de Martius (Memoriae J.B. de Spix) Monachii, xvi + i-ii + 138 p., 101 pl. (for dating, see Whitehead, P.J.P. & G.S. Myers, 1971, J. Soc. Biblphy nat. Hist., 5(6):478-497).

SPRINGER, S. 1959. A new shark of the family Squalidae from the Carolina continental slope. Copeia, 1959:30-33.

--. 1973. Odontaspididae, Mitsukurinidae, Lamnidae, Cetorhinidae, Alopiidae, Orectolobidae, Scyliorhinidae, p. 11-21, in Hureau, J.C. & Th. Monod (ed.). Check-list of the fishes of the north-eastern Atlantic and of the Mediterranean/Catalogue des poissons du nord-est Atlantique et de la Méditerranée (Clofnam). Paris, Unesco, 2 vol.

--. 1979. A revision of the catsharks, family Scyliorhinidae. NOAA tech. Rep. NMFS Circ., (422):1-152, 97 fig., 4 tab.

SPRINGER, S.; BULLIS, H. R. 1956. Collections by the "Oregon" in the Gulf of Mexico. Spec. scient. Rep. U.S. Fish. Wildl. Serv. (Fish.), (196):1-134.

SPRINGER, S.; COLLETTE, B. B. 1971. The Gulf of Guinea stingray Dasyatis rudis. Copeia, 1971(2):338-341, 3 fig.

SPRINGER, V. G. 1958. Systematics and zoogeography of the clinid fishes of the subtribe Labrisomini Hubbs. Publs Inst. mar. Sci., Univ. Tex., 5:417-492.

--. 1959. Cypselurus luetkeni (Jordan & Evermann) a synonym of Cypselu-

<u>rus</u> <u>heterurus</u> (Rafinesque)?. <u>Copeia</u>, 1959(2):166-167, 4 tab.

--. 1962. A review of the blenniid fishes of the genus <u>Ophioblennius</u> Gill. <u>Copeia</u>, 1962(2):426-433.

--. 1966. A new species of the blenniid fish genus <u>Entomacrodus</u> from the eastern Atlantic Ocean. <u>Atlantide Rep.</u>, (9):59-61, pl. 6 (fig. A-C).

--. 1967. Revision of the circumtropical shorefish genus <u>Entomacrodus</u> (Blenniidae: Salariinae). <u>Proc. U.S. natn. Mus.</u>, 122(3582):1-150.

--. 1968. Osteology and classification of the fishes of the family Blenniidae. <u>Bull. U.S. natn. Mus.</u>, (284):1-85, 16 fig., 11 pl.

SPRINGER, V. G.; GOMON, M. F. 1975. Variation in the western Atlantic clinid fish <u>Malacoctenus</u> <u>triangulatus</u> with a revised key to the Atlantic species of <u>Malacoctenus</u>. <u>Smithson. Contrib. Zool.</u>, (200):1-11.

SPRINGER, V. G.; HOESE, H. D. 1958. Notes and records of marine fishes from the Texas coast. <u>Tex. J. Sci.</u>, 10(3):343-348.

STAIGER, J. C. 1965. Atlantic flying-fishes of the genus <u>Cypselurus.</u> <u>Bull. mar. Sci.</u>, 15(3):672-725, 19 fig.

--. 1972. <u>Bassogigas</u> <u>profundissimus</u> (Pisces: Brotulidae) from the Puerto Rico Trench. <u>Bull. mar. Sci.</u>, 22(1):26-33.

STANDER, G. G.; Le ROUX, P. J. 1968. Notes on fluctuations of the commercial catch of the South African pilchard (<u>Sardinops</u> <u>ocellata</u>) 1950-1965. <u>Investl Rep. Div. Fish. Un. S. Afr.</u>, (65):1-14.

STARCK II, W. A. 1964. A contribution to the biology of the grey snapper, <u>Lutjanus</u> <u>griseus</u> (Linnaeus), in the vicinity of lower Matecumbe Key, Florida. Ph. D. thesis, Univ. Miami, xv + 258 p., 23 fig.

STARKS, E. C. 1908. The characteristics of <u>Atelaxia</u>, a new suborder of fishes. <u>Bull. Mus. comp. Zool. Harv.</u>, 52:15-22, 5 pl.

--. 1922. The specific differences between the chub mackerels of the Atlantic and Pacific Oceans. <u>Copeia</u>, 1922(103):9-11.

STARNES, W. C. 1981. Priacanthidae, <u>in</u>: Fischer, W., Bianchi G. & W.B. Scott (ed.), FAO species identification sheets for fishery purposes. Eastern Central Atlantic; fishing areas 34 and 47 (in part). Canada Funds-in Trust. Ottawa, Department of Fisheries and Oceans Canada, by arrangement with the Food and Agriculture Organization of the United Nations, vol. 1-7: pag. var.

STASSANO, E. 1890. La pesca sulle spiagge atlantiche del Sahara, Pesci. <u>Annali Agric.</u>:1-103. (also 1932, in <u>Boll. Pesca Piscic. Idrobiol.</u>, 8(2):219-265, 1 map).

STAUCH, A. 1965. Caractères morphologiques et biogéographiques de deux <u>Heteromycteris</u> africains (Pisces, Teleostei, Heterosomata, Soleidae). <u>Cah. ORSTOM. Océanogr.</u>, 3(2):5-10, 5 fig., map.

--. 1965. Sur la répartition géographique d'<u>Arnoglossus</u> <u>imperialis</u> (Raf., 1810) et description d'une espèce nouvelle, <u>Arnoglossus</u> <u>blachei</u> (Pisces, Teleostei, Heterosomata, Bothidae). <u>Bull. Mus. natn. Hist. nat.. Paris</u>, (2)37(2):252-260, 8 fig.

--. 1966. Quelques données sur les <u>Bothus</u> de l'Atlantique et description d'une espèce nouvelle <u>Bothus</u> <u>guibei</u> n. sp. (Pisces, Teleostei, Heterosomata). <u>Bull. Mus. natn. Hist. nat.. Paris</u>, (2)38(2):118-125, 8 fig.

--. 1967. Description d'une nouvelle espèce de Bothinae: <u>Arnoglossus</u> <u>entomorhynchus</u> n. sp. (Pisces, Teleostei, Heterosomata). <u>Bull. Mus. natn. Hist. nat.. Paris</u>, (2)39(4):660-664, fig.

STAUCH, A.; BLACHE, J. 1964. Contribution à la connaissance du genre <u>Ateleopus</u> Schlegel, 1846 (Pisces, Teleostei, Ateleopoidei, Ateleopidae) dans l'Atlantique oriental. <u>Cah. ORSTOM. Océanogr</u>, 2(2):47-54, 3 fig.

STAUCH, A.; BLANC, M. 1962. Description d'un sélacien rajiforme des eaux douces du Nord-Cameroun: <u>Potamotrygon</u> <u>garouaensis</u> n. sp. <u>Bull. Mus. natn. Hist. Nat.. Paris</u>, (2)34(2):166-171, 4 fig., 1 tab.

STAUCH, A.; CADENAT, J. 1965. Révision du genre _Psettodes_ Bennett, 1831 (Pisces, Teleostei, Heterosomata). _Cah. ORSTOM. Océanogr._, 3(4):19-30, 11 fig.

STEAD, D. G. 1908. New fishes from New South Wales. No. 1. Sydney, 12 p.

STEHMANN, M. 1970. Vergleichend morphologische und anatomische Untersuchungen zur Neuordnung der Systematik der nordostatlantischen Rajidae. _Arch. FischWiss._, 21(2):73-163, 1 fig., 27 pl.

--. 1971. _Raja (Leucoraja) leucosticta_ spec. nov. (Pisces, Batoidei, Rajidae), eine neue Rochenart aus dem Seegebiet des tropischen Westafrika; gleichzeitig zur Frage des Vorkommens von _Raja ackleyi_ Garman, 1881 im mittleren Ostatlantik. _Arch. FischWiss._, 22(1):1-16, fig. 1-12.

--. 1971. Ergebnisse der Forschungsreisen des FFS "Walther Herwig" nach Südamerika. XVII. _Raja (Raja) herwigi_ Krefft, 1965; ergänzende Untersuchungen zum subgenerischen Status der Art. _Arch. FischWiss._, 22(2): 85-97, fig. 1-8.

--. 1971. Untersuchungen zur Validität von _Raja maderensis_ Lowe, 1839, zur geographischen Variation von _Raja straeleni_ Poll, 1951, und zum subgenerischen Status beider Arten (Pisces, Batoidei, Rajidae). _Arch. FischWiss._, 22(3):175-199, fig. 1-13.

--. 1977. Ein neuer archibenthaler Roche aus dem Nordostatlantik, _Raja kreffti_ spec. nov. (Elasmobranchii, Batoidea, Rajidae), die zweite Spezies im Subgenus _Malacoraja_ Stehmann, 1970. _Arch. FischWiss._, 28 (2/3):77-93, fig. 1-9.

--. 1978. _Raja_ "bathyphila", eine Doppelart des Subgenus _Rajella_: Wiederbeschreibung von _R. bathyphila_ Holt & Byrne, 1908 und _Raja bigelowi_ spec. nov. (Chondrichthyes, Rajiformes, Rajidae). _Arch. FischWiss._, 29(1/2):23-58, fig. 1-14.

--. 1981. Platyrhinidae, Pristidae, Rhinobathidae, Rhynchobatidae, _in_: Fischer, W., G. Bianchi & W.B. Scott (ed.), FAO species identification sheets for fishery purposes. Eastern Central Atlantic; fishing areas 34 and 47 (in part). Canada Funds-in-Trust. Ottawa, Department of Fisheries and Oceans Canada, by arrangement with the Food and Agriculture Organization of the United Nations, vol. 1-7: pag. var.

STEHMANN, M.; BÜRKEL, D. L. 1984. Pristidae, _in_ Whitehead, P.J.P., Bauchot, M.L., Hureau, J.C., Nielsen, J. & E. Tortonese, Fishes of the North-eastern Atlantic and the Mediterranean/Poissons de l'Atlantique du Nord-Est et de la Méditerranée, Paris, Unesco, 1:153-155, 2 fig., 2 maps.

--. 1984. Torpedinidae, _in_ Whitehead, P.J.P., Bauchot, M.L., Hureau, J.C., Nielsen, J. & E. Tortonese, Fishes of the North-eastern Atlantic and the Mediterranean/Poissons de l'Atlantique du Nord-Est et de la Méditerranée, Paris, Unesco, 1:159-162, 3 + 3 fig., 3 maps.

--. 1984. Rajidae, _in_ Whitehead, P.J.P., Bauchot, M.L., Hureau, J.C., Nielsen, J. & E. Tortonese, Fishes of the North-eastern Atlantic and the Mediterranean/Poissons de l'Atlantique du Nord-Est et de la Méditerranée, Paris, Unesco, 1:163-196, 32 + 9 fig., 33 maps.

--. 1984. Chimaeridae, _in_ Whitehead, P.J.P., Bauchot, M.L., Hureau, J.C., Nielsen, J. & E. Tortonese, Fishes of the North-eastern Atlantic and the Mediterranean/Poissons de l'Atlantique du Nord-Est et de la Méditerranée, Paris, Unesco, 1:212-215, 3 fig., 3 maps.

--. 1984. Rhinochimaeridae, _in_ Whitehead, P.J.P., Bauchot, M.L., Hureau, J.C., Nielsen, J. & E. Tortonese, Fishes of the North-eastern Atlantic and the Mediterranean/Poissons de l'Atlantique du Nord-Est et de la Méditerranée, Paris, Unesco, 1:216-218, 2 fig., 2 maps.

STEHMANN, M.; LENZ, W. 1973. Ergebnisse der Forschungsreisen des FFS "Walther Herwig" nach Südamerika. XXVI. Systematik und Verbreitung der Artengruppe - _Seriolella punctata_ (Schneider, 1801), _S. porosa_

Guichenot, 1848, S. dobula (Günther, 1869) - sowie taxonomische Bemerkungen zu Hyperoglyphe Günther, 1859 und Schedophilus Cocco, 1839 (Osteichthyes, Stromateoidei, Centrolophidae). Arch. FischWiss., 23: 179-201.

STEIN, D. L. 1986. Cyclopteridae, in P.J.P. Whitehead, M.L. Bauchot, J.C. Hureau, J. Nielsen & E. Tortonese. Fishes ot the North-eastern Atlantic and the Mediterranean/Poissons de l'Atlantique du Nord-Est et de la Méditerranée, Paris, Unesco, 3:1269-1274, 5 fig.

STEIN, D. L.; ABLE, K. W. 1986. Liparidae, in P.J.P. Whitehead, M.L. Bauchot, J.C. Hureau, J. Nielsen & E. Tortonese. Fishes of the North-eastern Atlantic and the Mediterranean/Poissons de l'Atlantique du Nord-Est et de la Méditerranée, Paris, Unesco, 3:1275-1283, 11 fig.

STEIN, D.; BUTLER, J. 1972. First records of Bathysaurus mollis (Pisces: Bathysauridae) from the northeast Pacific Ocean. J. Fish. Res. Bd Can., 29(7):1093-1094.

STEINDACHNER, F. 1861. Beiträge zur Kenntniss der Gobioiden. Sber. Akad. Wiss. Wien, 42:283-292, pl.

--. 1861. Ichthyologische Mittheilungen. II. Verh. zool.-bot. Ges. Wien, 11:133-144, pl.

--. 1865. Vorläufiger Bericht über die an der Ostküste Tenerife's bei Santa Cruz gesammelten Fische. Sber. Akad. Wiss. Wien, (1)51:398-404.

--. 1866. Zur Fischfauna von Port Jackson in Australien. Sber. Akad. Wiss. Wien, (1)53:424-480, 7 pl.

--. 1867. Über einige Meeresfische aus der Umgebung von Monrovia in Westafrika. p. 517-527. In: Ichthyologische Notizen (IV). Sber. Akad. Wiss. Wien, (1)55:517-534, pl. 1-6.

--. 1867. Über eine neue Scopelus- und Monacanthus-Art aus China. p. 711-713. In: Ichthyologische Notizen (V). Sber. Akad. Wiss. Wien, (1)55: 701-717, pl. 1-3.

--. 1867. Ichthyologischer Bericht über eine nach Spanien und Portugal unternommene Reise. IV. Übersicht der Meeresfische an den Küsten Spaniens und Portugals. Sber. Akad. Wiss. Wien, (1)56:603-708, 9 pl.

--. 1868. Ichthyologischer Bericht über eine nach Spanien und Portugal untermommene Reise. V-VI. Übersicht der Meeresfische an den Küsten Spaniens und Portugals. Sber. Akad. Wiss. Wien, (1)57:351-424, 6 pl.; 667-738, 6 pl.

--. 1868. Ichthyologische Notizen (VII). Sber. Akad. Wiss. Wien, (1)57: 965-1008, pl. 1-5.

--. 1869-1870. Zur Fischfauna des Senegal. Sber Akad. Wiss. Wien, (1)60 (1869):669-714, 12 pl., 945-995, 8 pl.; (1)61(1870):533-583, 8 pl.

--. 1870. Ichthyologische Notizen (IX). Sber. Akad. Wiss. Wien, (1)60: 290-318.

--. 1877. Ichthyologische Beiträge. (V.) V. Über einige neue oder seltene Fischarten aus dem atlantischen, indischen und stillen Ocean. Sber. Akad. Wiss. Wien, (1)74:203-240.

--. 1878. Ichthyologische Beiträge. VI. Sber. Akad. Wiss. Wien, 77(1): 379-392, 3 pl.

--. 1880. Ichthyologische Beiträge. VIII. Sber. Akad. Wiss. Wien, (1)80: 119-191, 3 pl.

--. 1881. Ichthyologische Beiträge. X. Sber. Akad. Wiss. Wien, (1)83: 179-219, 8 pl.

--. 1881. Ichthyologische Beiträge. XI. Sber. Akad. Wiss. Wien, (1)83: 393-408, 1 pl.

--. 1881. Beiträge zur Kenntniss der Fische Afrika's und Beschreibung einer neuen Sargus-Art von den Galapagos-Inseln. Denkschr. Akad. Wiss.. Wien, (1)44:19-58, 10 pl.

--. 1882. Beiträge zur Kenntnis der Fische Afrika's und Beschreibung einer neuen Paraphoxinus-Art aus der Herzegowina. Denkschr. Akad.

Wiss., Wien, (1)45:1-18, 6 pl.

--. 1887. Ichthyologische Beiträge. XIV. Sber. Akad. Wiss. Wien, (1)96: 56-68, 4 pl.

--. 1891. Ichthyologische Beiträge. XV. Über einige seltene und neue Fischarten aus dem canarischen Archipel. Sber. Akad. Wiss. Wien, (1) 100: 343-374, 3 pl.

--. 1892. Über einige neue und seltene Fischarten aus der ichthyologischen Sammlung des k. k. Naturhistorischen Hofmuseums. Denkschr. Akad. Wiss., Wien, (1)59:357-385, 6 pl.

--. 1894. Ichthyologische Beiträge. XVII. Sber. Akad. Wiss. Wien, (1) 103: 443-464, 5 pl.

--. 1895. Vorläufige Mittheilung über einige neue Fischarten aus der ichthyologischen Sammlung des k. k. Naturhistorischen Hofmuseums in Wien. Anz. Akad. Wiss. Wien, 32:180-183.

--. 1895. Die Fische Liberia's. Notes Leyden Mus., 16:1-96, 4 pl.

--. 1903. Fische. p. 409-464, pl. 17-18. In: W. Kükenthal: Ergebnisse einer zoologischen Forschungsreise in den Molukken und in Borneo. Abh. senckenb. naturforsch. Ges., 25.

STEINDACHNER, F.; DÖDERLEIN, L. 1883. Beiträge zur Kenntniss der Fische Japan's (I). Denkschr. Akad. Wiss., Wien, 47:211-242, 7 pl.

--. 1884. Beiträge zur Kenntniss der Fische Japan's. (III). Denkschr. Akad. Wiss., Wien, 49:171-212, 7 pl.

--. 1887. Beiträge zur Kenntniss der Fische Japan's. (IV). Denkschr. Akad. Wiss., Wien, 53:257-296, 4 pl.

STEINDACHNER, F.; KNER, R. 1870. Über einige Pleuronectiden, Salmoniden, Gadoiden und Blenniiden aus der Decastris-Bay und von Viti-Levu. Sber. Akad. Wiss. Wien, (1)61:421-446, pl.

STEINITZ, H. 1949. Contribution to the knowledge of the Blenniidae of the eastern Mediterranean II. Istanb. Univ. Fen. Fak. Mecm., (B)14: 170-197.

--. 1950. On the zoogeography of the teleostean genera Salarias, Ophioblennius and Labrisomus. Archo zool. ital., 35:325-348.

STEPHENS, W. M. 1965. Summer cruise to the Sargasso Sea. Sea Front., 11: 108-123.

STERBA, G. 1962. Freshwater fishes of the world (translated and revised by D. W. Tucker). London, 878 p., 1193 fig., 192 pl.

STEVEN, G.A. 1947. Report on the fisheries of Sierra Leone. London, 66p.

STILES, C. W. 1912. Opinion 41. Athlennes vs. Ablennes. p. 94-95. In: Opinions rendered by the International Commission on Zoological Nomenclature, Opinions 38 to 51. Smithson. Inst. Publs, (2060):89-117.

STINTON, F. C. 1967. The otoliths of the teleostean fish Antigonia capros and their taxonomic significance. Bocagiana, (13):1-17, 2 pl.

STORER, D. H. 1839. Fishes of Massachusetts. p. 1-202, pl. 1-3. In: Reports on the fishes, reptiles and birds of Massachusetts. Boston, 439 p., 4 pl.

--. 1843. Description of a new species of Torpedo. Am. J. Sci., 45(1): 165-170, pl. 3.

--. 1853-67. A history of the Fishes of Massachusetts. Mem. Am. Acad. Arts Sci., (n.s.) 5(2), 1853:49-92, pl. 1-8; 122-168, pl. 9-16; 257-296, pl. 17-23; 6(2), 1858:309-372, pl. 24-29; 8(2), 1863:389-434, pl. 30-35; 9(1), 1867:217-256, pl. 36-39 (also book, Cambridge & Boston, 1867, 287 p.).

STOREY, M. H. 1939. Contributions toward a revision of the ophichthyid eels. 1. The genera Callechelys and Bascanichthys, with descriptions of new species and notes on Myrichthys. Stanford ichthyol. Bull., 1 (3):61-84.

STORM, V. 1881. Bidrag til kundskab om Trondhjemsfjordens fauna. III. K. norske Vidensk. Selsk. Skr. (1880), 4:73-83.

STRASBURG, D. W. 1964. Postlarval scombroid fishes of the genera _Acanthocybium, Nealotus_ and _Diplospinus_ from the central Pacific Ocean. _Pacif. Sci._, 18(2):174-185, 10 fig.

--. 1966. _Golem cooperae_, a new antennariid fish from Fiji. _Copeia_, 1966(3):475-477.

--. 1970. A report on the billfishes of the central Pacific Ocean. _Bull. mar. Sci._, 20(3):575-604, 20 fig.

STREETS, T. H. 1877. Contributions to the natural history of the Hawaiian and Fanning Islands and Lower California. Ichthyology. _Bull. U.S. natn. Mus._, 7:43-102.

STRUHSAKER, P. 1965. The whalefish _Barbourisia rufa_ (Cetunculi) from waters off southeastern United States. _Copeia_, 1965(3):376-377, 1 fig.

--. 1969. Demersal fish resources: composition, distribution, and commercial potential of the continental shelf stocks off southeastern United States. _Fishery ind. Res._, 4(7):261-300.

STRÖMMAN, P. H. 1896. Leptocephalids in the University Zoological Museum at Upsala. Upsala, vi + 53 p., 5 pl.

SUDA, A. 1973. Developing fisheries for non-conventional species. _J. Fish. Res. Bd Can._, 30(12):2121-2158.

SULAK, K. J. 1974. Morphological and ecological observations on Atlantic ipnopid fishes of the genus _Bathytyphlops. Copeia_, 1974(2):570-573.

--. 1975. _Talismania mekistonema_, a new Atlantic species of the family Alepocephalidae (Pisces: Salmoniformes). _Bull. mar. Sci._, 25(1):88-93, 5 fig.

--. 1977. _Aldrovandia oleosa_, a new species of the Halosauridae, with observations on several other species of the family. _Copeia_, 1977(1): 11-20.

--. 1977. The systematics and biology of _Bathypterois_ (Pisces, Chlorophthalmidae) with a revised classification of benthic myctophiform fishes. _Galathea Rep._, 14:49-108.

--. 1981. Aulopidae, Synodontidae, _in_ Fischer, W., Bianchi, G. & W.B. Scott (ed.), FAO species identification sheets for fishery purposes. Eastern Central Atlantic; fishing areas 34, 47 (in part). Canada Funds-in-Trust. Ottawa, Department of Fisheries and Oceans Canada, by arrangement with the Food and Agriculture Organization of the United Nations, vol. 1-7: pag. var.

--. 1984. Aulopidae, _in_ Whitehead, P.J.P., Bauchot, M.L., Hureau, J.C., Nielsen, J. & E. Tortonese, Fishes of the North-eastern Atlantic and the Mediterranean/Poissons de l'Atlantique du Nord-Est et de la Méditerranée, Paris, Unesco, 1:403-404, 1 fig., 1 map.

--. 1984. Synodontidae, _in_ Whitehead, P.J.P., Bauchot, M.L., Hureau, J.C., Nielsen, J. & E. Tortonese, Fishes of the North-eastern Atlantic and the Mediterranean/Poissons de l'Atlantique du Nord-Est et de la Méditerranée, Paris, Unesco, 1:405-411, 5 + 7 fig., 5 maps.

--. 1984. Chlorophthalmidae, _in_ Whitehead, P.J.P., Bauchot, M.L., Hureau, J.C., Nielsen, J. & E. Tortonese, Fishes of the North-eastern Atlantic and the Mediterranean/Poissons de l'Atlantique du Nord-Est et de la Méditerranée, Paris, Unesco, 1:412-420, 8 fig., 8 maps.

--. 1986. Halosauridae, _in_ P.J.P. Whitehead, M.L. Bauchot, J.C. Hureau, J. Nielsen & E. Tortonese. Fishes of the North-eastern Atlantic and the Mediterranean/Poissons de l'Atlantique du Nord-Est et de la Méditerranée, Paris, Unesco, 2:593-598, 5 fig.

--. 1986. Notacanthidae, _in_ P.J.P. Whitehead, M.L. Bauchot, J.C. Hureau, J. Nielsen & E. Tortonese. Fishes of the North-eastern Atlantic and the Mediterranean/Poissons de l'Atlantique du Nord-Est et de la Méditerranée, Paris, Unesco, 2:599-603, 4 fig.

SULAK, K. J.; CRABTREE, R. E.; HUREAU, J. C. 1984. Provisional review of the genus <u>Polyacanthonotus</u> (Pisces, Notacanthidae) with description of a new Atlantic species <u>Polyacanthonotus merretti</u>. <u>Cybium</u>, **8**(4): 57-68, 2 fig.

SULAK, K. J.; WENNER, C. A.; SEDBERRY, G. R.; GUELPEN, L. van. 1985. The life history and systematics of deep-sea lizard fishes, genus <u>Bathysaurus</u> (Synodontidae). <u>Can. J. Zool.</u>, **63**:623-642, 18 fig., 6 tab.

SUPINO, F. 1905. Il <u>Todarus brevirostris</u> Gr. e Cal. <u>Ric. Lab. Anat. norm. R. Univ. Roma</u>, **11**(3):255-259, 6 fig.

--. 1905. Il <u>Saurenchelys cancrivora</u> Peters. <u>Ric. Lab. Anat. norm. R. Univ. Roma</u>, **11**(1):1-7.

SVENSSON, G. S. O. 1933. Freshwater fishes from the Gambia river (British West Africa). Results of the Swedish Expedition, 1931. <u>K. svenska Vetensk. Akad. Handl.</u>, **13**(3):1-102.

SVETLOV, M. F. 1978. A record of <u>Xenocyttus nemotoi</u> Abe in the Southeast Pacific. <u>Vop. Ikhtiol.</u>, **18**:544-545 (Engl. transl.: <u>J. Ichthyol.</u>, **18**: 493-494).

SVETOVIDOV, A. N. 1952. Fauna USSR. Fishes, 2(1). Clupeidae. 331 p., 54 fig., 53 pl. (in Russian)(Engl. transl., Israel Program Scient. Transl., Jerusalem, 1963:374 p.).

--. 1955. Contribution to the classification of the needlefishes. (<u>Belone belone</u> (L.)(Pisces, Belonidae). <u>Trudy. zool. Inst.. Leningr.</u>, **18**: 343-345. (in Russian)(Engl. transl.: Bur. Comm. Fish. Ichthyol. Lab. transl. no 4).

--. 1958. The blenniid fishes of the Black Sea. <u>Zool. Zh..</u> **37**(4):584-592, fig. 1-3, tab. 1-3.

--. 1961. The European species of the family Ophidiidae and the functional significance of peculiarities in the structure of their swimbladders. <u>Vop. Ikhtiol.</u>, (17):3-13. (transl. no 41, Ichthyol. Lab., USNM).

--. 1963. Fauna of USSR, Fishes, 2(1), Clupeidae (Engl. transl. of Svetovidov, 1952). Israël Program Scient. Transl., Jerusalem, 428 p.

--. 1964. Ryby Chernogo Morya (Fishes of the Black Sea). <u>Opred. Faune SSSR</u>, **86**:1-550, 191 fig. (in Russian).

--. 1981. The Pallas fish collection and the Zoographia Rosso-Asiatica: an historical account. <u>Archs nat. Hist.</u>, **10**:45-64.

--. 1984. Acipenseridae, <u>in</u> Whitehead, P.J.P., Bauchot, M.L., Hureau, J.C., Nielsen, J. & E. Tortonese, Fishes of the North-eastern Atlantic and the Mediterranean/Poissons de l'Atlantique du Nord-Est et de la Méditerranée, Paris, Unesco, 1:220-225, 6 + 6, fig., 6 maps.

--. 1984. Salmonidae, <u>in</u> Whitehead, P.J.P., Bauchot, M.L., Hureau, J.C., Nielsen, J. & E. Tortonese, Fishes of the North-eastern Atlantic and the Mediterranean/Poissons de l'Atlantique du Nord-Est et de la Méditerranée, Paris, Unesco, 1:373-385, 11 + 4 fig., 10 maps.

--. 1986. Merluccidae, <u>in</u> P.J.P. Whitehead, M.L. Bauchot,J.C. Hureau, J. Nielsen & E. Tortonese. Fishes of the North-eastern Atlantic and the Mediterranean/Poissons de l'Atlantique du Nord-Est et de la Méditerranée, Paris, Unesco, 2:677-679, 3 fig.

--. 1986. Gadidae, <u>in</u> P.J.P. Whitehead, M.L. Bauchot, J.C. Hureau, J. Nielsen & E. Tortonese. Fishes of the North-eastern Atlantic and the Mediterranean/Poissons de l'Atlantique du Nord-Est et de la Méditerranée, Paris, Unesco, 2:680-710, 33 + 8 fig.

SWAEN, A., 1886. Etudes sur le développement de la torpille (<u>Torpedo ocellata. Archs Biol.. Paris</u>, 7:537-585, pl. 14-16.

SWAIN, J. 1883. A review of Swainson's genera of fishes. <u>Proc. Acad. nat. Sci. Philad.</u>, 34:272-284.

SWAINSON, W. 1838-39. The natural history and classification of fishes, amphibians and reptiles, or monocardian animals. London, 1, 1838:386 p., 100 fig.; 2, 1839:448 p., 135 fig.

SYLVA, D.P. (see de SYLVA, D.P.)

TALBOT, F. H. 1960. Additions to the South African Museum collections of marine fishes. Ann. S. Afr. Mus., 45(2):257-259.

TALBOT, F. H.; PENRITH, M. J. 1962. Spearing behavior in feeding in the black marlin Istiompax marlina. Copeia, 1962(2):468.

--. 1962. Tunnies and marlins of South Africa. Nature, Lond., 193:558-559.

--. 1963. The white marlin Makaira albida (Poey) from the seas around South Africa. S. Afr. J. Sci., 59(7):330-331, 1 fig.

--. 1963. Synopsis of biological data on species of the genus Thunnus (Sensu lato)(South Africa). FAO Fish. Rep., (6)2:608-646.

--. 1965. Ctenogobius cloatus Smith, 1960, a synonym of Ctenogobius saldanha (Barnard, 1927). Ann. S. Afr. Mus., 48:189-193, 1 pl., 2 fig.

--. 1968. The tunas of the genus Thunnus in South African waters. Ann. S. Afr. Mus., 52(1):1-41, 8 fig., 8 tab.

TANAKA, S. 1905. On two new species of Chimaera. J. Coll. Sci. imp. Univ. Tokyo, 20(11):1-14, pl. 1-2.

--. 1908. Notes on some Japanese fishes with descriptions of fourteen new species. J. Coll. Sci. imp. Univ. Tokyo, 23(7):1-54, 4 pl.

--. 1908. Notes on some rare fishes of Japan, with descriptions of two new genera and six new species. J. Coll. Sci. imp. Univ. Tokyo, 23 (13):1-24, 2 pl.

--. 1909. Descriptions of one new genus and ten new species of Japanese fishes. J. Coll. Sci. imp. Univ. Tokyo, 27(8):1-27, pl. 1.

--. 1911-1930. Figures and descriptions of the fishes of Japan, including Riukiu Islands, Bonin Islands, Formosa, Kurile Islands, Korea and Southern Sakhalin. Tokyo. 18, 1914:295-318, pl. 86-90; 19, 1915: 319-342, pl. 91-95; 27, 1918:475-494, pl. 131-135; 28, 1918:495-514, pl. 136-137; 29, 1918:515-538, pl. 138-139; 30-34, 1925:539-644, pl. 114-153; 35-41, 1927:645-808, pl. 154-171; 42-45, 1928:809-902, pl. 172-182; 46, 1929:903-924, pl. 183-184; 47-48, 1930:925-960, pl. 185-190.

--. 1916. Two new species of Japanese fishes. Zool. Mag., Tokyo, 28:257-258.

--. 1917. Eleven new species of fishes of Japan. Zool. Mag., Tokyo, 29: 7-12. (in Japanese).

TANING, A. V. 1918. Mediterranean Scopelidae (Saurus, Aulopus, Chlorophthalmus and Myctophum). Rep. Dan. oceanogr. Exped. Mediterr., 2 (Biol.)(A7): 1-154.

--. 1928. Synopsis of scopelids in the North Atlantic. Vidensk. Meddr dansk naturh. Foren., 86:49-69.

--. 1931-33. Teleostei. Iniomi-Myctophidae. Fiches 109-128. Faune ichthyol. Atlant. N., Cahier 7, 1931:109-110, 117-118, 121; Cahier 11, 1932:111-116, 119-120, 124-128; Cahier 12, 1933:122-123.

--. 1932. Notes on scopelids from the Dana Collections I. Vidensk. Meddr dansk naturh. Foren., 94:125-146.

--. 1955. On the breeding areas of the swordfish (Xiphias.). Deep Sea Res., 3(Suppl.): 438-450, 4 fig.

TAQUIN, A. 1903. Les îles Canaries et les parages de pêche canariens. Bull. Soc. belge r. Géogr., 27:37-99.

TARANETS, A. Ya.; ANDRIASHEV, A. P. 1934. On a new genus and species, Petroschmidtia albonotata (Zoarcidae, Pisces) from the Okhotsk Sea. Dokl. Akad. Nauk SSSR, 2(8):506-512.

TARDENT, P. 1959. Capture d'un Abudefduf saxatilis vaigiensis Q. et G. (Pisces, Pomacentridae) dans le Golfe de Naples. Revue suisse Zool.,

66(20):347-351, fig. 1-2.

TASSEL, J. L. van (see Van TASSEL, J. L.)

TAVOLGA, W. 1950. Development of the gobiid fish, _Bathygobius soporator._ J. Morph., **87**:467-492.

TAYLOR, W. R.; DYKE, G. van. 1981. Ariidae, _in_ Fischer, W., Bianchi, G. & W.B. Scott (ed.), FAO species identification sheets for fishery purposes. Eastern Central Atlantic; fishing areas 34, 47 (in part). Canada Funds-in-Trust. Ottawa, Department of Fisheries and Oceans Canada, by arrangement with the Food and Agriculture Organization of the United Nations, vol. 1-7: pag. var.

TCHERNAVIN, V. V. 1947. Six specimens of Lyomeri in the British Museum (with notes on the skeleton of Lyomeri). _J. Linn. Soc. (Zool.)_, **41** (279):287-350, 15 fig., pl. 2-3.

--. 1953. Summary of the feeding mechanisms of the deep-sea fish, _Chauliodus sloani_. London, 101 p., 10 pl., 53 fig.

TEE-VAN, J. 1940. Eastern Pacific Expeditions of the New-York Zoological Society. 17. A review of the American fishes of the family Cirrhitidae. _Zoologica, N. Y._, **25**(5):53-64, 4 fig., 1 pl.

TEMMINCK, C. J.; SCHLEGEL, H. 1843-1850. Pisces. _In_: P. F. von Siebold. Fauna Japonica. Lugduni Batavorum, 324 p., 144 pl. (for dating, see M.L. Bauchot, P.J.P. Whitehead & Th. Monod, 1982, _Cybium_, **6**(3): 67).

TEMPLEMAN, W. 1965. Rare skates of the Newfoundland and neighbouring areas. _J. Fish. Res. Bd Can._, **22**(2):259-279, fig. 1-13.

--. 1966. A record of _Bathypterois dubius_ Vaillant from the western North Atlantic, and a review of the status of the species. _J. Fish. Res. Bd Can._, **23**(5):715-722.

TEMPLEMAN, W.; SQUIRES, H. J. 1963. Three records of the black scabbardfish, _Aphanopus carbo_ Lowe, from the Canadian region of the western Atlantic. _J. Fish. Res. Bd Can._, **20**:273-278.

TENG, H. T. 1959. Studies on élasmobranch fishes from Formosa. Part II. A new carcharoid shark, _Carcharias yangi_, from Formosa. _Rep. Inst. Fish. Biol. natn. Taiwan Univ._, **1**(3):12-15, 1 fig.

TENORE, M. 1810. Sopra una nuova specie di Squadro pescato nelle acque della riviera di Chiaja del littorale di Napoli. _Memorie Accad. pont. Napoli_, **1**:241-265.

TESCH, F. W. 1981. Anguillidae, _in_ Fischer, W., Bianchi, G. & W.B. Scott (ed.), FAO species identification sheets for fishery purposes. Eastern Central Atlantic; fishing areas 34, 47 (in part). Canada Funds-in-Trust. Ottawa, Department of Fisheries and Oceans Canada, by arrangement with the Food and Agriculture Organization of the United Nations, vol. 1-7: pag. var.

THIAM, M. 1979. Biologie de la reproduction de _Cynoglossus canariensis_ (Steind.). _Docum. scient. Centre Rech. océanogr. Dakar-Thiaroye_, **68**: 205-213.

THIELEN, R. van. 1977. The food of juvenile _Sardinella aurita_ and of juvenile and adult _Anchoa guineensis_ in near shore waters off Ghana, West Africa. _Ber. dt. wiss. Kommn Meeresforsch._, (N.F.)**25**(1-2):46-53.

THIOLLIERE, V. J. 1857. Ichthyologie. p. 138-222. _In_: X. Montrouzier: Essai sur la faune de l'île de Woodlark ou Moiou. Lyon, 226 p.

THOMAS, L. J. 1953. _Botriocephalus abyssmus_ n. sp., a tapeworm from the deep-sea fish _Echiostoma tanneri_ (Gill) with notes on its development. p. 269-276. G.S. Thapar Commemoration Volume, Lucknow.

THOMINOT, A. 1883. Note sur le genre _Aplodon_, poisson de la famille des Sparidae, voisin des girelles. _Bull. Soc. philomath. Paris_, (7)**5**: 141-144.

THOMOPOULOS, A. 1954. Sur quelques oeufs planctoniques de téléostéens de la Baie de Villefranche. _Bull. Inst. océanogr. Monaco_, (1043):1-15, 18 fig.

THOMPSON, W. 1837. (Notes on the natural history of Ireland). Proc. zool. Soc. Lond., 1837:52-63.

--. 1838. Contributions towards a knowledge of the Crenilabri of Ireland, including new species. Mag. Zool. Bot., 2:442-450, 2 pl.

--. 1840. On a new genus of fishes from India. Mag. nat. Hist., (n.s.)4:184-186.

THOMPSON, W. F. 1916. Fishes collected by the United States Bureau of Fisheries Steamer "Albatross" during 1888, between Montevideo, Uruguay, and Tome, Chile on the voyage through the Straits of Magellan. Proc. U.S. natn. Mus., 50:401-476, pl. 2-6.

THOMPSON, W. W. 1914. Catalogue of fishes of the Cape Province. Mar. biol. Rep., Cape Town, (2):132-167.

THOMSON, C. W. 1877. The Voyage of "The Challenger". London., 2 vol.

THOMSON, J. M. 1981. Mugilidae, in Fischer, W., Bianchi, G. & W.B. Scott (ed.), FAO species identification sheets for fishery purposes. Eastern Central Atlantic; fishing areas 34, 47 (in part). Canada Funds-in-Trust. Ottawa, Department of Fisheries and Oceans Canada, by arrangement with the Food and Agriculture Organization of the United Nations, vol. 1-7: pag. var.

THUOC, P.; SZYPULA, J. 1973 (see PHAM-THUOC).

THORSON, T. B.; WATSON, D. E. 1975. Reassignment of the African freshwater stingray Potamotrygon garouaensis, to the genus Dasyatis, on physiologic and morphologic grounds. Copeia, 1975:701-712, 3 fig., 3 tab.

THERSHER, R. E. 1976. Field analysis of the territoriality of the three-spot damselfish, Eupomacentrus planifrons (Pomacentridae). Copeia, 1976(2):266-275.

THYS van den AUDENAERDE, D. F. E. 1965. List of the freshwater fishes presently known from the island of Fernando Poo. Bonn. zool. Beitr., 16(3/4):316-317.

--. 1967. The freswater fishes of Fernando Poo. Verh. K. vlaam. Acad. Wet., 29(100):1-167, 47 fig.

--. 1968. Addendum to the "freshwater fishes of Fernando Poo" on some fishes collected by Col. J.J. Scheel. Rev. Zool. Bot. afr., 78:123-128.

TICKELL, S. R. 1865. Description of a supposed new genus of the Gadidae, Arakan (Asthenurus atripinnis). J. Asiat. Soc. Beng., 34(2):32, 2 fig.

TIEWS, K. 1963. Synopsis of biological data on bluefin tuna Thunnus thynnus (Linnaeus 1758) (Atlantic and Mediterranean). FAO Fish. Rep., (6)2:422-481.

TILESIUS, W. G. 1809. Description de quelques poissons observés pendant son voyage autour du monde. Mem. Soc. nat. Moscou, 2:212-249, pl. 13-17.

--. 1812. (Opisanie...) Rech. théor. Acad. imp. Sci. St Pétersb., 3:278-303, pl. 13-17 (in Russian, latin diagn.).

TKACHENKO, V. A. 1969. On the changes with age of some morphological features of Sardinella aurita Val. on the northwestern coast of Africa. Trudy atlant. nauchno-issled. Inst. ryb. Khoz. Okeanogr., 22:103-112 (in Russian).

TKACHUK, L. P. 1979. New species of trematodes from the deepwater Zeidae of the Indian and Atlantic Oceans. Zool. Zh., 58:1290-1296 (in Russian).

TOBOR, J. A. 1969. Species of the Nigerian ariid catfish, their taxonomy, distribution and preliminary observations of the biology of one of them. Bull. Inst. fond. Afr. noire, (A)31(1):643-658, 3 fig.

TOMINAGA, Y. 1970. On the gland-like organ situated in front of the eye of Rondeletia loricata. Zool. Mag., Tokyo, 79 (11-12):368 (in Japane-

se).
TOMINAGA, Y.; KUBOTA, T. 1972. Records of the redmouth whalefish, <u>Ron-deletia</u> <u>loricata</u>, from Sagami Bay and Suruga Bay, Japan, with notes on the holotype. <u>Jap. J. Ichthyol.</u>, 19(3):181-185, 1 fig.

TOMIYAMA, I.; ABE, T.; TOKIOKA, T. 1958. Encyclopedia Zoologica. Illustrated in Colours. II. 1st ed. (Vertebrata, Pisces, Cyclostomata, Protochordata). 392 p., 1155 col. fig. (in Japanese).

TOMO, A. 1985. Gadidae, p. 284-285. <u>In</u>: Fischer, W. & Hureau, J.C. (ed.). FAO Species identification sheets for fishery purposes. Southern Ocean (Fishing Areas 48, 58 and 88). Prepared and printed with the support of the Commission for the Conservation of Antarctic Marine Living Resources (CCAMLR). Rome, vol. 2.

TOOLE, M. J. (see O'TOOLE, M. J.)

TORCHIO, M. 1958. Myctophidae (Pisces) presenti nel Mar Ligure. <u>Doriana</u>, 2(91):1-6.

--. 1961. <u>Arnoglossus</u> <u>moltoni</u>, nuova specie di Bothidae del Mediterraneo (Pisces, Pleuronectiformes). <u>Atti Soc. ital. Sci. nat.</u>, 100(1-2):213-224, 4 fig., pl. 15.

--. 1968. Sulla eventuale presenza in acque mediterranee di individui dei generi <u>Cephalopholis</u> Bl. Schn. e <u>Chaetodon</u> L. <u>Natura, Milano</u>, 59 (3-4):210-212, fig. 1, pl. 2 (col.).

TORTONESE, E. 1935. Elenco dei pesci italiani con annotazioni sistematiche. <u>Boll. Pesca Piscic. Idrobiol.</u>, 11:211-254.

--. 1938. Note di Ittiologia. I. Pesci rari o poco noti del Golfo di Genova. <u>Boll. Musei Zool. Anat. comp. R. Univ. Torino</u>, (3)46(74):73-102, 7 fig. (reprint p. 1-32).

--. 1938. Revisione degli Squali del Museo Civico di Milano. <u>Atti Soc. ital. Sci. nat.</u>, 77:283-318, 5 fig.

--. 1939. Risultati ittiologici del viaggio di circumnavigazione del globo della R. N. "Magenta" (1865-68). <u>Boll. Musei Zool. Anat. comp. R. Univ. Torino</u>, (3)47(100):177-421, 9 pl.

--. 1940. Sugli Scomberoidei mediterranei del genere <u>Tetrapturus</u>. <u>Boll. Musei Zool. Anat. comp. R. Univ. Torino</u>, 48:173-178.

--. 1942. Contributo allo studio dell'ittiofauna marina dell'Africa occidentale. <u>Boll. Musei Zool. Anat. comp. R. Univ. Torino</u>, (4)49 (125):151-173, 2 fig., 2 pl.

--. 1947. Richerche zoologiche nell'isola di Rodi (Mar Egeo). Pesci. <u>Boll. Pesca Piscic. Idrobiol.</u>, (n.s.)23(2):143-192, 9 fig.

--. 1949. Identificazione di due Sgombroidi (Pesci) accidentali nel Mediterraneo. <u>Boll. Zool.</u>, 16(1-3):61-66, 1 fig.

--. 1951. Revisione delle specie mediterranee della subfam. Sudinae (Pisces Iniomi). <u>Boll. Pesca Piscic. Idrobiol.</u>, (n.s.)6(2):177-186, 3 fig.

--. 1952. Un Percoide marino e batifilo nuovo per l'ittiofauna italiana (<u>Epigonus</u> <u>denticulatus</u> Dieuz.). <u>Boll. Pesca Piscic. Idrobiol.</u>, 28 (n.s.) 7:72-74.

--. 1952. Gli Hemirhamphidae del Mediterraneo (Pisces Synentognathi). <u>Boll. Ist. Mus. Zool. Univ. Torino</u>, 3(3):1-8.

--. 1953. Nota sui Centracanthidae del Mediterraneo (Pisces Perciformes). <u>Atti Soc. ital. Sci. nat.</u>, 92:24-29, 3 fig.

--. 1954. Spedizione subacquea italiana nel Mar Rosso. Ricerche zoologiche. VI. Plettognati. <u>Riv. Biol. colon.</u>, 14:73-86, 1 fig.

--. 1956. Iniomi, Plectognathi. <u>In</u>: Uova, larve e stadi giovanili di Teleostei (U. d'Ancona ed.). <u>Fauna Flora Golfo Napoli</u>, 38:889-977, fig. 786-821, pl. 51.

--. 1956. Leptocardia, Ciclostomata, Selachii. <u>Fauna Ital.</u>, 2:1-332, 163 fig.

--. 1957. Studi sui Plagiostomi. VIII. I <u>Pristis</u> del Museo Civico di

Genova. <u>Doriana</u>, 2(81):1-7, fig. A + B.

--. 1958. Elenco dei Leptocardi, Ciclostomi, Pesci cartilaginei ed ossei del Mare Mediterraneo. <u>Atti</u> <u>Soc.</u> <u>ital.</u> <u>Sci.</u> <u>nat.</u>, **97**(4):309-345.

--. 1958. Primo reperto ligure di un raro mictofide: <u>Diaphus</u> <u>metoplocampus</u> (Cocco)(Pisces Iniomi). <u>Annali</u> <u>Mus.</u> <u>civ.</u> <u>Stor.</u> <u>nat.</u> <u>Giacomo</u> <u>Doria</u>, 70:71-72.

--. 1960. Revisione dei Centrolophidae (Pisces Perciformes) del Mare Ligure (1). <u>Annali</u> <u>Mus.</u> <u>civ.</u> <u>Stor.</u> <u>nat.</u> <u>Giacomo</u> <u>Doria</u>, 71:57-82.

--. 1960. Su alcuni squali e sgombroidi del Atlantico orientale. <u>Doriana</u>, 3(109):1-5, 1 fig.

--. 1960. General remarks on the Mediterranean deep-sea fishes. <u>Bull.</u> <u>Inst.</u> <u>océanogr.</u> <u>Monaco</u>, (1167):1-14.

--. 1961. Mediterranean fishes of the family Istiophoridae. <u>Nature,</u> <u>Lond.</u>, **192**:80.

--. 1962. Osservazioni comparative intorno alla ittiofauna del Mediterraneo e dell'Atlantico occidentale (Florida e Isole Bahamas). <u>Natura,</u> <u>Milano</u>, **53**:1-20, 6 fig.

--. 1963. Catalogo dei tipi di pesci del Museo Civico di Storia Naturale di Genova. Parte III. <u>Annali</u> <u>Mus.</u> <u>civ.</u> <u>Stor.</u> <u>nat.</u> <u>Giacomo</u> <u>Doria</u>, 73:333-350.

--. 1963. Elenco riveduto dei Leptocardi, Ciclostomi, Pesci cartilaginei e ossei del Mare Mediterraneo. <u>Annali</u> <u>Mus.</u> <u>civ.</u> <u>Stor.</u> <u>nat.</u> <u>Giacomo</u> <u>Doria</u>, 74:156-185.

--. 1963. <u>Belone</u> <u>imperialis</u> (Raf.)(Pisces) nel Mediterraneo. <u>Doriana</u>, 3(129):1-6.

--. 1964. Contributo allo studio dei Cirrhitidae (Pisces Perciformes). <u>Doriana</u>, 3(148):1-8, 1 fig.

--. 1964. Comment on the proposed ruling on the type species of <u>Sciaena</u> Linnaeus, 1758, Z.N.(S.)850. <u>Bull.</u> <u>zool.</u> <u>Nom.</u>, 21:362.

--. 1965. I Pesci e i Cetaci del Mare Ligure con particolare riguardo alle specie economicamente importanti. Genova, 216 p., 150 fig., n. num., 4 phot. n. num., 4 col. pl. n. num. (inset).

--. 1966. Note sistematiche e nomenclatoriali intorno agli Aracanidi e agli Ostracionidi. <u>Annali</u> <u>Mus.</u> <u>civ.</u> <u>Stor.</u> <u>nat.</u> <u>Giacomo</u> <u>Doria</u>, 76:75-89, 1 fig.

--. 1967. Su alcuni pesci del Golfo di Genova. <u>Doriana</u>, 4(177):1-5, 2 fig.

--. 1970. Comparsa di <u>Tetrapturus</u> <u>albidus</u> Poey (Pesci Sgombroidi) nel Golfo di Genova. <u>Boll.</u> <u>Pesca</u> <u>Piscic.</u> <u>Idrobiol.</u>, 25(1):80-83, 1 fig.

--. 1970. Osteichthyes (Pesci ossei). Parte prima. <u>Fauna</u> <u>Ital.</u>, **10**:1-565, 198 fig.

--. 1970. First report of a zoarcid fish from the Mediterranean (<u>Melanostigma</u> <u>atlanticum</u> Koefoed). <u>Doriana</u>, 4(190):1-4, 2 fig.

--. 1972. Risultati ittiologici di alcune crociere nel Mediterraneo e nel vicino Atlantico (1970-71). <u>Annali</u> <u>Mus.</u> <u>civ.</u> <u>Stor.</u> <u>nat.</u> <u>Giacomo</u> <u>Doria</u>, 79:18-26, 1 fig.

--. 1972. On the affinities and systematic position of the genus <u>Callanthias</u> after a study of its type species <u>C.</u> <u>rubens</u> (Raf.). <u>Boll.</u> <u>Zool.</u>, 39(1):71-82, 2 fig.

--. 1973. Apogonidae, p. 365-367, Coryphaenidae, p. 385, Lobotidae, p. 389, Xiphiidae, p. 482, Molidae, p. 649-650, <u>in</u> Hureau, J.C. & Th. Monod (ed.). Check-list of the fishes of the north-eastern Atlantic and of the Mediterranean/Catalogue des poissons du nord-est Atlantique et de la Méditerranée (Clofnam). Paris, Unesco, 2 vol.

--. 1975. Osteichthyes (Pesci ossei). Parte Seconda. <u>Fauna</u> <u>Ital.</u>, 11:xviii + 1-639, 240 fig., 8 pl.

--. 1977. Observations sur un Uranoscopidae de l'Afrique occidentale. <u>Uranoscopus</u> <u>polli</u> Cadenat, 1953 (Pisces). <u>Bull.</u> <u>Off.</u> <u>natn.</u> <u>Pêch.</u>

Tunisie, 1(2):189-192, 2 fig.

--. 1986. Serranidae, in P.J.P. Whitehead, M.L. Bauchot, J.C. Hureau, J. Nielsen & E. Tortonese. Fishes of the North-eastern Atlantic and the Mediterranean/Poissons de l'Atlantique du Nord-Est et de la Méditerranée, Paris, Unesco, 2:780-792, 14 + 3 fig.

--. 1986. Moronidae, in P.J.P. Whitehead, M.L. Bauchot, J.C. Hureau, J. Nielsen & E. Tortonese. Fishes of the North-eastern Atlantic and the Mediterranean/Poissons de l'Atlantique du Nord-Est et de la Méditerranée, Paris, Unesco, 2:793-796, 3 fig.

--. 1986. Apogonidae, in P.J.P. Whitehead, M.L. Bauchot, J.C. Hureau, J. Nielsen & E. Tortonese. Fishes of the North-eastern Atlantic and the Mediterranean/Poissons de l'Atlantique du Nord-Est et de la Méditerranée, Paris, Unesco, 2:803-809, 7 fig.

--. 1986. Cepolidae, in P.J.P. Whitehead, M.L. Bauchot, J.C. Hureau, J. Nielsen & E. Tortonese. Fishes of the North-eastern Atlantic and the Mediterranean/Poissons de l'Atlantique du Nord-Est et de la Méditerranée, Paris, Unesco, 2:810-811, 1 fig.

--. 1986. Pomatomidae, in P.J.P. Whitehead, M.L. Bauchot, J.C. Hureau, J. Nielsen & E. Tortonese. Fishes of the North-eastern Atlantic and the Mediterranean/Poissons de l'Atlantique du Nord-Est et de la Méditerranée, Paris, Unesco, 2:812-813, 1 fig.

--. 1986. Rachycentridae, in P.J.P. Whitehead, M.L. Bauchot, J.C. Hureau, J. Nielsen & E. Tortonese. Fishes of the North-eastern Atlantic and the Mediterranean/Poissons de l'Atlantique du Nord-Est et de la Méditerranée, Paris, Unesco, 2:814, 1 fig.

--. 1986. Centracanthidae, in P.J.P. Whitehead, M.L. Bauchot, J.C. Hureau, J. Nielsen & E. Tortonese. Fishes of the North-eastern Atlantic and the Mediterranean/Poissons de l'Atlantique du Nord-Est et de la Méditerranée, Paris, Unesco, 2:908-911, 5 + 1 fig.

--. 1986. Kyphosidae, in P.J.P. Whitehead, M.L. Bauchot, J.C. Hureau, J. Nielsen & E. Tortonese. Fishes of the North-eastern Atlantic and the Mediterranean/Poissons de l'Atlantique du Nord-Est et de la Méditerranée, Paris, Unesco, 2:912-913, 2 fig.

--. 1986. Trachinidae, in P.J.P. Whitehead, M.L. Bauchot, J.C. Hureau, J. Nielsen & E. Tortonese. Fishes of the North-eastern Atlantic and the Mediterranean/Poissons de l'Atlantique du Nord-Est et de la Méditerranée, Paris, Unesco, 2:951-954, 4 fig.

--. 1986. Balistidae, in P.J.P. Whitehead, M.L. Bauchot, J.C. Hureau, J. Nielsen & E. Tortonese. Fishes of the North-eastern Atlantic and the Mediterranean/Poissons de l'Atlantique du Nord-Est et de la Méditerranée, Paris, Unesco, 3:1335-1337, 3 fig.

--. 1986. Monacanthidae, in P.J.P. Whitehead, M.L. Bauchot, J.C. Hureau, J. Nielsen & E. Tortonese. Fishes of the North-eastern Atlantic and the Mediterranean/Poissons de l'Atlantique du Nord-Est et de la Méditerranée, Paris, Unesco, 3:1338-1339, 2 fig.

--. 1986. Ostraciontidae, in P.J.P. Whitehead, M.L. Bauchot, J.C. Hureau, J. Nielsen & E. Tortonese. Fishes of the North-eastern Atlantic and the Mediterranean/Poissons de l'Atlantique du Nord-Est et de la Méditerranée, Paris, Unesco, 3:1340, 1 fig.

--. 1986. Tetraodontidae, in P.J.P. Whitehead, M.L. Bauchot, J.C. Hureau, J. Nielsen & E. Tortonese. Fishes of the North-eastern Atlantic and the Mediterranean/Poissons de l'Atlantique du Nord-Est et de la Méditerranée, Paris, Unesco, 3:1341-1345, 6 fig.

--. 1986. Diodontidae, in P.J.P. Whitehead, M.L. Bauchot, J.C. Hureau, J. Nielsen & E. Tortonese. Fishes of the North-eastern Atlantic and the Mediterranean/Poissons de l'Atlantique du Nord-Est et de la Méditerranée, Paris, Unesco, 3:1346-1347, 2 fig.

--. 1986. Molidae, in P.J.P. Whitehead, M.L. Bauchot, J.C. Hureau, J.

Nielsen & E. Tortonese. Fishes of the North-eastern Atlantic and the Mediterranean/Poissons de l'Atlantique du Nord-Est et de la Méditerranée, Paris, Unesco, 3:1348-1350, 3 + 2 fig.

TORTONESE, E.; ARBOCCO, G. 1958. Pesci dell'Africa occidentale raccolti dal Prof. Edoardo Zavattari. _Doriana_, 2(87):1-6, 2 fig.

TORTONESE, E.; CASANOVA QUEIROLO, L. 1970. Contributo allo studio dell' ittiofauna del Mar Ligure orientale. _Annali Mus. civ. Stor. nat. Giacomo Doria_, 78:21-46, fig. 8.

TORTONESE, E.; DEMIR, M. 1965. Rapport sur les travaux récents (1962-64) concernant les vertébrés marins et les céphalopodes de la Méditerranée et de ses dépendances. _Rapp. P.-v. Réun. Commn int. Explor. scient. Mer Méditerr._, 18(3):575-590.

TORTONESE, E.; FABIANO, H. 1975. On the species of _Notoscopelus_ living in the Italian seas (Pisces, Myctophidae). _Atti Soc. ital. Sci. nat._, 116:227-230.

TORTONESE, E.; HUREAU, J.C. (ed.) 1979. Supplément 1978 au Clofnam (Check-list of the fishes of the north-eastern Atlantic and of the Mediterranean/ Catalogue des poissons du nord-est Atlantique et de la Méditerranée). _Cybium_, 3(1):5-66 (also in second edition of _Clofnam_, 2:333-394).

TORTONESE, E.; TROTTI, L. 1950. Catalogo dei pesci del Mare Ligure. _Atti Accad. ligure_, 6(1):49-164.

TOWNSEND, C. H.; NICHOLS, J. T. 1925. Deep-sea fishes of the "Albatross" Lower California Expedition. _Bull. Am. Mus. nat. Hist._, 52:1-20, 4 fig., 4 pl.

TREWAVAS, E. 1932. A contribution to the classification of the fishes of the order Apodes, based on the osteology of some rare eels. _Proc. zool. Soc. Lond._, 1932 (2):639-659, fig. 1-9, pl. 1-4.

--. 1933. On the structure of the oceanic fishes, _Cyema atrum_ Günther and _Opisthoproctus soleatus_ Vaillant. _Proc. zool. Soc. Lond._, 1933 (2): 601-614,pl. 1-2.

--. 1962. A basis for classifying the sciaenid fishes of tropical West Africa. _Ann. Mag. nat. Hist._, (13)5:167-176.

--. 1963. _Sciaena_ Linnaeus, 1758 (Pisces): proposed validation under the plenary powers of the ruling given in Opinion 93 concerning the type species. Z.N.(S.)850. _Bull. zool. Nom._, 20:349-360.

--. 1964. The sciaenid fishes with a single mental barbel. _Copeia_, 1964 (1):107-117.

--. 1966. Comments on the type species of _Sciaena_ Linnaeus, 1758, Z.N. (S.)850. _Bull. zool. Nom._, 23:4-5.

--. 1977. The sciaenid fishes (croakers or drums) of the Indo-West-Pacific. _Trans. zool. Soc. Lond._, 33:253-541.

TREWAVAS, E.; INGHAM, S. E. 1972. A key to the species of Mugilidae (Pisces) in the north-eastern Atlantic and Mediterranean, with explanatory notes. _J. Zool._, 167:15-29.

TROADEC, J. P. 1964. Prises par unité d'effort des sardiniers de Pointe-Noire (Congo). Variations saisonnières de l'abondance des sardinelles (_Sardinella eba_ C.V. et _Sardinella aurita_ C.V.) dans les eaux congolaises (de 3°30'S à 5°30'S). _Cah. ORSTOM, Océanogr._, 2(4):17-25.

TROADEC, J. P.; GARCIA, S. (ed.). 1980. The fish resources of the eastern central Atlantic. Part one: the resources of the Gulf of Guinea from Angola to Mauritania. _FAO Fish. tech. Pap._, (186.1): vii + 1-166. (French version published in 1979).

TROADEC, J. P.; BARRO, M.; BOUILLON, P. 1969. Pêches au chalut sur la radiale de Grand Bassam (Côte d'Ivoire)(Mars 1966 - Février 1967). _Docums scient. provis. Centre Rech. océanogr., Abidjan_, (33):1-14, 11 tab., 86 pl. (mimeo).

TROIS, E. F. 1905. La femmina de _Coris julis. Atti Ist. veneto Sci._, 64

(2):197-198.

TROSCHEL, F. H. 1856. Bericht über die Leistungen in der Ichthyologie während des Jahres 1855. Arch. Naturgesch., 22(2):67-89.

--. 1866. Ein Beitrag zur ichthyologischen Fauna der Inseln des Grünen Vorgebirges. Arch. Naturgesch., 32(1):190-239, pl. 5.

TROTT, L. B.; OLNEY, J. E. 1986. Carapidae, in P.J.P. Whitehead, M.L. Bauchot, J.C. Hureau, J. Nielsen & E. Tortonese. Fishes of the Northeastern Atlantic and the Mediterranean/Poissons de l'Atlantique du Nord-Est et de la Méditerranée, Paris,Unesco, 3:1172-1176, 5 + 3 fig.

TROTTI, L. 1936. Contributo alla conoscenza di probabili organi luminosi nell' Hymenocephalus italicus Gigl. Annali Mus. civ. Stor. nat. Giacomo Doria, 59:160-170.

--. 1945. Comportamento del V et VII paio di nervi craniali nel Notacanthus bonapartei Risso. Annali Mus. civ. Stor. nat. Giacomo Doria, 62: 216-252.

TRUNOV, I. A. 1968. New data on the distribution of Electrona rissoi (Cocco) and Diaphus ostenfeldi (Tåning)(Myctophidae). Vop. Ikhtiol., 8(4=51):745-748 (in Russian).

--. 1968. The whalefish (Barbourisia rufa, fam. Barbourisiidae) and the frilled shark (Chlamydoselachus anguineus, fam. Chlamydoselachidae) from southwest African coastal waters. Probl. Ichthyol., 8(1):135-138, 3 fig. (Engl. transl. from Russian Vop. Ikhtiol.).

--. 1968. Some preliminary data on the composition and distribution of some fishes from southeastern Atlantic. Vop. Ikhtiol., 8:952-954, map.

--. 1969. Schedophilus huttoni (Centrolophidae), a species of fish new to the Atlantic Ocean. J. Ichthyol., 9:443-445.

--. 1976. New species and species recorded for the first time in the pelagic area of the tropical Atlantic of the families Serranidae, Emmelichthyidae and Ariommidae. Vop. Ikhtiol., 16(2):263-273, 7 fig. (in Russian; Engl. transl.: J. Ichthyol., 16(2):229-238, 7 fig.).

--. 1981. The ichthyofauna of the submarine Bank Valdivia (south-eastern Atlantic). Byull. mosk. Obshch. Ispyt. Prir., 86(5):51-64, 5 fig.

--. 1982. Zeiformes of the thalassobathyal of the southwestern Atlantic. Byull. mosk. Obshch. Ispyt. Prir., 87(2):41-53, 1 fig., 1 tab.

TRUNOV, I. A.; ISAREV, A. T. 1971. Guntherus altivela Osorio, 1917 (fam. Ateleopidae) from the southeastern Atlantic. J. Ichthyol., 11(1):115-177, 1 fig.

TSCHUDI, J. J. von. 1844. Ichthyologie: 1-35, pl. 1-6. In: Untersuchungen über die Fauna Peruana. St Gallen, 693 p.

TSUJIMOTO, M. 1932. The liver oils of elasmobranch fish. J. Soc. chem. Ind., Lond., 51:318T-322T.

TUCKER, D. W. 1950. The biology of Aphanopus carbo Lowe. Rep. Challenger Soc., 3(2):26.

--. 1953. The fishes of the genus Benthodesmus (family Trichiuridae). Proc. zool. Soc. Lond., 123:171-197.

--. 1954. The "Rosaura" Expedition 1937-38. Report on the fishes collected by S.Y. "Rosaura" in the North and Central Atlantic 1937-38. Part I. Bull. Br. Mus. nat. Hist. (Zool.), 2(6):163-214, 19 fig., 2 pl.

--. 1955. Studies on trichiuroid fishes (2). Benthodesmus tenuis (Günther) collected by the "Expedition Océanographique Belge dans les eaux côtières de l'Atlantique Sud (1948-1949)", with additional notes on the genus Benthodesmus. Bull. Inst. r. Sci. nat. Belg., 31(64):1-26, 1 pl.

--. 1956. Studies on the trichiuroid fishes. A preliminary revision of the family Trichiuridae. Bull. Br. Mus. nat. Hist. (Zool.), 4(3): 73-130, 1 pl., 23 fig.

TUCKER, D. W.; PALMER, C. 1949. New British records of two rare deep-sea fishes: <u>Oxynotus</u> <u>paradoxus</u> Frade and <u>Aphanopus</u> <u>carbo</u> Lowe. <u>Nature,</u> <u>Lond.</u>, **146**(4178):930-931, 1 fig.

TURNER, C. L. 1936. The absorptive processes in the embryos of <u>Parabro-</u> <u>tula</u> <u>dentiens</u>, a viviparous deep-sea brotulid fish. <u>J.</u> <u>Morph.</u>, **59**: 313-325.

TURON, J. M.; RUCABADO, J.; LLORIS, D.; MACPHERSON, E. 1986. Datos pes- queros de las expediciones realizadas en aguas de Namibia durante los anos 1981 a 1984 ("Benguela III" a "Benguela VII" y "Valdivia I"). <u>Datos</u> <u>informativos</u> <u>Inst.</u> <u>Investnes</u> <u>pesq.</u>, Barcelona, **17**:11-343.

TURON, W. 1807. The British Fauna, containing a compendium of the zoolo- gy of the British Islands; arranged according to the Linnean system. Swansea, London, 230 p.

TYLER, J. C. 1965. A synopsis of the four species of cowfishes (<u>Acan-</u> <u>thostracion</u>, Plectognathi) in the Atlantic Ocean. <u>Proc.</u> <u>Acad.</u> <u>nat.</u> <u>Sci.</u> <u>Philad.</u>, **117**(8):261-287, 20 fig.

UENO, T. 1971. List of the marine fishes from the waters of Hokkaido and its adjacent regions. <u>Scient.</u> <u>Rep.</u> <u>Hokkaido</u> <u>Fish.</u> <u>expl.</u> <u>Stn</u>, (13): 61-102.

UENO, T.; ABE, K. 1966. Studies on deep-water fishes from off Hokkaido and adjacent regions. VIII-IX. <u>Jap.</u> <u>J.</u> <u>Ichthyol.</u>, **14**(1/3):35-39, 2 fig.

--. 1966. On rare or newly found fishes from the water of Hokkaido (II). <u>Jap.</u> <u>J.</u> <u>Ichthyol.</u>, **13**(4-6):229-236, fig. 1-11.

UEYANAGI, S. 1960. On the larvae and the spawning areas of the shiroka- jiki, <u>Marlina</u> <u>marlina</u> (Jordan & Hill). <u>Rep.</u> <u>Nankai</u> <u>reg.</u> <u>Fish.</u> <u>Res.</u> <u>Lab.</u>, (12):85-96, 5 fig. (in Japanese).

--. 1960. Preliminary note on the larvae of the shortnosed spearfish, <u>Tetrapturus</u> <u>angustirostris</u> (Tanaka). <u>Rep.</u> <u>Nankai</u> <u>reg.</u> <u>Fish.</u> <u>Res.</u> <u>Lab.</u>, (12):97-98, 4 fig. (in Japanese).

--. 1962. On the larvae of the shortnosed spearfish, <u>Tetrapturus</u> <u>angus-</u> <u>tirostris</u> Tanaka. <u>Rep.</u> <u>Nankai</u> <u>reg.</u> <u>Fish.</u> <u>Res.</u> <u>Lab.</u>, (16):173-189, 16 fig. (in Japanese).

--. 1963. Methods for identification and discrimination of the larvae of five istiophorid species distributing in the Indo-Pacific. <u>Rep.</u> <u>Nan-</u> <u>kai</u> <u>reg.</u> <u>Fish.</u> <u>Res.</u> <u>Lab.</u>, (17):137-150, 6 fig. (in Japanese).

--. 1963. A study of the relationships of the Indo-Pacific istiophorids. <u>Rep.</u> <u>Nankai</u> <u>reg.</u> <u>Fish.</u> <u>Res.</u> <u>Lab.</u>, (17):151-165, 7 fig., 2 pl. (in Japanese).

--. 1964. Description and distribution of larvae of five istiophorid species in the Indo-Pacific. <u>Proc.</u> <u>Symp.</u> <u>Scombroid</u> <u>Fishes</u> <u>mar.</u> <u>biol.</u> <u>Ass.</u> <u>India</u>, **1**(1):499-528, 17 fig., 5 tab.

--. 1973. Present state of billfish larval taxonomy. p. 649-658, 6 fig. <u>In</u>: J.H.S. Blaxter (ed.). The early life history of fish. Berlin, Heidelberg, New-York, 1-765 p.

--. 1974. On an additional diagnostic character for the identification of billfish larvae with some notes on the variations in pigmentation. <u>NOAA</u> <u>tech.</u> <u>Rep.</u> <u>NMF</u> <u>SSRF</u>, (675)(2):73-78, 10 fig.

UEYANAGI, S.; WATANABE, H. 1965. On certain specific differences in the vertebrae of the istiophorids. <u>Rep.</u> <u>Nankai</u> <u>reg.</u> <u>Fish.</u> <u>Res.</u> <u>Lab.</u>, (22):1-8, 3 fig. 5 pl. (in Japanese).

UEYANAGI, S.; YABE, H. 1959. Larva of the black marlin (<u>Eumakaira</u> <u>nigra</u> Nakamura). <u>Rep.</u> <u>Nankai</u> <u>reg.</u> <u>Fish.</u> <u>Res.</u> <u>Lab.</u>, (10):151-169, 18 fig., 1 pl. (in Japanese).

--. 1960. On the larva possibly referable to <u>Marlina</u> <u>marlina</u> (Jordan & Hill). <u>Rec.</u> <u>oceanogr.</u> <u>Wks</u> <u>Japan</u>, **5**(2):167-173, 4 fig.

UEYANAGI, S.; IMAKOJI, G.; DOI, T. 1972. Ecological studies on the resources of the offshore pelagic fishes (Results of the year 1971). Division of Pelagic Fish Resources, Far Seas Fisheries Research Laboratory:15-19, fig. (in Japanese).

UEYANAGI, S.; MORI, K.; NISHIKAWA, Y. 1969. Tuna spawning ground and potential resources of pelagic fishes based on distribution of larvae. Division of Pelagic Fish Resources, Far Seas Fisheries Research Laboratory (Enyo Shigen Suisan Kenkyu-sho):1-9. (in Japanese).

UEYANAGI, S.; KIKAWA, S.; UTO, M.; NISHIKAWA, Y. 1970. Distribution, spawning and relative abundance of billfishes in the Atlantic Ocean. Bull. Far Seas Fish. Res. Lab., 3:15-55, 23 + 2 fig. (in Japanese).

UWATE, K. R. 1979. Revision of the anglerfish Diceratiidae with descriptions of two new species. Copeia, 1979(1):129-144.

VAHL, M. 1794. Beskrivelse af en nye fiskeslaegt (Caecula). Skr. Naturh. Selsk. Kjøbenhavn, 3(2):149-156.

--. 1798. Beskrivelse over 2de Arter af Fiske-Slaegten Polynemus. Skr. Naturh. Selsk. Kjøbenhavn, 4(2):158-168.

VAILLANT, L.; BOCOURT, M. 1915. Etudes sur les poissons. Observations sur les poissons de la région centrale de l'Amérique, p. 1-265, 20 pl. In: Milne Edwards M.H., Recherches zoologiques pour servir à l'histoire de la faune de l'Amérique centrale et du Mexique, Zool. 4è partie. Mission scientifique au Mexique et dans l'Amérique Centrale. Paris, 1869-1915.

--. 1882. Sur un poisson des grandes profondeurs de l'Atlantique, l'Eurypharynx pelecanoides. C. r. hebd. Séanc. Acad. Sci., Paris, 95: 1226-1228.

--. 1885. Sur les dimensions comparatives des adultes et des jeunes chez un poisson élasmobranche, l'Alopias vulpes. Bull. Soc. philomath. Paris, (7)10:41-42.

--. 1886. Considérations sur les poissons des grandes profondeurs, en particulier sur ceux qui appartiennent au sous-ordre des abdominales. C. r. hebd. Séanc. Acad. Sci., Paris, 103(25):1237-1239.

--. 1887. Remarques sur la construction du nid de l'Antennarius marmoratus Less. et G., dans la Mer des Sargasses. C. r. Séanc. Soc. Biol., (8)4:732-733.

--. 1888. Poissons. In: Expéditions scientifiques du "Travailleur" et du "Talisman" pendant les années 1880, 1881, 1882, 1883. Paris, Masson, 1-406 p., pl. 1-28.

--. 1889. Observations relatives à la montée de l'anguille sur les côtes de France. C. r. hebd. Séanc. Acad. Sci., Paris., 109:31-33.

--. 1894. Note sur les poissons de la famille des siluridés appartenant à la faune Madécasse et description d'une espèce nouvelle. Bull. Soc. philomath. Paris, 6:75-80.

--. 1901. Sur un griset (Hexanchus griseus L. Gm.) du Golfe de Gascogne. Bull. Mus. Hist. nat., Paris, 7:202-204.

--. 1919. In: L. Roule, Appendice: Liste des espèces déterminées par M. le professeur L. Vaillant. Résult. Camp. scient. Prince Albert I, 52: 129-135.

VALENCIENNES, A. 1824. Description du cernié, Polyprion cernium. Mém. Mus. Hist. nat., Paris, 11:265-269, pl. 17.

--. 1828-49. See G. CUVIER & A. VALENCIENNES, 1828-49.

--. 1832. Description de plusieurs nouveaux poissons du genre Apogon. Nouv. Annls Mus. Hist. nat., Paris, 1:51-60, fig.

--. 1837-44. Ichthyologie des îles Canaries, ou histoire naturelle des poissons rapportés par MM. Webb et Berthelot. In: P.B. Webb & S. Berthelot, Histoire naturelle de îles Canaries. Paris, 2(2):1-109, 26 pl. (Plates:1837: pl. 1, 3-4, 8, 11-12; 1838: pl. 7, 10, 13, 14, 17;

1839: pl. 5, 9, 15-16, 18; 1840: pl. 2; 1841: pl. 6; 1842: pl. 19; 1843: pl. 20-23; 1844: pl. 24-26. Text:1842, p. 1-8; 1843, p. 9-96; 1844, p. 97-109).

--. 1838-43. Les poissons. In: G. Cuvier, Le Règne animal... Ed. 3 (Ed. des Disciples) Paris, 4:1-392 p., 120 pl. (1838: p. 1-40; 1839:41-48; 1840:49-120; 1841:121-176; 1842:177-360; 1843:361-392).

--. 1838-43. Atlas des Poissons., In: G. Cuvier, Le Règne animal distribué d'après son organisation... Ed. 3 (ed. des Disciples), 120 pl. (?1836).

--. 1843. (Lophius hystrio). In: Dict. universel Hist. nat. (ed. Ch. d'Orbigny). Paris, 3:592.

--. 1847. Poissons. Catalogue des principales espèces de poissons rapportées de l'Amérique méridionale. In: A. D. d'Orbigny. Voyage dans l'Amérique méridionale. Paris, 5(2):1-11, pl. 1-16.

--. 1855. Ichthyologie, p. 297-351, pl. Poissons 1-10. In: A. du Petit-Thouars. Voyage autour du monde sur la frégate la "Vénus" pendant les années 1836-1839. Paris, 355 p., 79 pl.

VANDELLI, D. 1797. Florae et faunae Lusitanicae specimen. Mems R. Acad. Sci. Lisb., 1:37-79.

VAN DEN AUDENAERDE, D. F. E. (see THYS VAN DEN AUDENAERDE, D. F. E.)

VAN ECK, T. H. 1969. The South African hake: "Merluccius capensis" or "Merluccius paradoxus" ? S. Afr. Shipp. News Fishg Ind. Rev., 23(1): 124-135, 6 fig.

VAN OLST, J. C.; HUNTER, J. R. 1970. Some aspects of the organization of fish schools. J. Fish. Res. Bd Can., 27:1223-1238, 4 fig., 3 tab.

VAN PRAET, M.; GEISTDOERFER, P. 1980. Etudes des zymogrammes des tissus digestifs de poissons et invertébrés abyssaux. C. R. hebd. Séanc. Acad. Sci.. Paris, (D)290:1083-1086.

VAN TASSEL, J. L.; MILLER, P. J.; BRITO, A. 1988. A revision of Vanneaugobius (Teleostei: Gobiidae), with description of a new species. J. nat. Hist., 22:545-567, 11 fig.

VARAGNOLO, S. 1964. Calendario di comparse di uova di Teleostei marini nel plancton di Chioggia. Archo Oceanogr. Limnol., 13(2):249-279, 8 pl.

VASCONCELOS, M. de S.; PAES DA FRANCA, M. L. 1962. Contribuiçâo para um melhor conhecimento do género Aulostomus (Teleostei, Aulostomoidei). Notas mimeogr. Cent. Biol. pisc., (30):1-24, 16 fig., 1 map.

--. 1962. Contribution à une meilleure connaissance du genre Aulostomus (Teleostei Aulostomoidei). Mem. Jta Invest. Ultramar, (2)(36):123-145, 9 pl.

VERANY, J. B. 1847. Aggiunte al catalogo dei pesci della Liguria. Atti 8va Riun. Scienziati ital., (Genova 1846):492-494.

VERGARA, R. 1978. Lutjanidae, in Fischer, W. (ed.), FAO species identification sheets for fishery purposes. Western Central Atlantic (Fishing Area 31). FAO, Rome, vol. 1-7, pag. var.

VERIGHINA, I. A. 1979. The microscopic anatomy of the digestive tract in the deep-sea fish Bajacalifornia calcarata (Weber). Vop. Ikhtiol., 19(6):1139-1142.

VERIGHINA, I. A.; GOLOVAN, G. A. 1978. Peculiarities of the digestive tract structure in fishes of the family Alepocephalidae. Vop. Ikhtiol., 18(2):277-289.

VERRIER, M. L.; ESCHER-DESRIVIERES, J. 1937. Recherches sur la sensibilité lumineuse des poissons (Note préliminaire). Bull. Soc. zool. Fr., 62:126-136.

VIALLI, M. 1956. Sphyraenidae. In: Uova, larve e stadi giovanili di Teleostei. (ed. U. d'Ancona). Fauna Flora Golfo Napoli, 38:457-461.

VICK, N. G. 1963. A morphometric study of seasonal concentrations of the sailfish, Istiophorus albicans in the northern Gulf of Mexico, with notes on other Gulf istiophorids. Texas A. & M. Univ. Res. Found. Proj., 286D:1-41, 14 fig.

VILELA, A. J. 1923. Il pesca e industrias derivadas no distrito Mossamedes, 1921-1922. Porto, 426 p.

VILELA, H.; FRADE, F. 1963. Exposé synoptique sur la biologie du thon à nageoires jaunes Neothunnus albacora (Lowe) 1839 (Atlantique oriental). FAO Fish. Rep., (6)2:900-930.

VILELA, H.; MONTEIRO, R. 1959. Sobre atuns de Angola. I. Parte. Come se pesca e quanto se pesca. II. Parte. Caracteriçao morfologica de Neothunnus albacora (Lowe). Bolm Pesca, (64):11-54.

VILLEGAS, L. 1981. Postlarvas de Gobidos del Mar Cantabrico. Boln Inst. esp. Oceanogr., 6(1):19-45.

VILLEGAS, L.; BRETHES, J. C. 1976. Distribution et abondance relative de la bécasse dans l'Atlantique marocain. Trav. Docums Dev. Pêche Maroc, 17:1-19, 7 fig.

VILLIERS, L. 1980. Changes in predation by the juvenile goby Deltentosteus quadrimaculatus (Teleostei, Gobiidae). Netherl. J. Sea Res., 14: 362-373.

--. 1982. The feeding of juvenile goby Deltentosteus quadrimaculatus (Pisces: Gobiidae). Sarsia, 67:157-162.

VINCIGUERRA, D. 1879-1885. Appunti ittiologici sulle collezioni del Museo Civico di Genova. Annali Mus. civ. stor. nat. Genova, I, 1879, 14:384-397; II, 1879, 14:609-627, pl.; III, 1880, 15:430-453, 3 fig.; IV, 1880, 16: 161-182, 5 fig.; V, 1883, 18:651-660; VI, 1885, (2)2: 82-96; VII, 1885, (2)2:446-475.

--. 1883. Risultati ittiologici delle crociere del "Violante". Annali Mus. civ. Stor. nat. Genova, 18:465-590, 3 pl.

--. 1883. Le crociere dell'yacht "Corsario" del Capitano Armatore Enrico d'Albertis. III. Pesci. Annali Mus. civ. Stor. nat. Genova, 18:607-620.

--. 1890. Intorno ad alcune specie di pesci raccolti dal dottore Enrico Stassano presso la costa occidentale del Sahara. Annali Agric., 1890:61-103. (also 1932, in: Boll. Pesca Piscic. Idrobiol., 8(2):266-304, 1 map).

--. 1890. Appunti intorno ad alcune collezioni ittiologiche recentemente pervenute al Museo zoologico della R. Università di Roma. Spallanzani, 28(10-11):466-489 (= II. Pesci di Las Palmas).

--. 1893. Catalogo dei pesci delle Isole Canarie. Atti Soc. ital. Sci. nat. (1892), 34:293-334.

--. 1932. Del genere Hymenocephalus (Pesci Macruridi) e particolarmente della specie mediterranea (H. italicus Gigl.). Annali Mus. civ. Stor. nat. Giacomo Doria, 56:14-26, pl. 1.

VIVIANI, D. 1805. Phosphorescentia maris. Genuae 17 p.

VLADYKOV, V. D. 1984. Petromyzonidae, in Whitehead, P.J.P., Bauchot, M.L., Hureau, J.C., Nielsen, J. & E. Tortonese, Fishes of the Northeastern Atlantic and the Mediterranean/Poissons de l'Atlantique du Nord-Est et de la Méditerranée, Paris, Unesco, 1:64-67, 3 + 3 fig., 3 maps.

VLADYKOV, V. D.; McALLISTER, D. E. 1961. Preliminary list of the marine fishes of Quebec. Naturaliste can., 88(3):53-78.

VLADYKOV, V. D.; TREMBLAY, J. L. 1936. Nouvelles espèces de Lycodes (Pisces, Zoarcidae) du Saint-Laurent et révision de toutes les espèces du même genre de l'Atlantique occidental. Fauna Flora laurent., (1):45 p., 2 tab., 7 pl.

VODJANITZKI, W. A.; KAZANOVA, I. I. 1954. Opredelitel pelagiceski ikrino i licino ryb Cernogo Morea (Key for identification of the pelagic eggs and fish larvae in the Black Sea). Trudy vses. nauchno-issled. Inst. morsk. ryb. Khoz. Okeanogr., 28:240-327, 72 fig.

VOIGT, F. S., 1831-43. Das Thierreich, geordnet nach seiner Organisa-
tion. Als Grundlage der Naturgeschichte der Thiere und Einleitung in
die vergleichende Anatomie, von Baron von Cuvier,... nach der zwei-
ten, vermehrten Ausgabe übersetzt, und durch Zusätze erweitert von
F.S. Voigt. 6 vol., Leipzig, Reptilien und Fische, 2, 1832: xvi + 539
p.
VOLZ, W. 1907. Catalogue of the fishes of Sumatra. Natuurk. Tijdschr.
Ned.-Indië, 66:35-250.
VON BONDE, C. (see BONDE, C. von).
VON LENDENFELD, R. (see LENDENFELD, R. von).
VOSS, G. L. 1953. A contribution to the life history and biology of the
sailfish, Istiophorus americanus Cuv. and Val., in Florida waters.
Bull. mar. Sci. Gulf Caribb., 3(3):206-240.
--. 1954. The postlarval development of the fishes of the family Gempy-
lidae from the Florida current. I. Nesiarchus Johnson and Gempylus
Cuv. & Val. Bull. mar. Sci. Gulf Caribb., 4(2):120-154, 15 fig.
--. 1969. The pelagic mid-water fauna of the eastern tropical Atlantic
with special reference to the Gulf of Guinea. p. 91-99. In: Proc.
Symp. Oceanogr. Fish. Resources trop. Atlant. (20-28 October 1966,
Abidjan). Paris, 430 p.
VOSS, N. A. 1957. Fishes of the family Gempylidae collected by the Ber-
muda oceanographic expeditions, 1929, 1930, 1931 and 1934. Copeia,
1957(4):304-305.
VROLIK, A. J. 1873. Studien über die Verknöcherung und die Knochen des
Schädels der Teleostei. Niederl. Arch. Zool., 1:219-318.
VU-TÂN-TUÊ, 1964. Evolution de la denture et du régime alimentaire de
Boops boops (L.) au cours de la croissance. Vie Milieu, (Suppl. 17):
505-515, 5 fig.

WADE, C. B. 1946. Two new genera and five new species of apodal fishes
from the eastern Pacific. Allan Hancock Pacif. Exped., 9(7):181-202,
pl. 1-5.
--. 1949. Notes on the Philippine frigate mackerels, family Thunnidae,
genus Auxis. Fishery Bull. Fish. Wildl. Serv. U.S., 51:229-240.
WADE, R. A. 1962. The biology of the tarpon Megalops atlanticus, and the
ox-eye, Megalops cyprinoides, with emphasis on larval development.
Bull. mar. Sci. Gulf Caribb., 12, 4:545-622, 20 fig., 5 tab.
WAGNER, H. 1939. Biologische Beobachtungen an Antennarius marmoratus
Gthr. Zool. Anz., 126 (9/10):285-297, fig. 1-7.
WAITE, E. R. 1899. Scientific results of the trawling expeditions of
H.M.C.S. "Thetis". Mem. Aust. Mus., 4:27-128, 10 fig., 32 pl.
--. 1900. Additions to the fish-fauna of Lord Howe Island. Rec. Aust.
Mus., 3(7):193-209, 2 pl., 2 fig.
--. 1903. Notes on the zoology of Paanopa or Ocean Island and Naura or
Pleasant Island. Gilbert group. The fishes. Rec. Aust. Mus., 5(1):2-
3.
--. 1904. Additions to the fish-fauna of Lord Howe Island. No. 4. Rec.
Aust. Mus., 5:135-186, 32 fig., pl. 17-24.
--. 1904. Catalogue of the fishes of Lord Howe Island. Rec. Aust. Mus.,
5:187-230.
--. 1907. A basic list of fishes of New Zealand. Rec. Canterbury Mus.,
1(1):1-39.
--. 1910. A list of the known fishes of Kermadec and Norfolk Islands,
and a comparison with those of Lord Howe Island. Trans. N.Z. Inst.,
42:370-383, 2 pl.
--. 1909. Scientific results of the New Zealand government trawling
Expedition, 1907. Pisces. Part I. Rec. Canterbury Mus., 1(2):131-156,

pl. 14-23.

--. 1911. Scientific results of the New Zealand government trawling Expedition, 1907. Pisces. Part II, and Outcome of the Expedition. Rec. Canterbury Mus., 1(3):157-272, pl. 24-57.

--. 1912. Notes on New Zealand fishes: No 2. Trans. Proc. N. Z. Inst., 44:194-202, pl.

--. 1916. Fishes. Scient. Rep. Australas. Antarct. Exped., (C)3(1):1-92, 5 pl.

WAKIYA, Y. 1924. The carangoid fishes of Japan. Ann. Carnegie Mus., 15 (2/3):139-293, 38 pl.

WALBAUM, J. J. 1792. Petri Artedi sueci genera piscium in quibus systema totum ichthyologiae proponitur cum classibus, ordinibus, generum characteribus, specierum differentiis, observationibus plurimus. Ichthyologiae, Pars III (ed. 2). Grypeswaldiae, 723 p., 3 pl.

WALFORD, L. A. 1936. Contribution from the Fleischmann Expedition along the west coast of Mexico. I. New fishes from the west coast of Mexico. II. The groupers (Mycteroperca) of the Pacific coast of the America. Occ. Pap. S. Barbara Mus. nat. Hist., (4):1-8.

--. 1937. Marine game fishes of the Pacific coast from Alaska to the Equator. Berkeley, 205 p.

WALLACE, D. H.; WALLACE, E. M. 1942. Observations on the feeding habits of the white marlin, Tetrapturus albidus Poey. Md Dep. Res. Educ., (50):3-10, 1 fig.

WALLACE, J. H. 1967. The batoid fishes of the east coast of southern Africa, Part II: Manta, eagle, duckbill, cownose, butterfly and stingrays. Investl Rep. (S. Afr. Ass. mar. biol. Res.), (16):1-56.

WALTERS, V. 1961. A contribution to the biology of the Giganturidae with description of a new genus and species. Bull. Mus. comp. Zool. Harv., 125(10):297-319, 7 fig.

--. 1963. The trachipterid integument and an hypothesis on its hydrodynamic function. Copeia, 1963:260-270.

--. 1963. On two hitherto overlooked teleost families: Guentheridae (Ateleopodiformes) and Radiicephalidae (Lampridiformes). Copeia, 1963:455-457.

--. 1964. Order Giganturoidei. In: Fishes of the western North Atlantic. Mem. Sears Fdn. mar. Res., 1(4):566-576, fig. 152-155.

WALTERS, V.; FITCH, J. E. 1960. The families and genera of the lampridiform (Allotriognathi) suborder Trachipteroidei. Calif. Fish Game, 46 (4):441-451.

WAPENAAR, M. S.; TALBOT, F. H. 1964. Note on the rigidity of the pectoral fin of Makaira indica (Cuvier). Ann. S. Afr. Mus., 48:167-180, 7 fig., 1 pl.

WASSERSUG, R. J.; JOHNSON, R. K. 1976. A remarkable pyloric caecum in the evermannellid genus Coccorella, with notes on gut structure and function in alepisauroid fishes (Pisces, Myctophiformes). J. Zool., 179:273-289.

WATERMAN, T. H. 1939. Studies on deep-sea anglerfishes (Ceratioidea). I. An historical survey of our present state of knowledge. Bull. Mus. comp. Zool. Harv., 85(3):65-81.

WATTS, J. C. D. 1957. The chemical composition of West African fish. 1. The West African shad, Ethmalosa dorsalis (C. & V.), from Sierra Leone River estuary. Bull. Inst. fr. Afr. noire, (A)19(2):539-547.

--. 1963. A note on Ethmalosa fimbriata (Bowd.) from Sierra Leone. Bull. Inst. fr. Afr. noire, (A)25(1):235-237.

--. 1965. The preservation of fish in Sierra Leone. Bull. Inst. fr. Afr. Noire, (A)27(1):339-396.

WEBB, P. B.; BERTHELOT, S. 1845. Histoire naturelle des Iles Canaries. Paris. (see VALENCIENNES, A. 1837-44)

WEBER, M. 1897. Beiträge zur Kenntnis der Fauna von Süd-Afrika. Zool. Jb. (Syst.), 10:135-199, pl. 15.

--. 1910. Eine neue Art von Macrorhamphosus und Revision dieses Genus. Tijdschr. ned. dierk. Vereen., (2)11:71-79, pl. 4.

--. 1911. Die Fische der Aru- und Kei-Inseln. Ein Beitrag zur Zoogeographie dieser Inseln. Abh. senckenb. naturforsch. Ges., 34:1-49.

--. 1913. Die Fische der Siboga-Expedition. Siboga Exped., 57: i-xii + 1-710, 123 fig., pl. 1-12.

WEBER, M.; BEAUFORT, L. F. de. 1913. The fishes of the Indo-Australian Archipelago. Malacopterygii, Myctophoidea, Ostariophysi: I Siluroidea. Leiden, 2: xx + 404 p., 151 fig.

--. 1916. The fishes of the Indo-Australian Archipelago. Ostariophysi: II Cyprinoidea, Apodes, Synbranchi. Leiden, 3: xv + 455 p., 214 fig.

--. 1922. The fishes of the Indo-Australian Archipelago. Heteromi, Solenichthyes, Synentognathi, Percesoces, Labyrinthici, Microcyprini. Leiden, 4: xiii + 410 p., 103 fig.

--. 1929. The fishes of the Indo-Australian Archipelago. Anacanthini, Allotriognathi, Heterosomata, Berycomorphi, Percomorphi (Kuhliidae... Centropomidae). Leiden, 5: xiv + 458 p., 98 fig.

WEIHS, D.; MOSER, H. G. 1981. Stalked eyes as an adaptation towards more efficient foraging in marine fish larvae. Bull. mar. Sci., 31(1):31-36, fig. 1-3.

WEILER, W. 1968. Otolithi piscium. Fossilium Cat., (I) 117:1-196.

WEINLAND, D. F. 1858. A new division of the five species of flying-fish found along the coast of North America, which have hitherto all been referred to the genus Exocoetus. Proc. Bost. Soc. nat. Hist. (1856-1859), 6:385.

WEITZMAN, S. H. 1967. The origin of the stomiatoid fishes with comments on the classification of salmoniform fishes. Copeia, 1967(3):507-540.

--. 1967. The osteology and relationship of the Astronesthidae, a family of oceanic fishes. Dana Rep., (71):1-54, 31 fig.

WELLERSHAUS, S. 1963. Über zwei Tiefseefische in der marinbiologischen Sammlung des Instituts für Meeresforschung in Bremerhaven: Linophryne lucifer Collett und Trigonolampa miriceps Regan and Trewavas. Veröff. Inst. Meeresforsch. Bremerh., 8:163-166.

WELSH, W. W. 1923. Seven new species of fish of the order Malacopterygii. Proc. U.S. natn. Mus., 62:1-11, 10 fig.

WENNER, C. A. 1978. Making a living on the continental slope and in the deep-sea: life history of some dominant fishes of the Norfolk Canyon Area. Ph. D. diss., College of William & Mary. Williamsburg, Va., 294 p.

WENNER, C. A.; MUSICK, J. 1977. Biology of the morid fish Antimora rostrata in the western North Atlantic. J. Fish. Res. Bd Can., 34(12): 2362-2368.

WEYMOUTH, F. W. 1910. Notes on a collection of fishes from Cameron, Louisiana. Proc. U.S. natn. Mus., 38:135-145, fig. 1-2.

WHEELER, A. C. 1955. A preliminary revision of the fishes of the genus Aulostomus. Ann. Mag. nat. Hist., (12)8:613-623.

--. 1958. The Gronovius fish collection: a catalogue and historical account. Bull. Br. Mus. nat. Hist. (Hist.), 1(5):187-249, pl. 26-34.

--. 1963. The nomenclature of the European fishes of the subfamily Trachinotinae. Ann. Mag. Nat. Hist. (1962), (13) 5(57):529-540.

--. 1969. The fishes of the British Isles and North-West Europe. London, Melbourne and Toronto, i-xvii + 613 p., fig., 16 pl., map.

--. 1973. Aulostomidae, Macrorhamphosidae, p. 272-273, in: Hureau, J.C. & Th. Monod (ed.). Check-list of the fishes of the north-eastern Atlantic and of the Mediterranean/Catalogue des poissons du nord-est Atlantique et de la Méditerranée (Clofnam). Paris, Unesco, 2 vol.

--. 1975. Fishes of the world: an illustrated dictionary. New York, xiv + 366 p., 700 fig., 501 col. photos.

--. 1978. Key to the fishes of northern Europe. A guide to the identification of more than 350 species. London. xix + 380 p., fig.

--. 1981. Macrorhamphosidae, in Fischer, W., Bianchi, G., W.B. Scott (ed.), FAO species identification sheets for fishery purposes. Eastern Central Atlantic; fishing areas 34, 47 (in part). Canada Funds-in-Trust. Otttawa, Department of Fisheries and Oceans Canada, by arrangement with the Food and Agriculture Organization of the United Nations, vol. 1-7, pag. var.

--. 1985. The Linnaean fish collection in the Linnean Society of London. Zool. J. Linn. Soc., 84:1-76, fig. 1-5.

WHEELER, A.; DUNNE, J. 1975. Tripterygion atlanticus sp. nov. (Teleostei-Tripterygiidae) the first record of a tripterygiid fish in north-western Europe. J. Fish Biol., 7:639-649.

WHITE, E. G. 1937. Interrelationships of the elasmobranchs with a key to the order Galea. Bull. Am. Mus. nat. Hist., 74(2):25-138.

WHITEHEAD, P. J. P. 1962. The species of Elops (Pisces: Elopidae). Ann. Mag. nat. Hist., (13)5:321-329.

--. 1963. A revision of the recent round herrings (Pisces: Dussumieriidae). Bull. Br. Mus. nat. Hist. (Zool.), 10(6):305-380.

--. 1964. Herklotsichthys Whitley 1951 to replace Harengula Valenciennes, 1847 for Indo-Pacific species (Pisces: Clupeidae). Ann. Mag. nat. Hist., (13)6:273-284.

--. 1964. A redescription of the holotype of Clupalosa bulan Bleeker, and notes on the genera Herklotsichtys, Sardinella and Escualosa (Pisces: Clupeidae). Ann. Mag. nat. Hist., (13)7:33-47.

--. 1966. The elopoid and clupeoid fishes in Richardson's "Ichthyology of the seas of China and Japan", 1846. Bull. Br. Mus. nat. Hist. (Zool.), 14(2):15-54.

--. 1967. The clupeoid fishes described by Lacepède, Cuvier & Valenciennes. Bull. Br. Mus. nat. Hist. (Zool.), Suppl. 2: 180 p., 11 pl.

--. 1967. The West African shad, Ethmalosa fimbriata (Bowdich, 1825): synonymy, neotype. J. nat. Hist., 4:585-593.

--. 1967. The dating of the first edition of Cuvier's Le Règne Animal distribué d'après son organisation. J. Soc. Biblphy nat. Hist., 4(6): 300-301.

--. 1969. The clupeoid fishes described by Bloch and Schneider. Bull. Br. Mus. nat. Hist. (Zool.), 17(7):261-279.

--. 1970. The clupeoid fishes described by Steindachner. Bull. Br. Mus. nat. Hist. (Zool.), 20(1):1-46.

--. 1973. A synopsis of the clupeoid fishes of India. J. mar. biol. Ass. India, 14(1):160-256.

--. 1973. The clupeoid fishes of the Guianas. Bull. Br. Mus. nat. Hist. (Zool.), Suppl. 5: 227 p., 72 fig.

--. 1978. The Forster collection of zoological drawings in the British Museum (Natural History). Bull. Br. Mus. nat. Hist. (Hist.), 6(2):25-47.

--. 1978. Albulidae, Clupeidae, Elopidae, Engraulidae, Megalopidae, in W. Fischer, (ed.), FAO species identification sheets for fishery purposes. Western Central Atlantic (Fishing Area 31). Rome, FAO, vol. 1-6, pag. var.

--. 1981. Albulidae, Clupeidae, Elopidae, Engraulidae, Megalopidae, in W. Fischer, G. Bianchi & W.B. Scott (ed.), FAO species identification sheets for fishery purposes. Eastern Central Atlantic (Fishing areas 34 and part of 47). Dept. Fish. Oceans Canada, Ottawa, FAO, Rome, vol. 1-7, pag. var.

--. 1984. Clupeidae, in Whitehead, P.J.P., Bauchot, M.L., Hureau, J.C.,

Nielsen, J. & E. Tortonese, Fishes of the North-eastern Atlantic and the Mediterranean/Poissons de l'Atlantique du Nord-Est et de la Méditerranée, Paris, Unesco, 1:268-281, 13 + 8 fig., 13 maps.

--. 1984. Engraulidae, in Whitehead, P.J.P., Bauchot, M.L., Hureau, J.C., Nielsen, J. & E. Tortonese, Fishes of the North-eastern Atlantic and the Mediterranean/Poissons de l'Atlantique du Nord-Est et de la Méditerranée, Paris, Unesco, 1:282-283, 1 fig., 1 map.

--. 1986. FAO species catalogue. Clupeoid fishes of the world. An annotated and illustrated catalogue of the herrings, sardines, pilchards, sprats, anchovies and wolf-herrings. Part 1 - Chirocentridae, Clupeidae and Pristigasteridae. FAO Fish. Synop., (125)7(1):1-303.

--. 1986. The synonymy of Albula vulpes (Linnaeus, 1758)(Teleostei, Albulidae). Cybium, 10(3):211-230.

WHITEHEAD, P. J. P.; BAUCHOT, M. L. 1986. Catalogue critique des types de poissons du Muséum national d'Histoire naturelle (suite). Ordre des Clupeiformes (Familles des Clupeidae, Engraulididae et Denticipitidae). Bull. Mus. natn Hist. nat., Paris, (4)7A(4, suppl.):1-77.

WHITEHEAD, P. J. P.; MYERS, G. S. 1971. Problems of nomenclature and dating of Spix and Agassiz's Brazilian fishes (1829-1831). J. Soc. Biblphy nat. Hist., 5(6):478-497.

WHITEHEAD, P. J. P.; WHEELER, A. 1966. The generic names used for the seabasses of Europe and North America (Pisces: Serranidae). Annali Mus. civ. Stor. nat. Giacomo Doria, 76:23-41, 5 fig.

WHITEHEAD, P. J. P.; BOESEMAN, M.; WHEELER, A. C. 1966. The types of Bleeker's Indo-Pacific elopoid and clupeoid fishes. Zool. Verh., Leiden, (84):1-152.

WHITEHEAD, P. J. P.; NELSON, G. J.; WONGRATANA, T. 1988. FAO species catalogue. Clupeoid fishes of the world. An annotated and illustrated catalogue of the herrings, sardines, pilchards, sprats, anchovies and wolf-herrings. Part 2. Engraulididae. FAO Fish. Synop., (125)7(2): vii + 305-579.

WHITEHEAD, P.J.P.; BAUCHOT, M.L.; HUREAU, J.C.; NIELSEN, J.; TORTONESE, E. (ed.). 1984-1986. Fishes of the North-eastern Atlantic and the Mediterranean/Poissons de l'Atlantique du Nord-Est et de la Méditerranée. Paris, Unesco. 1, 1984:1-510, 863 fig., 427 maps; 2, 1986 (September):511-1008, 562 fig., 444 maps; 3, 1986 (December):1009-1473, 602 fig., 413 maps.

WHITLEY, G. P. 1928. Fishes from the Great Barrier Reef collected by Mr. Melbourne Ward. Rec. Aust. Mus., 16(6):294-304, 2 fig.

--. 1929. Some fishes of the order Amphiprioniformes. Mem. Qd Mus., 9 (3):207-247, fig. 1-4, pl. 27-28.

--. 1929. Studies in ichthyology. No 3. Rec. Aust. Mus., 17(3):101-143, 5 fig., pl. 30-34.

--. 1930. Ichthyological miscellanea. Mem. Qd Mus., 10(1):8-31, 1 fig., 1 pl.

--. 1930. Additions to the check-list of the fishes of New South Wales (No 3). Aust. Zool., 6:117-123.

--. 1930. Five new generic names for Australian fishes. Aust. Zool., 6: 250-251.

--. 1931. Studies in ichthyology. No 4. Rec. Aust. Mus., 18(3):96-133.

--. 1931. Studies in ichthyology. No 5. Rec. Aust. Mus., 18(4):138-160.

--. 1931. New names for Australian fishes. Aust. Zool., 6(4):310-334, pl. 25-27, 1 fig.

--. 1932. Fishes. Scient. Rep. Gt Barrier Reef Exped., 4:267-316, 4 pl., 5 fig.

--. 1932. Studies in ichthyology. No 6. Rec. Aust. Mus., 18:321-348, pl. 36-39, 3 fig.

--. 1933. Studies in ichthyology. No 7. Rec. Aust. Mus., 19:60-112, 4 fig., 5 pl.

--. 1933. Sunfishes. <u>Victorian</u> <u>Nat.</u>, 49:207-213, 7 fig.

--. 1935. Studies in ichthyology. No 9. <u>Rec. Aust. Mus.</u>, 19(4):215-250, 11 fig., pl. 18.

--. 1936. Ichthyological genotypes; some supplementary remarks. <u>Aust. Zool.</u>, 8(3):189-192.

--. 1937. Studies in ichthyology. No 10. <u>Rec. Aust. Mus.</u>, 20(1):3-24, fig. 1-5, pl. 2.

--. 1937. Further ichthyological miscellanea. <u>Mem. Qd Mus.</u>, 11(2):113-148.

--. 1937. The Middleton and Elizabeth Reefs, South Pacific Ocean. <u>Aust. Zool.</u>, 8(4):199-273.

--. 1938. Studies in ichthyology. No 11. <u>Rec. Aust. Mus.</u>, 20:195-199, 1 fig., pl. 21.

--. 1938. Ray's bream and its allies in Australia. <u>Aust. Zool.</u>, 9:191-194, pl. 19.

--. 1939. Ichthyological genotypes: Desmarest's designations, 1874. <u>Aust. Zool.</u>, 9(3):222-226.

--. 1939. Taxonomic notes on sharks and rays. <u>Aust. Zool.</u>, 9(3):227-262, fig. 1-18, pl. 20-22.

--. 1939. A new fish of the genus <u>Prionobatis</u>, from northern Australia. <u>Mem. Qd Mus.</u>, 11:296-298.

--. 1940. Illustrations of some Australian fishes. <u>Aust. Zool.</u>, 9(4): 397-428.

--. 1940. The Nomenclator Zoologicus and some new fish names. <u>Aust. Nat.</u>, 1940:241-243.

--. 1941. Ichthyological notes and illustrations. <u>Aust. Zool.</u>, 10(1):1-52, fig. 1-32, pl. 1-2.

--. 1941. A lanternfish from Macquarie Island. <u>Aust. Zool.</u>, 10(1):124.

--. 1943. Ichthyological notes and illustrations. (Part 2). <u>Aust. Zool.</u>, 10(2):167-187.

--. 1946. Australian marine eels. <u>Aust. Mus. Mag.</u>, 9:60-65.

--. 1947. New sharks and fishes from western Australia. Part 3. <u>Aust. Zool.</u>, 11:129-150.

--. 1948. Studies in ichthyology. No 13. <u>Rec. Aust. Mus.</u>, 22(1):70-94, fig. 1-11.

--. 1948. A list of the fishes of western Australia. <u>Fish. Bull. West. Aust.</u>, (2):1-35.

--. 1950. New fish names. <u>Proc. R. zool. Soc. N.S.W.</u>, (1948-49):44.

--. 1951. New fish names and records. <u>Proc. R. zool. Soc. N.S.W.</u>, (1949-50):61-68, 2 fig.

--. 1951. Studies in ichthyology. No 15. <u>Rec. Aust. Mus.</u>, 22(4):389-408.

--. 1953. Studies in ichthyology. No 16. <u>Rec. Aust. Mus.</u>, 23(3):133-138.

--. 1954. New locality records for some Australian fishes. <u>Proc. R. zool. Soc. N.S.W.</u>, (1952-53):23-30.

--. 1954. More new fish names and records. <u>Aust. Zool.</u>, 12(1):57-62, 1 pl.

--. 1955. The Australian Museum's marlins. <u>Aust. Mus. Mag.</u>, 11(9):292-297, 4 fig.

--. 1957. A new angler fish. <u>West. Aust. Nat.</u>, 5(7):207-209.

--. 1957. Ichthyological illustrations. <u>Proc. R. zool. Soc. N.S.W.</u>, (1955-56):56-71.

--. 1958. Descriptions and records of fishes. <u>Proc. R. zool. Soc. N.S.W.</u>, (1956-57):28-51.

--. 1964. Scombroid fishes of Australia and New Zealand. <u>Proc. Symp. Scombroid Fishes, mar. biol. Ass. India</u>, 1(1):221-253, 6 fig., 4 pl.

--. 1964. Presidential address. A survey of Australian ichthyology. <u>Proc. Linn. Soc. N.S.W.</u>, 84(1):11-127.

--. 1968. A check-list of the fishes recorded from the New Zealand region. _Aust. Zool._, 15(1):1-102, 2 fig.

WICTOR, K. 1970. Ikra i larwy palasza (_Trichiurus lepturus_ L.) w wodach szelfu Afryki polnocno-zachodniej. (Eggs and larvae of _Trichiurus lepturus_ L. in the shelf waters of north-western Africa). _Przegl. zool._, 14(4):366-370.

WILK, S. J. 1977. Biological and fisheries data on bluefish, _Pomatomus saltatrix_ (Linnaeus). _Tech. Sci. Rep. NMFS. NE Fish Center. Sandy Hook Lab._, (11):56 p.

WILLEM, V. 1950. Contributions à l'étude des organes respiratoires chez les téléostéens: _Dibranchus. Bull. Inst. r. Sci. nat. Belg._, 26(22): 1-5, 3 fig.

WILLIAMS, F. 1968. Report on the Guinean Trawling Survey. I. General Report. Lagos, ix + 828 p., 983 + 2 tab.

--. 1969. Review of the principal results of the Guinean Trawling Survey. p. 139-146, 2 fig. _In_: Proc. Symp. Oceanogr. Fish. Resources trop. Atlantic. (20-28 October 1966, Abidjan). Paris, 430 p.

WILLIS, A. C.; WOOD, E. M. 1973. Development of _Abudefduf luridus_ and _Chromis chromis_. Azores Expedition, _Rep. Exul Sub Aqua Club_, 1973:41-45.

WINTERBOTTOM, R. 1976. Notes on South African gobies possessing free upper pectoral fin rays (Pisces: Gobiidae). _Spec. Publs J.L.B. Smith Inst. Ichthyol._, (16):1-11, 7 fig.

WINTERBOTTOM, R.; TYLER, J. C. 1981. Balistidae, _in_ Fischer W., Bianchi, G. & W.B. Scott (ed.), FAO species identification sheets for fishery purposes. Eastern Central Atlantic; fishing areas 34, 47 (in part). Canada Funds-in-Trust. Ottawa, Department of Fisheries and Oceans Canada, by arrangement with the Food and Agriculture Organization of the United Nations, vol. 1-7, pag. var.

WIRTZ, P. 1976. A key to the European Blennioidea. _Vie Milieu_, (A)26: 145-156.

--. 1976. The otoliths of the Mediterranean Tripterygiidae. _Vie Milieu_, (A)26:293-298.

--. 1977. Welcher Blenniide ist das ? _Aquar. Mag._, 1977(5):204-215.

--. 1978. The behaviour of the Mediterranean _Tripterygion_ species. _Z. Tierpsychol._, 48:142-174.

--. 1980. A revision of the eastern Atlantic Tripterygiidae (Pisces, Blennioidei) and notes on some westafrican blennioid fish. _Cybium_, 4(4):83-101, fig. 1-8, tab. 1-2.

--. 1981. Clinidae, _in_ Fischer, W., Bianchi G. & W.B. Scott (ed.), FAO species identification sheets for fishery purposes. Eastern Central Atlantic; fishing areas 34, 47 (in part). Canada Funds-in-Trust. Ottawa, Department of Fisheries and Oceans Canada, by arrangement with the Food and Agriculture Organization of the United Nations, vol. 1-7, pag. var.

--. 1982. Range extension for _Hypleurochilus aequipinnis_ (Günther, 1861) in the eastern tropical Atlantic. _Cybium_, 6(3):74.

WIRTZ, P.; BATH, H. 1982. _Lipophrys bauchotae_ n. sp., a new blenniid fish from the eastern tropical Atlantic (Pisces: Blenniidae). _Senckenberg. biol._, 62:225-232.

WIRTZ, P.; ZANDER, C. D. 1986. Clinidae, _in_ P.J.P. Whitehead, M.L. Bauchot, J.C. Hureau, J. Nielsen & E. Tortonese. Fishes of the Northeastern Atlantic and the Mediterranean/Poissons de l'Atlantique du Nord-Est et de la Méditerranée, Paris, Unesco, 3:1117, 1 fig.

WISNER, R. L. 1962. Midwater trawl surveys. _Limnol. Oceanogr._, 7 (Suppl.):39-41.

--. 1963. A new genus and species of myctophid fish from the South-Central Pacific Ocean, with notes on related genera and the designation of a new tribe, Electronini. _Copeia_, 1963(1):24-28.

--. 1971. Descriptions of eight new species of myctophid fishes from the eastern Pacific Ocean. *Copeia*, 1971(1):39-54.

--. 1974. Descriptions of five new species of myctophid fishes from the Pacific, Indian and Atlantic Oceans. *Occ. Pap. Calif. Acad. Sci.*, (110):1-37.

--. 1976. The taxonomy and distribution of lanternfishes (Family Myctophidae) of the eastern Pacific Ocean. Navy Ocean Res. Dev. Activity (NORDA), Rep. (3):1-229.

--. 1978. Scomberesocidae, in Fischer, W. (ed.), FAO species identification sheets for fishery purposes. Western Central Atlantic (Fishing Area 31). Rome, FAO, vol. 1-7, pag. var.

--. 1981. Scomberesocidae, in Fischer, W., Bianchi, G. & W.B. Scott (ed.), FAO species identification sheets for fishery purposes. Eastern Central Atlantic; fishing areas 34, 47 (in part). Canada Funds-in-Trust. Ottawa, Department of Fisheries and Oceans Canada, by arrangement with the Food and Agriculture Organization of the United Nations, vol. 1-7: pag. var.

WITTENBERG, J. B.; HAEDRICH, R. L. 1974. A choroid rete mirabile of the fish eye. II. Distribution and relation to the pseudobranch and to the swimbladder rete mirabile. *Biol. Bull. mar. biol. Lab., Woods Hole*, 146:151-156.

WITTENBERG, J. B.; COPELAND, D. E.; HAEDRICH, R. L.; CHILD, J. S. 1980. The swimbladder of deep-sea fish: The swimbladder wall is a lipid-rich barrier to oxygen diffusion. *J. mar. biol. Ass. U.K.*, 60:263-276.

WITZELL, W. N. 1973. Gonostomatidae, p. 114-122, in Hureau, J.C. & Th. Monod (ed.). Check-list of the fishes of the north-eastern Atlantic and of the Mediterranean/Catalogue des poissons du nord-est Atlantique et de la Méditerranée (Clofnam). Paris, Unesco, 2 vol.

WOJOIECHOWSKI, J. 1972. Observations on biology of cutlassfish *Trichiurus lepturus* L. (Trichiuridae) of Mauritania shelf. *Acta Ichthyol. Piscatoria*, 2(2):67-75.

WOLLAM, M. B. 1969. Larval *Acanthocybium solanderi* (Cuvier) (Scombridae) from the Straits of Yucatan and Florida. *Leafl. Fla Dep. nat. Res.*, 4(1):1-7.

WONGRATANA, T. 1980. Systematics of clupeoid fishes of the Indo-Pacific region. Ph. D. thesis, Fac. of Science, Univ. of London, 2 vol., 432 p., 334 pl., 126 fig., 17 tab.

--. 1983. Diagnoses of 24 new species and proposal of a new name for a species of Indo-Pacific fishes. *Jap. J. Ichthyol.*, 29(4):385-407, 25 fig.

WOOD, S. D. 1837. The fishes (Pisces) of Britain, systematically arranged. *Analyst*, 5:204-215.

WOOD-MASON, J.; ALCOCK, A. W. 1891. Natural history notes from H.M. Indian Marine Survey Steamer "Investigator", Commander B.F. Hoskyn, R.N., commanding. Ser. 2, No 1: On the results of deep-sea dredging during the season 1890-91. *Ann. Mag. nat. Hist.*, (6)8:119-138, fig. 3-5, pl. 7-8.

WOODS, L. P.; KANAZAWA, R. H. 1951. New species and new records of fishes from Bermuda. *Fieldiana, Zool.*, 31(53):629-644, fig. 134-137.

WOODS, L. P.; SONODA, P. M. 1973. Order Berycomorphi (Beryciformes). In: Fishes of the western North Atlantic. *Mem. Sears Fdn mar. Res.*, 1(6): 263-396, fig. 1-66, tab. 1-12.

WOODWARD, A. S. 1902. The fossil fishes of the English Chalk. Part 1. *Palaeontogr. Soc. (Monogr.)*, 56:1-56.

WRIGHT, J. M. 1986. The ecology of fish occurring in shallow water creeks of a Nigerian mangrove swamp. *J. Fish Biol.*, 29:431-444.

WU, H. W. 1931. Notes on the fishes from the coast of Foochow region and

Ming River. <u>Contr.</u> <u>Biol.</u> <u>Lab.</u> <u>Sci.</u> <u>Soc.</u> <u>China</u> (Zool.), 7(1):1-64.

YABE, H. 1951. Larva of the swordfish, <u>Xiphias gladius</u>. <u>Jap.</u> <u>J.</u> <u>Ich-</u><u>thyol.</u>, 1(4):260-263, 1 fig.

YABE, H.; UEYANAGI, S.; KIKAWA, S.; WATANABE, H. 1959. Study on the life-history of the sword-fish, <u>Xiphias gladius</u> Linnaeus. <u>Rep.</u> <u>Nankai</u> <u>reg.</u> <u>Fish.</u> <u>Res.</u> <u>Lab.</u>, (10):107-150, 28 fig., 1 pl. (in Japanese).

YANG, H.-C.; NAKAMURA, I. 1973. A description of deep-sea fish obtained in the Pescadores Islands, Taiwan. <u>A.</u> <u>Rep.</u> <u>Sci..</u> <u>Taiwan</u> <u>Mus.</u>, 16:85-90. (Translation 73 from Systematics Laboratory, NOAA).

YARBERRY, E. L. 1965. Osteology of the zoarcid fish <u>Melanostigma</u> <u>pamme-</u><u>las.</u> <u>Copeia</u>, 1965(4):442-462, 9 fig.

YARRELL, W. 1832. (16 January 1832). (On the anatomy of the conger eel (<u>Conger</u> <u>vulgaris</u> Cuv.) and on the difference between the conger and freshwater eels). <u>Proc.</u> <u>Comm.</u> <u>Sci.</u> <u>Corresp.</u> <u>zool.</u> <u>Soc.</u> <u>Lond.</u>, 1:158-159.

--. 1836-39. A history of British fishes, illustrated by nearly 400 wood-cuts, in two volumes. London, 1, 1836: xxxvii + 408 p., fig. n. num.; 2, 1839:472 p., fig. n. num.; 1839, suppl.: vi + 48 p., fig.

--. 1859. A history of British fishes, illustrated by nearly 400 wood-cuts, in two volumes. (3rd ed.). London, 1:675 p.

YASUDA, F.; KOHNO, H.; YATSU, A.; IDA, H.; ARENA, P.; LI GRECI, F.; TAKI, Y. 1978. Embryonic and early larval stages of the swordfish, <u>Xiphias gladius</u>, from the Mediterranean. <u>J.</u> <u>Tokyo</u> <u>Univ.</u> <u>Fish.</u>, 65 (1):91-97, 2 fig.

YOSHINO, T. P.; NOBLE, E. R. 1973. Myxosporida in macrourid fishes of the North Atlantic. <u>Can.</u> <u>J.</u> <u>Zool.</u>, 51:745-752.

ZAHURANEC, B. J.; PUGH, W. L.; FARQUHAR, G. B. 1970. Biological sound scattering studies. Part I. Initial investigations in the Gulf of Mexico and western North Atlantic. <u>Tech.</u> <u>Rep.</u> <u>U.S.</u> <u>nav.</u> <u>oceanogr.</u> <u>Off.</u>, TR-224:35 p.

ZAMA, A.; YASUDA, F. 1979. An annotated list of fishes from the Ogasawa-ra Islands. Supplement I, with zoogeographical notes on the fish fauna. <u>J.</u> <u>Tokyo</u> <u>Univ.</u> <u>Fish.</u>, 66(2):139-163.

ZANDER, C. D. 1979. Morphologische und ökologische Untersuchung der Schleim-fische <u>Parablennius</u> <u>sanguinolentus</u> (Pallas, 1811) und <u>P.</u> <u>parvicornis</u> (Valenciennes, 1836)(Perciformes, Blenniidae). <u>Mitt.</u> <u>hamb.</u> <u>zool.</u> <u>Mus.</u> <u>Inst.</u>, 76:469-474.

--. 1981. Zur Nahrung kleiner Bodenfische des westafrikanischen Auf-triebsgebietes. <u>"Meteor"</u> <u>Forsch.-Ergebn.</u>, (D)(33):71-75.

--. 1982. Feeding ecology of littoral gobiid and blennioid fish of the Banyuls area (Mediterranean Sea). I. Main food and trophic dimension of niche and ecotope. <u>Vie</u> <u>Milieu</u>, 32:1-10.

--. 1986. Blenniidae, <u>in</u> P.J.P. Whitehead, M.L. Bauchot, J.C. Hureau, J. Nielsen & E. Tortonese. Fishes of the North-eastern Atlantic and the Mediterranean/Poissons de l'Atlantique du Nord-Est et de la Méditer-ranée, Paris, Unesco, 3:1096-1112, 22 + 1 fig.

--. 1986. Tripterygiidae, <u>in</u> P.J.P. Whitehead, M.L. Bauchot, J.C. Hu-reau, J. Nielsen & E. Tortonese. Fishes of the North-eastern Atlantic and the Mediterranean/Poissons de l'Atlantique du Nord-Est et de la Méditerranée, Paris, Unesco, 3:1118-1121, 3 + 3 fig.

ZANDER, C. D.; HEYMER, A. 1971. <u>Tripterygion</u> <u>tripteronotus</u> (Risso, 1810) und <u>Tripterygion</u> <u>xanthosoma</u> n. sp. eine ökologische Speziation (Pis-ces, Teleostei). <u>Vie</u> <u>Milieu</u> (1970), (A)21:363-394.

--. 1976. Morphologische und ökologische Untersuchungen an den speleophilen Schleimfischartigen *Tripterygion melanurus* Guichenot, 1850 und *T. minor* Kolombatovic, 1892. *Z. zool. Syst. & Evolutionsforsch.*, 14: 41-59.

--.1977. Analysis of ecological equivalents among littoral fish: p. 621-630, 7 fig. *In*: Keegan, B. F., P.O. Ceidigh & P.J. Boaden (ed.): Biology of benthic organisms. Oxford.

ZANDER, C. D.; JELINEK, H. 1976. Zur demersen Fischfauna der Grotte von Banjole (Rovinj/YU) mit Beschreibung von *Speleogobius trigloides* n. gen. n. sp. (Gobiidae, Perciformes). *Mitt. hamb. zool. Mus. Inst.*, 73: 265-280.

ZARUR, M. A. 1962. Algunas consideraciones geobiológicas de la laguna de Términos, Camp. *Revta Soc. mex. Hist. nat.*, 23:51-63.

ZEI, M. 1941. Studies on the morphology and taxonomy of the Adriatic species of Maenidae. *Acta adriat.*, 2(4):135-191, 21 fig., 4 pl.

--. 1949. Ova and developmental stages of *Maena smaris* and *Maena chryselis. Acta adriat.*, 4(5):3-19, 14 fig.

--. 1949. Typical sex reversal in teleosts. *Proc. zool. Soc. Lond.*, 119 (4):917-920.

--. 1951. Jadranske girice (Maenidae). Monografiska studija. *Slov. Akad. Znan. Umetu. Ljubljana*, (4)(3):1-127, 27 fig., 18 pl.

--. 1963. Vertebrata (Cyclostomata, Chondrichthyes, Osteichthyes): p. 406-565, pl. 176-215. *In*: R. Riedl, Fauna und Flora der Adria. Hamburg, 640 p., fig., 8 col. pl., 221 pl.

--. 1965. Some facts on *Sardinella aurita* Val. in relation to temperature. *Ghana J. Sci.*, 5:264-272.

--. 1969. The behaviour of *Sardinella aurita* Val., in relation to light and temperature. p. 469-475. *In*: Ben-Tuvia, A. & W. Dickson (1969).

--. 1969. Sardines and related species of the eastern tropical Atlantic, p. 101-108. *In*: Proc. Symp. Oceanogr. Fish. Resources trop. Atlant. (20-28 October 1968, Abidjan). Paris, 430 p.

ZHAROV, V. L.; ZHUDOVA, A. M. 1967. Some data on occurrence of scombroid larvae (Order Perciformes, Suborder Scombroidei) in the open waters of the tropical Atlantic. *Trudy AtlantNIRO*, 18:201-214. (in Russian, transl. by W.L. Klawe).

ZHUDOVA, A. M. 1969. Materials on the study of the eggs and larvae of some species of fish from the Gulf of Guinea. *Trudy atlant. nauchno-issled. Inst. ryb. Khoz. Oceanogr.*, 25:135-163, 54 fig., 2 tab. (Inter-Am. trop. Tuna Commn, La Jolla, 38 p., transl. by W.L. Klawe, 1971).

ZILANOV, V. K.; BOGDANOV, S. I. 1969. Results of research on *Scomberesox saurus* in the north-eastern Atlantic in 1968. *Annls biol.. Copenh.*, 25:252-255.

ZOLEZZI, G. 1938. I pesci della spedizione Carniglia al Giubi 1926. *Boll. Pesca Piscic. Idrobiol.*, 14:210-219.

ZURBRIGG, R. E.; SCOTT, W. B. 1976. *Diaphus hudsoni* (Pisces: Myctophidae), a new lanternfish from the South Atlantic Ocean. *Can. J. Zool.*, 54 (9):1538-1541.

ZUGMAYER, E. 1911. Poissons provenant des campagnes du yacht "Princesse Alice" 1901-1910. *Résult. Camp. scient. Prince Albert I*, 35:1-174, 6 pl.

--. 1911. Diagnoses de poissons nouveaux provenant des campagnes du yacht "Princesse Alice" (1901-1910). *Bull. Inst. océanogr. Monaco*, (193):1-14.

--. 1913. Diagnoses des stomiatidés nouveaux provenant des campagnes du yacht "Hirondelle II" (1911 et 1912). *Bull. Inst. océanogr. Monaco*, (253):1-7.

--. 1914. Diagnoses de quelques poissons nouveaux provenant des campagnes du yacht "Hirondelle II" (1911-13). *Bull. Inst. océanogr. Mona-*

<u>co</u>, (288):1-4.

--. 1933. Liste complémentaire des déterminations faites par M. Zugmayer. p. 79-85. <u>In</u>: Roule, L. & F. Angel. Poissons provenant des campagnes du Prince de Monaco. <u>Résult. Camp. scient. Prince Albert I</u>, 86:1-115, 4 pl.

--. 1940. Diagnoses des stomiatidés nouveaux provenant des campagnes du yacht Hirondelle II (1911 et 1912)(avec un tableau de détermination). <u>Résult. Camp. scient. Prince Albert I</u>, 103:201-205 (same as Zugmayer, 1913).

ZUPANOVIC, S.; CISSE, M. 1977. Quelques observations sur les sardinelles (<u>S. aurita</u> et <u>S. eba</u>) et balistes (<u>B. capriscus</u>) capturés au large des côtes de Guinée. UNDP(SF)/FAO, Project GUI/74/024, Dével. pêche marit., Conakry, 34 p., (mimeo).

Alphabetical index of scientific names
Index alphabétique des noms scientifiques